NATIONAL ACADEMIES
Sciences
Engineering
Medicine

NATIONAL ACADEMIES PRESS
Washington, DC

The Future of Water Quality in Coeur d'Alene Lake

Committee on the Future of Water Quality in Coeur d'Alene Lake

Water Science and Technology Board

Division on Earth and Life Studies

Consensus Study Report

NATIONAL ACADEMIES PRESS 500 Fifth Street, NW Washington, DC 20001

This activity was supported by contracts between the National Academy of Sciences and the Idaho Department of Environmental Quality, Kootenai County, and the U.S. Environmental Protection Agency. Any opinions, findings, conclusions, or recommendations expressed in this publication do not necessarily reflect the views of any organization or agency that provided support for the project.

International Standard Book Number-13: 978-0-309-69041-6
International Standard Book Number-10: 0-309-69041-2
Digital Object Identifier: https://doi.org/10.17226/26620

This publication is available from the National Academies Press, 500 Fifth Street, NW, Keck 360, Washington, DC 20001; (800) 624-6242 or (202) 334-3313; http://www.nap.edu.

Copyright 2022 by the National Academy of Sciences. National Academies of Sciences, Engineering, and Medicine and National Academies Press and the graphical logos for each are all trademarks of the National Academy of Sciences. All rights reserved.

Printed in the United States of America.

Cover image credit: Content is the intellectual property of Esri and is used herein with permission. Copyright © 2022 Esri and its licensors. All rights reserved.

Suggested citation: National Academies of Sciences, Engineering, and Medicine. 2022. *The Future of Water Quality in Coeur d'Alene Lake.* Washington, DC: The National Academies Press. https://doi.org/10.17226/26620.

The **National Academy of Sciences** was established in 1863 by an Act of Congress, signed by President Lincoln, as a private, nongovernmental institution to advise the nation on issues related to science and technology. Members are elected by their peers for outstanding contributions to research. Dr. Marcia McNutt is president.

The **National Academy of Engineering** was established in 1964 under the charter of the National Academy of Sciences to bring the practices of engineering to advising the nation. Members are elected by their peers for extraordinary contributions to engineering. Dr. John L. Anderson is president.

The **National Academy of Medicine** (formerly the Institute of Medicine) was established in 1970 under the charter of the National Academy of Sciences to advise the nation on medical and health issues. Members are elected by their peers for distinguished contributions to medicine and health. Dr. Victor J. Dzau is president.

The three Academies work together as the **National Academies of Sciences, Engineering, and Medicine** to provide independent, objective analysis and advice to the nation and conduct other activities to solve complex problems and inform public policy decisions. The National Academies also encourage education and research, recognize outstanding contributions to knowledge, and increase public understanding in matters of science, engineering, and medicine.

Learn more about the National Academies of Sciences, Engineering, and Medicine at **www.nationalacademies.org**.

Consensus Study Reports published by the National Academies of Sciences, Engineering, and Medicine document the evidence-based consensus on the study's statement of task by an authoring committee of experts. Reports typically include findings, conclusions, and recommendations based on information gathered by the committee and the committee's deliberations. Each report has been subjected to a rigorous and independent peer-review process and it represents the position of the National Academies on the statement of task.

Proceedings published by the National Academies of Sciences, Engineering, and Medicine chronicle the presentations and discussions at a workshop, symposium, or other event convened by the National Academies. The statements and opinions contained in proceedings are those of the participants and are not endorsed by other participants, the planning committee, or the National Academies.

Rapid Expert Consultations published by the National Academies of Sciences, Engineering, and Medicine are authored by subject-matter experts on narrowly focused topics that can be supported by a body of evidence. The discussions contained in rapid expert consultations are considered those of the authors and do not contain policy recommendations. Rapid expert consultations are reviewed by the institution before release.

For information about other products and activities of the National Academies, please visit www.nationalacademies.org/about/whatwedo.

COMMITTEE ON THE FUTURE OF WATER QUALITY IN COEUR D'ALENE LAKE

SAMUEL N. LUOMA (*Chair*), Institute of the Environment, University of California, Davis
ROBERT L. ANNEAR, Geosyntec Consultants, Portland, OR
WILLIAM A. ARNOLD, University of Minnesota, St. Paul
MICHAEL T. BRETT, University of Washington, Seattle
JAMES J. ELSER (NAS), Flathead Lake Biological Station, University of Montana, Polson
SCOTT E. FENDORF, Stanford University, Stanford, CA
ALEJANDRO N. FLORES, Boise State University, Idaho
PRIYA M. GANGULI, California State University, Northridge[*]
ROBERT M. HIRSCH, U.S. Geological Survey (retired), Reston, VA
LYNN E. KATZ, University of Texas, Austin
JAMES G. MOBERLY, University of Idaho, Moscow
S. GEOFFREY SCHLADOW, Tahoe Environmental Research Center and University of California, Davis

National Academies of Sciences, Engineering, and Medicine Staff

LAURA J. EHLERS, Senior Program Officer, Water Science and Technology Board
RACHEL SILVERN, Program Officer, Board on Atmospheric Sciences and Climate
CALLA ROSENFELD, Senior Program Assistant, Water Science and Technology Board
ERIC EDKIN, Manager of Internal Communications and Program Support, Division on Earth and Life Studies

---

[*] Resigned from the committee in May 2022.

WATER SCIENCE AND TECHNOLOGY BOARD

CATHERINE L. KLING (NAS) (*Chair*), Cornell University, Ithaca, NY
NEWSHA AJAMI, Lawrence Berkeley National Laboratory, Berkeley, CA
PEDRO J. ALVAREZ (NAE), Rice University, Houston, TX
JONATHAN D. ARTHUR, American Geosciences Institute, Washington, DC
RUTH L. BERKELMAN (NAM), Emory University, Atlanta, GA
JORDAN R. FISCHBACH, The Water Institute of the Gulf, Pittsburgh, PA
ELLEN GILINSKY, Ellen Gilinsky, LLC, Seattle, WA
ROBERT M. HIRSCH, U.S. Geological Survey (retired), Reston, VA
VENKATARAMAN LAKSHMI, University of Virginia, Charlottesville
MARK W. LeCHEVALLIER, Dr. Water Consulting, LLC, Morrison, CO
CAMILLE PANNU, Columbia University, New York, NY
DAVID L. SEDLAK (NAE), University of California, Berkeley
JENNIFER TANK, University of Notre Dame, Notre Dame, IN
DAVID WEGNER, Woolpert Engineering, Tucson, AZ

Water Science and Technology Board Staff

DEBORAH GLICKSON, Board Director
LAURA EHLERS, Senior Program Officer
STEPHANIE JOHNSON, Senior Program Officer
CHARLES BURGIS, Associate Program Officer
MARGO REGIER, Associate Program Officer
JONATHAN TUCKER, Associate Program Officer
JEANNE AQUILINO, Finance Business Partner
EMILY BERMUDEZ, Program Assistant
PADRAIGH HARDIN, Program Assistant
MILES LANSING, Program Assistant
OSHANE ORR, Program Assistant

Preface

Coeur d'Alene Lake (or the Lake) in northern Idaho is an invaluable recreational, economic, and natural resource asset for residents of Idaho, eastern Washington, and the nation. The issues that confront the Lake today reflect the history of other water bodies in the western United States. As the United States expanded westward, during the era of "manifest destiny," what were once tribal homelands were confiscated across the west. In this case, it left the Coeur d'Alene Tribe with a fraction of its original homeland, including only the southern part of Coeur d'Alene Lake and remnants of what was once its breadbasket in the north. Although more recent agreements have granted some lands back to the Coeur d'Alene Tribe, those include wetlands and inshore areas so contaminated with lead and arsenic that the Tribal Authority has had to ban traditional practices like harvesting water potatoes in those parts of the northern Lake.

When the mineral extraction boom began in the western United States in the last half of the 1800s, usually after discoveries of gold, the Coeur d'Alene basin was one of the centers of activity. Mining, milling, and smelting dominated the valley from 1880 through the 1960s over a large area 60 kilometers upstream from the Lake. Like elsewhere in the west, great riches were extracted from mineral deposits in the basin (dominated in this case by extraction of lead and silver). What was left behind seemed like a moonscape[1] of unvegetated floodplain soils and mountainsides: lands so contaminated that little vegetation could grow. Riverbanks and riverbed sediments were heavily contaminated with lead, zinc, arsenic, and cadmium from wastes released directly to the South Fork of the Coeur d'Alene River, and barren mountainsides reflected the atmospheric fallout from smelter operations. Contaminated runoff made its first detectable impact on the sediments of Coeur d'Alene Lake around 1900. Contaminated waters and sediments continue to be deposited in the Lake today, reflecting the legacy of nearly 100 years of mineral extraction.

In 1983, the U.S. Environmental Protection Agency (EPA) listed the Bunker Hill Mining and Metallurgical Complex in northern Idaho as a Superfund site on the National Priorities List and remediation activities began shortly thereafter. In 1998, EPA began applying Superfund requirements beyond the original Bunker Hill boundaries to areas throughout the 1,500-square mile Coeur d'Alene River basin project area but did not select a remedy to address Coeur d'Alene Lake.

The latest phase in the development of the west includes increased population growth in the more rural communities, especially exceptionally scenic areas such as Coeur d'Alene Lake. In 2021, the *Wall Street Journal*

---

[1] My personal observation from visits to the region in the 1970s and 1980s.

declared that Coeur d'Alene represented the "hottest" real estate market in America. With population growth came concern about nutrient inputs and eutrophication that, in other water bodies, have resulted in a loss of ecosystem services critical to development, including loss of lake clarity, eutrophication, nuisance algal blooms, and loss of native fisheries. In 2002, EPA began plans for remedial actions to address mine waste contamination in the lower basin of the Coeur d'Alene River up to the Lake. But rather than select a remedy for the Lake itself, EPA agreed to have the state of Idaho and the Coeur d'Alene Tribe develop and implement a Lake Management Plan (LMP) outside of the Superfund process. The LMP was published in 2009. The goal of the plan was to manage nutrient loads entering the basin in an effort to sustain adequate oxygen levels in the Lake's water column and minimize the potential for metals mobilization.

The issues of concern today represent the nexus of the issues that developed over this history. More than two decades of scientific studies of the Coeur d'Alene basin have built a body of data useful to evaluating the intersection of these issues. But geographic differences in jurisdictions appear to be one factor limiting collaborative syntheses of these studies. EPA-sponsored studies and ongoing U.S. Geological Survey (USGS) stream monitoring are concentrated in the watershed. The Tribal Authority conducts monitoring and special studies, mostly in the Lake south of the confluence of the Coeur d'Alene River. The Idaho Department of Environmental Quality (IDEQ) conducts monitoring and studies north of the confluence. Monitoring data are publicly available, data reports describing trends were joint authored by the Tribal Authority and the IDEQ until recently, and an impressive array of process studies were published by USGS and academia over a 10-year period (1995–2005) but have continued at a much slower pace since then.

Nevertheless, concerns remain about questions critical to the future of the Lake. Part of the concern appears to stem from differences in interpretations of the existing data, perhaps influenced by collisions among the interests of different constituencies (not surprising in a situation where the problem is complicated and the stakes are high). Thus, critical questions remain. Is remediation of the damage from mineral extraction in the upstream landscape benefitting the Lake? Are nutrient inputs from increasing development and urbanization affecting lake clarity, and will they ultimately result in eutrophication? Are there interactions between declining metal inputs to the Lake, as remediation proceeds, and the types of lake productivity that affect clarity and other signs of eutrophication? Can the native fisheries and wetland resources that once provided the Tribe's breadbasket and traditional practices be recovered? Hence, the request was made to the National Academies of Sciences, Engineering, and Medicine to establish an *ad hoc* consensus committee to analyze available data and information about Coeur d'Alene Lake water quality and provide recommendations to address identified issues of concern.[2]

The study was commissioned specifically to evaluate the future of water quality in Coeur d'Alene Lake by three of the parties with closest connections to the well-being of the Lake: the IDEQ, EPA, and Kootenai County, Idaho. The remit was restricted to analyses relevant to the status and future of the Lake. The committee appointed by the National Academies began its task in January 2021, in the middle of the COVID-19 pandemic. The pandemic limited travel throughout the study period and forced the committee to meet virtually for five of its six meetings—an unprecedented approach for such panels. Although the committee was unable to visit the Lake as a group, we were aided greatly by a virtual course from the University of Idaho on the history of the Coeur d'Alene Tribe and a virtual film tour of key sites in the valley put together by Jamie Brunner of IDEQ and Ed Moreen of the EPA. The Coeur d'Alene Tribe, IDEQ, EPA, and USGS went out of their way, over and over, to find the data and reports we needed for the investigation. Our efforts included reviews of the relevant literature, analyses of raw data provided as above, and statistical analysis of trends. Details of analytical methods are presented in the report's appendixes. This is an unusual report for the National Academies in that it includes original analysis and key conclusions that resulted from those analyses. In many cases, the conclusions stand alone and no recommendations were necessarily warranted.

The study was established under the auspices of the Water Science and Technology Board of the National Academies. The Committee on the Future of Water Quality in Coeur d'Alene Lake included 11 individuals whose joint expertise covered the diversity of disciplines relevant to the study. The committee heard from many local experts about ongoing water quality monitoring and modeling in Coeur d'Alene Lake, climate in the Coeur d'Alene

---

[2] The formal study statement of task is found in the Summary.

region, and updates on the Bunker Hill Superfund site. I would like to thank the following individuals for giving numerous informative presentations to the committee: Dan McCracken, Jamie Brunner, Craig Cooper, and Robert Steed, IDEQ; Ed Moreen, Cami Grandinetti, and Kim Prestbo, EPA; Dale Chess, Rebecca Stevens, and Phil Cernera, the Coeur d'Alene Tribe; Lauren Zinsser, Dan Wise, and Chris Mebane, USGS; Chris Fillios, County of Kootenai; Tyler Jantzen, Jacobs; Guillaume Mauger, University of Washington; and Erin Brooks, University of Idaho. The committee also thanks the many individuals who spoke during open-mic sessions or submitted written comments to the committee during the course of the study.

                                  Samuel N. Luoma, *Chair*
                                  Committee on the Future of Water Quality in Coeur d'Alene Lake

Acknowledgments

This Consensus Study was reviewed in draft form by individuals chosen for their diverse perspectives and technical expertise. The purpose of this independent review is to provide candid and critical comments that will assist the National Academies of Sciences, Engineering, and Medicine in making each published report as sound as possible and to ensure that it meets the institutional standards for quality, objectivity, evidence, and responsiveness to the charge. The review comments and draft manuscript remain confidential to protect the integrity of the process.

We thank the following individuals for their review of this proceedings:

James B. Cotner, University of Minnesota
Joseph L. Domagalski, USGS Sacramento
James N. Galloway (NAS), University of Virginia
Matthew Ginder-Vogel, University of Wisconsin
K. David Hambright, University of Oklahoma
Dennis P. Lettenmaier (NAE), University of California at Los Angeles
Ann S. Maest, Buka Environmental
Jerome O. Nriagu, University of Michigan
Kimberly J. Van Meter, Pennsylvania State University.

Although the reviewers have provided many constructive comments and suggestions, they were not asked to endorse the conclusions or recommendations nor did they see the final draft of the report before its release. The review of this report was overseen by Joan B. Rose (NAE), Michigan State University, and Richard G. Luthy (NAE), Stanford University. Appointed by the National Academies, they were responsible for making certain that an independent examination of this report was carried out in accordance with institutional procedures and that all review comments were carefully considered. Responsibility for the final content of this report rests entirely with the authoring Committee and the institution.

Contents

SUMMARY   1

1 INTRODUCTION   11
Introduction to the CDA Region, 13
Mining History and the Superfund Site, 33
Federal, Regional, and Local Authority and Oversight, 38
Current Water Quality Conditions, 40
National Academies Study, 42
References, 44

2 LONG-TERM MONITORING OF COEUR D'ALENE LAKE AND ITS WATERSHED   47
Lake Monitoring, 48
Coeur d'Alene River Sampling, 53
Published Summaries of Long-Term Monitoring Data, 54
Data Availability, 55
References, 56

3 ANALYSIS OF INPUTS TO COEUR D'ALENE LAKE   59
Sources of Metal Input to Coeur d'Alene Lake, 59
The Superfund Remedy and Its Effects on Metal Inputs, 64
Sources of Phosphorus to Coeur d'Alene Lake, 72
Analysis of Inputs to Coeur d'Alene Lake, 80
Conclusions and Recommendations, 119
References, 121

4 IN-LAKE PROCESSES: HYDRODYNAMICS   125
Introduction to Lake Processes, 125
Lake Hydrodynamics, 127
Sediment Transport, 140

Reservoir Modeling of Coeur d'Alene Lake, 145
Conclusions and Recommendations, 149
References, 150

5   IN-LAKE PROCESSES: DISSOLVED OXYGEN AND NUTRIENTS   153
Eutrophication, Productivity, and Oxygen Depletion, 153
Analyses of Dissolved Oxygen Trends, 155
In-Lake Nutrient Analyses, 167
In-Lake Productivity Analyses, 177
Conclusions and Recommendations, 184
References, 186

6   IN-LAKE PROCESSES: METALS   189
In-Lake Processes Relevant to Metals, 189
Analysis of Trends in Dissolved Zinc, 191
Analysis of Trends in Dissolved Cadmium, 201
Analysis of Trends in Total Lead, 204
Conclusions and Recommendations, 212
References, 213

7   LAKE BED PROCESSES   215
Introduction to Biogeochemical Processes in Lake Sediments, 215
Metal(loid) Dynamics Within CDA Lake Sediments, 219
Key Issues About Sediment Processes Under Changing Conditions, 233
Conclusions and Recommendations, 235
References, 238

8   GAPS IN LAKE AND WATERSHED MONITORING   241
Improvements to River Monitoring, 241
Improvements to Lake Monitoring, 245
Institutional Considerations, 253
Conclusions and Recommendations, 253
References, 255

9   RISKS OF METALS CONTAMINATION IN COEUR D'ALENE LAKE   259
Human Health Risks, 259
Ecological Health, 273
Conclusions and Recommendations, 293
References, 294

10   FUTURE WATER QUALITY CONSIDERATIONS   303
Climate Change, 303
Will Climate Change, Population Growth, and Land Use Reverse Water Quality Trends in CDA Lake?, 321
Future Water Quality Scenarios, 327
Conclusions and Recommendations, 335
References, 336

ACRONYMS 339

APPENDIXES

A Coeur d'Alene Watershed Analysis Methodology for Metals and Nutrients 341
B In-Lake Analysis Methodology for Metals, Nutrients, and Dissolved Oxygen 361
C Biographical Sketches of Committee Members 367

Summary

Coeur d'Alene Lake (CDA Lake, or the Lake) in northern Idaho is an invaluable natural, recreational, and economic resource to residents of Idaho and eastern Washington. The 3,740-mi^2 watershed that drains into the Lake can be divided into five major subbasins: the North and South Forks of the CDA River (collectively called the upper basin), the lower CDA River basin, the St. Joe River basin, and the nearshore subwatersheds directly surrounding CDA Lake. The entire watershed lies within the homeland of the Coeur d'Alene Tribe (the CDA Tribe). Starting in the late 1880s, the area along the South Fork of the CDA River (the Silver Valley) was mined for lead, silver, and zinc. Mineral extraction and beneficiation wastes, laden with heavy metals, were discharged to the South Fork of the CDA River and flowed downstream, subsequently contaminating more than 75 million metric tons of CDA Lake sediments with lead, cadmium, arsenic, and zinc. Although mining activities have declined significantly and metal inputs to CDA Lake have also declined, metal concentrations in CDA Lake and its sediments remain at or above ambient water quality standards set by the State of Idaho and the CDA Tribe, with arsenic, cadmium, lead, and zinc concentrations orders of magnitude higher than in most lakes in the United States.[1]

Coincident with the diminution of mining, in 1983 the Bunker Hill mining district in the Silver Valley was designated as a regulated hazardous waste site under the nation's Superfund law, and remediation of the affected parts of the CDA basin began. Although all mineral extraction activities were located ~60 km or more upstream, the Lake continues to be a repository for wastes heavily contaminated with metals. Nonetheless, CDA Lake itself was not included as a target of remediation under Superfund. Rather, protection of water quality in the Lake was left to a Lake Management Plan to be implemented by the CDA Tribe, which owns the bed and banks of the southern third of the Lake, and the State of Idaho, which controls the northern two-thirds. The Plan was based on the assumption that increased nutrient loading from lakeshore development, land use changes in the basin, and other dynamics might pose a potential new threat to the Lake that could promote anoxic conditions in the bottom waters and release metals bound to Lake sediments that would then pose a threat to ecosystems and human health.

Figure S-1 shows a map of CDA Lake, its major inflows (the CDA River and the St. Joe River) and outflow (the Spokane River), and the monitoring stations where water quality data have been collected variously over the past 30 years by the U.S. Geological Survey (USGS), the Idaho Department of Environmental Quality (IDEQ), the CDA Tribe, the U.S. Environmental Protection Agency (EPA), independent researchers, and others. Despite sustained efforts to collect data on various water quality parameters in the Lake and watershed, comprehensive analyses of the monitoring data and explicit testing of hypotheses are rare. Given the uncertainties in the data

---

[1] Text here and throughout the report was modified after report release to clarify that mining has not completely ceased in the CDA watershed.

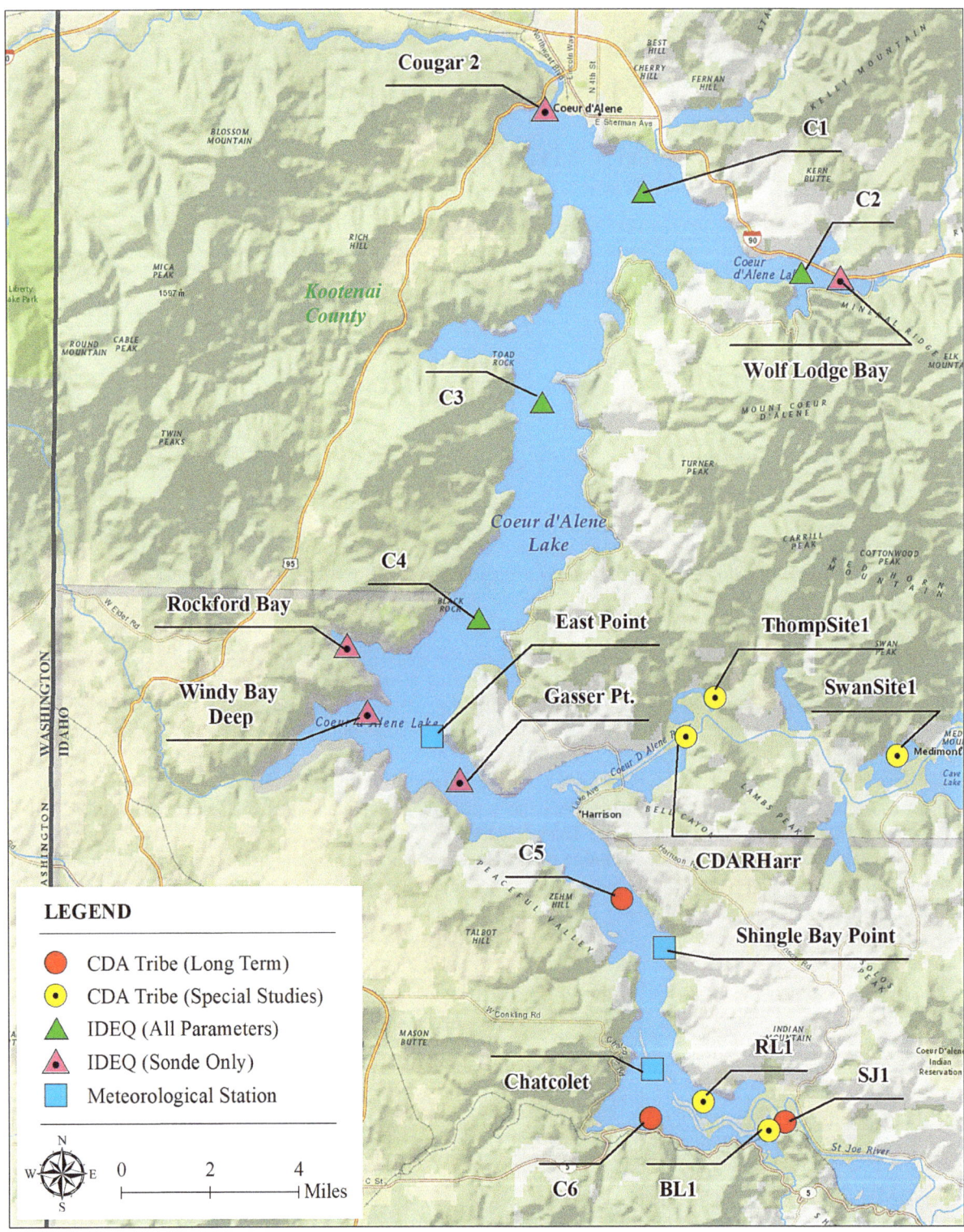

FIGURE S-1 IDEQ and CDA Tribe long-term monitoring locations in CDA Lake. SOURCE: Generated by the Committee using data from IDEQ, CDA Tribe, and the USGS National Map and associated datasets.

> **BOX S-1**
> **The Future of Water Quality in Coeur d'Alene Lake: Statement of Task**
>
> An *ad hoc* consensus committee appointed by the National Academies of Sciences, Engineering, and Medicine will assess whether Coeur d'Alene Lake is at near-term risk of going anoxic and releasing toxic metals back into the water column. This will involve reviewing historical and recent water quality data, and any available modeling efforts, stemming from the 2009 Lake Management Plan and other available information, with the goal of determining what future water quality conditions in the lake will be. More specifically, the study will
>
> - Evaluate current water quality in the lake, lower rivers and lateral lakes with a focus on observed trends in nutrient loading and metals concentrations, while also considering how changes in temperature, precipitation, and streamflow could affect those trends.
> - Consider the impacts of current summertime anoxia on the fate of the metals and nutrients.
> - Consider whether reduced levels of zinc entering the lake as a result of the upgrade to the Central Treatment Plant and other upstream activities are removing an important control on algal growth.
> - Discuss whether metals currently found in lake sediments will be released into the lake if current trends continue. If sufficient data are not available to result in a high level of confidence in its conclusions, the National Academies will identify the additional data that are required to achieve an appropriate level of confidence.
> - Discuss the relevance of metals release in the lake to human and ecological health risks.

analysis activities, along with concerns of various stakeholder groups, in late 2020, IDEQ, EPA, and Kootenai County requested that the National Academies of Sciences, Engineering, and Medicine analyze available data and information about CDA Lake water quality and provide insights about future Lake conditions. The statement of task for the National Academies' study is found in Box S-1.

Chapters 1 and 2 introduce CDA Lake and its watershed and current long-term monitoring efforts. Chapter 3 reviews the watershed's historical mining activities, the Superfund remedy, and land use changes, and it analyzes 30 years of data on metal and nutrient inputs to the Lake from the watershed. The next three chapters evaluate in-lake data relevant to the committee's charge to assess trends over time, including physical data (Chapter 4), dissolved oxygen and nutrient data (Chapter 5), and metals data (Chapter 6). Chapter 7 reviews studies done to understand processes occurring in the Lake sediments to determine if deteriorating water quality conditions in the Lake might lead to metals release from the sediments. Given the analyses of the previous four chapters, Chapter 8 describes gaps in long-term monitoring that should be filled in order for water quality in CDA Lake to be adequately and continuously assessed into the future. Chapter 9 reviews the human health and ecological risks posed by metals in CDA Lake. Finally, Chapter 10 discusses the future of water quality in CDA Lake, including climate change and how it may alter the trends observed in Chapters 3 through 7. Each chapter ends with conclusions and (where appropriate) recommendations that synthesize more technical and specific statements found within the body of each chapter. Because of the extensive data analyses in this report, not every conclusion leads to a recommendation. Several key conclusions and recommendations are compiled in this summary.

ANALYSIS OF INPUTS TO COEUR D'ALENE LAKE

Inputs of lead, cadmium, and zinc to CDA Lake reflect the century-long legacy of mine waste deposition in the Lake's watershed. The frequent floods that transport mining wastes downstream and the minimal dilution that occurs between the source of the primary contamination and the Lake contribute to ongoing metal loading to the Lake. In Chapter 3, the committee analyzes inputs to the Lake, focusing on the two major river systems that contribute 84 percent of the inflow—the CDA River and the St. Joe River. The trends observed in the river monitoring network data over the past 30 years provide clues about the effectiveness of the Superfund cleanup efforts in the upper basin and can help prioritize and plan for future remediation efforts. Although the Superfund remediation has reduced metal inputs from the upper basin via stabilization of the landscape, capping, and sequestration activities, the lower basin comprises an immense stockpile of metal-enriched particulates poised for transport to CDA Lake. The following conclusions about trends in inputs to CDA Lake are found in Chapter 3.

Cadmium, lead, and zinc concentrations and loads into the mainstem CDA River from the South Fork have declined over the past 30 years, and Superfund activities have likely contributed to this decline. For the South Fork of the CDA River at Elizabeth Park, fluxes of the three metals have declined since the early 1990s, with 2020 values being 40 to 50 percent of their maximum. Similarly, at the South Fork near Pinehurst, just downstream of the Box,[2] fluxes of zinc and cadmium have declined more than 60 percent since 1992, while the decline in lead flux has been about 80 percent. Stabilization of the landscape, capping, and sequestration activities have likely been effective at reducing fluxes of particle-associated lead. For zinc, remedial activities in the upper basin and particularly in the Box, including the continuous improvements at the Central Treatment Plant, have helped substantially lower concentrations and fluxes.

Reductions of total lead fluxes from the South Fork of the CDA River were offset by processes in the lower basin that released lead between 2000 and 2010, such that present-day inputs to the Lake are still substantial. Overall, lead flux to the Lake at Harrison was still 1.3 times higher in 2020 compared to the 1990s because of the increase in fluxes between 2000 and 2010. In 2020, lead fluxes into the lower basin at Pinehurst were only 2.6 percent of lead flux to CDA Lake at Harrison, demonstrating that there are large reservoirs of metals in the river sediments and floodplains of the lower basin. Future decreases in lead fluxes into the Lake will be determined more by evolving storage and release mechanisms in the lower basin than by further efforts to reduce lead flux from the South Fork watershed. The committee's analysis of total lead in high-flow discharges at Harrison shows that lead concentrations in these flows have decreased over time, suggesting that remediation is having a beneficial effect. Remediation of the lower basin will require careful planning so as not to remobilize metals and increase their transport to the Lake.

There has been a downward trend in both zinc and cadmium concentrations and fluxes throughout the CDA basin, and fluxes of both metals to the Lake (measured at Harrison) were lower in 2020 than in 1992 (by 63 and 45 percent, respectively). At the CDA River at Harrison, the cadmium and zinc fluxes leveled off during 2000–2010 but are declining again in the most recent decade. Unlike total lead, as of 2020 the inputs of total cadmium and total zinc from the South Fork of the CDA River to the lower CDA River were 43 and 44 percent (respectively) of the outputs of the lower CDA River to the Lake. This suggests that further reductions in cadmium and zinc coming from the South Fork *are* likely to be important to reducing inputs of these metals to the Lake. Targeted studies and trend data show that the primary sources of cadmium and zinc are now base flow, presumably coming from the groundwater system. The Central Treatment Plant is now a minor source of zinc.

Over the past decade, total phosphorus fluxes and concentrations at monitoring sites in the CDA River, the St. Joe River, and the Spokane River below the Lake outlet have all been declining (typically 20–30 percent reductions during the 2010–2020 decade). In the case of the CDA River, this is a reversal of the trend observed over the prior decade. Like lead, total phosphorus flux to the Lake in 2020 was higher (by 2.3 times) than in the early 1990s. Projecting future trends of phosphorus in Lake CDA will require a sustained effort at monitoring and regular data synthesis for phosphorus across the whole watershed, with monitoring efforts closely connected to research aimed at understanding the reasons for this current decline. Without a better understanding of the history of phosphorus transport in the whole watershed, there is no basis for projecting future phosphorus transport or the potential for future increases in phosphorus loading to the Lake.

IN-LAKE PROCESSES: HYDRODYNAMICS

Seasonal and long-term water quality trends within a lake are the result of the interactions among the key physical, chemical, and biological processes and the associated process drivers that alter inputs, outputs, and internal dynamics within the lake. The main high-level processes relevant within CDA Lake include the lake hydrodynamics, sediment transport, biogeochemical processes, and primary production—with the first two being covered in Chapter 4. The CDA River provides the main source of particulate and dissolved metals, while the cleaner St. Joe River dilutes metal concentrations in the southern reach of the Lake during the spring. Internal lake processes, including in-lake hydrodynamics, biogenic recycling, benthic flux from the Lake sediment to bottom waters, and physical processes

---

[2] The Box is an area of the Superfund site 3 miles (4.8 km) wide and 7 miles (11.2 km) long, from Kellogg on the eastern end to Pinehurst on the western end, that had the most intense mining activity.

such as sediment transport all may contribute to elevated concentrations later in the year. Internal hydrodynamics have redistributed particulate metal contamination through the northern and southern Lake, as evidenced by widespread sediment contamination. In Chapter 4, the committee analyzed physical data from the watershed and CDA Lake, along with special studies on modeling and sediment transport in CDA Lake, and makes the following conclusions.

The limited data available on specific conductivity in the river inflows and CDA Lake showed that water from the St. Joe River entered CDA Lake as an interflow, meaning that, because of its density relative to the stratification of the Lake, it intrudes below the thermocline but above the Lake bottom. The CDA River inflow, on the other hand, is classified as an overflow, intruding into the epilimnion above the thermocline. **Hence, the two major river inflows are not likely to play a major role in resuspending sediment within CDA Lake.** Sediment resuspension could, however, occur in the littoral zones of the Lake due to other factors. The identification of those areas most subject to sediment resuspension, together with the relevant water quality impacts, can only be ascertained with a nearshore monitoring program, something that currently does not exist.

The committee's analysis of conductivity data found that a St. Joe River inflow of 1,000 cubic feet per second (cfs) is a threshold below which thermal stratification commences and internal lake processes can predominate. At river inflows above 1,000 cfs, which occur during winter and early spring, river discharge controls water quality (dissolved oxygen and pH) in CDA Lake, and the Lake behaves like a run-of-the-river system with little opportunity for biogeochemical processes to become established. At St. Joe River inflows below 1,000 cfs, which generally occur in June or later, internal dynamics and thermal stratification become important for CDA Lake, especially at site C5. It is during this period that water column processes such as nutrient uptake, phytoplankton proliferation, and decomposition can happen.

There is a critical lack of deterministic model usage for CDA Lake—for both heuristic and predictive purposes—despite having invested in the first stages of developing such a model (ELCOM–CAEDYM).[3] Development of a powerful 3-D hydrodynamic and water quality model began over 15 years ago, but little has been done to use the model to better understand key processes within the Lake, the evolution of changes within the Lake, and the likely trajectory of the Lake under future climate changes.

IN-LAKE PROCESSES: DISSOLVED OXYGEN AND NUTRIENTS

Chapter 5 analyzes water column data from CDA Lake over the past 30 years to reveal trends in dissolved oxygen, nutrients, and lake productivity. The chapter also specifically addresses the item in the statement of task that asks whether reduced levels of zinc are removing an important control on algal growth. The following conclusions about trends in nutrient and dissolved oxygen concentrations in CDA Lake are found in Chapter 5.

For the most recent decade, the data indicate declines in total phosphorus concentration in CDA Lake, although this decline was not statistically significant in the northern lake. Declines in total phosphorus in the Lake are consistent with the declines in total phosphorus from the two major rivers entering the Lake. The ability to project future changes in phosphorus inputs and concentrations in the Lake will depend on research aimed at better understanding the causative mechanisms of these recent trends in phosphorus delivery from the watershed to the Lake. Trends in total nitrogen concentration in the Lake are more complex than those for total phosphorus. Gaining a better understanding of the nitrogen cycle, and analyzing trends of nitrogen concentration in the Lake and rivers, is an important topic that needs to be included in the research agenda supporting adaptive management of lake trophic state.

The evidence that dissolved oxygen concentrations are worsening in the bottom waters of the C1, C4, C5, and C6 stations is equivocal at best. Rather, there is an increasing dissolved oxygen concentration trend particularly evident at C5.[4] **Coupled with the recent trends in phosphorus concentrations in the Lake and phosphorus loading to the Lake, low dissolved oxygen is not a current problem in the main body of the Lake,[5] nor is it expected to become a problem if current trends continue.** However, if climate change were to strengthen

---

[3] ELCOM-CAEDYM is the Estuary, Lake and Coastal Ocean Model–Computational Aquatic Ecosystem DYnamics Model.

[4] This conclusion was edited after report release to reflect only the dissolved oxygen trend analyses shown in the report, which were for bottom waters.

[5] As discussed in Chapter 1, the main body of the Lake is characterized by the primary long-term monitoring stations C1 through C5; this footnote was added after report release.

thermal stratification substantially in the future, dissolved oxygen concentrations at the sediment–water interface would likely decrease.

The available field evidence does not support the concept that the current high zinc concentrations in the Lake suppress chlorophyll *a*. CDA Lake supports lower amounts of chlorophyll *a* per unit phosphorus than generally observed for lakes worldwide. Although this is consistent with zinc suppression of phytoplankton biomass, beyond this observation there is little evidence from more detailed analysis of field data that the current high zinc concentrations suppress chlorophyll *a*. For example, chlorophyll *a* levels are disproportionately lower than expected in the southern part of the Lake in both zinc-enriched and zinc-poor locations. Consistent with this, multiple regression analysis relating chlorophyll to total phosphorus and total zinc finds a strongly positive and statistically significant association for total phosphorus but a highly nonsignificant association for total zinc. Further research involving field experimentation would help develop greater confidence in predicting the response of Lake chlorophyll concentrations and particular taxa of phytoplankton to potential reductions in legacy metal contamination in the Lake.

IN-LAKE PROCESSES: METALS

Chapter 6 focuses on the concentration of heavy metals of concern, particularly lead, zinc, and cadmium, in CDA Lake. In addition to revealing trends in metals concentrations over the past 30 years, the committee attempted to elucidate seasonal trends that are indicative of various processes occurring in the water column, such as hydrodynamics, benthic flux, and biogenic cycling. The following conclusions about trends in metals concentrations in CDA Lake are found in Chapter 6.

Downward trends in dissolved zinc concentration in CDA Lake at sites C1 and C4 over the past 30 years are highly significant. Furthermore, zinc concentrations at these sites are decreasing at a similar rate to declines in zinc inputs from the CDA River. Differences in dissolved zinc concentrations between surface waters and bottom waters during stratification suggest inputs to bottom waters from internal sources are occurring. But internal cycling of zinc has not detectably slowed the response of the northern Lake to changes in zinc input driven by remediation in the CDA basin. The dynamics of zinc at C5 (where zinc concentrations are declining more slowly) are more complex than at C1 and C4 and reflect spring dilution from the St. Joe River, summer/fall hydrodynamic redistribution of metals from north to south, and other internal processes.

Dissolved cadmium concentrations at C1 and C4 declined from 2004 to 2020, with virtually all of the decline occurring after 2014. Dissolved cadmium concentrations in deeper water (> 20 m) are greater than those in shallower water, by about 15 percent, suggesting that internal processes affect dissolved cadmium concentrations (as they do for zinc). Lake cadmium trends are highly responsive to changes in cadmium inputs from the CDA River.

From about 2003 to 2012, total lead concentrations in CDA Lake rose slowly but they have declined over the past eight to ten years. Most lead enters the Lake during periods of high discharge, such that the future trajectory of Lake lead concentrations will depend strongly on the extent of scouring of legacy sediment from the bed and banks of the CDA River. Despite thermal stratification in summer, there were only small differences between total lead concentrations in surface and bottom waters at C1 and C4. This suggests minimal flux of lead from internal sources and is consistent with strong lead binding to particulates and the lower bioavailability of lead to phytoplankton compared to zinc and cadmium.

LAKE BED PROCESSES

Chapter 7 examines the (bio)geochemical reactions that control the partitioning of lead, cadmium, zinc, and arsenic between Lake sediments and porewater, ultimately leading to the specific conditions of CDA Lake. Although data available to analyze were sparse, the committee examined metal deposition and migration within the Lake sediments and the operative processes controlling dissolved porewater metal(loid) concentrations and possible entry into Lake waters. It also assessed the potential for eutrophication and/or pH changes to release metals from Lake sediments into the water column. The following conclusions about processes in Lake sediments, including possible metals release to the overlying water column, are found in Chapter 7.

Iron(III) (hydr)oxides are an abundant and dominant control on metal and arsenic concentrations in the Lake sediments (at the sediment–water interface). With the pH conditions of the sediments, the Fe(III) oxides serve as principal adsorbents of arsenic, cadmium, lead, and zinc that regulate dissolved concentrations as shown through measurement and modeling. Due to the abnormally high metal concentrations (particularly iron, which makes up more than 10 percent of the solids) within the sediments, sulfur occurs in quantities insufficient to control the full suite of heavy metal concentrations and may exert local and selective controls, but not universal control, on metal retention in the sediments.

The greatest threat of enhanced anoxia, if it were to occur, is release of arsenic into the Lake water column; however, there is no evidence that anoxia is getting worse. With As(V) adsorption on Fe(III) oxides having a dominant control on arsenic concentrations within the porewaters of Lake sediments, decreased oxygen concentrations leading to anoxia in bottom water could promote their reductive dissolution within the upper sediments and arsenic release into overlying Lake waters. A second threat of bottom water anoxia is release of phosphorus. Although not redox active itself, phosphate is largely bound to Fe(III) oxides, and anoxia leads to the reductive dissolution of the Fe(III) phases and thus the concomitant release of phosphate. If the Lake is phosphorus limited, it is possible that such release of phosphorus from the Fe(III) oxides could create a feedback loop that further promotes biological productivity, anoxia, and arsenic release from the sediment.

Because the adsorption edge for zinc on iron oxides occurs in the pH range of 6.0–7.0 (typical of CDA Lake), local lowering of pH can cause release of dissolved zinc from the sediment. A decline in pH to less than 7 occurs in bottom waters of CDA Lake in some locations and some years, with the onset of stratification. Of the metal(loid) contaminants, zinc has the highest dissolved concentration within the sediment porewater, while cadmium is present at much lower concentrations than zinc. Upward diffusion of zinc from porewaters and desorption from surficial layers of the sediment is possible at pH less than 7. Less of a concern for zinc release would be an alteration in oxygen concentrations within the bottom waters and top few centimeters of the sediment. Lead partitions strongly to Lake sediments both under oxygenated and anoxic conditions and is less likely than zinc or cadmium to be released from sediments into the water column.

GAPS IN WATERSHED MONITORING

Chapter 8 considers improvements in several aspects of the Lake and river monitoring programs, such as where monitoring occurs, which analytes should be monitored (including questions of detection and precision), how samples should be collected and how many, and when samples should be collected. The program would benefit by adding monitoring for physical and ecological parameters and taking additional sediment cores (including porewater) as well as by improving methods for sampling of phosphorus and metals. The following conclusions and recommendations about long-term monitoring are found in Chapter 8.

Understanding the water quality of CDA Lake would be improved by increasing the spatial and temporal intensity of sampling in the Lake and the rivers. With respect to Lake assessments, the Lake program would benefit by expanding to encompass littoral areas, where the greatest interactions with the public occur and watershed impacts manifest. A strategy to increase temporal resolution in Lake sampling can be achieved by implementation of carefully chosen sensors targeting physical (temperature), chemical (oxygen), and biological (chlorophyll, fluorescence) parameters of interest. For the rivers, the sampling strategy needs increased frequency, particularly at the upstream and downstream ends of the lower basin, with continued attention to sampling high-flow events. Use of new continuous monitoring strategies, particularly for turbidity, can be of great value to the estimation of transport into and out of the lower CDA River reach.

An efficient sampling strategy designed to better understand inputs of nutrients from the lakeshore tributaries would benefit Lake management as the population in nearshore areas grows. Since the time this study began, the IDEQ has initiated a strategy to fill this information gap. It will be crucial after a few years of data have been collected to undertake and publish a synthesis of what can be learned from this monitoring of the Lake's tributaries, and to use that to improve the monitoring network and ongoing analyses. The goal would be to use these new data and geospatial landscape and development data to produce estimates of nutrient fluxes across the majority of the previously unmonitored lakeshore watersheds.

Important ecological components of the Lake are understudied. Targeted expansion of ecological monitoring beyond the phytoplankton community could help identify how ecological processes in the Lake are responding to changing metal and nutrient concentrations. Obvious questions relate to the effects of the legacy of metal enrichment on the pelagic and benthic food webs as well as the status of food webs in parts of the Lake with different exposures to metals.

The agencies involved in data collection are encouraged to provide a mechanism to make the relevant data available to the wider community of stakeholders, agencies, and scientists. The river data are already available through such a system, but the Lake data are not. Furthermore, to succeed at adaptively managing the Lake for decades into the future, **a scientific and institutional structure for carrying out data synthesis, coordinated among jurisdictions and interest groups, is needed.** The required synthesis tasks include regularly evaluating mass balances, relating concentration trends to Lake processes and inputs, generating hypotheses about system drivers, and periodically evaluating the ecological and human health implications of the findings.

RISKS OF METALS CONTAMINATION IN COEUR D'ALENE LAKE

Human exposures and ecological risks associated with CDA Lake itself have not been the subject of comprehensive study compared to the systematic evaluation of risks in the basin upstream of the Lake. Trends in the basin indicate that human exposures to metals like lead have declined, but further assessment of human health risks from occupational, recreational, and subsistence-living exposure to lead and arsenic in the Lake could benefit the region. To better assess ecological risk of metals in CDA Lake, it is important to advance the body of knowledge across multiple levels of biological organization. Identifying present-day ecological implications of the legacy of metal contamination in CDA Lake is a first step toward addressing the future viability of ecosystem services, such as biodiversity, ecological functions, fisheries, wildlife, and support for activities ranging from recreation to a subsistence life style. The following conclusions about better understanding human health and ecological risk from CDA Lake metals are found in Chapter 9.

Expansion of existing monitoring to include a few sensitive nearshore environments could provide an early warning system for the onset of harmful algal blooms and expansion of nuisance-attached algae and of invasive plants. Although on a lake-wide basis the Lake remains oligotrophic with some mesotrophy in the south, experience elsewhere suggests the first signs of changes in trophic status can occur in nearshore, local waters in the form of blooms of attached algae. Expanded lakeshore monitoring could aid in detecting those changes before they become widespread.

Systematically developing a body of knowledge on how CDA Lake food webs are influenced by the legacy of mineral extraction will inform decisions about remediation and efforts to maximize ecosystem services. High priorities for better understanding ecological processes in the Lake include (1) expanded characterization of benthic and pelagic food webs; (2) evaluation of metal exposures in key components of the food web, and (3) experiments with benthic and water column mesocosms to identify thresholds below which the Lake ecosystem will improve.

FUTURE WATER QUALITY CONSIDERATIONS

Chapter 10 reviews the recent climate history of the CDA region, examines climate projections that have been made through 2100, and considers how future changes in climate, population, and land use could affect the trends noted in Chapters 3–7. The chapter also projects long-term trends in metal enrichment in the water column of the Lake (assuming the rate of change in the past decade applies to the future) in order to provide a context for the progress that has been made to date in reducing inputs.

A major impact of climate change likely to affect water quality in the CDA region is air temperature warming as much as 2.5–3°C (4.5–5.4°F) by the year 2050, depending on the month. Data from the CDA region over the past 30 years show warming of about 0.4°F per decade. Increases in air temperature are expected to increase lake temperature and increase fire risk across the region.

Although there are no apparent precipitation trends in the CDA region over the past 30 years, studies in the greater Pacific Northwest suggest that extreme precipitation events will become 5–34 percent more intense by 2080. A shift is expected in the percentage of precipitation that falls as snow versus rain, such that by 2080 the peak snow water equivalent could decrease by an average of 73 percent. Finally, in the Pacific Northwest, the center of timing of the annual hydrograph is predicted to shift from April 15 in the 1980s to as early as March 4 by 2040, although there is not yet evidence of this shift over the past 30 years in the CDA region.

Lake water temperatures have been increasing in surface water at station C4 over the past 30 years (although not in bottom water). It is reasonable to assume that this warming is related to the general global trend toward rising air temperatures, and this can be expected to increase in the future based on ongoing greenhouse gas forcing of the climate. Responses in the Lake to such warming will likely include increased rates of ecosystem metabolism and lengthening of the duration of lake stratification, which may extend periods of dissolved oxygen consumption in the bottom waters and sediments of the Lake as well as lower pH in the bottom waters.

Future climate change may slow or reverse the trends in metals and phosphorus loading to CDA Lake (discussed in Chapter 3) and the trends in dissolved oxygen, phosphorus, and metals concentrations within CDA Lake (discussed in Chapters 5 and 6), and it may increase the potential for metals release from Lake sediments (discussed in Chapter 7). The changes in climate considered by the committee were (1) increased frequency and magnitude of large runoff events, (2) a forward shift in the timing of flow to the Lake, (3) warming of Lake water, and (4) increased frequency and size of fires, along with increases in lakeshore populations. Although they are predicted to occur for the Pacific Northwest, there is not yet evidence in the CDA region of climate effects (1), (2), and (4)—emphasizing the importance of monitoring, data analysis, and further process studies into the future.

Zinc concentrations in surface waters are at or approaching the Lake Management Plan (LMP) target of 36 μg/L in some months and at some locations. If trends from the past decade continue into the future, it will take bottom waters 10 to more than 100 years to reach that target. The slowest changes are occurring in the southern Lake, where the response to declining inputs appears to be buffered by internal inputs from bottom sediments and hydrodynamic inputs from the northern Lake. Dissolved lead concentrations in CDA Lake are already below the LMP target of 0.54 μg/L in the measured Lake locations, with the exception of Site C4 during the spring. Cadmium concentrations are also already below the LMP target of 0.25 μg/L in the three measured Lake locations (C1, C4, and C5).

***

In summary, CDA Lake is beginning to recover from the waste inputs of more than a century of mineral extraction and beneficiation in its watershed. Data from the past 25 years do not show evidence of increasing eutrophication in the main body of the Lake. However, conditions are undocumented in shallower waters of nearshore areas where early signs of deteriorating water quality might first appear. The evidence from available field data does not support the concept that declining zinc concentrations have enhanced eutrophication over the past 15 years; detailed monitoring and experiments could provide further clarification. Better understanding of how metals are affecting the Lake ecosystem is necessary to address the question of whether further reductions in concentrations of zinc and cadmium in the water column of the Lake are more likely to result in ecosystem benefits than in greater risks to the ecosystem.

The mainstem CDA River and its watershed, along with the sediments of the Lake, contain an immense reservoir of sediment-bound metals. If changes in climate, population growth, or remediation activities result in greater metal inputs to the Lake or metal releases from Lake sediments, recovery of the Lake could be slowed or reversed. The processes controlling metal and nutrient inputs, flux from sediments, and cycling within the Lake are generally understood, but they are complex, they differ among contaminants, and important details are missing. The Chapter 10 projections of future water quality in CDA Lake, which assume that trends from the recent past carry forward, suggest that recovery from the legacy of mining will be a decades-long project at best. The committee recognizes that such projections provide an invaluable perspective on the task that lies ahead, but are rarely completely correct. The best preparation for the future will involve fortification and expansion of monitoring to provide an early warning of deteriorating conditions, regular syntheses of data, and targeted studies—all coordinated among interest groups—followed by application of those results to managing the Lake as the future unfolds.

1

Introduction

Coeur d'Alene Lake (hereafter CDA Lake, or the Lake) in northern Idaho is an invaluable natural, recreational, and economic resource to residents of Idaho and eastern Washington. The CDA basin that drains into the Lake encompasses approximately 3,740 mi^2 (9,690 km^2), and ranges in elevation from 2,000 to 6,850 ft above sea level (Woods and Beckwith, 2008; Zinsser, 2018). It is located within the 6,680 mi^2 (17,300 km^2) Spokane River basin, which is a subbasin of the Columbia River basin (Figure 1-1).

The CDA basin encompasses much of Kootenai, Shoshone, and Benewah Counties in Idaho. As seen in Figure 1-2, the CDA basin is composed of five major subbasins that are mentioned repeatedly in this report: the North Fork of the CDA River, the South Fork of the CDA River, the lower CDA River basin, the St. Joe River basin, and the so-called nearshore watershed directly surrounding CDA Lake. The most populated city in the Lake's watershed is the city of Coeur d'Alene near the Lake's northern outlet, home to about 166,000 people and a rapidly growing economic center in the region (Wall Street Journal, 2021). Other population centers around the basin include Harrison, Pinehurst, Smelterville, Kellogg, Wallace, Mullan, and St. Maries. Discussed in greater detail later, land cover within the CDA basin consists of approximately 74 percent forested land, 15 percent shrubland, and the remaining 11 percent a mix of agriculture, barren land, small water bodies, wetlands, and urban development. This entire area lies within the homeland of the Coeur d'Alene Tribe (hereafter CDA Tribe), for whom the entire Lake holds immense cultural, cosmological, and natural importance.

Starting in the late 1880s, the area along the South Fork of the CDA River was mined for gold and silver (and hence was named the Silver Valley). Waste from these mining activities was historically discharged to the South Fork of the CDA River, which flowed downstream and subsequently contaminated over 75 million metric tons (MT) of CDA Lake sediments with metals such as lead, cadmium, arsenic, and zinc. There are also about 32 million MT of contaminated sediments in the floodplain of the lower CDA River (Bookstrom et al., 2001) that could potentially make their way to the Lake in the future. Although metal inputs to CDA Lake have declined since the diminution of mining in the 1980s, metal concentrations in CDA Lake and its sediments remain above ambient water quality standards set by the State of Idaho and the CDA Tribe.

In 1983 the Bunker Hill mining district in the Silver Valley was designated as a regulated hazardous waste site under the nation's Comprehensive Environmental Response, Compensation, and Liability Act of 1980 (CERCLA, or Superfund Act). Administered by the U.S. Environmental Protection Agency (EPA), the Superfund Act requires the cleanup of listed hazardous waste sites to prevent risks to human health and the environment. Remediation at the Bunker Hill Superfund site has focused primarily on the basin of the South Fork of the CDA River, with plans to begin work along the main stem of the CDA River (i.e., the lower basin)

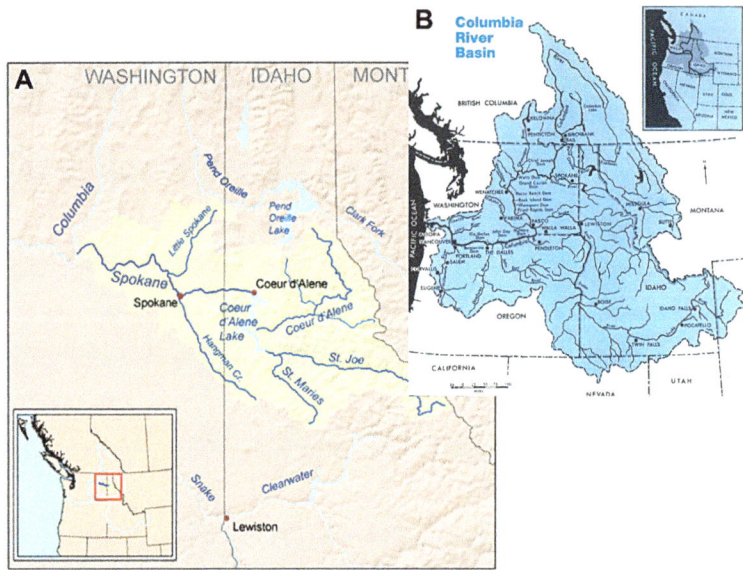

FIGURE 1-1 (A) Map of the Spokane River drainage basin showing CDA Lake and its major tributaries (CDA and St. Joe Rivers). The CDA Lake watershed is a subbasin of the Spokane River basin. (B) Map of the Columbia River drainage basin, which spans seven states and British Columbia, Canada. The Spokane River watershed is a subbasin of the Columbia River basin. The yellow star southeast of Spokane, Washington indicates the location of CDA Lake. SOURCES: Figure A from Kmusser (2008), Figure B from USGS (1981).

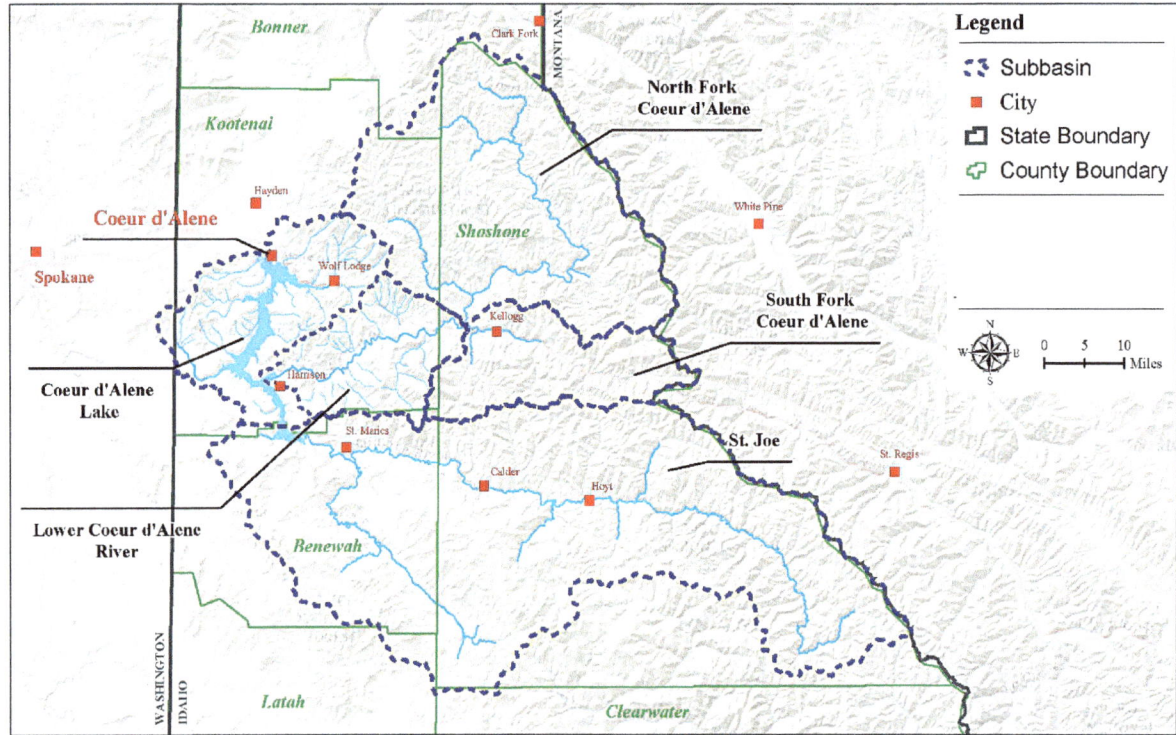

FIGURE 1-2 CDA basin and subbasin map. SOURCE: Generated by the committee using data from the Idaho Department of Environmental Quality (IDEQ), CDA Tribe, and the U.S. Geological Survey (USGS) National Map and associated datasets.

in the coming years. Although heavily contaminated with metals from mining waste, CDA Lake itself was not included as a target of remediation under Superfund. Rather, protection of water quality in the Lake was left to a Lake Management Plan (LMP) to be implemented by the CDA Tribe, which owns the bed and banks of the southern third of the Lake, and the State of Idaho, which controls the northern two-thirds. The LMP, which was made necessary by the sociopolitical complexities of the early 2000s, emphasizes control of nutrient loading to the Lake in order to maintain adequate dissolved oxygen throughout the water column. It was based on the assumption that increased nutrient loading to the Lake could lead to anoxic conditions, which might remobilize metals from the Lake sediments that would then pose a threat to ecosystems and human health. In addition, there was concern that Superfund remedial activities in the basin could lower the Lake's zinc concentration, which was hypothesized to suppress primary productivity and concomitant anoxia (Kuwabara et al., 2006, 2007).

Water quality data from both the Lake and the CDA basin that could shed light on the Lake's current status and potential future conditions have been collected variously over the past 30 years by the U.S. Geological Survey (USGS), the Idaho Department of Environmental Quality (IDEQ), the CDA Tribe, EPA, independent researchers, and other stakeholder groups. Monitoring data have been reported annually to comply with the LMP, and summarizing statements are released periodically (e.g., "total phosphorus levels in the lake are steadily increasing and the rate may be accelerating"—IDEQ, 2020a). Comprehensive analyses of the monitoring data and explicit testing of hypotheses, however, are rare. Given the uncertainties in the data analysis activities, along with divergent concerns of various stakeholder groups, in late 2020, IDEQ, Kootenai County, and EPA requested that the National Academies of Sciences, Engineering, and Medicine (the National Academies) establish an *ad hoc* committee to analyze available data and information about CDA Lake and provide recommendations that address the observed trends in water quality.

INTRODUCTION TO THE CDA REGION

CDA Lake, located in the northern panhandle of Idaho, is the second largest naturally occurring lake in Idaho (Horowitz et al., 1993a; Woods and Beckwith, 2008). It lies between the Selkirk and Coeur d'Alene Mountains, extending from the mouth of the St. Joe River to the headwaters of the Spokane River near the city of Coeur d'Alene. The lake formed 18,000 to 13,000 years ago when ice dams on glacial Lake Missoula breached, releasing catastrophic flood waters and coarse sediment downstream (Horowitz et al., 1992, 1995; Horowitz et al., 1993b; Kahle and Bartolino, 2007). As a result of these floods, the Spokane River was dammed, forming CDA Lake. Figure 1-3 shows the primary hydrologic features of CDA Lake, including its major inputs (the St. Joe River and the CDA River) and output (the Spokane River), the names of locations of interest with CDA Lake that are extensively studied, and the main lateral lakes that lie east of CDA Lake along the CDA River. This diagram is used throughout the report to denote major features of the Lake.

Within the CDA Basin, most of the population lies within Kootenai County, which is also where there is the most extensive lakeshore development and active recreational use of the Lake (e.g., swimming, boating, and fishing). Indeed, the CDA Lake region appeals as a resort area for vacationers and to those looking to relocate to a rural and relatively inexpensive location (Wall Street Journal, 2021). Much of the economic activity in the basin is found in the city of Coeur d'Alene, whose future job growth over the next decade is predicted to be 44.7 percent—higher than the U.S. average of 33.5 percent.[1] Construction is also growing in Kootenai County as indicated by the more than 45 percent increase in building permits in 2021 compared to 2020 (Hardy, 2021).

Nearly all of the shoreline around the Lake is occupied by campgrounds, marinas, or single-family houses (Maupin and Weakland, 2009). Not surprisingly, the growth in population and economic prosperity of the CDA region has been tied to the water resources in the area; for example, Liao et al. (2016) found that property values were positively associated with Secchi depth (a measure of water clarity) and negatively associated with the presence of the invasive aquatic plant watermilfoil.

---

[1] https://www.bestplaces.net/economy/city/idaho/coeur_d'alene

FIGURE 1-3 CDA Lake. The map shows major inputs and outputs to the Lake, some of the closer lateral lakes, neighboring mountain ranges, and several points of interest (black dots). SOURCE: Generated by the committee using data from IDEQ, CDA Tribe and the USGS National Map and associated datasets.

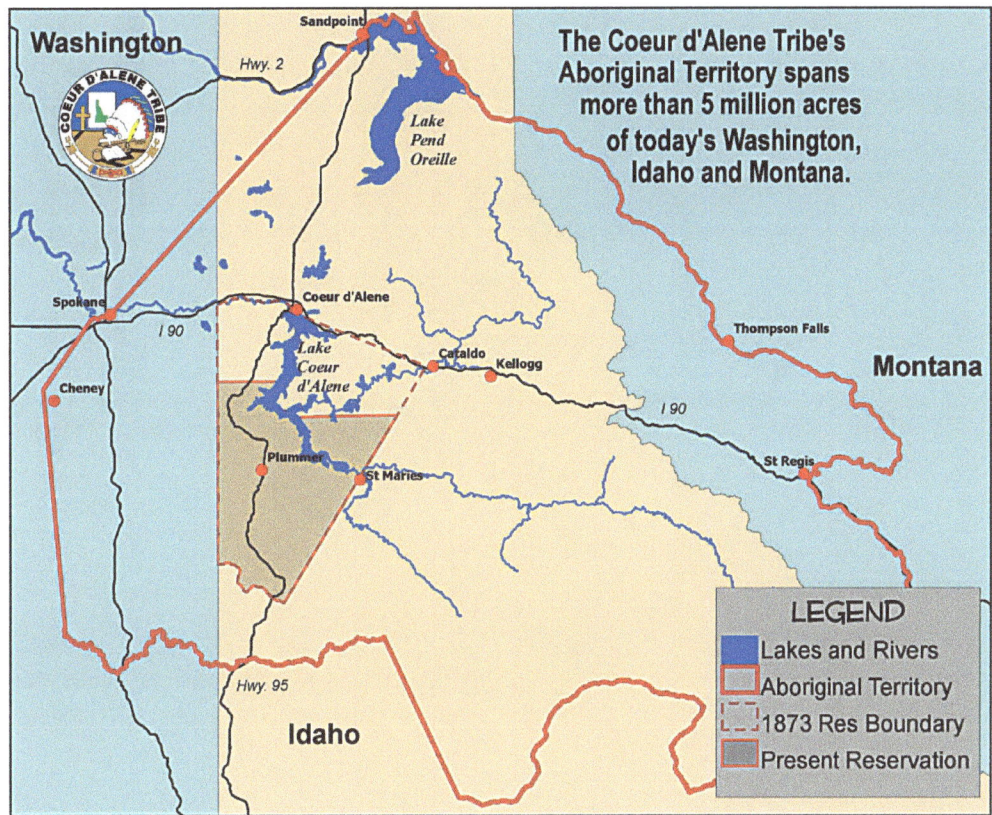

FIGURE 1-4 Map showing the boundaries of the Aboriginal Territory of the CDA Tribe, the 1873 Reservation, and the present reservation. SOURCE: NCAI Partnership for Tribal Governance (2016).

The CDA Tribe's Aboriginal Territory originally encompassed all of CDA Lake and covered more than 5 million acres of land across the states of Idaho, Washington, and Montana (see Figure 1-4). The current boundaries of the CDA Tribe's reservation now span 345,000 acres, with about one-third of the southernmost region of the Lake under tribal jurisdiction (NCAI Partnership for Tribal Governance, 2016; Woods and Beckwith, 2008).

The following sections describe several features of the CDA region that play a role in controlling water quality in the Lake. These include climate (particularly temperature and precipitation), hydrology of the region and hydraulics of the Lake, land use in the CDA basin, and population. These descriptions are not intended to provide a comprehensive overview of the physical, chemical, and biological characteristics of the Lake and its watershed. The reader is referred to the LMP (IDEQ and CDA Tribe, 2009), National Research Council (NCR, 2005), and numerous EPA reports on the Superfund site (e.g., the final Remedial Investigation/Feasibility Study—URS Greiner, Inc., and CH2M Hill, 2001) for relevant background information.

Climate

The watershed of CDA Lake typically has mild and dry summers followed by cold and wet winters, but the region's steep topography leads to wide variations in temperature and precipitation across the watershed.

Precipitation

Mean annual precipitation in the CDA watershed during the period of 1991–2020 was estimated to be 1,090 mm/yr (Daly et al., 2008, 2015). Orographic effects cause significant spatial variation to the mean annual

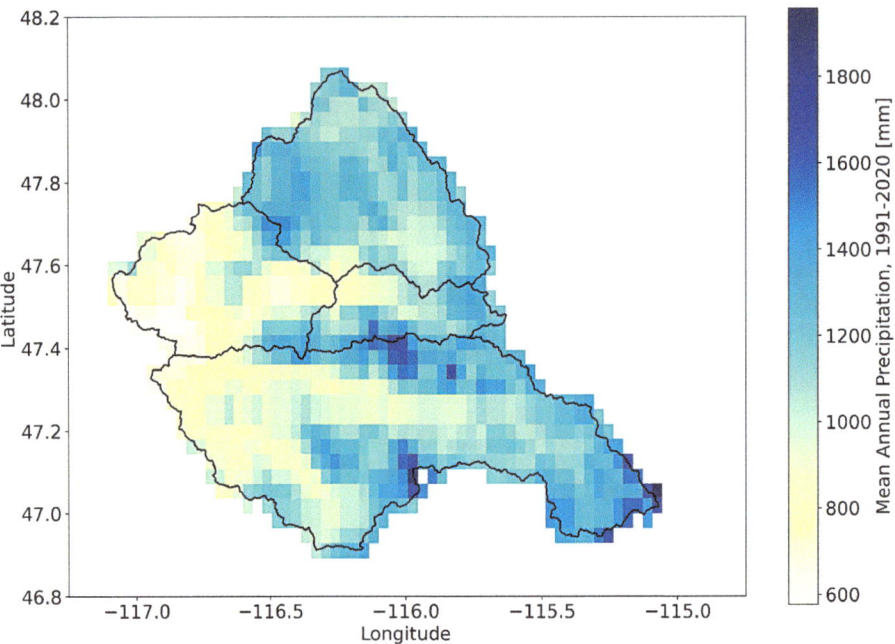

FIGURE 1-5 Mean annual precipitation across the CDA basin, 1991–2020. SOURCE: Generated by the committee with data from the PRISM Climate Group at Oregon State University. https://developers.google.com/earth-engine/datasets/catalog/OREGONSTATE_PRISM_AN81m#citations.

precipitation within the watershed, such that at lower elevations within the watershed (i.e., in the vicinity of the Lake) mean annual precipitation is as low as 600 mm/yr (see Figure 1-5). Conversely, at the highest elevation locations within the Lake's watershed, mean annual precipitation for the 1991–2020 period exceeds 1,800 mm/yr.

Mean annual precipitation varies only slightly between the St. Joe River (1,129 mm/yr), the North Forks of the CDA River (1,162 mm/yr), and South Fork of the CDA River (1,169 mm/yr) watersheds (Figure 1-6). The collection of small catchments that drain directly to CDA Lake (white bars in Figure 1-6) are associated with a significantly lower mean annual precipitation (876 mm/yr), owing primarily to a lower distribution of elevations than any of the tributaries.

Precipitation timing throughout the watershed varies significantly throughout the year (Maupin and Weakland, 2009; Zinsser, 2018). Based on analysis of the 30-year period between 1991 and 2020, the months of November through January are typically associated with the highest precipitation throughout the year, with the watershed receiving on average more than 140 mm/month of precipitation in these months. Indeed, approximately 70 percent of the yearly precipitation occurs between October and April, with about half falling as snow (Maupin and Weakland, 2009; Zinsser, 2018). Also during the winter, heavy precipitation events associated with storms that come in from the Pacific Ocean can result in rain-on-snow events (Woods and Beckwith, 2008), inducing snow melt and higher discharge in local rivers and streams. Precipitation is lowest in the summer months of July and August, with monthly average precipitation values of less than 40 mm/month in each of these months during the 30-year period from 1991 to 2020. As shown in Figure 1-6, the variability in long-term average monthly precipitation between subwatersheds is smaller than the seasonal variability.

The annual volume of precipitation within the CDA Lake watershed exhibits significant interannual variability. Over the 30-year period between 1991 and 2020, the annual total precipitation has varied from a low of 824 mm in 2019 to as high as 1,505 mm in 1996 within the watershed, with the interannual variability in annual precipitation volume being consistent in timing across the subwatersheds (data not shown). Furthermore, there are no strong trends in the total amount of precipitation over the past 40 years (Figure 1-7).

INTRODUCTION

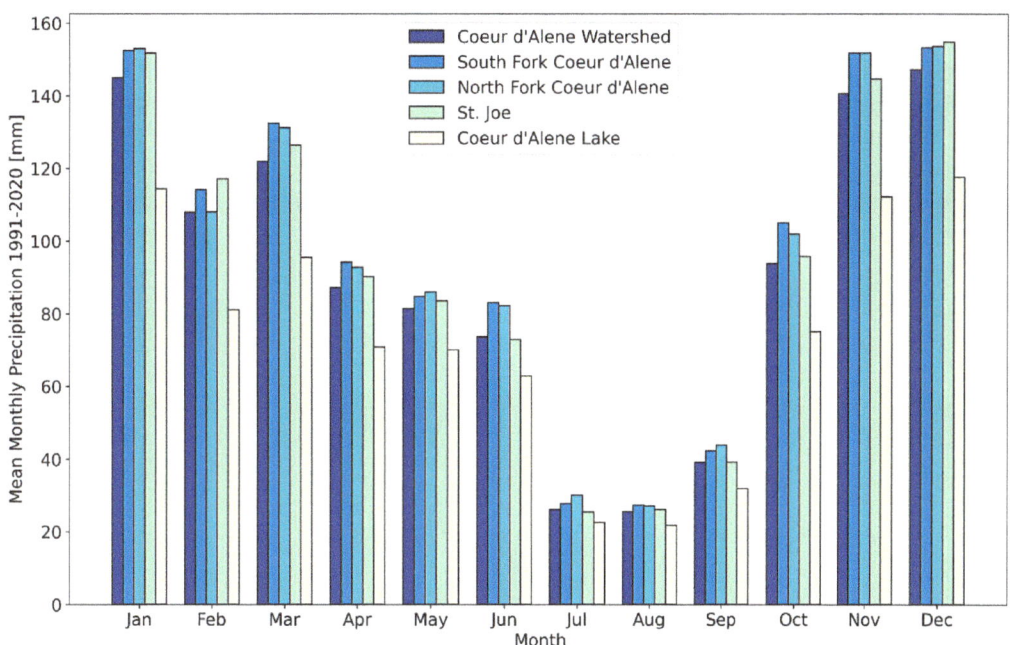

FIGURE 1-6 Mean monthly precipitation across the CDA Basin, 1991–2020. SOURCE: Generated by the committee with data from the PRISM Climate Group at Oregon State University. https://developers.google.com/earth-engine/datasets/catalog/OREGONSTATE_PRISM_AN81m#citations.

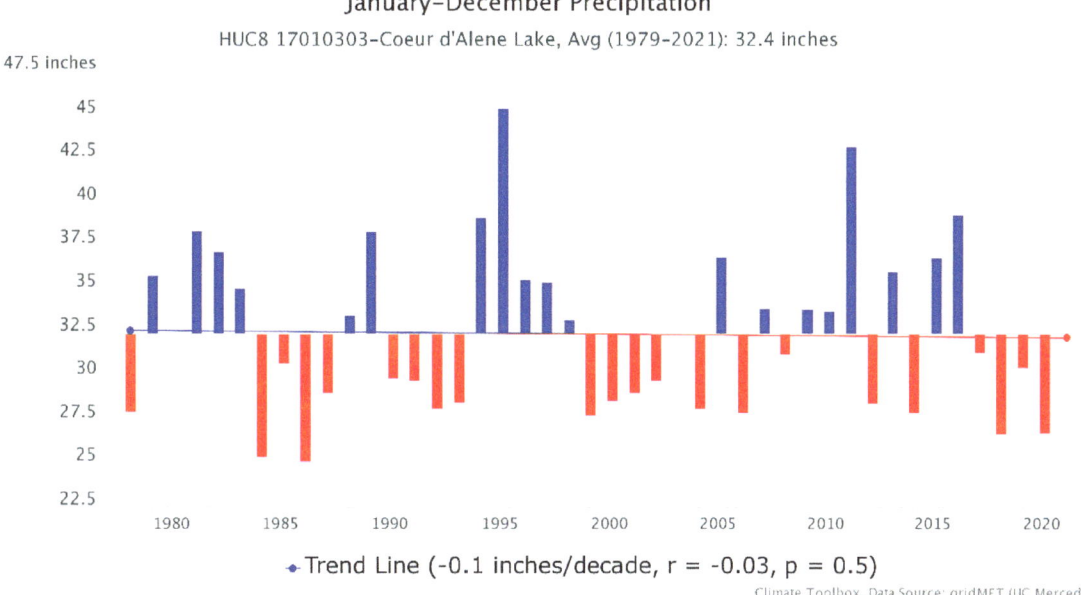

FIGURE 1-7 Trends in annual precipitation for CDA Lake region, 1979–2021. SOURCE: Hegewisch, K.C. and Abatzoglou, J.T. Historical Climate Tracker web tool. Climate Toolbox (https://climatetoolbox.org/) https://climate.northwestknowledge.net/NWTOOLBOX/historicalTracker.php#collapseDOWNLOAD.

Temperature

In the CDA region, average summer temperatures are typically around 60°F (15°C), with temperatures as much as 30°F warmer during the day compared to night. The warmest averages occur in July and can approach 70°F (21°C) (Maupin and Weakland, 2009). Monthly wintertime temperatures are around 29°F (−2°C), with less variability between day and night conditions. The coldest month is typically January. Although temperatures are often below freezing in the winter, the Lake typically only freezes at the shallower southern end (see Figure 1-8).

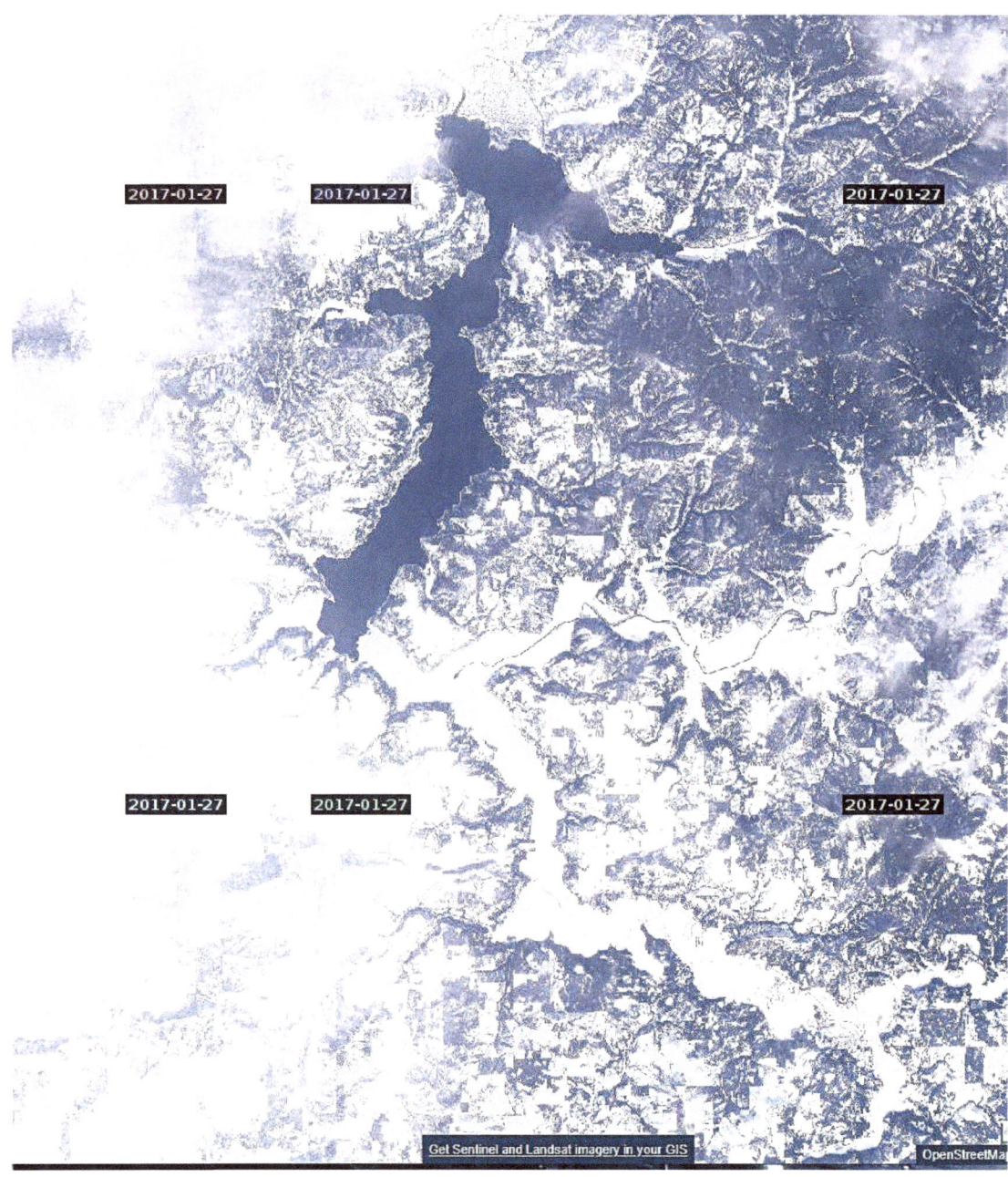

FIGURE 1-8 Image of ice cover over southern CDA Lake on January 27, 2017. SOURCE: Chess (2021).

INTRODUCTION

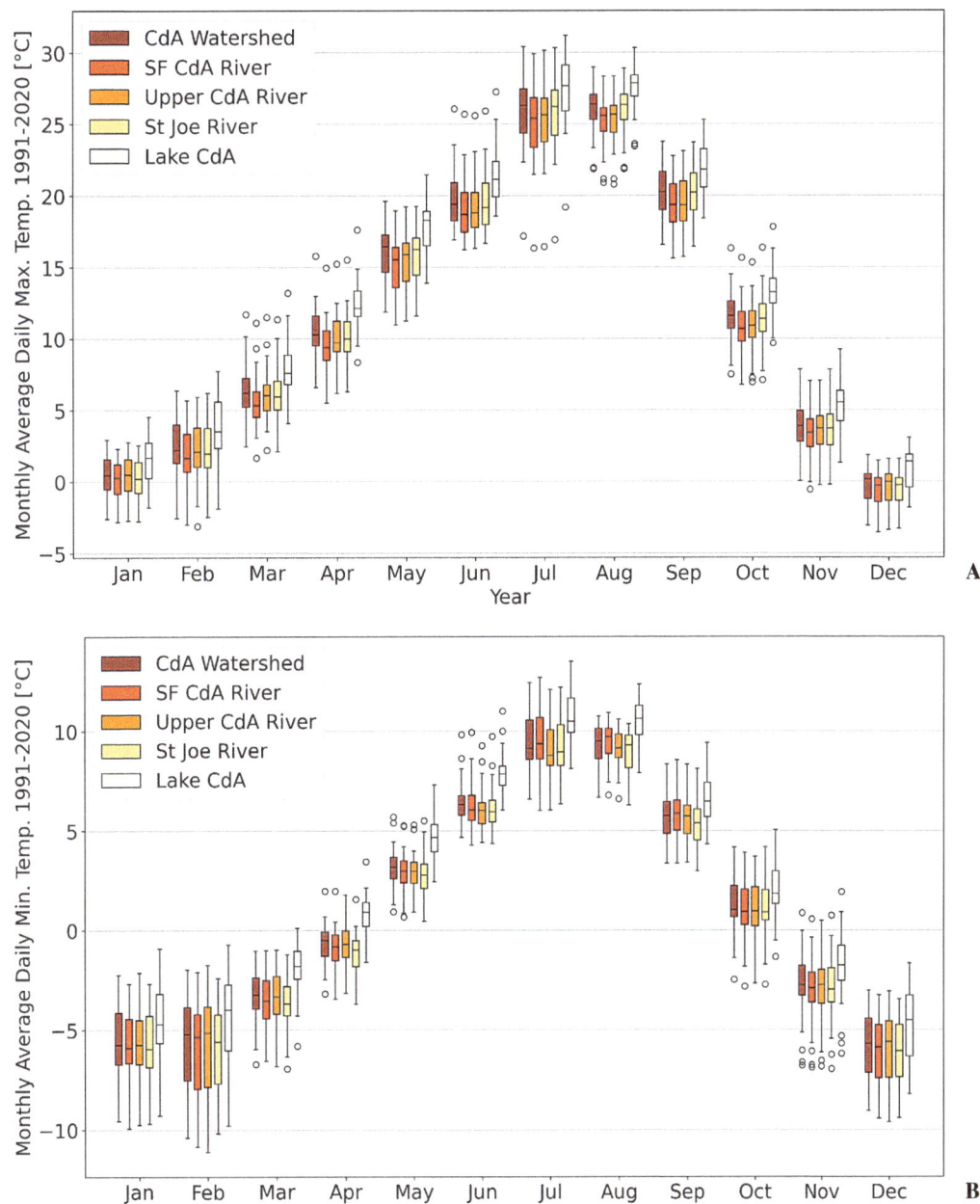

FIGURE 1-9 Mean monthly daily maximum temperature (A) and daily minimum temperature (B), 1991–2020, showing interannual variability. SOURCE: Generated by the committee with data from the PRISM Climate Group at Oregon State University. https://developers.google.com/earth-engine/datasets/catalog/OREGONSTATE_PRISM_AN81m#citations.

As with precipitation, the CDA watershed is also associated with large seasonal variations in temperature (Figure 1-9). The watershed-average daily maximum temperature ranges from slightly less than 32°F (0°C) in January to approximately 77°F (25°C) in July and August. Conversely the watershed-average daily minimum temperature ranges from slightly less than 23°F (−5°C) in January to approximately 48°F (9°C) in July and August.

Monthly average temperatures at the watershed scale can also vary significantly from year to year. From 1991 to 2020, the range of monthly average daily maximum temperatures varied by up to 5°C in December and

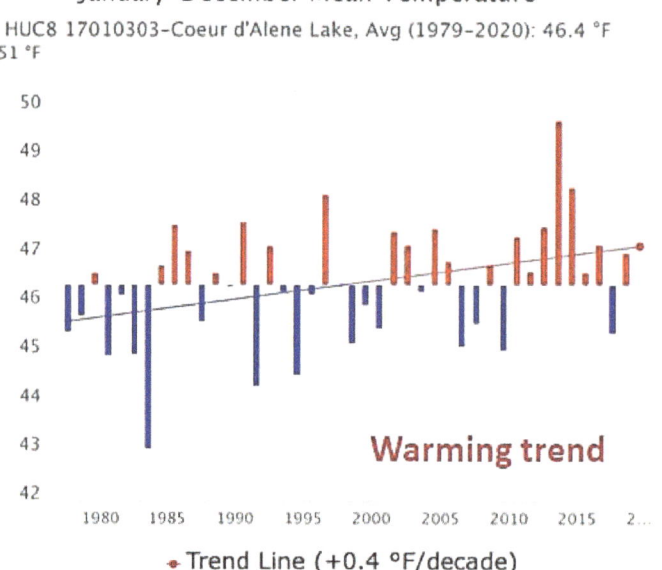

FIGURE 1-10 Trends in annual mean temperature for CDA Lake region, 1979–2020. Note: r = 0.40, p = 0.01. SOURCE: Hegewisch, K.C. and Abatzoglou, J.T. Historical Climate Tracker web tool. Climate Toolbox (https://climatetoolbox.org/) https://climate.northwestknowledge.net/NWTOOLBOX/historicalTracker.php#collapseDOWNLOAD.

January and up to approximately 8°C in July. Monthly average daily minimum temperature, by contrast, exhibits the greatest variability in the winter months of December through February and the least variability in the spring months of April through June. The large elevation gradient within the CDA watershed gives rise to significant spatial heterogeneity in temperatures in all seasons.

Figure 1-10 shows the trend in air temperature for the subwatershed around CDA Lake from 1979 to 2020, showing an average increase of 0.4°F per decade during that period. Although the warming trend is consistent with the climate warming in the western United States and within the broader region, a degree of caution should be exercised when interpreting the magnitude of this trend. Because the PRISM data underlying Figure 1-10 are based on geostatistical interpolations of available surface climate observations and because the density of observing sites and the length of records at those sites varies through time, there is some degree of uncertainty in the ability to accurately compute trends from the data.

Rain versus Snow

The observed seasonal patterns in precipitation and temperature within the CDA Lake watershed result in the bulk of precipitation delivered to the watershed arriving as snow. The U.S. Department of Agriculture (USDA) Natural Resources Conservation Service (NRCS) monitors snowpack conditions in the CDA watershed through a network of Snow Telemetry (SNOTEL) automated measurement stations and manual snow course measurement sites. Through this network of sites, the NRCS develops a basin scale analysis of snow water storage estimates for the Spokane River watershed, for which the CDA River watershed comprises the primary tributary. Based on the 30-year period between 1991 and 2020, the NRCS estimates that the median peak snow water equivalent (SWE) in the Spokane River Watershed is approximately 700 mm and the median date of maximum SWE occurs around April 1 (see Figure 1-11). In general, the snowpack builds from late November to as late as the beginning of June. There is, however, significant interannual variability in the magnitude and timing of peak SWE conditions. Over the 30-year period ending in 2020, peak SWE in the Spokane River watershed has been as low as approximately 250 mm and occurred as early as March 1. At the other extreme, peak SWE has exceeded 1,200 mm and the date of maximum SWE has been as late as May 3.

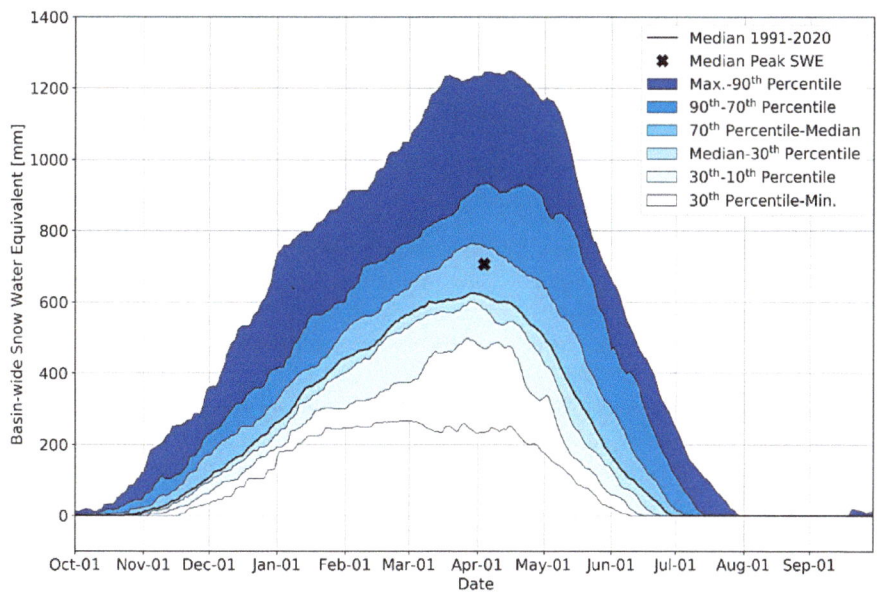

FIGURE 1-11 Historical snow storage as snow water equivalent (SWE) in the Spokane River basin, showing the historical range of variability in snow storage and the median peak snow water content and the historical median date of peak SWE April 4. SOURCE: Generated by the committee with data from the USDA NRCS National Water and Climate Center. https://www.nrcs.usda.gov/Internet/WCIS/AWS_PLOTS/basinCharts/POR/WTEQ/assocHUC6/170103_Spokane.html.

Over the 1983–2021 period of record for which Spokane River SWE estimates are available, there has been a decrease in the peak SWE within the Spokane River System of approximately 36 mm/decade (see Figure 1-12). Although the observed trend fails a traditional Mann-Kendall test for significance, it was found to be statistically significant when tested via a non-parametric Mann-Kendall test modified for effective sample size ($p < 0.001$; Yue and Wang, 2004). Conversely, there was no significant trend in the date on which the peak SWE storage within the

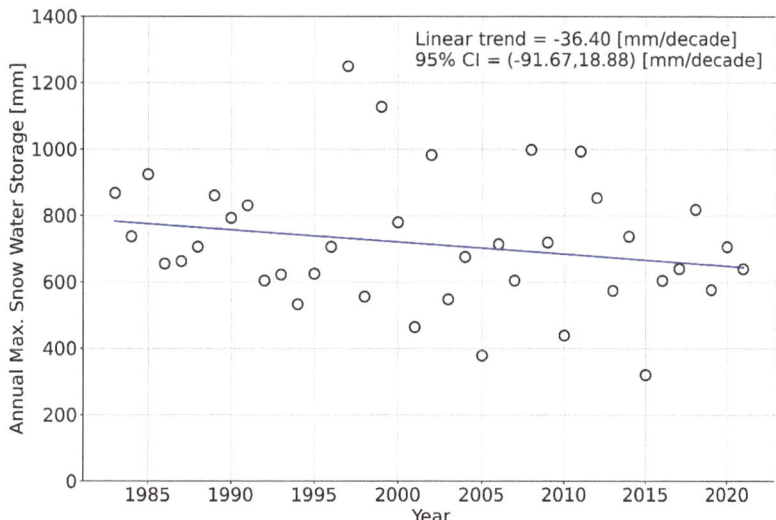

FIGURE 1-12 Annual maximum SWE in the Spokane River basin, 1983–2021. SOURCE: Generated by the committee with data from the USDA NRCS National Water and Climate Center. https://www.nrcs.usda.gov/Internet/WCIS/AWS_PLOTS/basinCharts/POR/WTEQ/assocHUC6/170103_Spokane.html.

Spokane River occurred over the period of record (not shown). The trend of decreasing peak maximum SWE in the Spokane River basin is broadly consistent with other declines in SWE metrics observed in the broader western United States (e.g., Mote et al., 2018; Hamlet et al., 2005).

Hydrology and System Hydraulics

CDA Lake is a 31,875-acre (129-km^2) lake in northern Idaho that drains the CDA River, St. Joe River, and other smaller tributaries around the lake's shoreline. When CDA Lake is full, it holds 2.3 million acre-feet of water (2.79 km^3) (Woods and Beckwith, 2008). The north-south trending Lake is roughly 25 miles (35 km) long and its width ranges from about 1 to 3 miles (1.6–4.8 km). The Lake's shoreline is approximately 135 miles (217 km) long (Horowitz et al., 1993a; Maupin and Weakland, 2009). The maximum depth of the Lake is 210 feet (64 m) (in the central pool), with a mean depth of 72 feet (22 m).

More than 94 percent of the water that flows into CDA Lake comes from the CDA and St. Joe Rivers, which enter along the eastern and southeastern shorelines, respectively. The St. Joe River has an average annual discharge slightly higher than that of the CDA River (2.45 billion m^3/yr compared to 2.28 billion m^3/yr). Because it drains the part of the basin containing the major mining operations of the last century, the CDA River and its floodplain are heavily contaminated with metals that represent a continuing source of metals contamination to CDA Lake. In comparison to the CDA watershed, the St. Joe watershed has been used primarily by the logging industry, has supported no major mining activities, and has been reported to be relatively uncontaminated with trace elements (Mink et al., 1971; Rabe and Flaherty, 1974; Bender, 1991). Smaller prominent tributaries to CDA Lake include Wolf Lodge Creek, Carlin Creek, Plummer Creek, and Fighting Creek (Maupin and Weakland, 2009). After flowing predominantly from south to north, water in CDA Lake discharges to the Spokane River in the northernmost part of the Lake, near the city of Coeur d'Alene. A small amount of Lake water also recharges the Spokane Valley–Rathdrum Prairie (SVRP) aquifer (discussed below).

Because the bulk of the precipitation that is delivered to the watershed arrives as snow, the hydrology of the lake's major contributing watersheds is characterized by a snowmelt-dominated regime and hence varies significantly with season. The rate and timing of snow accumulation, retention, and release is directly reflected in the hydrographs of the rivers that contribute to CDA Lake (Figure 1-13), which show that flow is largely a spring phenomenon, followed by dry summers and falls. Indeed, it is not uncommon for the inflows in April and May to each be roughly half of the entire volume of the Lake (Figure 1-14). Conversely, in the drier months (August, September, and October) the monthly inflows are typically less than 5 percent of the volume of the Lake.

Based on CDA Lake bathymetry (see Figure 1-15), there are roughly four "pool areas" in the Lake, each characterized by relatively deep areas. As a result, these pools are likely to be more depositional, where sediments settle out. In addition, tributary inflow loads from smaller tributaries (i.e., not the CDA or St. Joe rivers) tend to deposit either in the immediate bay areas and the confluence of the tributary with the Lake or in these pool areas. In the case of the CDA River, sediment sampling indicates that fine sediments are pervasive throughout the Lake, with lower loads in the small embayment areas than along the axis of the Lake.

The four main pool areas of CDA Lake, shown in Figure 1-15, are characterized by deep water, exceeding 20–60 m depending on the location. The main pool areas of the Lake experience a range of flows dictated by the CDA and St. Joe Rivers, with water moving south to north through the Lake and past the Post Falls dam, especially during high discharge.[2] The main body of the Lake is characterized by the primary long-term monitoring locations C1 through C5 (shown in Figure 2-1 and discussed in greater detail in Chapter 2). The southernmost long-term monitoring site, C6, near Chatcolet Lake, is somewhat of an outlier in that it is more characteristic of local conditions rather than the main body of the larger Lake. Chatcolet Lake is moderately deep, at roughly 10 m, and is separated from the main Lake by shallow vegetated sills 1–2 m in depth, which restricts water exchange between this basin and the main Lake. C6 reflects influence from the St. Joe River and its watershed, which was not as affected by mining. Hence, sediments in Chatcolet Lake are much less contaminated by legacy metals compared to the main body of CDA Lake.

---

[2] Hydrodynamics become more complex as discharge recedes (see Chapter 4).

INTRODUCTION

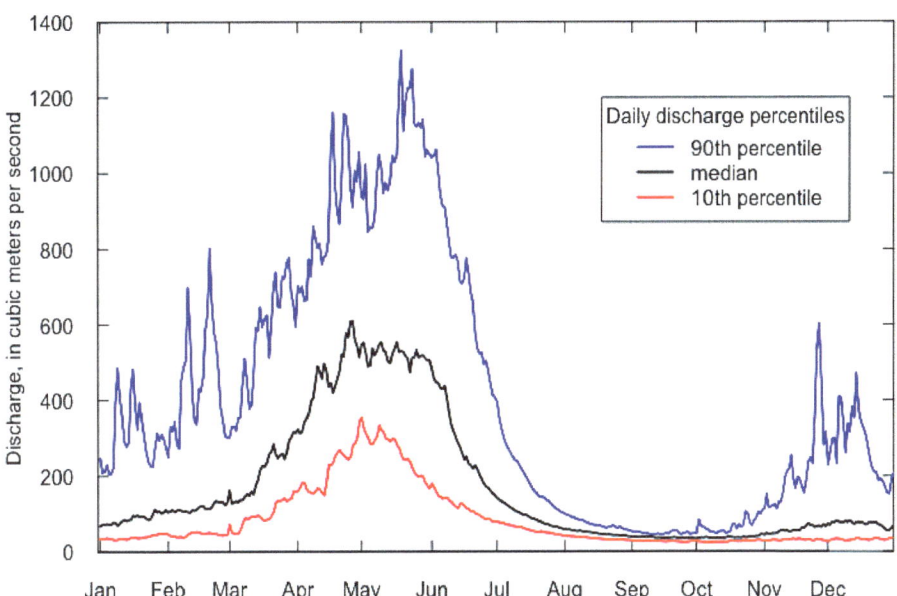

FIGURE 1-13 Inflows to CDA Lake based on water years 1987–2020, showing estimated inflow to the Lake by day of the year; 10th, 50th, and 90th percentiles are shown for each day. Note: As described in Chapter 3, discharge is the combined flows of the CDA and St. Joe rivers, multiplied by 1.1036 to represent the total inflow including all of the drainage area to the Lake not included in the CDA or St. Joe watersheds. SOURCE: Data from USGS analyzed by the committee.

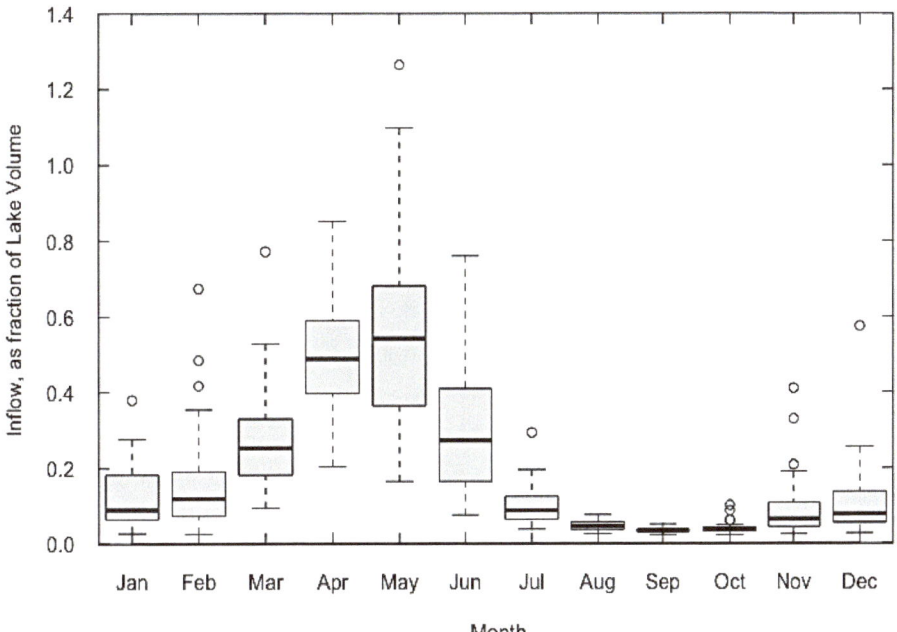

FIGURE 1-14 Total monthly inflow to CDA Lake, based on water years 1987–2020, showing estimated total inflow for the month as a fraction of the total volume of the Lake. Note: As described in Chapter 3, inflow is combined flows of the CDA and St. Joe Rivers, multiplied by 1.1036 to represent the total inflow including all of the drainage area to the Lake not included in the CDA or St. Joe watersheds. SOURCE: Data from USGS analyzed by the committee.

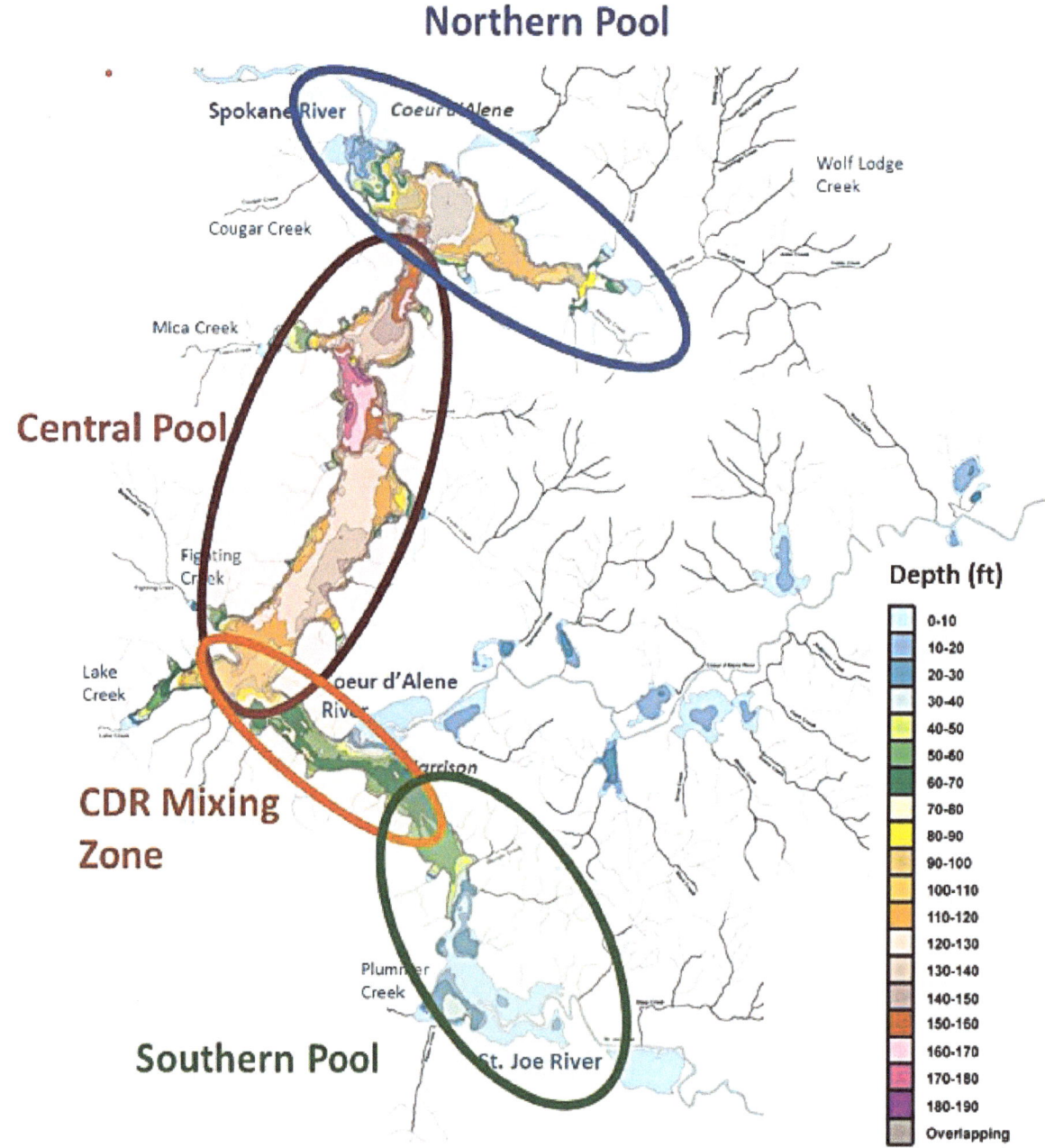

FIGURE 1-15 Bathymetry and pool areas in CDA Lake. SOURCE: Cooper (2021a).

Several lateral lakes are adjacent to the CDA River upstream of CDA Lake (see Figure 1-3). These lakes have been monitored irregularly over the past 30 years as part of special studies, but they are not part of the ongoing long-term monitoring programs like C1 through C6. The lateral lakes have much smaller drainage areas contributing flow to them, they are shallow and warmer than the main Lake, and in flood conditions they may be inundated by flood water from the CDA River.

CDA Lake's mean retention (or residence) time is 0.42 years, calculated based on the Lake volume and the mean discharge presented in Chapter 3 of 209 m^3/s. Lakes generally span a range of residence times from months to decades, depending on the volume of the lake and its watershed, so CDA Lake is at the lower end of the range. There is significant interannual variability in the mean residence time of CDA Lake. For years of high inflow to the Lake (such as 1996), the mean residence time is about 0.25 years. In years with the lowest inflows, the mean residence time can be as high as 0.9 years. Thus, even in the driest years that have been observed, enough water flows into the Lake to replace its volume each year.

Although broadly speaking CDA Lake has about a five-month mean retention time, the hydraulics of the Lake vary considerably throughout the year and by location because of the regulation of Lake levels by the Post Falls Dam, which backs up water into the Lake and further upstream on the CDA River during the summer. Seasonally, average hydraulic residence times vary from 90 days in the February–April period to about 1,100 days in the summer (Cooper, 2021b). During the winter, significant inflows pass through the Lake, bringing sediments laden with metals and other materials. A large proportion of these sediments settle out in the Lake once inflows enter the Lake and flow velocities decrease. In early summer as river inflows decrease, Post Falls Dam operations raise the water level (and volume) in the Lake, increasing the retention time significantly and allowing sediments that entered the Lake to spread out and settle to the bottom. These late spring and summer hydraulic conditions along with warmer temperatures lead to thermal stratification, during which the Lake develops three distinct layers of water: the epilimnion or the top warm layer, the thermocline or middle layer where a temperature gradient is observed, and the colder hypolimnion that extends to the Lake floor. Thermal stratification is caused by the differing density of water of different temperatures. As discussed in detail in Chapter 4, there is ample evidence of thermal stratification in CDA Lake, along with decreases in dissolved oxygen in the hypolimnion at certain locations. Due to the low flows, increased retention time, and increased volume characteristic of CDA Lake in the summer, in-lake physical (i.e., sedimentation), chemical, and biological processes dominate Lake dynamics during this season. Stratification lasts until late summer or early fall, when dam operations allow outflows to increase and disturb the stratification.

Post Falls Dam

Located downstream of CDA Lake on the Spokane River, the Post Falls Dam was constructed in the 1890s to divert the Spokane River's flow in order to provide power to a saw mill at the same location.[3] Following the destruction of the sawmill in a fire in 1902, Washington Water Power purchased the site and developed a hydroelectric facility that would provide power to mines nearly 100 miles away, via the longest high-voltage transmission line in the world. The dam is now operated by Avista Utilities, employing six turbine-generating units that are a source of electricity for mines, mills, businesses, factories, railways and cities in northern Idaho and eastern Washington.

To allow for increased recreational opportunities on CDA Lake, the Post Falls Dam is operated to hold the Lake at a high elevation during the summer months. According to CDA Tribe and Avista (2017), the dam is used to control CDA Lake elevation during the part of the hydrograph when flows are decreasing, typically in June, and then it maintains the Lake level at the summer full pool elevation of 2,128 ft (648.6 m) until the first Tuesday following Labor Day (see Figure 1-16). Following Labor Day, Avista draws the lake down approximately one foot in September, then 1.5 ft per month in October, November, and December until the Lake is at approximately 2,122.5 ft. Once the Lake reaches 2,122.5 ft (646.9 m), which is typically by the end of December, the natural channel restriction at the outlet of the Lake controls the lake's elevation.

The effects of the Post Falls Dam in summer extend well beyond the Lake into the watershed. In the lower reaches of both the CDA and St. Joe Rivers, elevated Lake levels lead to the creation of large, shallow open water areas that increase habitat for aquatic macrophyte growth and alter the natural levees, lateral lakes, and wetlands ecosystems in the lower reaches of both rivers. Indeed, the shallow southern lakes (e.g., Benewah, Chatcolet, Hidden, and Round Lakes—see Figure 1-3) used to be mostly emergent wetlands before the Post Falls Dam was constructed. During the summer, the dam decreases the gradient of the CDA River to 0.19 m/km such that backwater and substantial quantities of sediment are deposited in the main channel and along the banks and

---

[3] Myavista.com

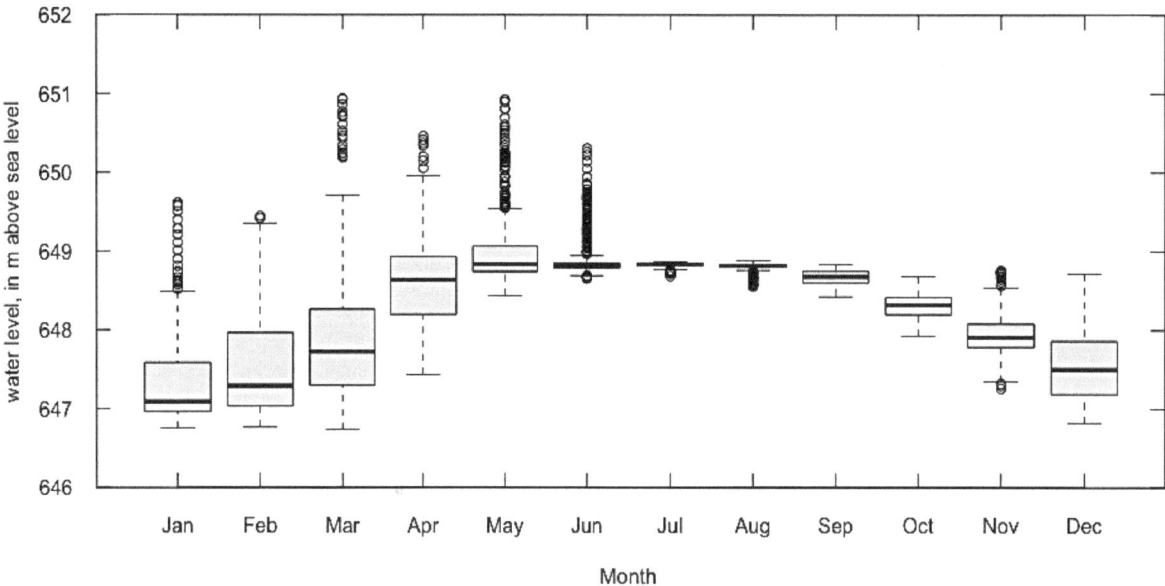

FIGURE 1-16 Daily water level of CDA Lake by month for water years 2008–2020. SOURCE: Data, plotted by the committee, are USGS lake level data for CDA Lake at Post Falls (USGS Gage 12415500).

floodplain (Bender, 1991). During high flow (e.g., spring snowmelt and runoff), these sediments can be resuspended, transported, and redeposited both in the river and in the Lake (Bender, 1991).

Spokane Valley-Rathdrum Prairie Aquifer

The Upper Spokane River watershed overlies the SVRP aquifer, discovered in 1895 (MacInnis et al., 2009). The aquifer underlies about 370 mi^2 of land, trending northeast to southwest from Bonner County, Idaho, near Lake Pend Oreille to CDA Lake in Kootenai County before extending west into Spokane County, Washington, until it ends near Nine Mile Falls, northwest of Spokane (Boese et al., 2015; Kahle et al., 2005) (Figure 1-17). The aquifer holds about 10 trillion gallons (3.78×10^{10} m^3) of water and is the sole source of drinking water to more than 500,000 residents in Kootenai County, Idaho, and Spokane County, Washington, making it one of the most valuable natural resources in the region (Kahle et al., 2005).

The sediments that comprise most of the SVRP aquifer are highly permeable, allowing groundwater to flow as fast as ~50 feet per day (15.2 m per day) in some portions (Boese et al., 2015). This rapid flow rate combined with the SVRP aquifer's large storage capacity, make it one of the fastest flowing and most productive aquifers in the country (MacInnis et al., 2009). It is also an unconfined aquifer, meaning it has a direct connection to surface water and is recharged by water from local lakes including CDA Lake, streams, rivers, and precipitation (Kahle and Bartolino, 2007; MacInnis et al., 2009). As a result, the SVRP aquifer is highly susceptible to contamination associated with human activities (e.g., urban runoff, septic systems, industrial discharge, mining waste, fertilizer and pesticide use) (MacInnis et al., 2009). In 1978, the SVRP aquifer became the second groundwater reservoir in the nation to be designated a "Sole Source Aquifer" (i.e., sole source of drinking water) by the EPA (Kahle and Bartolino, 2007), a designation that heightened public awareness of this critical resource and prompted resource agencies to protect the aquifer by establishing environmental management practices, such as septic tank removal and stormwater treatment.

The daily water budget for the SVRP aquifer is approximately 1 billion gallons. The Spokane River comprises 43 percent of recharge to the aquifer with minor inflows from irrigation water and septic systems. Recharge from lakes accounts for 28 percent of the inflow, with CDA Lake contributing a moderate amount to that value (Figure 1-18).

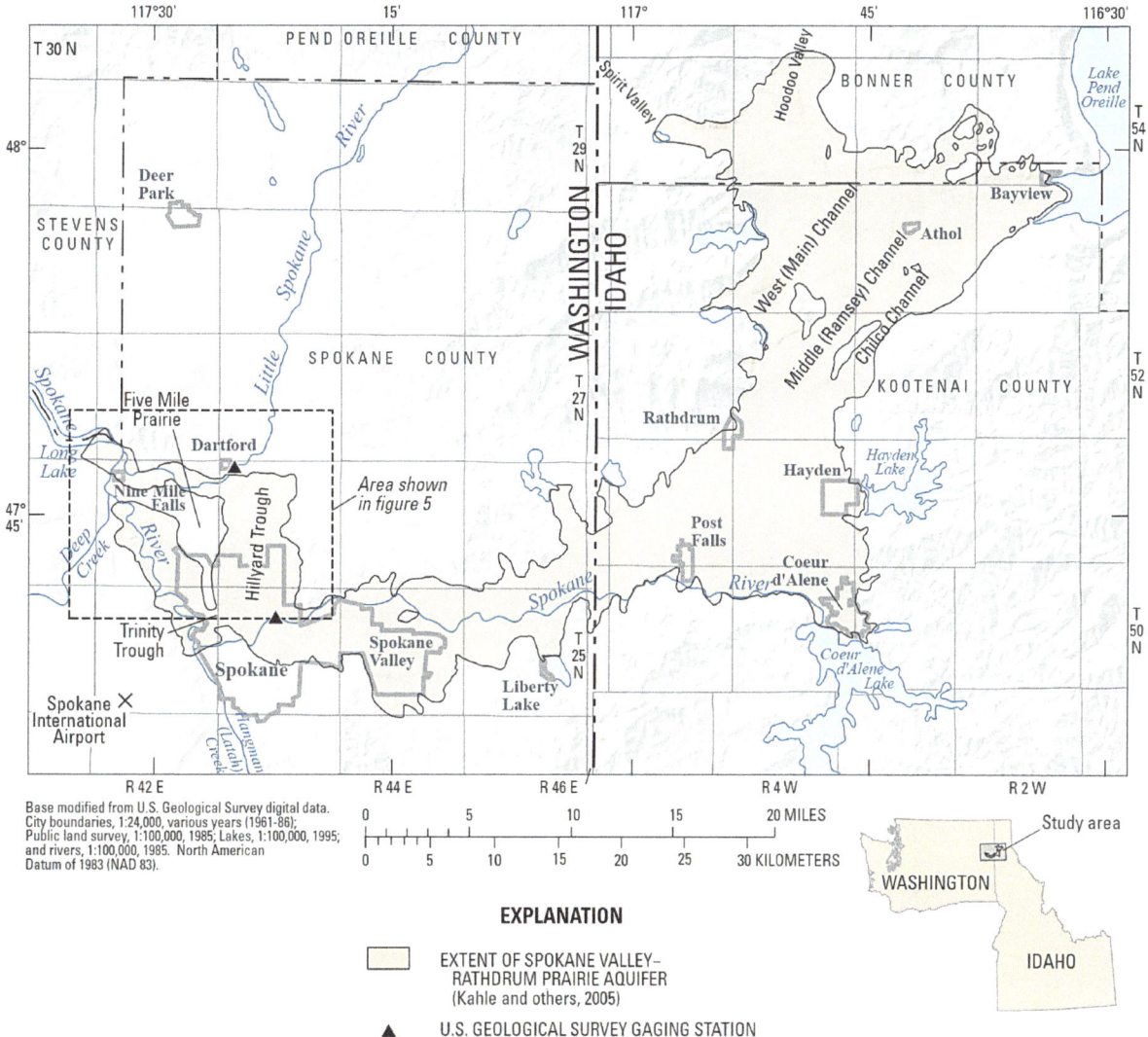

FIGURE 1-17 The SVRP aquifer (shaded tan) underlies Spokane County in Washington state and Bonner and Kootenai Counties in Idaho. SOURCE: Kahle and Bartolino (2007).

Land Uses in the CDA Watershed

Land use trends in the CDA watershed, such as increased development, are expected to affect sediment and nutrient inputs to CDA Lake. The three primary drainage areas in the watershed (CDA River basin, St. Joe River basin, and the nearshore catchments) vary with respect to geologic, historic, and land-use features and the extent to which they act as sources of contaminants to CDA Lake. The 2001 and 2019 land cover maps for these three areas are shown in Figures 1-19 and 1-20, respectively.

Table 1-1 shows similar land cover data for the CDA basin, but in percentages, with 30-m resolution data from the National Land Cover Database. The land cover categories can be grouped into eight main categories and additional subcategories (see Table 1-2). The year 2001 was selected by the committee as the baseline for evaluating land use trends in the CDA basin because the classification system used between 2001 and 2019 was unchanged, and land use data are available from 2001 to 2019 that facilitated using the Land Cover Change Index. Comparison of the 2001 and 2019 land cover maps shows relatively few changes in developed areas and

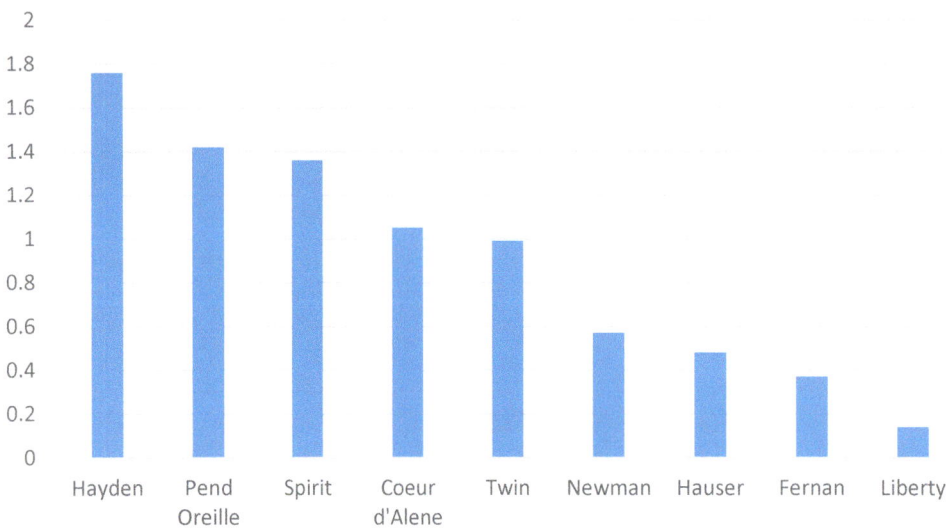

FIGURE 1-18 Estimates of groundwater recharge contributions from lakes near the SVRP aquifer. Values on y-axis are in units of cubic meters per second based on average conditions between 1990 and 2005. SOURCE: Kahle et al. (2005).

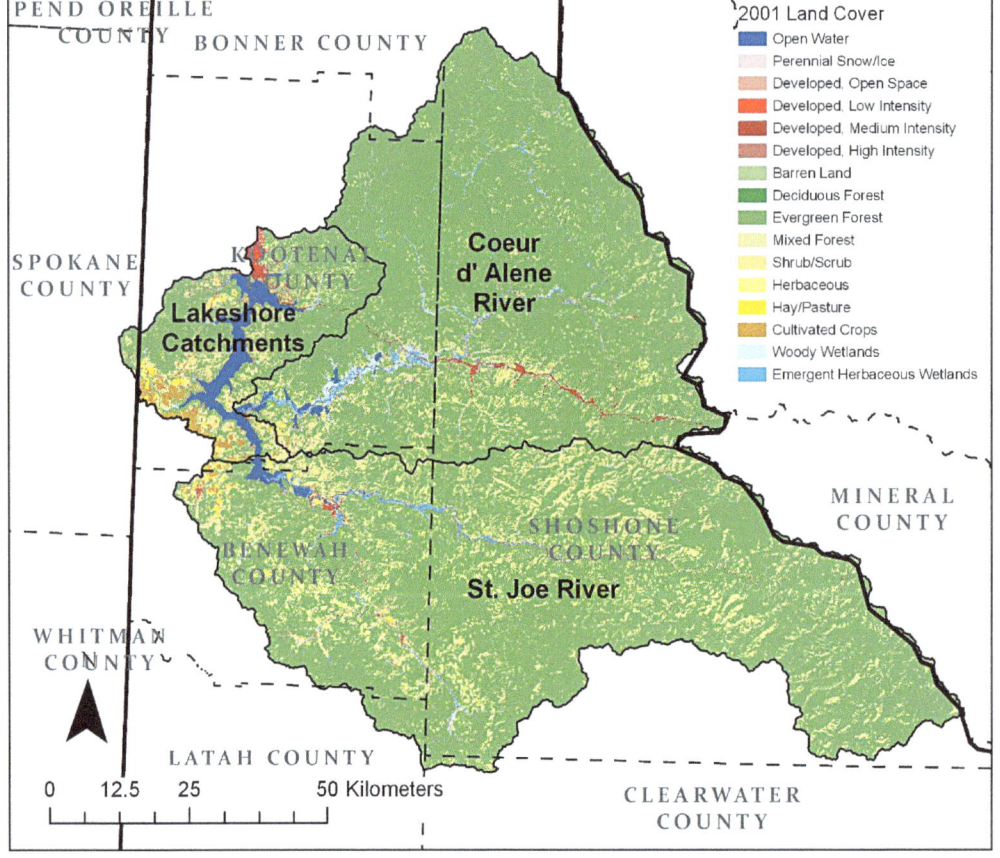

FIGURE 1-19 CDA basin land-cover data in 2001. SOURCE: Data plotted by the committee from the National Land Cover Database. https://www.mrlc.gov/data?f%5B0%5D=category%3ALand%20Cover&f%5B1%5D=region%3Aconus.

TABLE 1-1 Comparison of CDA Basin Land Use Category Percentages, 2001 and 2019

Main Category	2001 Percentage of Land Cover	2019 Percentage of Land Cover	Percentage Change in Land Cover (2019–2001)
Water	1.73%	1.60%	−0.13%
Developed	1.33%	2.17%	0.84%
Barren	0.19%	0.41%	0.23%
Forest	79.09%	74.27%	−4.82%
Shrubland	12.79%	14.58%	1.79%
Herbaceous	2.46%	3.07%	0.61%
Planted/Cultivated	0.29%	0.55%	0.25%
Wetlands	1.56%	1.24%	−0.32%

SOURCE: National Land Cover Database; https://www.usgs.gov/centers/eros/science/national-land-cover-database.

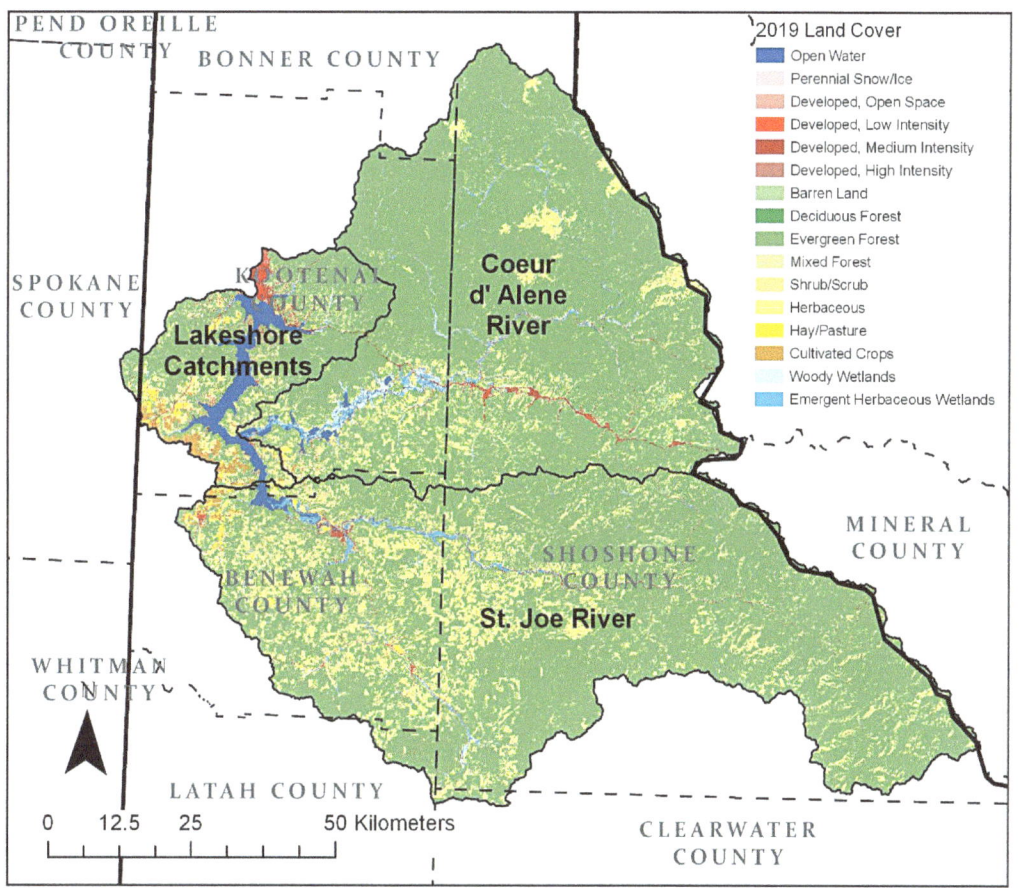

FIGURE 1-20 CDA basin land-cover data in 2019. SOURCE: Data plotted by the committee from the National Land Cover Database. https://www.mrlc.gov/data?f%5B0%5D=category%3ALand%20Cover&f%5B1%5D=region%3Aconus.

TABLE 1-2 Land Cover Categories Are Based on the Modified Anderson Level II Classification System

Main Category	Sub Class Number	Sub Category
Water	11	Open Water
Developed	21	Developed Open Space
	22	Developed Low Intensity
	23	Developed Medium Intensity
	24	Developed High Intensity
Barren	31	Barren Land
Forest	41	Deciduous Forest
	42	Evergreen Forest
	43	Mixed Forest
Shrubland	52	Shrub/Scrub
Herbaceous	71	Grassland/Herbaceous
Planted/Cultivated	81	Pasture/Hay
	82	Cultivated Crops
Wetlands	90	Woody Wetlands
	95	Emergent Herbaceous Wetlands

SOURCE: National Land Cover Database; https://www.mrlc.gov/data/legends/national-land-cover-database-class-legend-and-description.

herbaceous and planted/cultivated land cover, and yet significant changes in forested land and shrubland. As shown in Table 1-1, the percentage of land that was forested land decreased by nearly 5 percent over the 18-year time period whereas the percentage of shrubland showed the largest increase (1.79 percent). Developed land increased by less than one percent over the 18-year time period.

A transition in land use to more developed land is evident within the lakeshore catchments, as shown in Figure 1-21 (note the pink-colored pixels). Because agricultural and residential/urban lands can have significantly higher phosphorus loading than undeveloped land (Tong and Chen, 2002), there is potential for increased nutrient loadings due to changing land use and population growth. Changes in land surface permeability associated with deforestation, landscaping, road building, and development can also lead to reduced groundwater infiltration, floodplain alteration, flooding, run-off, and nutrient loading during run-off (Rogger et al., 2017; Hogan and Walbridge, 2007).

Population Growth

A growing population in the watershed presents water quality risks for CDA Lake due to potential increases in both point and nonpoint source pollutants of anthropogenic origin. Both stormwater and wastewater inputs of contamination to the Lake may increase as population grows unless appropriate treatment and management are put in place. The LMP (IDEQ and CDA Tribe, 2009) noted about a 20 percent increase in population between 2000 and 2006 in both the city of Coeur d'Alene and Kootenai County, and these trends seem to be continuing within the cities of Coeur d'Alene, Hayden, and Post Falls, and in numerous smaller communities along the lakeshore. In September 2021, Coeur d'Alene was identified as the fifth fastest growing small metro area in the United States, with a 14.6 percent increase in population between 2015 and 2020.[4] Recent analyses have also reported that over 75 percent of the population growth in Kootenai County was due to people who migrated to the area.[5] The statistics also show that the percentage increase in the over-65 demographic is more than double the percentage increase in the overall population since 2001.

---

[4] https://lacrossetribune.com/lifestyles/the-15-fastest-growing-metropolitan-areas-in-the-us/collection_b1762b54-e25f-540a-b873-8150c4c50dbb.html#19

[5] https://www.krem.com/article/news/local/what-does-north-idaho-growth-mean-for-kootenai-county/293-ffafd844-e778-4839-8087-afdefffbaf45

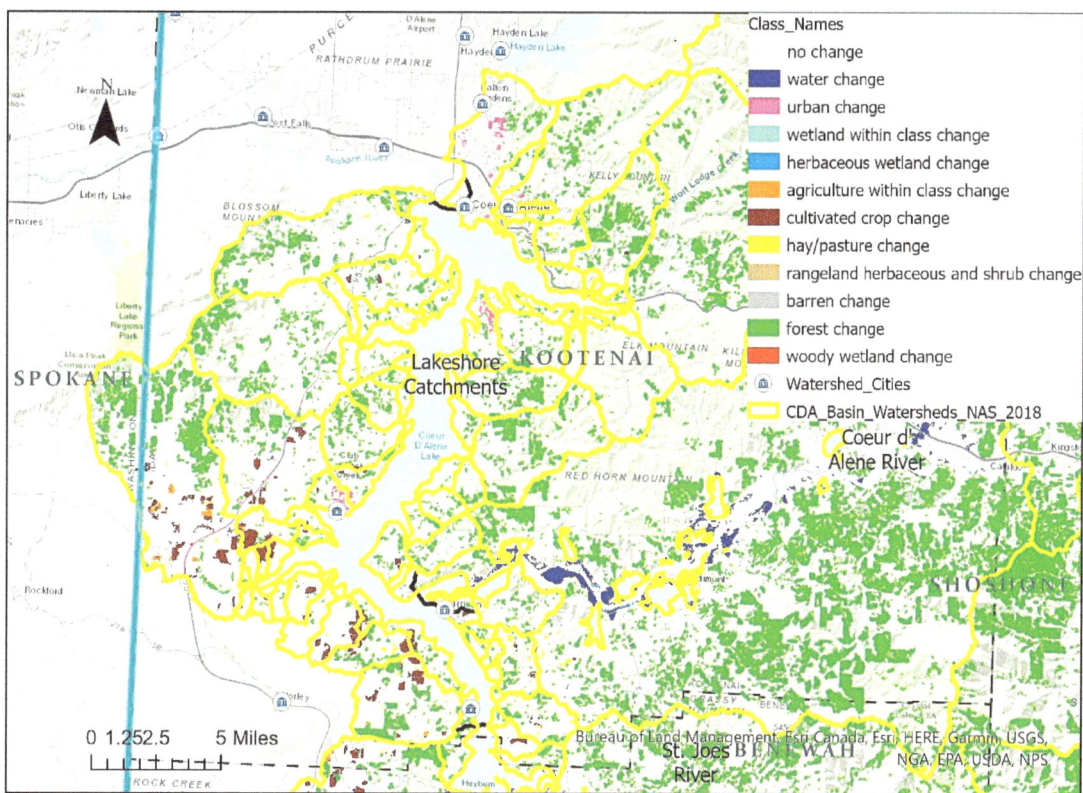

FIGURE 1-21 Land cover change index for the lakeshore catchments between 2001 and 2019. Although these data represent changes, the direction of change is not included in the continental U.S. dataset (Table 1-1 suggests that the changes are *increases* in developed land cover). SOURCE: Data plotted by the committee from the National Land Cover Database. https://www.mrlc.gov/data?f%5B0%5D=category%3ALand%20Cover&f%5B1%5D=region%3Aconus.

One of the issues identified within the LMP, the more recent Total Phosphorus Nutrient Inventory (IDEQ and CDA Tribe, 2020), and the Kootenai County Land Use Comprehensive Plan (Kootenai County, 2020) is the uncertain impact of rapid growth and urbanization of nearshore regions of CDA Lake. The Kootenai County Land Use documents reported an estimated population increase from 132,000 in 2006 to 174,673 in 2021, which corresponds to a 2.64 percent annual growth rate[6] (and is consistent with the previous estimate from the LMP). The Kootenai County Comprehensive Plan (Kootenai County, 2020) assumed growth over the next 10 years at approximately 2 percent per year, with estimates of the Kootenai County's population doubling in 35 years. The 2021 annual growth rates for the other counties in the CDA basin are much smaller (see Table 1-3). These values have been relatively consistent for the past several years.

The committee conducted its own analysis of population growth in the CDA basin using ArcGIS and 2021 Census data. The average annual population growth from 1990 to 2010 was 1.35 percent for the entire CDA Lake watershed, 0.96 percent for the St. Joe River basin, and 2.43 percent for the lakeshore watersheds (see first two columns of Table 1-4). In contrast, the CDA River watershed saw a population *decline* of 0.7 percent during the 1990–2010 period. More recent data show that population in each watershed has increased since 2010, with the greatest percent increase within the lakeshore watersheds, as indicated in the final column of Table 1-4.

The significant differences between the county and watershed populations are due to the presence of populated cities that fall outside of the watershed boundaries. For example, as shown in Figure 1-22, it is evident that Post Falls,

---

[6] https://worldpopulationreview.com/us-counties/id/kootenai-county-population

TABLE 1-3 2021 Population Growth in Counties That Include Parts of the CDA Watershed

County	Population	Growth (number)	Percent Annual Growth
Kootenai County	174,673	4,488	2.57
Benewah County	9,458	80	0.85
Latah County	40,666	279	0.69
Shoshone County	13,066	92	0.704

SOURCE: https://worldpopulationreview.com/us-counties/id.

TABLE 1-4 Watershed Population Data Comparison among Watersheds

Watershed	1990	2010	2021	Percent Change from 2010 to 2021
St. Joe River Basin	6,956	8,294	9,538	15.00
Coeur d'Alene River Basin	13,804	13,708	14,600	6.51
Lakeshore Watersheds	20,032	29,781	35,599	19.54
Coeur d'Alene Lake Watershed	40,792	51,783	59,737	15.36

SOURCE: 1990 and 2010 data are from Wise (2021); 2021 data are from the Census using ESRI methodology. https://doc.arcgis.com/en/esri-demographics/reference/data-allocation-method.htm.

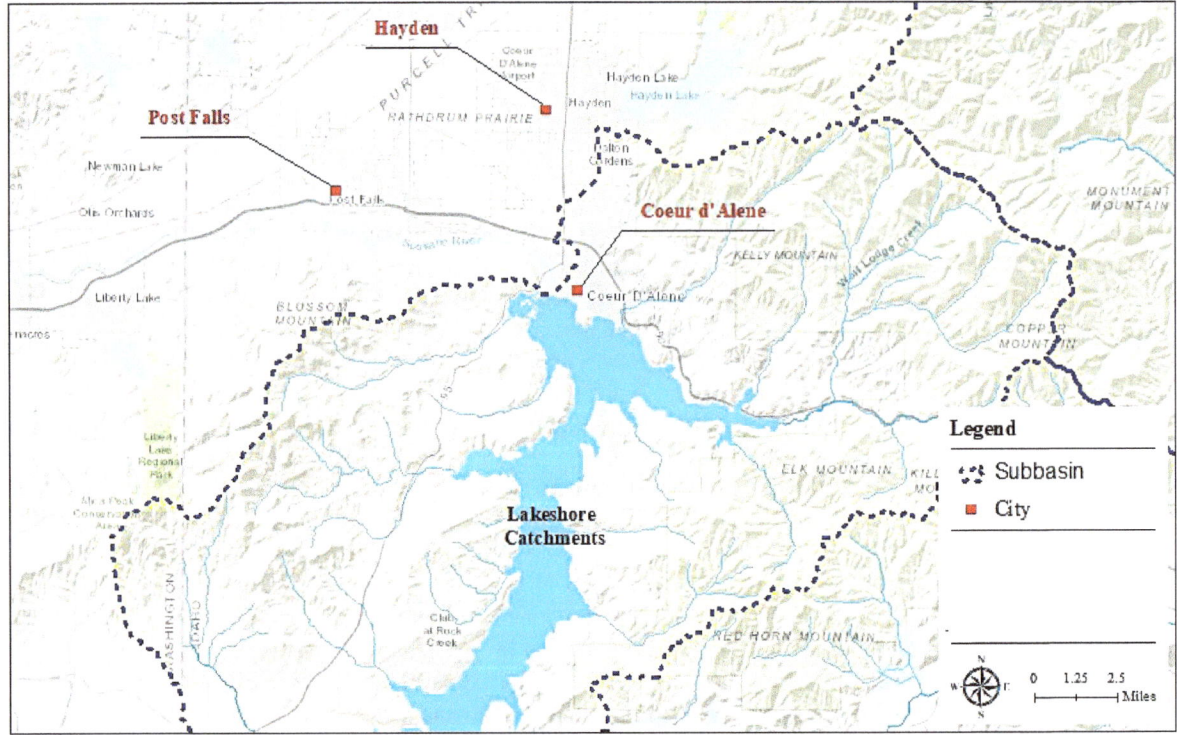

FIGURE 1-22 Northern part of the lakeshore watershed boundary (blue dash) and neighboring cities (shown in red). SOURCE: Generated by the committee using data from IDEQ, CDA Tribe, and the USGS National Map and associated datasets.

Hayden, and large portions of the city of Coeur d'Alene fall outside of the watershed. These cities are the three largest cities in the county, with 2021 populations of 30,160, 7,360 and 47,840, respectively. The data suggest that there is significant growth occurring in areas near the Lake, but these areas do not necessarily lie within the watershed boundary. Centralized wastewater systems servicing those three cities discharge outside of the CDA basin.

Between 2011 and 2019, land use changes within the entire CDA watershed have been modest, with the greatest change associated with deforestation (less than a 5 percent change in forest land area between 2001 and 2019). Population growth watershed-wide has increased by approximately 15 percent since 2010, with the most substantial urbanization and population growth occurring in the lakeshore watersheds within Kootenai County. Continued development within these areas may have some impact on land use and could lead to increases in total solids and nutrient loadings within nearshore areas, particularly for residences relying on septic systems (see Chapter 3).

MINING HISTORY AND THE SUPERFUND SITE

Over the past 150 years, CDA Lake has received a massive influx of heavy metals that emanated from mineral extraction in the upstream watershed, predominantly within the Coeur d'Alene mining district known as the Silver Valley (Figure 1-23). Mining operations began in the CDA mining district in the mid-1880s along the South Fork of the CDA River and its major tributaries. More than 100 mines and ore processing units within the region produced and processed about 130 million MT of ore containing silver, lead, zinc, and other metals during the first century of operation (Long, 1998b). Total production records highlight the magnitude of operations within the district, which has ranked among the top 1 percent of world producers for silver and lead and among the top 10 percent for zinc (Long, 1998a,b). The Bunker Hill Mine and Smelting Complex, located in Kellogg, Idaho, was the largest mining and ore processing company in the region and the largest smelter in the world at the time it was built.

Over the past three decades, detailed descriptions of the history of mining operations within the CDA region have been published (e.g., Woods and Beckwith, 1997; NRC, 2005). Regarding the CDA watershed between 1883 and 1888, Woods and Beckwith (1997) say, "Perhaps nowhere in the history of Euro-American settlement of western North America was an area so rapidly and drastically transformed." The focus of this section is to summarize aspects of the mining activities that have had a significant impact on water quality.

History of Mining Contamination

Previous reports describing the ore deposits within the district provide an understanding of the range of contaminants released from the site (Balistrieri et al., 2002). The host rocks are primarily quartzite and argillite

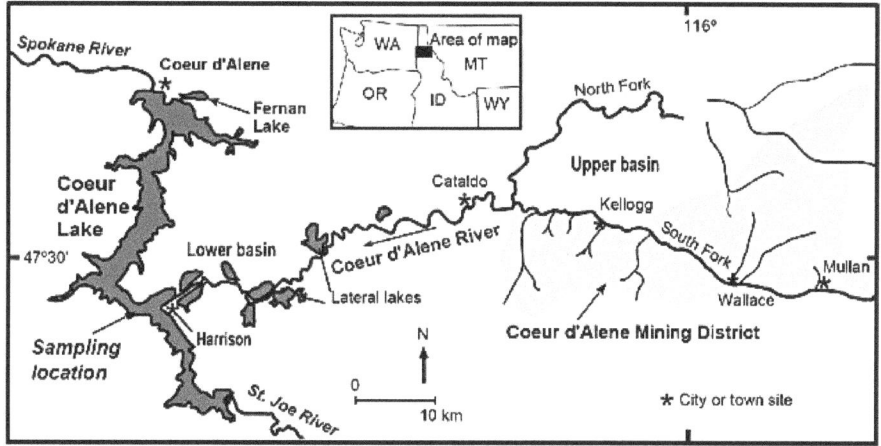

FIGURE 1-23 The Silver Valley, the CDA mining district, shown in the light gray shaded area along the South Fork of the CDA River and its major tributaries. SOURCE: Langman et al. (2020).

with carbonate bearing rocks interbedded. Quartz and siderite veins contain stratigraphically controlled Pb-Zn-Ag ore shoots. Veins rich in lead and zinc are comprised of argentiferous galena (PbS) and sphalerite (ZnS), while veins rich in silver contain argentiferous tetrahedrite [$(Cu, Ag)_{10}(Fe, Zn)_2(As, Sb)_4S_{13}$] with minor amounts of galena and sphalerite. In addition to minerals containing lead, copper, zinc, iron, silver, and arsenic, the veins contain pyrite (FeS_2), chalcopyrite ($CuFeS_2$), and minor amounts of arsenopyrite (FeAsS) and pyrrhotite ($Fe_{(1-x)}S$).

Mining and milling operations conducted over time led to dispersal of more than 100 million tons of contaminated materials spread over thousands of acres (EPA, 2015; Figure 1-24). The principal sources of the most severe metals contamination were air emissions from smelter operations and the tailings resulting from beneficiation (concentration) of the ore. The latter were released without containment directly into the South Fork of the Coeur d'Alene River and its tributaries. Flood events spread particulate contamination across the CDA River basin. At the same time, increased clogging of channels with tailings from mining activities raised stream levels to the point that annual overbank flooding drove flood waters to higher levels, eventually altering stream channels (NRC, 2005).

Increased population growth associated with the success of the mining industry also contributed to spreading contamination across the basin while altering the region's hydrology. Deforestation from both mining activities and development and road building increased runoff and promoted more rapid movement of contaminants. These impacts did not go unnoticed by residents and settlers in the area. Early attempts to satisfy farmers who complained that tailings were poisoning crops, livestock, dogs, and chickens led to placement of small wooden impounding dams (e.g., Canyon, Osburn, and Pine Creeks shown in Figure 1-24). However, the structures routinely failed, especially during floods. High-flow events deposited contaminated tailings as lenses of tailings or as tailing/sediment mixtures in the bed, banks, floodplains, and lateral lakes of the CDA River basin and in CDA Lake (EPA, 2015).

In response to concerns associated with the lack of fish, macroinvertebrates and planktonic organisms along the 50-mile long South Fork of the CDA River, a floating suction dredge was operated every summer starting in 1931 for 45 years on Mission Flats near Cataldo. The dredged material was allowed to settle in a series of impoundments and the supernatant was returned to the river. The discharge of tailings directly to the South Fork of the

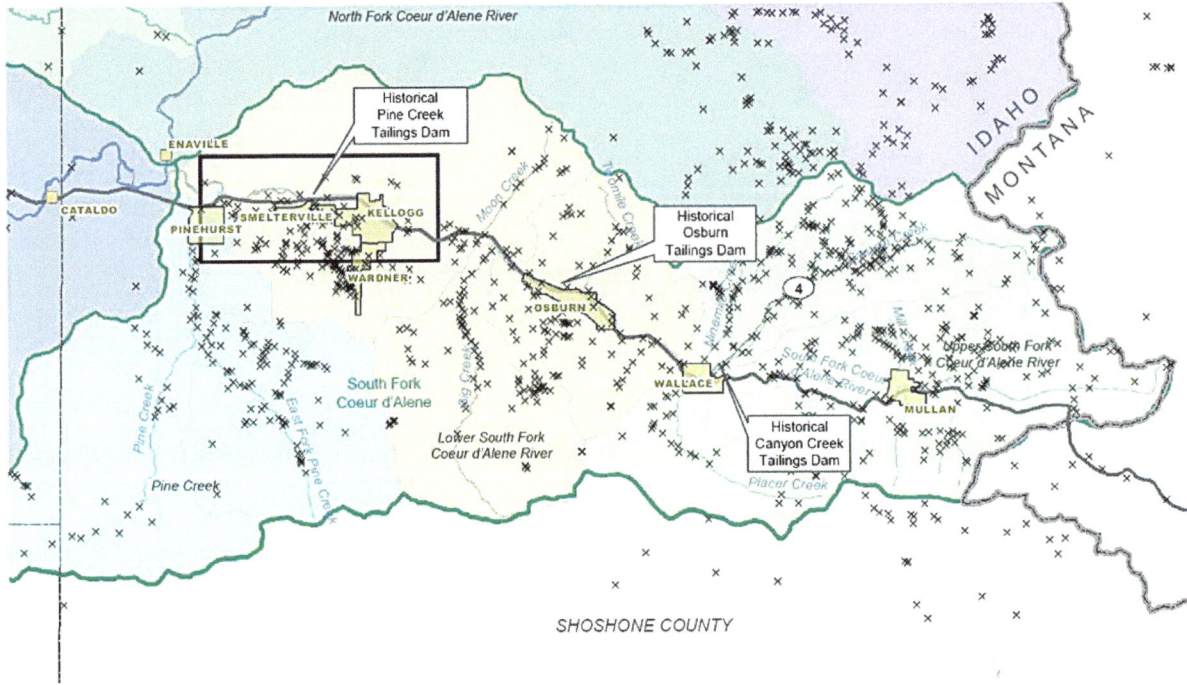

FIGURE 1-24 Extent of mining operations in Silver Valley (x's represent mining features) including the locations of wooden impounding dams constructed to reduce the potential for downstream metals transport. SOURCES: Moreen (2021) and https://astswmo.org/files/Meetings/2018/CaBs_Symposium/Presentations/Urban-Lead.pdf.

INTRODUCTION

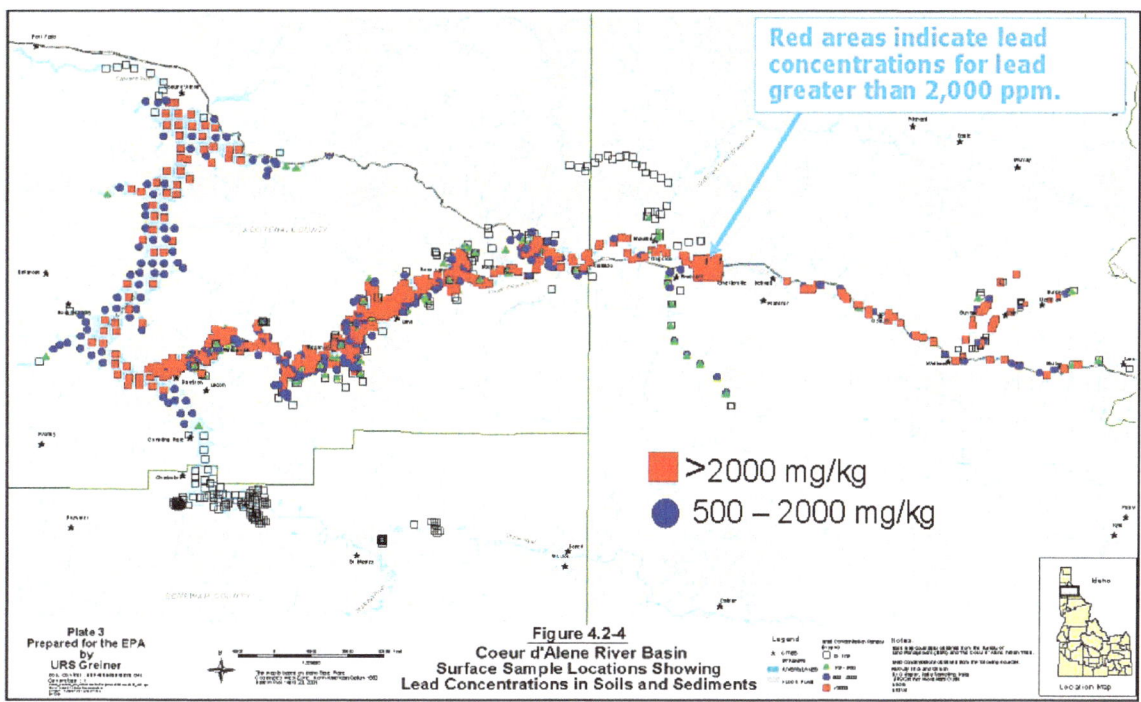

FIGURE 1-25 Nature and extent of sediments with significant concentrations of lead in the CDA River basin and Lake.
SOURCE: URS Greiner and CH2M Hill (2001), Moreen (2021).

CDA River and its tributaries continued until tailings impoundments were installed per 1968 regulations. Overall estimates by Long (1998b) indicate that 56 million MT of mill tailings were discharged, including 2,200 MT of silver, 800,000 MT of lead, and 650,000 MT of zinc. Substantial deposits remain in the river system and continue to be transported downstream. Figure 1-25 shows the profoundly high lead concentrations along the CDA River and within the Lake as of 2001.

The Central Impoundment Area (CIA) was built at the Bunker Hill Complex in the 1920s. This impoundment, as well as others, was an attempt to contain the tailings from the mining activities. The CIA was built on mine waste rock and other materials and enclosed in a ring dike structure (EPA, 1992). It served as an unlined repository for flotation tailings from the Bunker Hill ore concentration mills. Over time, its size increased to 200 acres, and it was divided into three major cells: the east cell for mine wastes and tailings, a gypsum pond, and a slag pile. It also handled acid mine drainage from the Bunker Hill Mine.

Air emissions from ore processing facilities in Kellogg (operated between 1917 and 1981) and Smelterville also provided significant dispersal of contaminants including lead, zinc and sulfur dioxide. Between 1886 and 1917, smelting operations were conducted offsite, but in 1917 the first blast furnace began producing lead, cadmium, and silver and alloys of these metals. Expansion and modification of the smelter continued, reaching a capacity of more than 300 tons of metallic lead per day at the time of its closure (EPA, 1992). Prior to the fall of 1973, typical releases of particulates from the air pollution control system ranged from 10 to 20 tons per month. After fire damaged the air control system baghouse in September 1973, continued metal production led to releases of 160 tons per month of particulate emissions containing 50 to 70 percent lead (TerraGraphics, 1990). These emissions contaminated nearby areas, led to high blood lead levels in residents, and contributed to denuding of surrounding hillsides. Smelter operations ceased in 1981.

Prior to 1928, the Bunker Hill Complex operations focused on lead, cadmium, and silver. In 1928, an electrolytic zinc plant was put into operation; sulfuric acid plants were added in 1954, 1966, and 1970; and a phosphoric acid plant and fertilizer plant were added in 1960 and 1965, respectively. These operations continued until 1981. Limited mining and milling operation took place from 1983 until 1986 and from 1988 until 1991.

The Superfund Remedy

The widespread contamination generated by the mining activities within the Silver Valley and the documented evidence of human health impacts and ecological damage led to the region's labeling in 1983 as one of the largest and most complex hazardous waste sites in the United States. That year the Bunker Hill Mining & Metallurgical Complex was designated as a Superfund site based on the high blood lead levels in children and contamination of the local environment by lead, arsenic, cadmium, and zinc. As part of the Superfund process, EPA studied an area that spans 1,500 mi^2 (3,884 km^2) and 166 river miles (266 km) of northern Idaho and eastern Washington and includes the CDA River, associated tributaries, CDA Lake, and the Spokane River that drains from CDA Lake (Figure 1-26). The final project operating units for remediation efforts include only portions of this study area, as described in greater detail in Chapter 3.

During the remedial investigation, EPA concluded that ambient water quality criteria were exceeded throughout the basin, both Ninemile and Canyon Creeks were devoid of fish and other aquatic life, and habitat fragmentation and destruction was widespread. The most significant contamination was associated with the Box (Figure 1-26)—a rectangle 3 miles (4.8 km) wide and 7 miles (11.2 km) long, from Kellogg on the eastern end to Pinehurst on the western end, that had seen the most intense mining activity. Because mine tailings had been deposited in downstream beds, banks, and floodplains, and because acid mine drainage had discharged to surface waters including the South Fork of the CDA River for over a century, the contamination extended from the eastern end of the watershed, through the entire upper basin along the South Fork, into the lower basin of the CDA River, into CDA Lake, and deposited as far as the Spokane River. To this day, flooding continues to redistribute contaminants across the region and groundwater contamination is widespread.

The long-term remedy for the Superfund site is described in detail in Chapter 3 of this report and in many other reports (e.g., EPA, 2021). Since 1992, the remedial actions in the CDA watershed have attempted

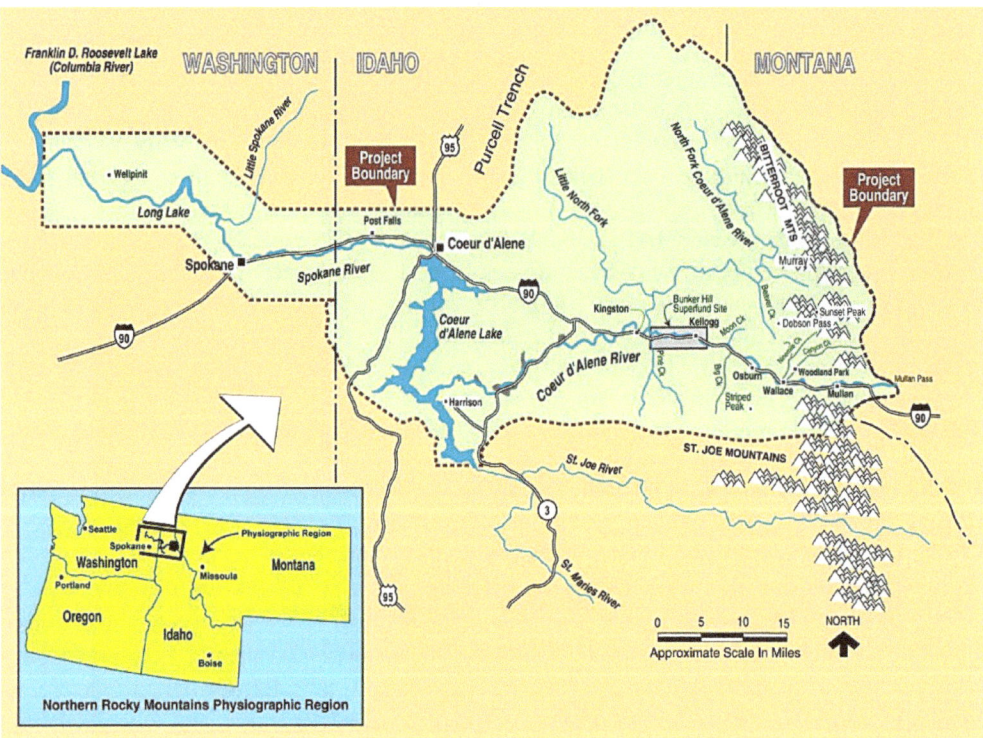

FIGURE 1-26 Map of the CDA River basin and the 3,884 km^2 (1,500 mi^2) area studied during the EPA's Remedial Investigation. The gray box marks the Bunker Hill Superfund Site. SOURCE: URS Greiner and CH2M Hill (2001).

to break exposure pathways for humans and ecological receptors, control source areas, or both. Actions include demolishing mining facilities and waste pits, building treatment plants for mine wastes and contaminated source waters, removing lead-contaminated soil in residential and commercial areas, installing protective barriers (capping), revegetating denuded landscapes and restoring other critical habitats, and remediating and reconstructing floodplains and streams. A 2012 amendment to the remedy extended the activities to consolidation and isolation of sources at upper basin mine and mill sites, construction and maintenance of waste repositories, remediation of road surfaces, protection of remedies from erosion and recontamination, upgrades to the Central Treatment Plant, construction of a groundwater collection system within the Box, and groundwater remediation in Ninemile and Canyon Creek watersheds.

Monitoring of physical, chemical, and biological parameters, including human blood lead levels, has been conducted throughout to determine the effectiveness of the various actions. The 30-year interim remedial plan includes benchmarks for both human health exposure to contaminated soils and ecological protection. However, the remedy is not expected to achieve ambient water quality criteria under the Clean Water Act at all locations, nor is it expected to achieve the applicable and relevant or appropriate requirements for groundwater (i.e., drinking water standards) under the Safe Drinking Water Act at all locations (EPA Region 10, 2012).

Coeur d'Alene Lake Management Plan

CDA Lake was omitted from the Superfund remedy for both technical and sociopolitical reasons. Rather, EPA deferred selecting remedial actions for the Lake, pending implementation of a Lake Management Plan (LMP) along with ongoing evaluation of the effect that current cleanup work in the lower basin, the Box, and the upper basin will have on water quality in CDA Lake (EPA, 2020). Finalized in 2009, the LMP was developed by the State of Idaho and the CDA Tribe with the goal of protecting and improving water quality by limiting basin-wide nutrient inputs that impair water quality or impact metal ion solubility (IDEQ and CDA Tribe, 2009). A low nutrient, high hypolimnetic dissolved oxygen condition was thought to be the optimal state for immobilization and net burial of residual metals in the Lake sediments and necessary to prevent harmful algal blooms.

The primary responsibilities outlined in the LMP for the CDA Tribe and IDEQ were to develop partnerships with community groups, businesses, and other governmental agencies; coordinate the different authorities involved in management of the basin; and seek funds for implementing activities to achieve the goals of the LMP. A core component of the LMP that was recently completed is the development of a basin-wide nutrient inventory (finalized for phosphorus; IDEQ and the CDA Tribe, 2020; reviewed in Chapter 3). Also within the LMP are activities meant to reduce nutrient loading to CDA Lake, as described in a nutrient management action plan.

Progress made on implementing the LMP is reported on annually by the Basin Environmental Improvement Project Commission (BEIPC), a group of representatives from the State of Idaho, the three Idaho counties in the basin, the CDA Tribe, the State of Washington, and the United States. BEIPC is broadly responsible for coordinating environmental cleanup to address heavy metal contamination, natural resource restoration, and water quality in the CDA basin, including the Superfund remedy. In its most recent annual report (2020), the BEIPC listed the major programs with the LMP as (1) the Science Core Program which covers routine lake monitoring by the CDA Tribe and IDEQ staff; (2) the Education and Outreach Core Program; and (3) the Nutrient Inventory and Nutrient Reduction Core Program. The 2020 annual report is available on the IDEQ's lake management web page.[7] There is no regular comprehensive accounting of specific actions taken to reduce nutrient loadings to CDA Lake done under the auspices of the LMP (personal communication, Jamie Brunner, IDEQ, January 2022).

As of the summer of 2018, the CDA Tribe determined that the LMP was inadequate, in itself, as an effective tool to protect water quality in the Lake. The CDA Tribe withdrew their support as a signatory government to the LMP in 2019. In 2020, the State of Idaho, Kootenai County and EPA sponsored the contract with the National Academies to conduct a neutral third-party review of the Lake data—an action that the CDA Tribe supported.

---

[7] http://www.basincommission.com/wp-content/uploads/2021/03/2020-Draft-Annual-Report.pdf

FEDERAL, REGIONAL, AND LOCAL AUTHORITY AND OVERSIGHT

The administrative structure for management of water quality in CDA Lake and its basin is complex, involving federal, state and tribal jurisdictional and authoritative bodies, especially EPA Region 10, IDEQ, and the Lake Management Department of the CDA Tribe. These three organizations are primarily responsible, respectively, for operating the Superfund site remedy; assessing compliance with water quality standards and identifying water quality limited waterbodies; and maintaining authority over most water quality issues within tribal boundaries. Additional oversight and support for water quality management come from the U.S. Army Corps of Engineers (ACOE) and USDA; Washington Department of Ecology; Panhandle Health District 1; Idaho's Soil Conservation Commission, Transportation Department, Department of Water Resources (IDWR), and Department of Lands (IDL); and city and county agencies within the basin. These entities, along with the CDA Tribe and IDEQ, are responsible for overseeing wastewater treatment operations and discharges, stormwater management, construction site erosion control, and water-related activities such as dredging, excavation and fill, streambed alterations, nuisance organisms, water levels and outflow rates, and land use activities (IDEQ and CDA Tribe, 2009). County governments in the basin use their authority under state law to promulgate zoning ordinances that regulate private land uses that can affect water quality conditions in the Lake. Federal and state resource agencies also exercise authorities over upland activities that may influence water quality conditions in tributary waters and the Lake.

Federal Authority

EPA's direct involvement in the region has been heavily weighted toward Comprehensive Environmental Response, Compensation, and Liability Act of 1980 (CERCLA) implementation and enforcement at the Bunker Hill Superfund site. In addition to this large role, EPA reviews and approves water quality standards that have been established by the State of Idaho and Tribes under the 1972 federal Clean Water Act (CWA). Idaho water quality standards are reviewed every three years.

Until 2018, EPA also was the agency responsible for administering the National Pollutant Discharge Elimination System permits for facilities across state and tribal waters. Authority was passed to Idaho for municipal and pretreatment permits on July 1, 2018; industrial permits on July 1, 2019; general permits on July 2, 2020; and federal facilities, stormwater, and biosolids permits on July 1, 2021.[8] An Idaho Pollution Discharge Elimination System permit authorizes point source pollutant discharges into Idaho waters of the United States, except on tribal land.

State Authority

Although IDEQ is the designated agency responsible for implementing state water quality standards to restore and maintain designated beneficial uses of streams, lakes, and other surface waters (Idaho Code §39-3601), the Coeur d'Alene Regional Office of IDEQ has local responsibility for the five-county Panhandle Area that includes the basin. This IDEQ office also receives and responds to water quality complaints and issues water quality certifications for federal permits under Section 401 of the CWA, reviews permits under the Idaho Joint Application for Permits (which includes those from IDL, IDWR, and ACOE), reviews engineering plans and specifications for water and wastewater systems, issues permits for wastewater reuse, and performs analyses and administers rules for groundwater. As part of its authority under the CWA, IDEQ establishes the total maximum daily load (TMDL) of a contaminant that can be discharged into a waterbody or segment for the sub-basins, identifies impairments within water segments or lakes, and lists those segments on a 303(d) list of impaired or threatened waters. IDEQ prepares integrated reports of current conditions of all state waters [per 305(b) requirements], including an assessment of status and trends of publicly owned lakes and listing of those waters that are impaired (not supporting their intended use) and need a TMDL. The most recent report, Idaho's 2018/2020 Integrated Report (IDEQ, 2020b), is also presented as an Integrated Report StoryMap.[9]

---

[8] https://www.deq.idaho.gov/water-quality/wastewater/permit-options/

[9] https://storymaps.arcgis.com/stories/740e317eebc546d0b3ebbed5419aba79

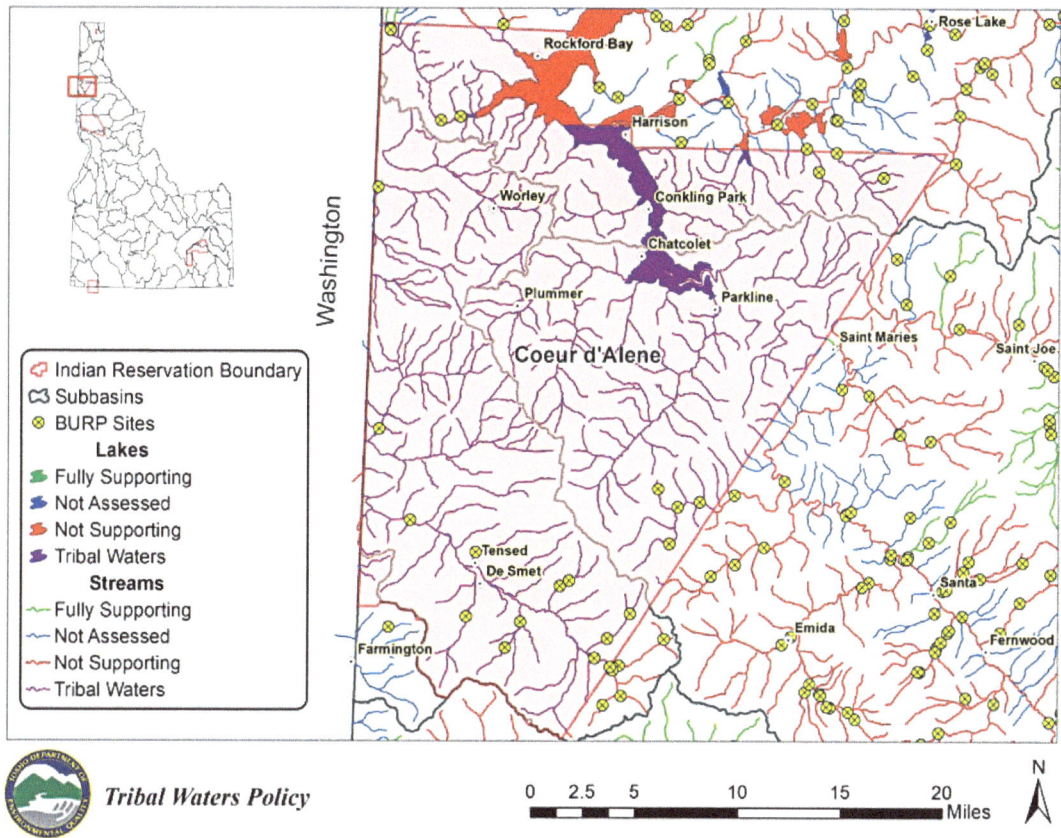

FIGURE 1-27 Map showing tribal waters in and around CDA Lake. These waters are not classified under the IDEQ TMDL process; therefore, they are not classified as to whether they are supporting or assessed under the TMDL nomenclature. SOURCE: IDEQ (2020b).

Coeur d'Alene Tribal Authority

The CDA Tribe has authority for setting water quality standards and enhancing and protecting water quality within their reservation. It has established a Lake Management Department to implement parts of the federal CWA and manage recreation, water resources, and hazardous wastes. As such, the Tribe also receives and responds to water quality complaints and issues water quality certifications for federal permits, reviews permits under the Idaho Joint Application for Permits, develops and enforces encroachment standards, regulates dredge and fill activities, performs groundwater analyses and characterization, and implements wellhead protection activities. Figure 1-27 provides a map showing the tribal waters in and surrounding CDA Lake.

Other Authorities Including Land Use Authorities

Water quality within CDA Lake is under the purview of a number of additional agencies and commissions beyond those already discussed. The IDL manages Idaho's public trust lands and the lands beneath Idaho's navigable waterways, including riverbeds and the beds and banks of Idaho's navigable lakes.[10] The IDL regulates encroachments (e.g., docks, marinas, shoreline stabilization) on navigable lakes in order to balance the protection of property, navigation, fish and wildlife habitat, aquatic life, recreation, aesthetic beauty and water quality with navigational or economic necessities (Idaho Code §58-1301).

---

[10] https://www.idl.idaho.gov/lakes-rivers/encroachments/#

Shoshone, Kootenai, Benewah, and Latah Counties and their respective cities have enacted ordinances to manage upland development and other land use activities in the basin. Kootenai County's (2020) Comprehensive Plan outlines goals, policies and objectives in light of projected future population growth and land use, but it is not enforceable land use regulation. Nonetheless, the Comprehensive Plan recognizes a **shoreline** designation that "includes lands within 500 feet of bodies of water . . . directly adjacent to shorelines and lands where a portion of the parcel is within the 500-foot boundary." Recognizing that development demand in these areas is high, the purpose of this designation is to guard against water quality degradation by managing erosion and surface water runoff, reducing impervious surfaces in developments, and implementing setbacks from shorelines and surface water corridors. The 2016 ordinances[11] provide detailed information about high water mark elevations, prohibited activities such as application of fertilizer or storage of chemicals, permitted activities including infrastructure development, site improvement, vegetation modification and erosion control; and development exceptions.

CURRENT WATER QUALITY CONDITIONS

The current water quality conditions in CDA Lake demonstrate the legacy of more than 100 years of mining in the watershed. Table 1-5 shows the water quality standards laid out by both the State of Idaho and the CDA Tribe, depending on the location in the Lake. In addition, the table lists the LMP water quality triggers for key metals and nutrients (total phosphorus, dissolved oxygen, chlorophyll *a*, lead, zinc, and cadmium), which in some cases default to the state and tribal water quality standards, but in other cases are slightly different.

Because the toxicity of metals to humans and ecological receptors is discussed in greater detail in Chapter 9, and the current concentrations of metals and nutrients are evaluated in great depth in Chapter 4, only cursory remarks are made here, to highlight the extent of the metals contamination in CDA Lake. CDA Lake is in violation of Idaho's water quality standards for some of the metals of concern in the northern half of the Lake, with zinc concentrations routinely well over the water quality standard of 36 µg/L and cadmium concentrations hovering around the standard of 0.25 µg/L. The CDA Tribe's water quality standards for these same metals are being violated at the more southern C5 location, but not at the southernmost monitoring location, C6. The Lake is also in violation of water quality standards for mercury, based on fish tissue concentrations. The IDEQ 2018–2020 Integrated Report (IDEQ, 2020b) and the most recent LMP status update on water quality (IDEQ, 2020a) report current metals concentrations and violations of standards in the northern half of the Lake. As would be expected given its drainage area, violations of metals standards are also evident along the extent of the CDA River.

Despite these violations, there are no current enforceable requirements to return metals concentrations to levels that meet water quality standards in CDA Lake (via the creation of a TMDL under the CWA). In 2000, EPA and Idaho established a metals TMDL for CDA Lake, but the TMDL was challenged in the Idaho Supreme Court and vacated for failure to follow state rulemaking provisions.[12] Nonetheless, there are TMDLs in place for certain metals (particularly mercury) in some segments of the CDA River (see Idaho's 2018/2020 Integrated Report—IDEQ, 2020b). The addition of mercury impairment for the CDA River segments in 2019 resulted from fish tissue samples that exceeded the water quality criterion (0.3 mg Hg/kg fresh weight tissue) for designated uses of primary contact recreation, cold water aquatic life, and salmonid spawning (Essig, 2010). The priorities for developing TMDLs in the CDA basin (as reflected in Idaho's 2018/2020 Integrated Report) range from low to high, with the South Fork of the CDA River and the Spokane River being the highest priority for the IDEQ.

Although there have been concerns expressed among IDEQ, the CDA Tribe, and stakeholders with respect to nutrient loading, CDA Lake is not currently listed as impaired for nutrients.

Dissolved Metal Concentrations in CDA Lake Compared to Other Lakes

Although annual mean concentrations of dissolved zinc, cadmium, lead, and arsenic have continued to exceed water quality standards in at least some locations in CDA Lake in recent years (IDEQ, 2020a), the reports on these

---

[11] https://www.kcgov.us/DocumentCenter/View/1506/Ordinance-493-Land-Use-Development-Code-Amended-PDF
[12] https://www.idl.idaho.gov/wp-content/uploads/sites/116/2020/10/ASARCO-v-State-of-Idaho-69P3d139_2003.pdf

TABLE 1-5 Water Quality Standards (WQS) and LMP Triggers for Contaminants of Concern in CDA Lake

	C1		C4		C5		C6	
	Idaho WQS	LMP Trigger	Idaho WQS	LMP Trigger	CDA Tribe Standard	LMP Trigger	CDA Tribe Standard	LMP Trigger
Total P	narrative	8 µg/L	narrative	8 µg/L	narrative	8 µg/L	narrative	9 µg/L
DO	Hypolimnion exempt	min < 6 mg/L	Hypolimnion exempt	min < 6 mg/L	min > 8 mg/L	min < 8 mg/L	min > 6 mg/L	Exceedance of WQS
chl a	narrative	3 µg/L annual geo mean; 5 µg/L max	narrative	3 µg/L annual geo mean; 5 µg/L max	narrative	3 µg/L annual geo mean; 5 µg/L max	narrative	3 µg/L annual geo mean; 5 µg/L max
Pb	0.54 µg/L	Already exceeds WQS	0.54 µg/L	Already exceeds WQS	0.39–0.56 µg/L	Already exceeds WQS	0.45–0.55 µg/L	Exceedance of WQS
Zn	36 µg/L	Already exceeds WQS	36 µg/L	Already exceeds WQS	26–37 µg/L	Already exceeds WQS	26–37 µg/L	Exceedance of WQS
Cd	0.25 µg/L	Already exceeds WQS	0.25 µg/L	Already exceeds WQS	0.19–0.26 µg/L	Already exceeds WQS	0.2–0.27 µg/L	Exceedance of WQS

Notes: C1 through C6 denote Lake locations; see Chapter 2. P = phosphorus, DO = dissolved oxygen, chl a = chlorophyll a, Pb = lead, Zn = zinc, and Cd = cadmium.
SOURCES: C1 and C4 values from IDEQ (2020a); C5 and C6 values from IDEQ and CDA Tribe (2017).

data have not established a context for these concentrations relative to other lakes. Surveys of dissolved metal concentrations in the waters of lakes comparable in size to CDA Lake are not especially common (Luoma and Rainbow, 2008). Nonetheless, there are enough data available to clearly illustrate that **dissolved concentrations of zinc, cadmium, and lead in CDA Lake are more than an order of magnitude higher than those measured in other large lakes,** many with urbanized or industrialized watersheds (Table 1-6). The water column concentrations of lead and especially zinc and cadmium in CDA Lake are unusually high for such a large lake. This comparison illustrates the challenges of bringing metal enrichment in CDA Lake to a level commensurate with other lakes surrounded by human development and industry, and it is also useful for better understanding the potential for impacts on ecological and human health.

In CDA Lake, the annual mean or median dissolved lead concentration never exceeds 1 µg/L and only exceeds the target of 0.54 µg/L occasionally (IDEQ, 2020a). Nevertheless, it is important to recognize that dissolved concentrations of lead are typically extremely low in lake waters because of the high affinity of lead for particulate material. For example, Coale and Flegal (1989), using ultra-clean methodologies, found that dissolved lead concentrations across Lake Erie and Lake Ontario varied from 0.002 to 0.058 µg/L in the mid-1980s, with the highest concentrations near the urban developments and industrial activities on the shoreline. They showed that a primary source of that lead was atmospheric input originating from leaded gasoline. Use of leaded gasoline contributed to a global elevation of lead concentrations in large lakes and oceans throughout the 1980s. Industrial emissions of lead have declined since the 1980s, lead is no longer added to gasoline, and overall lead concentrations have declined in many environments (Luoma and Rainbow, 2008). Yet, present-day annual mean dissolved lead concentrations at the C4 monitoring location within CDA Lake would have to decline to less than 10 percent of the 2017–2019 concentrations to even reach the levels typical of shallow waters in the Great Lakes during the 1980s.

Annual mean or median dissolved cadmium and especially zinc concentrations within CDA Lake exceed the LMP targets more frequently than did dissolved lead (IDEQ and CDA Tribe, 2020). In the largest lakes listed in Table 1-6 other than CDA Lake, dissolved zinc concentrations do not exceed 1 µg/L. Higher concentrations were observed in the somewhat smaller lakes listed in Table 1-6 (i.e., Lake Greifen and a study of 28 lakes in Quebec

TABLE 1-6 Range of Dissolved Arsenic (As), Cadmium (Cd), Zinc (Zn), and Lead (Pb) Concentrations in Selected Large Lakes

Selected Lakes	As µg/L	Cd µg/L	Zn µg/L	Pb µg/L	Source
Great Lakes		0.002–0.0045	0.087–0.280	0.003–0.011	Nriagu et al., 1995
"Baseline" lakes	0.3–1.9				Smedley and Kinniburgh, 2002
Lake Tahoe	1.12	0.0032–0.008		0.0005–0.0052	Anderson and Bruland, 1991; Romero et al., 2013; Chien et al., 2019
Lake Erie, Ontario	0.52	0.002–0.011	0.005–0.11	0.002–0.058	Anderson and Bruland, 1991; Coale and Flegal, 1989
Lake Sammamish WA		0.0008–0.005	0.12–0.60		Balistrieri et al., 1992
Lake Greifen*			0.65–2.6		Xue et al., 1995
28 Quebec & Ontario lakes			0.13–0.97; 2.0–2.7		Tessier et al., 1989
Coeur d'Alene Lake					
C1: 2017–2020*	0.25–0.45	0.10–0.27	25–70	< 0.05–3.0	IDEQ, 2020a
C4: 2017–2020*	0.28–0.55	0.12–0.46	27–76	0.2–64	IDEQ, 2020a
C5: 2017–2020*	< 0.2–0.57	< 0.05–0.15	< 5–90	0.07–0.9	CDA Tribe data

*Range includes surface and bottom waters.

and Ontario, Canada, that include mine-influenced lakes with pH > 6.0). Lake Greifen is a eutrophic lake affected by sewage inputs, and many of the Canadian lakes are mine-influenced. The maximum zinc concentrations in the smaller contaminated lakes is 2.7 µg/L (5 of the 28 high pH lakes had [Zn] > 1 µg/L). The point to be made is that the annual mean and the maximum concentrations of zinc in CDA Lake far exceed concentrations typical of even enriched high-pH lakes. Dissolved zinc concentrations would have to decline to about 1 percent of annual mean or median zinc concentrations at the C4 monitoring location of CDA Lake to approach concentrations in a typical larger lake.

On the other hand, arsenic concentrations at C4 only occasionally exceed values typically seen in larger lakes and are far lower than extreme concentrations observed in alkaline lakes or systems where sediments regularly become anoxic (Smedley and Kinniburgh, 2002).

NATIONAL ACADEMIES STUDY

Despite the diminution of mining and the implementation of the Bunker Hill Superfund remedy, CDA Lake does not meet water quality standards for metals throughout the Lake and during all seasons. Exposure of humans as well as fish and wildlife in the region to unsafe levels of metals is still a concern in parts of the Lake and the CDA watershed, and the Superfund remedy is ongoing. Other uses of the Lake (for recreation and tourism) have continued to thrive despite the massive metals loading to the Lake over the past 100 years.

In the past 15 years, some stakeholders have become concerned about development along the lake shorelines, land use changes within the larger basin, and other dynamics that might pose a potential new threat in the form of increased loading of nutrients (i.e., phosphorus and nitrogen) that could lead to the release of metals bound to lake sediments. Indeed, the premise of the LMP is that an increase in phosphorus loading to the Lake, combined with decreases in zinc loading due to Superfund remedial actions, *could* pose new problems to CDA Lake if these changes lead to increased growth of algae in the Lake, associated production of algal toxins, and depression of oxygen levels (IDEQ and CDA Tribe, 2009). The hypothesis of concern is that a decrease in oxygen levels in Lake sediments could promote the release of metals from the Lake sediments into the Lake water column, potentially leading to even worse water quality (IDEQ and CDA Tribe, 2009).

These issues are complicated by the fact that funds for the Superfund remedy cannot currently be spent to study or remediate CDA Lake. Rather, the State of Idaho and the CDA Tribe are implementing the LMP outside of the Superfund process, primarily to manage nutrient loads entering the basin in an effort to maintain adequate oxygen levels in the Lake. Thirteen years since the writing of the LMP, reports such as IDEQ (2020a) express concern about water quality conditions in the Lake. Hence, IDEQ, EPA, and Kootenai County requested the involvement of the National Academies to analyze available data and information about CDA Lake water quality and provide insights about future conditions in the Lake.

In late 2020, the National Academies assembled a committee to review historical and recent water quality data, and available modeling efforts, stemming from the 2009 LMP and other available information, with the goal of determining what near-future water quality conditions in the Lake might be. More specifically, the committee was asked to do the following:

- Evaluate current water quality in the Lake, lower rivers, and lateral lakes with a focus on observed trends in nutrient loading and metals concentrations (Chapters 3 through 6), while also considering how changes in temperature or precipitation could affect those trends (Chapter 10).
- Consider the impacts of current summertime anoxia on the fate of the metals and nutrients (Chapters 5 and 6).
- Consider whether reduced levels of zinc entering the Lake as a result of the upgrade to the Central Treatment Plant and other upstream activities are removing an important control on algal growth (Chapters 5 and 9).
- Discuss whether metals currently found in Lake sediments will be released into the Lake if current trends continue (Chapters 7 and 10). If sufficient data are not available to result in a high level of confidence in its conclusions, the National Academies will identify the additional data that are required to achieve an appropriate level of confidence (Chapter 8).
- Discuss the relevance of metals release in the Lake to human and ecological health (Chapter 9).

The committee met six times over a 15-month period. Given the ongoing COVID-19 pandemic, these meetings were held entirely in a virtual format, with the exception of the final closed-session meeting. Nonetheless, the committee benefited from ongoing dialogues with a large stakeholder group, including the IDEQ, the CDA Tribe, EPA, USGS, county representatives, researchers from the University of Idaho, and other parties.

Organization of Report

This report begins with two introductory chapters that provide context for understanding the future of water quality in CDA Lake. Chapter 1 has already introduced the Lake and the region, including its climate, hydrology, current water quality, population, land use, and regulatory environment. Chapter 2 discusses the physical, chemical, and biological monitoring of the Lake, including a discussion of the data sets used by the committee to elucidate water quality trends and how monitoring could be improved in the future.

Chapter 3 discusses inputs of water, metals, and nutrients to CDA Lake from the watershed, including the committee's analysis of trends in contaminant concentrations and loading over time. Similarly, Chapters 4, 5, and 6 evaluate the water quality of the Lake itself, including the committee's analyses of trends over time in hydrodynamics (Chapter 4), dissolved oxygen and nutrients (Chapter 5), and metals (Chapter 6). Chapter 7 discusses processes occurring in the Lake sediments that could lead to release of metals from the sediments into the water column, including analyses of bottom water data and limited sediment core data for dissolved oxygen, nutrients, and metals.

With the benefit of the completed data analyses, Chapter 8 returns to the subject of Chapter 2 by making recommendations for improving the water quality monitoring in CDA Lake. Chapter 9 reviews the human health and ecological risks posed by metals in CDA Lake. Finally, Chapter 10 discusses the role of climate change and how it may alter the trends observed in Chapters 3 through 7. Each chapter ends with conclusions and (when appropriate) recommendations that synthesize more technical and specific statements in the body of each chapter. The most important conclusions and recommendations from Chapters 3 through 10 are compiled in the report summary.

REFERENCES

Anderson, L. C., and K. W. Bruland. 1991. Biogeochemistry of arsenic in natural waters: the importance of methylated species. *Environ. Sci. Technol.* 25(3):420–427.

Balistrieri, L. S., J. W. Murray, and B. Paul. 1992. The biogeochemical cycling of trace metals in the water column of Lake Sammamish, Washington: Response to seasonally anoxic conditions. *Limnol. Oceanogr.* 37(3):529–548.

Balistrieri, L., S. Box, A. Bookstrom, R. Hooper, and J. Mahoney. 2002. Impacts of Historical Mining in the Coeur d'Alene River Basin. Chapter 6 *In:* Pathways of Metal Transfer from Mineralized Sources to Bioreceptors: A Synthesis of the Mineral Resources Program's Past Environmental Studies in the Western United States and Future Research Directions. USGS Bulletin 2191. L. S. Balistrieri and L. L. Stillings, eds.

Bender, S. F. 1991. Investigation of the chemical composition and distribution of mining wastes in Killarney Lake, Coeur d'Alene area, northern Idaho. Doctoral dissertation, University of Idaho.

Boese, R., G. Stevens, A. Hess, J. Jenkins, R. Barlow, A. Clary, J. Covert, J. Ekins, M. Galante, T. Hanson, L. Laumatia, M. LaScuola, R. Lindsay, S. Phillips, and L. Schmidt. 2015. The Spokane Valley-Rathdrum Prairie Aquifer Atlas, 2015 Update. City of Spokane.

Bookstrom, A. A., S. E. Box, J. K. Campbell, K. I. Foster, and B. L. Jackson. 2001. Lead-rich sediments, Coeur d'Alene River Valley, Idaho: area, volume, tonnage, and lead content. U.S. Geological Survey Open-File Report, 01-140.

Chien, C. T., B. Allen, N. T. Dimova, J. Yang, J. Reuter, G. Schladow, and A. Paytan. 2019. Evaluation of atmospheric dry deposition as a source of nutrients and trace metals to Lake Tahoe. *Chemical Geology* 511:178–189.

Chess, D. 2021. Water Quality Data Summary. Presentation to the NASEM Committee. February 26, 2021.

Coale, K. H., and A. R. Flegal. 1989. Copper, zinc, cadmium and lead in surface waters of Lakes Erie and Ontario. *Science of the Total Environment* 87:297–304.

Cooper, C. 2021a. Coeur d'Alene Lake Management Plan Water Quality Data and Trends. Presentation to the NASEM Committee. February 26, 2021.

Cooper, C. 2021b. Conceptual Model of CDA Lake. Presentation to the NASEM Committee. February 26, 2021.

Coeur d'Alene Tribe and Avista Corporation. 2017. Coeur d'Alene Reservation Five-Year Synthesis 2011–2015 Monitoring Data 4(E) Condition No. 5 Spokane River Hydroelectric Project FERC Project No. 2545. February 2017.

Daly, C., M. Halbleib, J. I. Smith, W. P. Gibson, M. K. Doggett, G. H. Taylor, J. Curtis, and P. P. Pasteris. 2008. Physiographically sensitive mapping of climatological temperature and precipitation across the conterminous United States. *International Journal of Climatology: a Journal of the Royal Meteorological Society* 28(15):2031–2064.

Daly, C., J. I. Smith, and K. V. Olson. 2015. Mapping atmospheric moisture climatologies across the conterminous United States. *PloS One* 10(10):e0141140.

EPA (U.S. Environmental Protection Agency). 1992. Superfund Record of Decision: Bunker Hill Mining & Metallurgical Complex. EPA ID: IDD048340921 OU 02 Smelterville, ID, September 1992. https://semspub.epa.gov/work/HQ/188517.pdf.

EPA Region 10. 2012. Interim Record of Decision (ROD) Amendment, Upper Basin of the Coeur d'Alene River, Bunker Hill Mining and Metallurgical Complex Superfund Site. https://semspub.epa.gov/work/10/664107.pdf.

EPA. 2015. Bunker Hill Mining and Metallurgical Complex, Idaho, Superfund Case Study (EPA 542-F-15-001; p. 11). https://www.epa.gov/sites/production/files/2018-03/documents/bunker_hill_eco_case_study_final_feb2015.pdf.

EPA. 2020. Optimization Review Report Remedial Process Optimization Study Lake Coeur d'Alene Bunker Hill Mining and Metallurgical Site Operable Unit 03 Coeur d'Alene, Kootenai County, Idaho.

EPA. 2021. Fifth Five-Year Review Report for the Bunker Hill Mining and Metallurgical Complex Superfund Facility (Bunker Hill Superfund Site) Shoshone, Kootenai, and Benewah Counties in Idaho Spokane County in Washington.

Essig, D. A. 2010. Arsenic, Mercury, and Selenium in Fish Tissue and Water from Idaho's Major Rivers: A Statewide Assessment. Idaho DEQ, March 2010. https://www2.deq.idaho.gov/admin/LEIA/api/document/download/15208.

Hamlet, A. F., P. W. Mote, M. P. Clark, and D. P. Lettenmaier. 2005. Effects of temperature and precipitation variability on snowpack trends in the western United States. *Journal of Climate* 18(21):4545–4561.

Hardy, M. 2021. Census confirmation: Our county is booming. CDA/Post Falls Press. August 21, 2021. https://cdapress.com/news/2021/aug/22/unsurprising-growth.

Hogan, D. M., and M. R. Walbridge. 2007. Best management practices for nutrient and sediment retention in urban stormwater runoff. *Journal of Environmental Quality* 36(2):386–395.

Horowitz, A. J., K. A. Elrick, and R. B. Cook. 1992. Effect of mining-related activities on the sediment-trace element geochemistry of Lake Coeur d'Alene, Idaho, USA. Part 1, Surface sediments. U.S. Geological Survey Open-File Report 92-109.

Horowitz, A. J., K. A. Elrick, and R. B. Cook. 1993a. Effect of mining and related activities on the sediment trace element geochemistry of Lake Coeur d'Alene, Idaho, USA. Part I: surface sediments. *Hydrol. Process.* 7:403–423.

Horowitz, A. J., K. A. Elrick, J. A. Robbins, and R. B. Cook. 1993b. The effect of mining and related activities on the sediment-trace element geochemistry of Lake Coeur d'Alene, Idaho. Part II, Subsurface sediments. U.S. Geological Survey Open-File Report 93-656.

Horowitz, A. J., K. A. Elrick, J. A. Robbins, and R. B. Cook. 1995. A summary of the effects of mining and related activities on the sediment-trace element geochemistry of Lake Coeur d'Alene, Idaho, USA. *J. Geochem. Explor.* 52(1–2):135–144.

IDEQ and Coeur d'Alene Tribe. 2009. Coeur d'Alene Lake Management Plan. State of Idaho Department of Environmental Quality.

IDEQ and Coeur d'Alene Tribe. 2017. Coeur d'Alene Lake Management Plan: Coeur d'Alene Lake Status Update, 2015.

IDEQ and CDA Tribe. 2020. Coeur d'Alene Lake Management Program: Total Phosphorus Nutrient Inventory, 2004–2013.

IDEQ. 2020a. Coeur d'Alene Lake Management Plan: Coeur d'Alene Lake Status Update, 2015–2018. May 2020.

IDEQ. 2020b. Idaho's 2018/2020 Integrated Report, Final. October 2020.

Kahle, S. C., and J. R. Bartolino. 2007. Hydrogeologic Framework and Ground-Water Budget of the Spokane Valley-Rathdrum Prairie Aquifer, Spokane County, Washington, and Bonner and Kootenai Counties, Idaho. U.S. Geological Survey Scientific Investigations Report 2007–5041.

Kahle, S. C., R. R. Caldwell, and J. R. Bartolino. 2005. Compilation of geologic, hydrologic, and ground-water flow modeling information for the Spokane Valley-Rathdrum Prairie Aquifer, Spokane County, Washington, and Bonner and Kootenai Counties, Idaho. U.S. Geological Survey Scientific Investigations Report 2005-5227.

Kmusser. 2008. Map of the Spokane River watershed—Western United States. https://commons.wikimedia.org/wiki/File:Spokanerivermap.png.

Kootenai County. 2020. Kootenai County Comprehensive Plan Update. February 2020. https://www.kcgov.us/DocumentCenter/View/13543/2020-Comp-Plan-Update.

Kuwabara, J. S., B. R. Topping, P. F. Woods, J. L. Carter, and S. W. Hager. 2006. Interactive effects of dissolved zinc and orthophosphate on phytoplankton from Coeur d'Alene Lake, Idaho. USGS Scientific Investigations Report 2006-5091. http://pubs.usgs.gov/sir/2006/5091.

Kuwabara, J. S., Topping, B. R., Woods, P. F., and J. L. Carter. 2007. Free zinc ion and dissolved orthophosphate effects on phytoplankton from Coeur d'Alene Lake, Idaho. *Environ. Sci. Technol.* 41(8):2811–2817. https://doi.org/10.1021/es062923l.

Langman, J. B., J. D. Ali, A. W. Child, F. M. Wilhelm, and J. G. Moberly. 2020. Sulfur species, bonding environment, and metal mobilization in mining-impacted lake sediments: column experiments replicating seasonal anoxia and deposition of algal detritus. *Minerals* 10(10):849.

Liao, F. H., F. M. Wilhelm, and M. Solomon. 2016. The effects of ambient water quality and Eurasian watermilfoil on lakefront property values in the Coeur d'Alene area of northern Idaho, USA. *Sustainability* 8(1):44. https://doi.org/10.3390/su8010044.

Long, K. R. 1998a. Grade and tonnage models for Coeur d'Alene type polymetallic veins. U.S. Geological Survey Open-File Report 98-583. https://doi.org/10.3133/ofr98583.

Long, K. R. 1998b. Production and disposal of mill tailings in the Coeur d'Alene mining region, Shoshone County, Idaho; Preliminary Estimates. U.S. Geological Survey Open-File Report 98-595. https://doi.org/10.3133/ofr98595.

Luoma, S. N., and P. S. Rainbow. 2008. Metal Contamination in Aquatic Environments: Science and Lateral Management. Cambridge: Cambridge University Press.

MacInnis, J. D., Jr., B. B. Lackaff, R. M. Boese, G. Stevens, S. King, and R. C. Lindsay. 2009. The Spokane Valley-Rathdrum Prairie Aquifer Atlas 2009 Update. City of Spokane.

Maupin, M. A., and R. J. Weakland. 2009. Water Budgets for Coeur d'Alene Lake, Idaho, Water Years 2000–2005. USGS Scientific Investigations Report 2009-5184. U.S. Geological Survey and Idaho Department of Water Resources.

Mink, L. L., R. E. Williams, and A. T. Wallace. 1971. Effect of industrial and domestic effluents on the water quality of the Coeur d'Alene River basin. Idaho Bureau of Mines and Geology Pamphlet 149, 30 p.

Moreen, E. 2021. Bunker Hill Mining and Metallurgical Complex Superfund Site Overview. Presentation to NASEM Committee. February 24, 2021.

Mote, P. W., S. Li, S., D. P. Lettenmaier, M. Xiao and R. Engel. 2018. Dramatic declines in snowpack in the western U.S. *npj Clim Atmos Sci* 1:2. https://doi.org/10.1038/s41612-018-0012-1.

National Research Council (NRC). 2005. Superfund and Mining Megasites: Lessons from the Coeur d'Alene River Basin. Washington, DC: National Academies Press. https://doi.org/10.17226/11359.

NCAI Partnership for Tribal Governance. 2016. Workforce Development: Building the Human Capacity to Rebuild Tribal Nations: Coeur d'Alene Tribe. National Congress of American Indians. https://www.ncai.org/PTG_Innovation_Spotlight_-_Coeur_dAlene_Workforce_Development_FINAL.pdf.

Nriagu, J. O., G. Lawson, H. K. T. Wong, and V. Cheam. 1995. Dissolved trace metals in Lakes Superior, Erie, and Ontario. *Environ. Sci. Technol.* 30(1):178–187.

Rabe, F. W., and D. C. Flaherty. 1974. The river of green and gold. Idaho Research Foundation. Inc., Natural Resources Series, (4). Moscow, ID, 98 pp.

Rogger, M., M. Agnoletti, A. Alaoui, J. C. Bathurst, G. Bodner, M. Borga, V. Chaplot, F. Gallart, G. Glatzel, J. Hall, J. Holden, L. Holko, R. Horn, A. Kiss, S. Kohnová, G. Leitinger, B. Lennartz, J. Parajka, R. Perdigão, S. Peth, L. Plavcová, J. N. Quinton, M. Robinson, J. L. Salinas, A. Santoro, J. Szolgay, S. Tron, J. J. H. van den Akker, A. Viglione, and G. Blöschl. 2017. Land use change impacts on floods at the catchment scale: Challenges and opportunities for future research. *Water Resources Research* 53(7):5209–5219.

Romero, I. C., N. J. Klein, S. A. Sanudo-Wilhelmy, and D. G. Capone. 2013. Potential trace metal co-limitation controls on N_2 fixation and NO_3^- uptake in lakes with varying trophic status. *Frontiers in Microbiology* 4(54). https://doi.org/10.3389/fmicb.2013.00054.

Smedley, P. L., and D. G. Kinniburgh. 2002. A review of the source, behaviour and distribution of arsenic in natural waters. *Applied Geochemistry* 17(5):517–568.

TerraGraphics. 1990. Risk assessment data evaluation report (RADER) for the populated areas of the Bunker Hill Superfund site. Moscow, ID: Prepared for Region 10 EPA, Superfund Section, 1990.

Tessier, A., R. Carignan, B. DuBreuil, and F. Rapin. 1989. Partitioning of zinc between the water column and the oxic sediments in lakes. *Geochimica et Cosmochimica ACTA* 53:1511–1522.

Tong, S. T., and W. Chen. 2002. Modeling the relationship between land use and surface water quality. *Journal of Environmental Management* 66(4):377–393.

U.S. Geological Survey (USGS). 1981. River Basins of the United States: The Columbia. U.S. Geological Survey. https://doi.org/10.3133/70039373.

URS Greiner, Inc., and CH2M Hill. 2001. Final (Revision 2) Remedial Investigation Report, Remedial Investigation Report for the Coeur d'Alene Basin Remedial Investigation/Feasibility Study, Vol. 1. Part 1. Setting and Methodology. URSG DCN 4162500.6659.05a. Prepared for U.S. Environmental Protection Agency, Region 10, Seattle, WA, by URS Greiner, Inc., Seattle, WA, and CH2M Hill, Bellevue, WA. September 2001.

Wall Street Journal. April 2021. Why Is Coeur d'Alene America's Hottest Housing Market?

Wise, D. 2021. Landscape Changes and Phosphorus Delivery to Coeur d'Alene Lake. Presentation to the NAQSEM Committee. May 4, 2021.

Woods, M. S., and M. A. Beckwith. 2008. Coeur d'Alene Lake, Idaho: Insights gained from limnological studies of 1991–92 and 2004–06. USGS Scientific Investigations Report 2008-5168. U.S. Geological Survey.

Woods, P. F., and M. A. Beckwith. 1997. Nutrient and trace-element enrichment of Coeur d'Alene Lake, Idaho. USGS Water Supply Paper 2485. https://doi.org/10.3133/wsp2485.

Xue, H. B., D. Kistler, and L. Sigg. 1995. Competition of copper and zinc for strong ligands in a eutrophic lake. *Limnol. Oceanogr.* 40(6):1142–1152.

Yue, S., and C. Wang. 2004. The Mann-Kendall test modified by effective sample size to detect trend in serially correlated hydrological series. *Water Resources Management* 18:201–218. https://doi.org/10.1023/B:WARM.0000043140.61082.60.

Zinsser, L. M. 2018. Coeur d'Alene Basin Environmental Monitoring Program, surface water, northern Idaho—Annual data summary, water year 2017. USGS Open-File Report 2018-1113. U.S. Geological Survey.

2

Long-Term Monitoring of Coeur d'Alene Lake and Its Watershed

This introductory chapter summarizes the various water quality data collected throughout the Coeur d'Alene (CDA) watershed, including the Lake and its tributaries, as well as the CDA and St. Joe rivers, by the U.S. Geological Survey (USGS), the Idaho Department of Environmental Quality (IDEQ), and the Coeur d'Alene Tribe (CDA Tribe). The generated data provide valuable information related to water quality, sediment quality, and biota, and have been used to drive numerous special studies (see Appendix A) and to model physical, chemical, and biological processes in the CDA watershed and the Lake (see Chapter 4). Data from these ongoing, long-term monitoring networks are analyzed by the Committee in Chapters 3 through 6 to reveal trends and highlight particular issues of concern. Subsequent to those analyses and building off of their limitations, Chapter 8 describes gaps in the long-term monitoring networks and in the evaluation of monitoring data, which should be filled in order for water quality in CDA Lake to be adequately and continuously assessed into the future.

The long history of research and associated monitoring activities in CDA Lake started with the U.S. Environmental Protection Agency (EPA) sampling the Lake as part of the National Eutrophication Survey (EPA, 1977). The USGS conducted comprehensive river and lake sampling in 1991–1992 as part of its baseline study of limnological conditions (Woods and Beckwith, 1997), which was used in the development of the initial Lake Management Plan in 1996. USGS and the CDA Tribe did an additional baseline study from October 2003 to August 2006 (Wood and Beckwith, 2008). During the 1990s and early 2000s, USGS published a number of studies characterizing water quality and sediments in the Lake (e.g., Horowitz et al., 1993, 1995; Woods and Beckwith, 1997; Kuwabara et al., 2000, 2003, 2006). Since then, USGS has monitored the rivers primarily in consultation with EPA as part of the Basin Environmental Monitoring Program, which began in October 2003, to evaluate the long-term effects of the Bunker Hill Superfund site remediation.

After efforts to develop a revised Lake Management Plan were unsuccessful, an assessment was completed in 2006 as part of an Alternative Dispute Resolution process among the CDA Tribe, IDEQ, and EPA. A key finding of the resolution process was a need for the CDA Tribe, IDEQ, and EPA to develop a consistent description of current water quality conditions in the Lake (Harty, 2007). Hence, one of the first objectives of the 2009 Lake Management Plan was to "improve scientific understanding of lake conditions through monitoring, modeling, and special studies" (IDEQ and CDA Tribe, 2009). Strategies to achieve this objective included the establishment of water quality triggers and routine monitoring of the Lake and rivers. In 2007, the CDA Tribe and IDEQ began a regular monitoring program that was a continuation of the baseline monitoring and some of the special studies previously done by USGS at key sites.

LAKE MONITORING

The long-term monitoring efforts of IDEQ, the CDA Tribe, and USGS capture different parts of the multifaceted CDA Lake and river system, as shown in Table 2-1 and Figure 2-1. IDEQ maintains long-term monitoring sites in the northern and central pools of the Lake, known as C1, C2, C3, and C4. Measurements at sites C1, C2, and C4 are done for all physical and water quality parameters, while less comprehensive assessment is done at the other IDEQ sites (see Table 2-1). The CDA Tribe maintains long-term monitoring sites in the southern part of the lake (called C5 and C6) and site SJ1 in the lower St. Joe River. It also maintains three meteorological stations and has measured many of the lateral lakes along the CDA River as part of special studies (Figures 2-1 and 2-2).

Throughout the Lake, water column sampling is done at discrete depths as well as several depth categories defined in the 2009 Lake Management Plan (IDEQ and CDA Tribe, 2009). These include the photic zone composite (five equally spaced samples from 1 m below the surface to the depth where underwater photosynthetically active radiation [PAR] is 1 percent of the light incident on the surface, composited into a churn splitter), discrete sampling at 20 and 30 m for northern pool stations, and 1 m above the Lake bottom or near bottom (a discrete sample with sampling depth determined from the Hydrolab profile). In general, physical parameters are measured continuously by profilers at discrete depths, while metals and nutrients are measured in the depth zones described above. Starting in 2017, the CDA Tribe also began measuring an epilimnion composite for metals and nutrients defined as grab samples from five equal depth increments from 1 m to the bottom of the epilimnion (composited in a churn splitter).

TABLE 2-1 Long-Term Monitoring Sites within CDA Lake

Agency	Site Number (colloquial name)	Type of Monitoring	Latitude (°N)	Longitude (°W)	Max Depth (m)
IDEQ	C1 (Tubbs)	all WQ parameters	47.65000	116.75833	40
	C2 (Wolf)	all WQ parameters	47.62500	116.68300	30
	C3 (Driftwood)	WQ parameters (sonde only)	47.58330	116.80600	52
	C4 (University)	all WQ parameters	47.51500	116.83500	40
	Cougar 2	WQ parameters (sonde only)	47.67433	116.45639	~20
	Wolf Lodge Bay	WQ parameters (sonde only)	47.62361	116.66417	~10
	Rockford Bay	WQ parameters (sonde only)	47.50589	116.89806	~15.5
	Windy Bay Deep	WQ parameters (sonde only)	47.48500	116.88778	~30
	Gasser Pt.	WQ parameters (sonde only)	47.46408	116.84306	~28
CDA Tribe	C5 (Chippy Pt.)	all WQ parameters	47.42717	116.76476	18
	C6 (Chatcolet)	all WQ parameters	47.35815	116.74987	11
	SJ1	all WQ parameters	47.35744	116.68634	21
	CDARHarr	All WQ parameters	47.478448	116.735051	10.5
	ThompSite1	Phytoplankton and physical parameters	47.490894	116.721537	7
	SwanSite1	Phytoplankton and physical parameters	47.473211	116.634575	6.1
	East Pt.	meteorological station	47.477470	116.856526	
	Shingle Bay Pt.	meteorological station	47.411881	116.744303	
	Chatcolet	meteorological station	47.373530	116.749151	

Note: WQ = water quality.

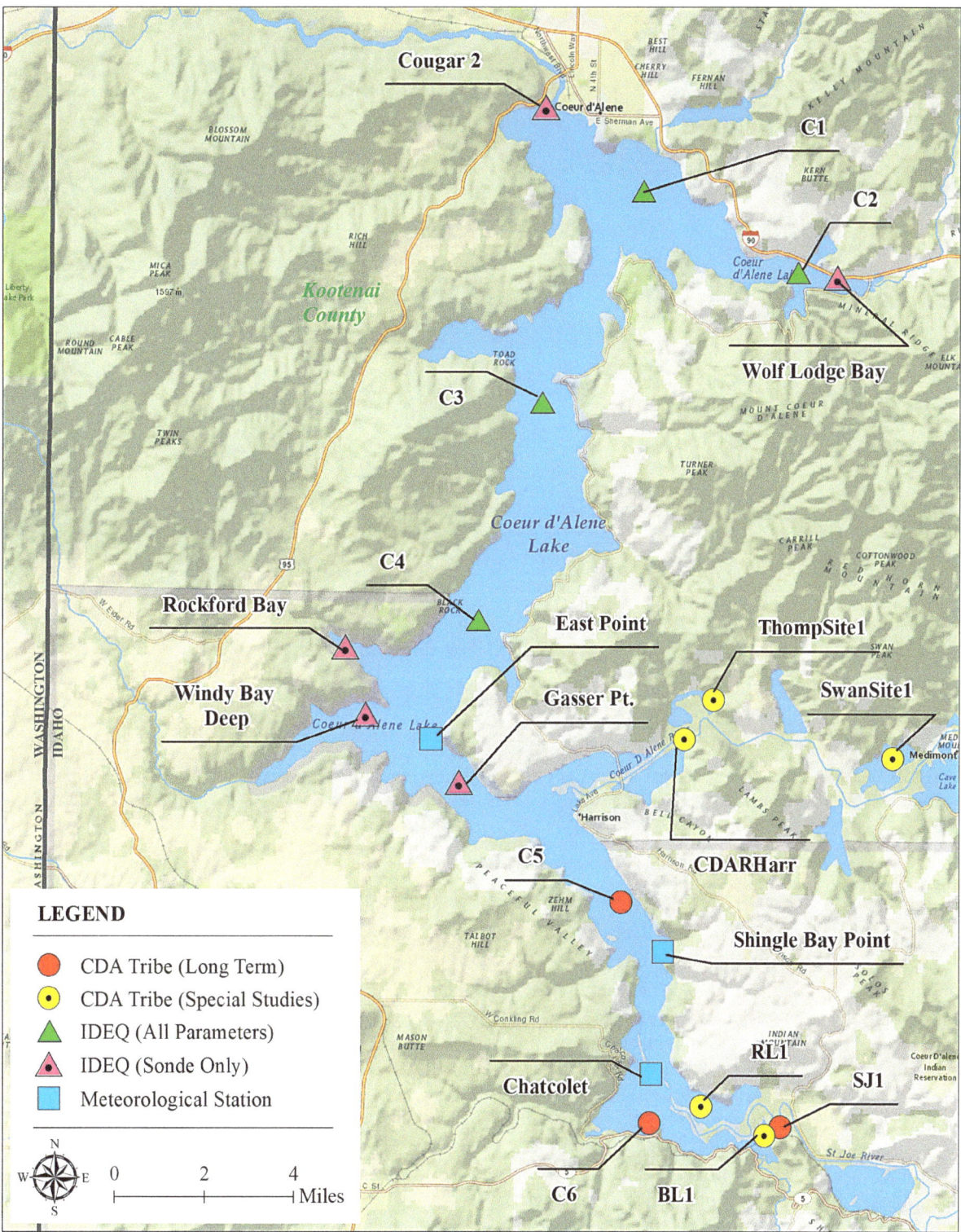

FIGURE 2-1 IDEQ and CDA Tribe long-term monitoring locations in CDA Lake. SOURCE: Generated by the committee using data from IDEQ, CDA Tribe and the USGS National Map and associated datasets.

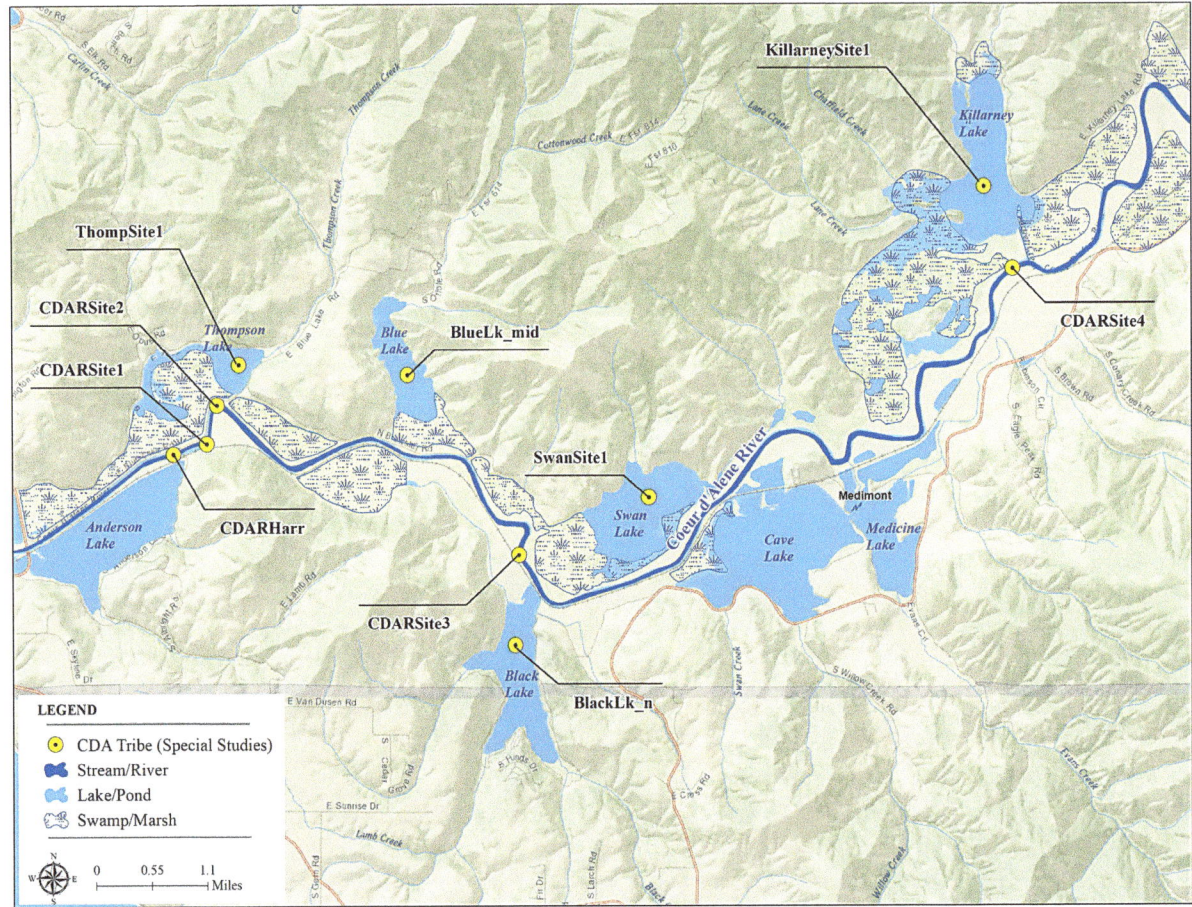

FIGURE 2-2 Lateral lake sites sampled by the CDA Tribe in the southern Lake. Parameters measured at these sites are described in Table 2-1 above. SOURCE: Generated by the committee using data from IDEQ, CDA Tribe, and the USGS National Map and associated datasets.

Constituents Measured and Methods

Tables 2-2 to 2-5 provide an overview of water quality and physical parameters measured in CDA Lake and some river sites by IDEQ and the CDA Tribe, organized by the constituents measured.[1] The parameters measured are in part guided by the Lake Management Plan, which includes triggers for zinc (Zn), lead (Pb), cadmium (Cd), phosphorus (P), phytoplankton, and dissolved oxygen (DO) as indicators of lake health. Table 2-2 shows the metals measured in the Lake; it is apparent that the focus is on those heavy metals commonly found in the mining waste. In almost all cases, both total and dissolved metal fractions are measured. Nutrients and chlorophyll a, which are important to understanding the productivity of CDA Lake and the potential for future problems with nutrient enrichment, are the subject of Table 2-3. Table 2-4 shows the biological parameters related to lake productivity that are measured by the CDA Tribe in the southern portion of CDA Lake as well as in the lateral lakes Swann Lake and Thompson Lake. Finally, Table 2-5 lists the physical parameters measured throughout the Lake, including light, dissolved oxygen, pH, and other parameters.

Prior to reinitiating long-term monitoring in 2007, IDEQ and the CDA Tribe jointly prepared a Quality Assurance Project Plan approved by EPA, which was designed to address quality assurance/quality control issues and

---

[1] This paragraph was edited after report release to more accurately describe the nutrient datasets provided to the committee.

TABLE 2-2 Monitoring of Metals and Other Water Quality Parameters in CDA Lake

Site Name	Sample Dates[a]	Sample Depth	Number of Samples	Parameters Measured
C5, C6, SJ1	2007–2020	Photic zone and one meter above bottom	~200–300 per site	Filtered and total: As, Ca, Cd, Fe, Mg, Mn, Pb, Zn; total hardness
CDARHarr	2013, 2015, 2018–2020	Photic zone	37	Filtered and total: As, Ca, Cd, Fe, Mg, Mn, Pb, Zn
C1	1991–1992, 1995–2020	varies	varies	Total and dissolved: As, Ca, Cd, Fe, Mg, Mn, Pb, Si, Zn, hardness
C2	2010–2020			
C3	1991–1992, 1995–2020			
C4	1991–1992, 1995–2020			
C5	1991–1992, 1995–1998, 2003–2006, 2013–2019			
C6	1991–1992, 2003–2006, 2015–2018			
Cougar 2	2013–2020			
Gasser Pt	2014–2020			

[a]Colors denote the agency conducting the monitoring: CDA Tribe, IDEQ, USGS.

TABLE 2-3 Monitoring of Nutrients and Chlorophyll *a* in CDA Lake

Site	Sample Dates[a]	Sample Depth	Number of Samples	Parameters Measured
C1	1991–1992, 1995–2020	Photic-near bottom	~700	Phosphorus[b], nitrogen[c], chlorophyll *a*
C2	2010–2019		~150	
C3	1991–1992, 1995–2006, 2010–2015		~400	
C4	1991–1992, 1995–2020		~700	
C5	1991–1992, 1995–1998, 2003–2006, 2015–2019		~200	
C6	1991–1992, 2003–2006		~200	
Cougar 2	2013–2015		<100	
Gasser Pt	2014–2015		<50	
C5, C6, SJ1	2007–2020	Photic zone, one meter above bottom	~300/site	Phosphorus (total, phosphate), nitrogen (total, ammonia, nitrate), chlorophyll *a*
CDARHarr	2013, 2015, 2018–2020	Photic zone	37	Phosphorus (total, phosphate), nitrogen (total, ammonia, nitrate), sulfate, TOC, DOC
SJ1	2007–2020	Photic zone	~100	

[a]Colors denote the agency conducting the monitoring: CDA Tribe, IDEQ, USGS.
[b]Phosphorus speciation when measured: total, dissolved total, dissolved reactive, total reactive, total acid-hydrolyzable, total reactive+acid-hydrolyzable, total organic.
[c]Nitrogen speciation when measured: total, ammonia, nitrate, nitrite, nitrate + nitrite.
Note: TOC and DOC = total organic carbon and dissolved organic carbon.

TABLE 2-4 Phytoplankton Monitoring in CDA Lake and Lateral Lakes by the CDA Tribe

Site	Sample Dates	Sample Depth	Number of Samples per Site	Parameters Measured
BL1, RL1	2011–2019	Photic zone	~1,200	Phytoplankton taxa identified to level of genus (usually), natural counting units (e.g., cells or colonies), biovolume
C5, C6, SJ1	2007–2020		~2,300	
ThompSite1 SwanSite1	2015	Epilimnion	~90	

TABLE 2-5 Physical Parameters and Light Monitoring in CDA Lake and the Lateral Lakes

Site	Sample Dates[a]	Sample Depth (m)	Number of Samples	Parameters Measured
C1 Tubbs	1991–1992, 1995–2021	1–40m	~5,000	Temperature, conductivity, DO, pH, turbidity, pressure, photic zone depth, extinction coefficient, percent light, Secchi depth
C2 Wolf	2010–2021	0–30 m	~2,000	
C3 Driftwood	1991–1992, 1995–2006, 2010–2021	1–50+ m	~4,000	
C4 University	1991–1992, 1995–2021	0–40 m	~5,000	
C5 Chippy Pt	1991–1992, 1995–1998, 2003–2019	0–20 m	~1,000	
C6 Chatcolet	1991–1992, 2003–2006	1–10 m	~500	
Cougar 2	2013–2021	0–20 m	~2,000	
Gasser Pt	2014–2021	0–20 m	~1,300	
Windy Deep	2016–2021	0–30 m	~1,300	
C5	2011, 2015, 2019	0.5- to 1-m resolution	67,600 over 3 years	Temperature, specific conductance, pH, DO
C5, C6, SJ1	2007–2020	1-m resolution	2,000–3,000 per site	Temperature, specific conductance, pH, DO, PAR, fluorescence
BlackLk_n	2015, 2017	0–10 m	200–300 per site	Temperature, specific conductance, pH, DO, PAR, fluorescence
BlueLk_mid	2015			
Cataldo	2015			
CDARHarr	2013, 2015, 2018–2020			
CDARSite1	2013, 2015, 2018			
CDARSite2	2013, 2015–2016, 2018			
CDARSite3	2013, 2018			
CDARSite4	2018			
KillarneySite1	2015			
SwanSite1	2015, 2019			
ThompSite1	2015, 2017, 2019			

[a] Colors denote the agency conducting the monitoring: CDA Tribe, IDEQ, USGS.

serve as an initial plan for the 2007 monitoring season (IDEQ and CDA Tribe, 2007). Currently, IDEQ and the CDA Tribe use different analytical laboratories, which affects the comparisons done across the Lake and data analysis of long-term trends. Samples collected by IDEQ are sent to EPA-certified labs for analysis (IDEQ, 2020). The CDA Tribe collects total recoverable and 0.45 µm filtered metals (arsenic, calcium, cadmium, iron, magnesium, manganese, lead, zinc), which are analyzed by the EPA Manchester Laboratory; total and speciated nitrogen and phosphorus, chlorophyll *a*, and organic carbon (2019–2020), which are analyzed by the Tshimakain Creek Laboratory; and phytoplankton, which is analyzed by Advanced Eco-Solutions, Inc. (Chess, 2021). Furthermore, different methods have been used to measure chlorophyll *a*, such that lab measurements of the same water yield different results (Cooper, 2021). Nutrient and chlorophyll *a* concentrations are also often near or below measurement detection limits, presenting challenges for the analytical methods (Cooper, 2021).

The Committee was sensitive to the possibility that comparability issues can affect using data from different laboratories and that methodologies change over time as technologies improve. Individual analyses found in Chapters 3–7 were usually associated with a single data source (e.g., watershed data were all from the quality-controlled databases of USGS). The report notes those instances where data were censored (e.g., metals time series were restricted to periods where quality assurance/quality control were reported or caveats are made about older data).

With respect to physical parameters, IDEQ measures Secchi depth and sonde profiles of light, temperature, conductivity, pH, turbidity and dissolved oxygen (Table 2-5). The CDA Tribe measures profiles of physical parameters at 1-m resolution coincident with their water quality measurements. The multi-parameter sonde measures temperature, dissolved oxygen, pH, specific conductance, fluorescence, and PAR (Table 2-5). An automated buoy profiler system has occasionally been deployed in the summertime at sites C5 (2011, 2015) and C6 (2014, 2018) to collect temperature, dissolved oxygen, pH, and specific conductance. In 2018, the Tribe began collecting profiles of temperature, specific conductance, and salinity at every site when they sampled water quality. The CDA Tribe also has three meteorological stations—East Point (2009–2019), Shingle Bay (2009–2019), and Chatcolet (2016–2020)—that log data every five minutes and measure temperature, relative humidity, solar radiation, PAR, wind speed, and wind direction (Chess, 2021).

Sampling Frequency

The frequency and timing of sampling varies by site and year, but generally sampling coincides with lake and river flow conditions of interest. There has historically been less monitoring during the winter when the Lake is cold and unstratified. Sampling tends to pick up in the spring, when there is peak runoff and influence from the rivers and it continues throughout the strong thermal stratification that develops in earlier summer (July) and weakens in the fall (October). Hence, IDEQ has traditionally sampled seven to nine times per year, including two rain-on snow events in February–April, monthly water quality measurements from May to September, fall lake turnover, and winter measurements in November–December, with sonde-only measurements in January–February (Cooper, 2021). From 2007 to 2020, the CDA Tribe's monitoring of their three long-term sites has been most concentrated from April to November (10–14 samples per site per month), with fewer samples (0–6 samples per site per month) during the winter months when portions of the southern Lake freeze over, making sampling conditions difficult (Chess, 2021). Although there is less sampling during winter, there has been a focus on sampling rain-on-snow events, which may bias wintertime samples.

COEUR D'ALENE RIVER SAMPLING

USGS measures discharge and water quality at many gages throughout the Spokane River basin (Figure 2-3). The oldest gages in the system to measure discharge came online in 1911; more recent sites were added in the late 1980s on the South Fork of the CDA River associated with the Superfund site. Although sampling timing and frequency was variable prior to the late 1990s, in 2004 it became unified under EPA's Basin Environmental Monitoring Program (EPA, 2004). Hence, the USGS data currently collected are concentrated in the CDA River basin and at the mouth of the St. Joe River as well as in the Spokane River where water flows out of CDA Lake. The water quality measurements are primarily of metals and nutrients. Water quality measurements from USGS date back to 1971, but the longest well-sampled periods date back to only 1990. There are 13 monitoring sites with sufficient data to be used in calculations of average concentrations and fluxes and to evaluate trends (shown

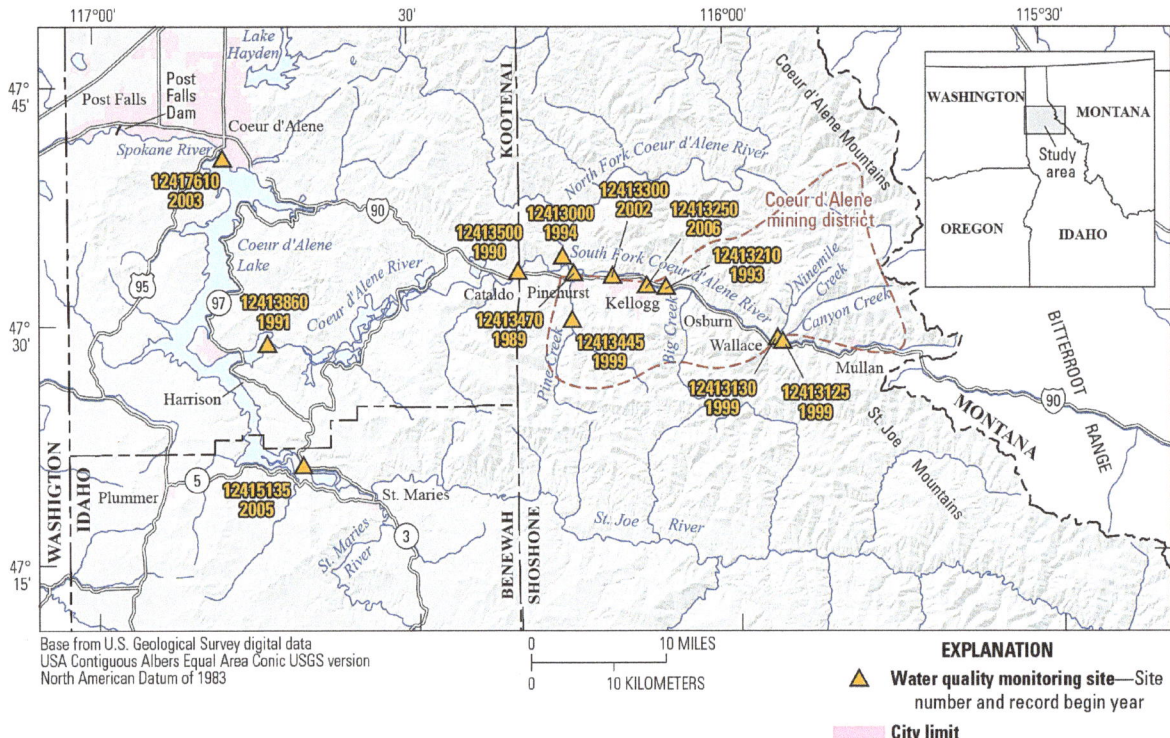

FIGURE 2-3 USGS water quality monitoring sites (triangles) in the Spokane River basin used in Zinsser's (2020) long-term trend analysis. Site number and year the record began are given for each site. SOURCE: Zinsser (2021).

in Figure 2-3 and documented in Zinsser, 2020). Chapter 3 uses data from 6 of these 13 sites in its analysis (Site 12417610: Spokane River below lake outlet; Site 12413860: CDA River near Harrison; Site 12415135: St. Joe River at Ramsdell; Site 12413000: North Fork CDA River at Enaville; Site 12413470: South Fork CDA River near Pinehurst; and Site 12413210: South Fork CDA River at Elizabeth Park).

Water quality samples follow standardized collection methods and are isokinetic and width- and depth-integrated such that samples represent water quality across the entire river cross-section. USGS measures total and filtered metals, total and filtered nutrients, major cations, and suspended sediment concentrations. Samples are collected approximately five times per year during a variety of hydrologic conditions with an emphasis on high discharge events. All of these water quality monitoring sites are co-located at USGS stream gages, so there is a complete record of daily mean discharge available to accompany these water quality records. The discharge and water quality data are all available from the USGS (2020) National Water Information System (NWIS) web interface.

PUBLISHED SUMMARIES OF LONG-TERM MONITORING DATA

Agencies and consultants have regularly analyzed the data described above and produced results in reports and papers. IDEQ has regularly produced reports on CDA Lake status and trends in response to the 2009 Lake Management Plan. The most recent report, published in 2020, explored trends over the years 2015–2018 (IDEQ, 2020). Earlier such reports were produced in collaboration with the CDA Tribe, while the most recent report was produced by IDEQ independently. Most recently, in 2017 the Tribe produced a five-year synthesis of their water quality monitoring data from 2011 to 2015 (CDA Tribe and Avista, 2017). USGS produces both USGS reports and articles in the peer-reviewed literature. One recent example analyzing USGS river water quality data was Zinsser (2020), which analyzed trends in metals and nutrients from 1990 to 2018. Beyond its collaboration with USGS, EPA separately contracts with consultants to collect data in the upper basin to determine the effectiveness of the Superfund remedy.

DATA AVAILABILITY

Availability of long-term monitoring data to all interested parties is important for several reasons. There may be a wide range of scientists, regulators, and stakeholders who should have easy access to the relevant data. Given current data storage and data sharing capabilities, it should be possible for users to make requests for the data without needing assistance from the staff of the agencies that collected the data. Making the data widely and easily available could be of real benefit to all parties involved. Such a repository under the umbrella of the Lake Management Plan could be a constructive way to involve all stakeholders in developing a unified understanding of changes in the Lake into the future. When data are easily accessed, it becomes possible for more individuals and groups to analyze the data, develop new hypotheses, or test existing hypotheses. When these analyses are then published, it becomes possible for there to be an open debate about conclusions drawn, which, in the long run, will provide insights that can improve the strategies for future management of CDA Lake.

In addition, open sharing of datasets can be an important component of the quality assurance process. If multiple parties can view the data in a variety of ways, data that are likely to be in error may be recognized, and those who control the datasets can then evaluate the accuracy of data values and either delete or flag them. In short, the more a dataset is used, the more likely it is to improve over time. In the case of the Coeur d'Alene system, the river-related data (flow and chemistry) are stored in the USGS NWIS system (USGS, 2020), which has a national capability for visual examination of the data and direct download of the data in versions that are compatible with open source statistical software. For example, the R packages "dataRetrieval" (De Cicco et al., 2018) and "EGRET" (Hirsch and De Cicco, 2015) both provide simple automatic access to the USGS river system data.

There is no such system for the physical or chemical data that are routinely collected from CDA Lake; rather, data used by the Committee were provided by the two primary agencies (IDEQ and the CDA Tribe) as Excel spreadsheets. The spreadsheets of the two agencies were similar, but not identical, making it somewhat cumbersome to do unified consistent analyses across all of the sampling locations. Developing institutional arrangements and technical protocols to bring all data of a given type (e.g., lake chemistry data) into a unified data management and data sharing system is not a trivial matter, but experience in other systems has shown that the net result of the enhanced data transparency and delivery is worth the effort. One such example is the Non-Tidal Network for the Chesapeake Bay watershed, which includes data from 123 river monitoring sites sampled by 10 different agencies (see Box 2-1).

BOX 2-1
River Monitoring in the Chesapeake Bay Watershed

The Chesapeake Bay estuary in the eastern United States covers an area of 166,534 km^2 (about 100 times the size of CDA Lake) and drains a watershed of about 167,000 km^2 (about 38 times the size of the CDA watershed). The Chesapeake Bay has been adversely affected by nutrient and sediment enrichment, and reductions in the loading of nutrients and sediment are required to improve conditions in the estuary and meet water quality standards set by the federal-state partners involved in the Chesapeake Bay Program. Within that partnership there is a river monitoring program called "The Non-Tidal Network," operated by a consortium of 10 agencies (federal, state, and interstate), that conducts river monitoring at 123 locations by a standard set of protocols and provides its data to a central data repository. As the program developed, research was carried out to establish the impact of various sampling design strategies on the ability of the system to estimate loads of the target constituents (nutrients and sediment) and to evaluate trends. Trends are evaluated on both a short-term basis (most recent ten years) and a long-term basis (typically 30 to 35 years) and published online once every two years. For a high-priority subset of these monitoring locations (the nine sites that are near the point where the river enters the estuary) trend results are updated annually.[a] The sampling protocol is to collect routine samples monthly as well as additional storm-event samples, for a total of 20 samples per year representing a range of discharge and loading conditions. The decision to establish this protocol was the result of extensive research and evaluation that resulted in a number of publications mentioned in the body of this report. The storm-event sampling is crucial to the effectiveness of the monitoring program because, for most of the constituents of interest, large fractions of the annual loads are delivered at high flows, but monitoring of low and moderate flows, throughout the entire year is also important because they provide a clearer picture of conditions driven by point sources and groundwater baseflow to the rivers.

---

[a] https://www.sciencebase.gov/catalog/item/60d37347d34e12a1b0097243. This was published in 2021 and provides trend results through the end of water year 2020

In addition to unified data availability, a system such as the Coeur d'Alene should have a process for regularly producing data synthesis products that make interpretations about the spatial and temporal characteristics of the data, with particular attention to trends. Such synthesis products should be based on data that are readily available to other scientists, as different researchers will have different approaches to data synthesis and there is strength in that diversity of approaches. The synergy generated by multiple approaches requires readily available data sets and publicly documented data analysis protocols.

***

The data generated by the long-term monitoring programs for the CDA watershed and CDA Lake form the basis for the analyses presented in the next three chapters. Those analyses revealed gaps and limitations in the monitoring program that form the basis of Chapter 8, which describes how the long-term monitoring programs for the CDA system can be improved.

REFERENCES

Chess, D. 2021. Water Quality Data Summary. Presentation to the NASEM Committee. February 26, 2021.

Coeur d'Alene Tribe and Avista Corporation. 2017. Coeur d'Alene Reservation Five-Year Synthesis 2011–2015 Monitoring Data 4(E) Condition No. 5 Spokane River Hydroelectric Project FERC Project No. 2545. February 2017.

Cooper, C. 2021. Coeur d'Alene Lake Management Plan Water Quality Data and Trends. Presentation to the NASEM Committee. February 26, 2021.

De Cicco, L. A., R. M. Hirsch, D. Lorenz, and W. D. Watkins. 2018. dataRetrieval: R packages for discovering and retrieving water data available from Federal hydrologic web services. doi:10.5066/P9X4L3GE.

EPA (U.S. Environmental Protection Agency). 1977. National Eutrophication Survey—Report on Coeur d'Alene Lake, Benewah and Kootenai Counties, Idaho, EPA Region X, Working Paper No. 778.

EPA. 2004. Basin Environmental Monitoring Plan, Bunker Hill Mining and Metallurgical Complex Operable Unit 3. Prepared by URS Group and CH2M Hill, for EPA Region 10, Seattle, WA.

Harty, J. M. 2007. Assessment Report on Prospects for Mediated Negotiation of a Lake Management Plan for Coeur d'Alene Lake. Harty Conflict Consulting & Mediation, Davis CA, prepared for U.S. Institute for Environmental Conflict Resolution.

Hirsch, R. M., and L. A. De Cicco. 2015. User guide to Exploration and Graphics for RivEr Trends (EGRET) and dataRetrieval: R Packages for Hydrologic Data (version 2.0, February 2015). U.S. Geological Survey Techniques and Methods Book 4, Chap. A10. doi:10.3133/tm4A10.

Horowitz, A. J., K. A. Elrick, and R. B. Cook. 1993. Effect of mining and related activities on the sediment trace element geochemistry of Lake Coeur d'Alene, Idaho, USA. Part I: surface sediments. *Hydrol. Process.* 7:403–423.

Horowitz, A. J., K. A. Elrick, J. A. Robbins, and R. B. Cook. 1995. Effect of mining related activities on the sediment trace element geochemistry of Lake Coeur d'Alene, Idaho. Part II—subsurface sediments. *Hydrol. Process.* 9:35–54. https://doi.org/10.1002/hyp.3360090105.

IDEQ (Idaho Department of Environmental Quality) and Coeur d'Alene Tribe. 2007. Quality Assurance Project Plan for Continued Monitoring of Water Quality Status and Trends in Coeur d'Alene Lake, Idaho. IDEQ, Coeur d'Alene, Idaho and CdA Tribe, Plummer, Idaho.

IDEQ and Coeur d'Alene Tribe. 2009. Coeur d'Alene Lake Management Plan. State of Idaho Department of Environmental Quality.

IDEQ. 2020. Coeur d'Alene Lake Management Plan: Coeur d'Alene Lake Status Update, 2015–2018. May 2020.

Kuwabara, J. S., B. R. Topping, P. F. Woods, J. L. Carter, and S. W. Hager. 2006. Interactive effects of dissolved zinc and orthophosphate on phytoplankton from Lake Coeur d'Alene, Idaho. U.S. Geological Survey Scientific Investigations Report 2006-5091. http://pubs.usgs.gov/sir/2006/5091.

Kuwabara, J. S., P. F. Woods, W. M. Berelson, L. S. Balistrieri, J. L. Carter, B. R. Topping and S. V. Fend. 2003. Importance of sediment-water interactions in Coeur d'Alene Lake, Idaho, USA: management implications. *Environmental Management* 32:348–359. https://doi.org/10.1007/s00267-003-0020-7.

Kuwabara, J. S., W. M. Berelson, L. S. Balistrieri, P. F. Woods, B. R. Topping, D. J. Steding, and D. P. Krabbenhoft. 2000. Benthic flux of metals and nutrients into the water column of Lake Coeur d'Alene, Idaho: Report of an August 1999 pilot study. U.S. Geological Survey Water-Resources Investigations Report 2000-4132. https://doi.org/10.3133/wri004132.

U.S. Geological Survey (USGS). 2020. National Water Information System (NWIS): U.S. Geological Survey, web interface, accessed January 2022. https://waterdata.usgs.gov/nwis.

Wood, M. S., and M. A. Beckwith. 2008. Coeur d'Alene Lake, Idaho: insights gained from limnological studies of 1991–92 and 2004–06. U.S. Geological Survey Scientific Report 2008-5168.

Woods, P. F., and M. A. Beckwith. 1997. Nutrient and trace-element enrichment of Coeur d'Alene Lake, Idaho. U.S. Geological Survey Water-Supply Paper 2485.

Zinsser, L. M. 2020. Trends in Concentrations, Loads, and Sources of Trace Metals and Nutrients in the Spokane River Watershed, Northern Idaho, Water Years 1990–2018. USGS Scientific Investigations Report 2020-5096. https://doi.org/10.3133/sir20205096.

Zinsser, L. M. 2021. U.S. Geological Survey Water-Quality Data and Analyses in the Spokane River Basin. Presentation to the NASEM Committee. February 16, 2021.

3

Analysis of Inputs to Coeur d'Alene Lake

The committee was tasked to evaluate current water quality in Coeur d'Alene (CDA) Lake, the lower rivers, and the lateral lakes, with a focus on observed trends in nutrient loading and metals concentrations. This task necessitates looking at the water quality not only in the Lake itself (see Chapters 4, 5, and 6), but also in the major inputs to the Lake. Because CDA Lake is a surface water–dominated system, this chapter focuses on the inputs of the two major river systems: the CDA River and the St. Joe River. The trends observed in the river monitoring network data over the past 30 years provide clues about the effectiveness of the Superfund cleanup efforts that have already taken place in the watershed. They also provide a window to help prioritize and plan for future remediation efforts and to better understand the role of potential changes in land use and population in the watershed.

This chapter begins with three narratives that describe (1) how metals from historical mining activities have moved through the landscape, (2) how the Superfund activities begun in the late 1990s have affected sources of metals in the watershed, and (3) ongoing sources of phosphorus in the watershed. The chapter then analyzes data on water flows and metal and phosphorus concentrations and fluxes collected over the past 30 years in the major river systems to reveal the most important trends in the inputs of these constituents from the watershed to CDA Lake.

SOURCES OF METAL INPUT TO COEUR D'ALENE LAKE

Ore deposits are composed of primary minerals (or mixtures of minerals) from which metal(loid)s can be profitably extracted. Mineral extraction beneficiation[1] and processing of these deposits inherently create massive quantities of wastes, all of which are enriched to one degree or another with the potentially toxic trace elements characteristic of the ore deposit. The most important contaminants in the wastes from the CDA activities included silver (Ag), lead (Pb), zinc (Zn), cadmium (Cd), arsenic (As), and antimony (Sb) (NRC, 2005).

The release of large volumes of tailings until 1968 was one of the most important factors in the widespread metal contamination in the CDA watershed. Once wastes are released or escape the immediate physical area of primary activities, hydrologic and geochemical processes distributed them widely over the basin. Particularly important in this case is the relatively limited dilution of contaminated particulates that occurs in the CDA watershed before the mine wastes are ultimately deposited in a large, deep lake of great societal value. This section describes the processes that contribute to metal inputs to the Lake, including the sources and stockpiles of metals

---

[1] Beneficiation is the process by which the mineral-rich ore is concentrated, usually involving flotation in these types of mines.

established up until 1968 (when the effective tailings impoundments were completed), and the physical and biogeochemical processes that determine the fate of those metal(oid)s.

Character of Mine Wastes in Different Basins

Mining, milling, and smelting in the CDA basin created the primary, secondary and tertiary contamination typical of mineral extraction (Moore and Luoma, 1990). Historic mining operations typically scattered the immediate wastes of the operation (primary contamination) across the local landscape in an "ill-defined patchwork of waste rock, mill tailings, furnace slag and flue dust" (Moore and Luoma, 1990). In simplistic terms, waste rock is the material that is removed to uncover the ore deposits. Mill tailings are mostly fine-grained particulate wastes that result from the combination of physical and chemical processing used to separate and concentrate the metals targeted for extraction from the raw ore. A high percentage of the ore materials that are processed end up discarded as highly contaminated fine-grained tailings (Downs and Stocks, 1977). The concentrate from the milling process is refined by smelting, which produces flue dust and slag in which metal contaminants are even more concentrated than in tailings. Smelter complexes typically created intensely contaminated soils and groundwater, often in close proximity to larger population centers (Moore and Luoma, 1990). Fine particulates and gases are also emitted into the atmosphere by smelting for distribution over tens to hundreds of square kilometers.

Primary wastes, by definition, create localized contamination issues. Constraining wastes to the footprint of the operation can limit risks to downstream waterways, floodplains, and resources like lakes (Luoma and Rainbow, 2008). Unfortunately, during the western U.S. mining boom between 1850 and 1950, environmental considerations like constraining waste distribution were of little or no concern.

Secondary particulate contamination is created when the discarded tailings and other mine wastes are washed downstream, mixed with other sediments in transport, and re-deposited in lower areas. Bookstrom et al. (2013) found lead-rich sediments (defined as more than 1,000 µg Pb/g dw sediment) throughout the CDA basin, co-enriched with zinc, silver, copper, cadmium, iron, and manganese (and in some cases, arsenic, antimony, and mercury). Floods can disperse these tailings across the floodplain, with more intense floods moving larger sized particles. During the U.S Environmental Protection Agency's (EPA's) remedial investigations in 2001, 1,080 mining-related areas on the CDA floodplains were identified as sources of either primary or secondary wastes, varying in size from less than an acre to hundreds of acres (NRC, 2005). Tertiary contamination occurs as a result of biogeochemical reactions within the secondary deposits.

Upper Basin

The upper basin, and specifically the South Fork of the CDA River upstream from Elizabeth Park, is dominated by primary waste deposits, including waste rock and small mill deposits. These can be seen amidst the workings at the mouths of the underground mines (deposits visible in Figure 3-1). Most of the mining in the basin was done in steep terrain through a complex of 200 underground mines and a few very small pits. Thus, the deposits are large in number but limited in volume and area.

A common source of metal-contaminated water in the upper basin are adits, which are horizontal passages leading into a mine for the purposes of access or drainage. Where adits cut through aquifers in the region, adit waters characterized by low pH and high dissolved metal concentrations are created—a consequence of oxygenated aquifer water reacting with sulfide mineral ores.

The section of the upper basin along the South Fork of the CDA River between Elizabeth Park and Pinehurst is commonly known as "the Box" and includes the Bunker Hill smelter, the largest ore-processing facilities, and associated population centers (Figure 3-2). Due to the intense contamination typical of smelting activities, both primary and secondary wastes are deposited in this part of the upper basin. The river gradient here is more moderate and the floodplains are broader than upstream. The basin is bordered by steep valley walls, such that primary wastes from the mining facilities were deposited directly onto the floodplains or released into streams. Facilities and housing were built on top of the vast amounts of mine tailings deposited in this region (NRC, 2005).

The efficiency of metal removal from the ore evolved over time, which affected the nature of the wastes. Fine-grained wastes deposited at Cataldo early in the mining history contained 10,000–30,000 µg Pb/g dry weight

FIGURE 3-1 Standard Mammoth Mine and milling complex on Canyon Creek during the era of active mining (no date); a mining operation typical of many in the western United States. Waste is scattered across the landscape. Here, steep topography limits the area available for storage of tailings. The upper buildings are resting on small waste tips constrained by fencing, but there is little other room for waste storage. The lack of trees on the mountainside is typical of such areas, probably as a result of sulfurous gases emitted by the milling process. Dying and dead trees around such mining/milling operations ultimately resulted in forest fires, which helped denude the landscape. Atmospheric metal (e.g., Cd and As) and acid deposition limited recovery of the forest once it burned. Note also the apparent silty (or perhaps acidic) nature of the creek and the outhouses built on the creek. SOURCE: Idaho State Historical Society (2019) (see also Moore and Luoma, 1990).

(1–3 percent lead; Bookstrom et al., 2013). The finer, more easily mobilized wastes from beneficiation created in later operations (NRC, 2005) contained 10,000–14,000 μg Pb/g dry weight at Cataldo. But whatever the era, deposits of tailings have consistently contained lead concentrations up to and more than 100 times natural levels.[2] Box et al. (2001) estimated that much of the floodplain in the Box was covered with metal-contaminated alluvium with an average thickness of approximately 1.3 m.

As of 2005, the upper basin was a continued source of metals to the river system and carried 20,000 to 70,000 metric tons (MT) of particulates downstream per year (Bookstrom et al., 2013). Of the approximately 800,000 MT of mined lead historically lost directly or indirectly to streams, Bookstrom et al. (2001) estimated that 24 percent (200,000 ± 100,000 MT) resided in the upper basin. These wastes have been the focus of remediation to date.

---

[2] Based largely upon sediments obtained from cores that extended to the pre-mining era, Bookstrom et al. (2013) calculated pre-mining Pb soil concentrations were 31 ± 19 μg/g dw. Estimates from several studies reviewed by the NRC (2005) and from CDA lake cores (Horowitz et al., 1993) are within one standard deviation of this estimate. Median concentrations of Pb in fine-grained floodplain sediments in the lower basin are 2,900 μg/g dw.

A few attempts were made at constraining wastes from the upper basin in the first 70 years of mineral extraction. For example, at least two in-stream tailings ponds were built between 1900 and 1933, but these both failed during large floods, leaving behind thick deposits of contaminated sediments over the floodplain the dams had engulfed (Bookstrom et al., 2013). A few unlined tailing-settling ponds were built off-stream between 1928 and 1968, including the Central Impoundment Area (CIA; Box 3-1) in the Box. But the large number of unregulated operations scattered over a wide area of the watershed limited the overall effectiveness of these efforts (Bookstrom et al., 2013; NRC, 2005). Other mining areas in the West of comparable size to the Silver Valley began employing effective settling basin technologies in the 1930s (Bookstrom et al., 2013). But it was not until 1968, more than 80 years after the first mining claim discharged tailings into the waterways, that all operating mills and mines in the Silver Valley had effectively impounded their tailings (Morra et al., 2015).

Lower Basin

The lower basin includes the main stem of the CDA River extending from Cataldo Landing to CDA Lake. This area is characterized by broad alluvial floodplains that range in width from 300 m at Cataldo, below the confluence of the North and South Forks of the CDA Rivers, to 5 km at the river mouth at CDA Lake (NRC, 2005). The lower basin is now a source of secondary contamination, created when the discarded tailings and other wastes from the upper basin were washed downstream, mixed with other sediments in transport, and re-deposited in the lower basin.

Floods have been instrumental in dispersing tailings from the upper basin across the floodplains of the lower basin, with more intense floods moving larger sized particles. Channel aggradation was caused by sedimentation of mine wastes and reduced forest cover during the early phases of mining. This increased the extent and severity of overbank flooding as the mineral extraction progressed (Bookstrom et al., 2013; NRC, 2005). Between 1893 and 2004, at least 40 discharge episodes peaked at 17,000 cfs (481 m^3/s) or more, inundating the floodplain of the CDA River valley, such that "on average, much of the valley floor was flooded about every 2.5 years" (Bookstrom et al., 2013). Floods with peak discharge of about 70,000 cfs (1,982 m^3/s) or more occurred in 1933, 1974, and 1996. By 1900, mill tailings deposited in the upper reaches of the CDA basin had reached CDA Lake and had affected as much as 25,000 acres (10,117 hectares) along the South Fork and main stem of the CDA River (Long, 1998). Thus, the accumulation of mine wastes that dominate the sediments of CDA Lake began 120 years ago and continues today.

The river channel in the lower 44 km of the lower basin has a meandering pattern and, for most of the year, has essentially a zero gradient (NRC, 2005). After floods and spring runoff subside, the channel of the CDA River in the lower part of the lower basin is backflooded in summer/early fall (as controlled by the Post Falls Dam, see Chapter 1). This has created delta-like floodplains in the lower basin that are metal-enriched and are now permanent features.

The CDA River in the lower basin is about 60 to 100 m wide and 5 to 15 m deep, with a bottom composed of sandy lead-rich sediments averaging about 3 m thick (Bookstrom et al., 2013). These riverbed sediments contain about 51 percent of the lead in all sediments on the floor of the CDA River valley (Bookstrom et al., 2001). The other 50 percent of the lead is scattered across the floodplain. Bookstrom et al. (2013) estimated that mill tailings covered about 60 km^2 of the 80-km^2 floor of the main stem of the CDA River valley. As river channels naturally meander and cut new banks, metal-rich deposits on the floodplains are remobilized into the river—yet another process that is accentuated by floods.

Hence, the lower basin comprises an immense stockpile of metal-enriched particulates poised for transport to CDA Lake. Indeed, in the 1999–2000 water year, the South Fork of the CDA River delivered only about 20 percent of the total lead load to CDA Lake, with the remaining 80 percent derived from erosion of the bed and along the course of the main stem of the CDA River below the confluence of the North Fork (Clark, 2003, Fig. 12).

Absence of Watershed Scale Dilution

An important issue with secondary wastes in the CDA basin is minimal dilution of particulate material with uncontaminated sediments. When metals in primary waste deposits enter streams and rivers as solutes and particulates, dilution is determined by the area of the watershed (Helgen and Moore, 1995). For example, metal contamination in sediments can be detected 650 km downstream from the Clark Fork mining complex in Montana, but the concentrations of those metals progressively decline downstream from their source by as much as 100-fold as a result of inputs of unenriched particles and waters from numerous tributaries (Axtmann and Luoma, 1991; Helgen and Moore, 1995). The CDA basin differs from this situation in that the river reach between the mineral extraction operations (the Superfund Site) and the Lake is relatively short (~80 km) (Bookstrom et al., 2013). The North Fork of the CDA River is the only large unenriched tributary to enter the CDA River below the mining district (and ~60 km from CDA Lake).

The dilution effect of tributaries can be reversed if the bed and banks of the river below the tributary are contaminated by secondary and tertiary wastes (Hornberger et al., 2009; Axtmann and Luoma, 1991). On average, the North Fork of the CDA River has about twice the discharge of the South Fork and about five times the sediment load—both largely uncontaminated, which NRC (2005) estimated should result in dilution of South Fork metal concentrations by 25–35 percent. The expected dilution, however, is not evident in sediments or soils from the levees, lateral lakes, or the river bottom in the lower basin. Bookstrom et al. (2013) found that lead concentrations between 1988 and 1993 in river-side levee sediments of the lower basin (below the North Fork confluence) varied from 3,300 µg/g dw on riverbanks to 3,800 µg/g dw on levee back-slope uplands, similar to the average upstream. In lateral floodplains, median lead concentration increased with water depth, from 1,900 µg/g dw in lateral marshes, to 2,100 µg/g dw in littoral margins of lateral lakes, and 4,400 µg/g dw on the bottoms of lateral lakes. These sources of contamination in the lower basin will continue to contribute metals to CDA Lake until they are isolated from the river.

Floodplain Geochemistry and Dissolved Metal Inputs

Biogeochemical reactions affect the distribution, partitioning, mobilization and bioavailability of metals in river valleys contaminated by fluvial deposition of mine wastes (Nimick and Moore, 1991). In the CDA basin, the primary mined ore minerals included mostly metal and metalloid sulfides with some carbonates mixed with the sulfides (NRC, 2005). These were formed in anoxic environments deep within the earth's crust. When these deposits are exposed to oxygen in air or water, the minerals are oxidized, releasing the metals along with the secondary products abundant in such ores (including reduced iron, manganese, sulfate, and hydrogen ions). The sulfuric acid produced by the oxidation reactions results in waters with pH < 2. High concentrations of iron, manganese, aluminum, and other metals and metalloids are common in the low-pH waters generated by mining sulfide deposits. Acidic groundwater and overland flow (runoff) then moves from un-remediated, mining contaminated floodplains to nearby rivers and streams. Inputs of dissolved metals from groundwater are an especially important source of metal enrichment of rivers and streams in low-flow conditions, while particulate-bound metals in overland flow dominate inputs to rivers during high-flow conditions (Moore and Luoma, 1990).

When the low pH waters emanating from mining activities are neutralized (e.g., by mixing with oxidized, higher pH waters containing alkalinity), the reduced iron and manganese precipitate as various oxide and hydrous oxide phases. Most of these form multi-phase complex particulates that also include clays and organic material (Jenne and Luoma, 1975). Colloidal or nanophase iron and manganese, which take longer to aggregate and settle, are also common in groundwaters and surface waters of the CDA region (Langman et al., 2020). Because metal cation sorption to clays and clay minerals such as iron, aluminum, and manganese oxides increases with increasing pH, dissolved metal cations present at low pH become particulate-bound as pH increases, with the strength of binding to the particulate phases determining the equilibrium distribution of metals between particulates and solution. Binding strength to iron oxides among metals of interest in the CDA basin follows the order Pb >> Zn > Cd. In anoxic environments, like subsurface wetland sediments and lateral lake bottom sediments, particulate metal sulfides are formed under neutralized conditions where sufficient sulfide exists (Balistrieri et al., 2002; see Chapter 6). Thus, the fate of metals in the CDA valley involves mineral dissolution, metal mobilization and migration, and mineral re-precipitation and metal adsorption that are controlled by both pH and redox conditions. Although pH and the availability of oxygen will determine which processes dominate spatially and temporally, the permeability of the soils or sediments, organic content, and microbial activity of the depositional environment are also important (Balistrieri et al., 2002).

The distribution of metals between particulate and dissolved phases differs depending on the metal and the location's redox status. For example, Balistrieri et al. (2002) noted that riverbanks and natural levees of the CDA River are more oxidizing environments and have lower zinc than lead concentrations because zinc is preferentially leached from these sediments when they are inundated. During the major rain-on-snow flood of 1996, they observed that zinc/lead ratios of suspended sediment were less than 1, indicating that most of the suspended sediment was mobilized from oxidizing environments (i.e., surface soils). In contrast, permanently inundated (and reduced) bed sediments in the river channel are relatively more enriched in zinc because zinc is not mobilized from the sulfide-rich sediments. During the major 1997 spring flood, zinc/lead ratios of suspended sediment were greater than 1, indicating that fines winnowed out of riverbed sediment predominated over sediment from oxidizing environments.

Groundwater can play an important role in delivering mining contaminants to river systems, especially during summer and fall, when the rivers gain from the groundwater. The groundwater below contaminated floodplains of the CDA valley can be of low pH. Complex reactions occur when acidic groundwaters contact less acidic waters in the river or in the hyporheic zone (the subsurface zone where groundwater first mixes with river water). In particular, Paulson (2001) proposed that dissolved Fe(II) oxidizes to Fe(III) and precipitates within the oxygenated hyporheic zone of the stream channel, absorbing lead and reducing its transport to the river. Because of their lower binding intensity, significant concentrations of zinc and cadmium remain in solution in the drainage that then enters surface waters. This was supported by the observation that dissolved zinc was essentially conservative in the free-flowing reach of the South Fork of the CDA River between the Bunker Hill Superfund Site and Cataldo despite significant dilution and a slight increase in pH. The release of zinc from particulates in acidic groundwaters and its limited re-adsorption when acidic groundwaters are neutralized (see Chapter 6) is one justification for the high priority given to remediating acid mine drainage from the upper basin.

Barton (2002) calculated metal inputs to the South Fork of the CDA River during summer and fall (July, September, and October 1999) when the river was gaining groundwater. He found that dissolved zinc loads from tributaries were less than 10 percent of the dissolved zinc entering the river; hence, groundwater dominated zinc inputs. There was not a concomitant gain of dissolved lead from groundwater, probably because lead was retained in the hyporheic zone by the precipitating oxides as described above. Zinsser (2019) confirmed that groundwater inputs remain the main source of loading of zinc and cadmium to the CDA River during periods of low flow.

THE SUPERFUND REMEDY AND ITS EFFECTS ON METAL INPUTS

This section describes the remedy at the Superfund site because of the likely significant impacts of the remedy on both metals and nutrient loading to CDA Lake. In general, the chosen remedies have attempted to control sources of contamination in the watershed and/or reduce exposures of humans and wildlife to contaminated material in the river basin. This might suggest that metal fluxes to CDA Lake would have declined over the entire period of remedy implementation; indeed, as shown later in this chapter, lead inputs to CDA Lake have been declining for about the past ten years, but cadmium and zinc inputs to CDA Lake have been declining for at least 30 years. Nonetheless, the water quality impacts of remedy implementation have been highly variable, with periods of both increasing and decreasing fluxes of lead into the Lake from the watershed. This narrative discusses the timing of specific remedial activities, including an analysis of land use changes coincident with the implementation of the remedy. Remedial activities are then revisited later in this chapter's quantitative analysis of metals and phosphorus loading to the Lake, to further examine what impacts the Superfund cleanup has had, and will continue to have, on the quality of water entering CDA Lake via the CDA River.

Operable Units and Remediation Activities

Operable Units 1 and 2

The Superfund remedy for the Bunker Hill site has been divided both in space and time into three distinct operable units (OUs). As described in Chapter 1, OU-1 and OU-2 are encompassed within the 3 × 7 mile Box, from the vicinity of Kellogg on the eastern end to Pinehurst on the western end (Figure 3-2). Remedial actions

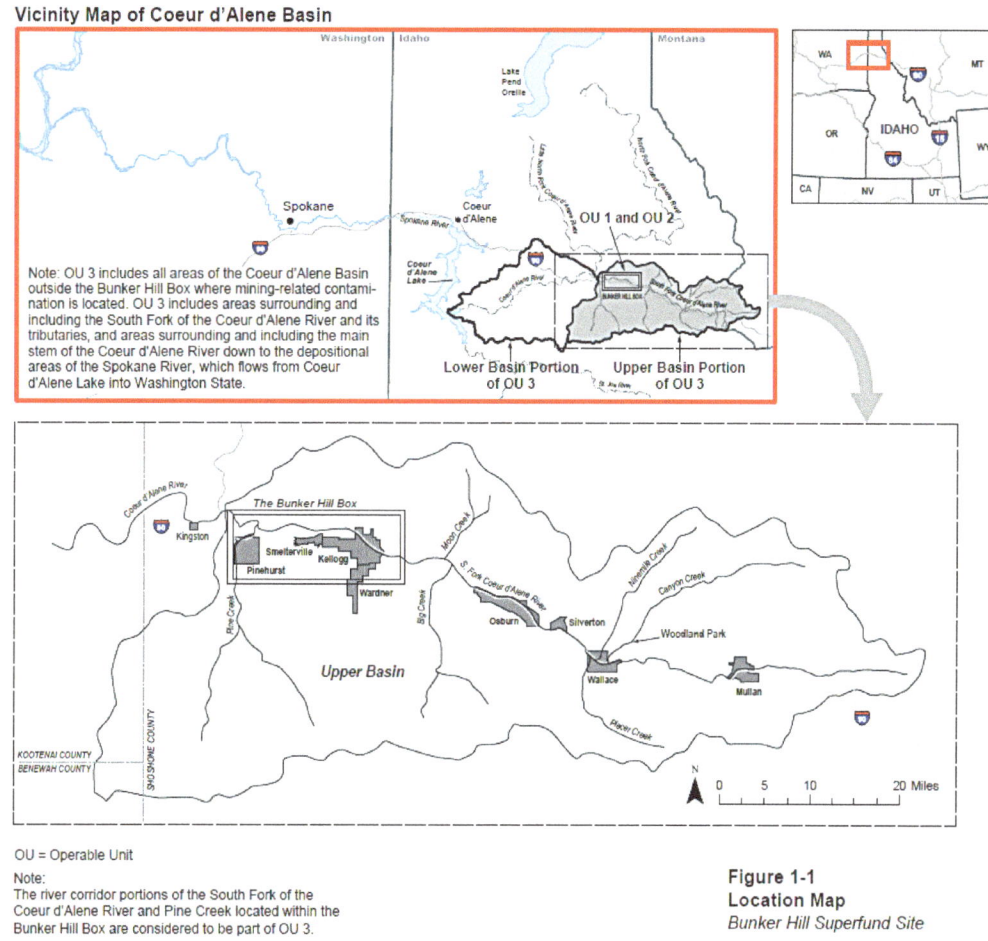

FIGURE 3-2 Delineation of operable units at the Bunker Hill Superfund Site. SOURCE: EPA Region 10 (2021).

began in 1986 when public areas in OU-1 and OU-2 were targeted for "fast-track" cleanup. Active remediation within OU-1 primarily focused on managing lead-contaminated soil in residential areas.

OU-1 remedies were designed to ensure that less than 5 percent of children have blood lead levels of 10 micrograms per deciliter (μg/dL) or greater; and that less than 1 percent of children have blood lead levels of 15 μg/dL or greater. Remediation was achieved through a number of actions, including removing soil to a depth of 12 inches (30 cm) that had greater than 1,000 milligrams of lead per kilogram soil (mg/kg), achieving a geometric mean yard soil lead concentration of less than 350 mg/kg for each residential community, and installing protective soil/vegetation barriers, culverts, and retaining walls to prevent runoff from re-contaminating remediated properties. Because removal of soil was conducted to a depth of 12 inches, contaminated material likely remains at greater depths.

The 1992 Record of Decision (ROD) for OU-2 also addressed non-populated areas, commercial areas, and other common-use areas including (1) the former industrial complex and Mine Operations Area in Kellogg; (2) the Smelterville Flats (the floodplain of the South Fork of the CDA River in the western half of OU-2); (3) hillsides, creeks, and gulches; (4) the CIA; and (5) the Bunker Hill Mine and associated acid mine drainage (EPA Region 10, 2015). Major surface remedial activities included demolition of the smelter and fertilizer plants (1995–1998); removal of contaminated material from hillsides and gulches and placement in constructed repositories (1995–2000); removal of contaminated material from the Smelterville Flats (1995–2005); and revegetation, stabilization of the land area, and creek reconstruction in the gulches and on the hillsides after removal of contaminated materials.

Remedial actions related to surface contamination in OU-1 and OU-2 were largely complete by 2008, with institutional controls allowing site use and reuse. The Central Treatment Plant (CTP), originally constructed in 1974 to treat acid mine drainage from the Bunker Hill Mine, has been upgraded to expand capacity and improve metals removal from groundwater over time. Most recently (2018–2020), a slurry wall and extraction wells have been installed to capture contaminated groundwater that was reaching the river, and this water is now being treated at the CTP (see Box 3-1 for details).

BOX 3-1
The Central Treatment Plant: Improvements and Performance over Time

The CTP was built in 1974 to treat acid mine drainage (pH = 2.8) from the Bunker Hill Mine and process water from various industrial facilities that were highly contaminated with metals. The EPA took over operation of the CTP in 1996. Improvements to the CTP prior to 2012 provided a direct gravity-fed pipeline for acid mine drainage to the CTP, refurbished and improved the thickener, increased hydraulic capacity, provided a new waste sludge line from the CTP to the sludge disposal pond at the Central Impoundment Area (CIA), replaced the lime storage and feed system with a quick lime system, constructed a new control building and plant control system, and updated the electrical system.

The 2012 ROD amendments called for additional upgrades to the CTP. In particular, the goal was to intercept metals from contaminated shallow groundwater in contact with tailings and waste rock along the valley floor and adit discharges derived from fractured bedrock aquifers that discharged into old underground workings. The groundwater collection system selected for implementation in OU-2 was an extraction well/cutoff wall system. Construction consisted of an 8,000-ft long, 30-ft deep, soil-bentonite groundwater cutoff wall along the north side of the CIA between the CIA slope and Interstate 90 and around the west end of the former slag pile area. Nine groundwater extraction wells were placed just south of the slurry bentonite groundwater cutoff wall and sized to provide water level control behind the wall and collection of between 1,500 and 2,500 gpm. This option was selected for its ability to accomplish a number of objectives, including minimizing overall groundwater collection flow rates, minimizing potential for groundwater ponding on the ground surface, maximizing hydraulic isolation for the CIA, minimizing infiltration of surface water into the groundwater collection system, minimizing the potential for groundwater discharge to Bunker Creek, minimizing operation and maintenance costs, and maximizing system flexibility (CH2M Hill for EPA, 2013). For OU-2, additional actions included construction of effluent conveyance via a pipeline to the South Fork of the CDA River to prevent recontamination. The upgrades also included expansion and upgrading of the CTP to achieve discharge limits and to allow for reduction of waste sludge volume. For OU-3, the ROD amendment specified additional expansion and upgrades to the CTP. The necessary conveyances from OU-3 were ultimately not installed due to costs and right of way/property rights issues (personal communication, E. Moreen, EPA, October 2021).

At the time of the 2012 ROD amendments, the CTP was able to effectively meet discharge requirements under normal flows, but zinc effluent limitations were exceeded during high flows. The acid mine drainage from the Kellogg Tunnel contained 100–500 mg/L of zinc depending on the flow. The potential OU-2 and OU-3 sources are significantly lower, ranging from 4 to 24.5 mg/L of zinc (EPA, 2013). Improvements in the ROD amendment included a granular media filter system to allow the plant to operate in a high-density sludge mode to improve metals removal effectiveness and reduce sludge volume. Other improvements to the CTP involved replacing the existing rapid mix tank and aeration basin, providing an automatic polymer feed system, removing the polishing pond, replacing sludge recycle and waste pumps, and adding an influent meter.

In 2016, the U.S. Army Corps of Engineers awarded a $48 million contract for even more upgrades to the CTP. Completion of the upgrades and installation of the groundwater cutoff wall and extraction pond appear to provide improved removal of lead, zinc and cadmium from the groundwater and mine flows. As shown by Moreen (2021a), effluent concentrations met discharge requirements over a five-month monitoring period between August and December 2020. Between 99.7 and 99.9 percent of the zinc was removed, leading to capture of 300–450 kg of zinc per day (110–165 MT/yr) within the system and preventing further contamination of the South Fork of the CDA River. Prior to the upgrade, effluent zinc concentrations were 0.15–0.3 mg/L, while after the upgrade the performance showed average levels < 0.04 mg/L. If removal is 99.9 percent, the CTP is removing 116 MT/yr on average, and discharging ~0.12 MT/yr.

Although not a contaminant of primary concern for the Superfund remedy, the average phosphorus removal from August 2020 to November 2021 was 98.4–98.8 percent. With a maximum effluent loading of 0.045 kg/day, the CTP is discharging less than 0.016 MT/yr of total phosphorus.

Operable Unit 3

OU-3 extends over a much larger land area (Figure 3-2) and includes the upper and lower basins of the CDA River. In the upper basin portion of OU-3, remedial actions are occurring, while for the lower basin portion, investigations and feasibility studies are ongoing, with pilot actions forthcoming. Remediation in the lower basin is complicated by the fact that metal contamination is present in the lower river beds, river banks, and floodplains, and in the lateral lakes; flooding continues to redistribute contaminants across the region; and groundwater contamination is widespread.

The ecological risk assessment (CH2M Hill and URS Corp, 2001) identified nine chemicals of potential concern in surface water, including antimony, arsenic, cadmium, copper, lead, manganese, mercury, silver, and zinc. Lead and zinc raised the greatest level of concern to EPA and were felt to serve as models for two distinct transport mechanisms. That is, lead is typically present in particulate form, whereas zinc is more likely to be present in dissolved form, except during floods. Across the upper basin, significant deposits of tailings (millions of tons) had built up on floodplains above the dams constructed in the early 1900s. Poor maintenance and flooding not only transported these contaminated materials downstream to the lower basin, but significant amounts of tailings remained behind the dams. There was no attempt to characterize groundwater sources of contamination, which is of concern since much of the dissolved zinc load that enters the CDA River during low discharge derives from groundwater.

The ROD for OU-3, issued in 2002, focuses on a large area that extends from near the Idaho-Montana border west through the Idaho Panhandle into the State of Washington. It includes communities, floodplains, rivers, tributaries and lakes, as well as Pine Creek and the portion of the South Fork of the CDA River within the Bunker Hill Box. OU-3 also includes areas where mine wastes were used for road building, fill, or construction. The ROD established a 30-year interim remedial plan that included final cleanup for human health exposure to residential/community soils and interim cleanup for ecological protection based on benchmarks. It is important to note that the ROD does not meet protectiveness standards or applicable or relevant appropriate requirements, it did not include groundwater or CDA Lake, and it did not allow for practice of tribal or subsistence lifestyles.

In 2012 an interim ROD amendment (EPA, 2012) was created to clarify and modify water collection and treatment actions from the initial RODs for OU-2 and OU-3. Relevant to OU-3, the amendment included consolidation/ isolation source control at upper basin mine and mill sites, construction and maintenance of repositories, remediation of road surfaces, protection of remedies from erosion and recontamination, and groundwater remediation in Ninemile and Canyon Creek watersheds. These remediation actions are complex, involving multiple actions over many years. For example, 19 mine and mill sites in Ninemile Creek have seen remedial activity that commenced in 1992 and continues today (see Box 3-2).

Future Remedial Activities

The remedial efforts associated with OU-3 have largely been implemented in the upper basin, and many remediation challenges remain, particularly along the South Fork of the CDA River, which provides 90 percent of the metals input to the main stem of the CDA River. Zinc has been the primary driver of cleanup activities in this area, and remediation activities to address zinc in the Ninemile and Canyon Creek watersheds have been extensive. Based on data collected at Harrison and analyzed by the committee, the combination of improved treatment at the CTP and remediation of major zinc sources in the South Fork of the CDA River has lowered zinc inputs to the Lake by about 45 percent, a reduction of 384 MT/yr. Note that the reduction simply due to the removal of zinc by the CTP has been approximately 116 MT/yr (see Box 3-1). The effect of remedial activities on lead, which is particle-associated, is more difficult to discern, but analysis of data from the U.S. Geological Survey (USGS) and by the committee (*vide infra* analysis) indicate that removal of materials during remediation may have either exposed materials with higher lead concentrations or led to other disturbances of the river and floodplain of the lower CDA River that temporarily increased total lead fluxes into the CDA River from 2005 to 2010, but that fluxes have been decreasing since. This information will be important to consider as remedial activities are undertaken in the lower basin.

> **BOX 3-2**
> **Remediation of Ninemile Creek**
>
> The Ninemile Creek basin within OU-3 has been a major source of zinc and contains multiple mining sites and remedial actions occurring for 30 years. Prior to 1998 and EPA involvement, remediation was undertaken by stakeholders. The following is a list of actions taken at the site.
>
> - **1992–1993:** Waste rock pile at Interstate Mill site regraded and moved out of floodplain.
> - **1993:** Relocation of 1,600 ft (488 m) of creek, waste rock regraded away from channel, capping of tailings pile, and grade control placement at the Success Mine Complex. Installation of upgradient surface water drainage and groundwater interceptor drains at two adits. Infiltration galleries at the Success Complex along waste rock piles were built to intercept drainage. A pilot permeable reactive barrier was tested for groundwater treatment.
> - **1994:** Contaminated tailings removed from two creek segments (0.5 and 0.75 miles, or 0.8 and 1.2 km, long), followed by riparian stabilization and revegetation. Material (150,000 yd^3 or 114,683 m^3) was placed in repository and capped. A fish pond near the confluence of two branches of Ninemile Creek was reconstructed.
> - **1998:** Waste rock from the Interstate Mill site was placed in an unlined repository. Tailings, debris, and contaminated sediments from the Interstate Mill site and the Creek were placed in a repository.
> - **2002–2007:** Consolidation, debris removal, and realignment of creek at Rex No.2/Sixteen-to-One Mine.
> - **2013–2019:** Road infrastructure and repairs for further remediation. The waste consolidation area (WCA) begins operation.
> - **2014–2015:** Waste rock excavated and removed from Interstate-Callahan (IC) Rock Dumps and placed in the WCA. Channel lining was reconstructed at the Rex No.2/Sixteen-to-One Mine. The riparian area at the IC Rock Dump was stabilized.
> - **2016–2019:** Waste rock excavated and removed from the Success Complex was moved to the WCA. The riparian area is stabilized.
> - **2019–2020:** Mine waste from the Interstate Mill site moved to the WCA, and infrastructure improvements were made. Creek reconfiguration at the Interstate Mill site.
>
> SOURCE: EPA (2019).

Compared to the upper basin, more limited remedial activities have been conducted in the lower basin. In addition, the lower basin has different and more complex hydrology, including large areas of wetlands and floodplains as well as river bed materials that are highly enriched in metals that present additional remediation challenges. Slowing of river flow at mile 160 near Cataldo causes much of the sediment that enters the lower river to deposit in the lower basin. Although the levels of metals entering this stretch of river have been reduced by remedial activities in the upper basin (particularly in the Box), the now contaminated river sediments, poorly stabilized river banks, and beaches in this stretch represent exposure risks and a continuing source of metals to CDA Lake because contaminated sediments are continuously transported downstream. EPA data (Prestbo, 2021a) and the committee's analysis of lead data later in this chapter indicate that the main stem of the CDA River downstream of Cataldo is a lead source, likely due to mobilization/transport of riverbed sediments, with < 15 percent of the load being derived from bank erosion. In addition to the river itself, the lower basin covers 18,000 acres (7,284 hectares), much of which has been contaminated during flooding events. Remediation challenges in this area include complex hydrology due to the fluctuation of lake levels, changing sediment supplies and streamflow conditions, limited access, and multiple stakeholders. The lower basin has large areas of waterfowl habitat that are contaminated above acute (80 percent of area) or chronic (95 percent of area) levels for lead (EPA, 2020).

Ongoing remedial efforts in the lower basin are shown in Box 3-3, with a focus on lead contamination and potential human and wildlife exposure risk. Efforts near the Cataldo Mission and recreation area from 1995 to 2002 focused on removal of contaminated tailings, erosion control, and capping. In 2003–2004, remediation near the Rose Lake boat launch occurred (Moreen, 2021b; EPA, 2010), which included bank stabilization to prevent erosion

> **BOX 3-3**
> **Decisions and Representative Remedial Actions in OU-3**
>
> **1992** Initial cleanup begins
> **1994** Tailings removal/stream bank stabilization at Elizabeth Park
> **1995–2002** Tailings and materials removed, erosion control, and capping at Cataldo Mission and recreation areas along lower river
> **1996–2000** Tailings and materials removed from Pine Creek mine and mill locations
> **1997** Tailings and materials removed from Canyon Creek and Osburn Flats
> **1998** Soil and tailings removal from South Fork floodplain
> **1998** Tailings, debris, and sediment removed from Interstate Mill site
> **1998–2000** Tailings, debris, and sediment removed from Silver Crescent and Charles Dickens Mines
> **2000** Water treatment systems at Gem Portal and Ninemile Creek
> **2000** Sediment and tailings removed from Elk Creek Pond
> **2000** Removal action and construction of Trail of the Coeur d'Alenes begins
> **2002** ROD for OU-3
> **2003–2004** Remediation of Rose Lake boat launch
> **2004** Trail of the Coeur d'Alenes opens
> **2012** Interim ROD Amendment for OU-1, OU-2, and upper basin OU-3
> **2015** Robinson Creek agriculture to wetland conversion
> **2017–2018** Multi-objective decision analysis to prioritize projects
> **2017** Thin layer capping trial at Lane Marsh
> **2019** Planning for Gray's Meadow restoration

and capping to limit exposure of recreational users. Construction of the Trail of the Coeur d'Alenes (2000–2004) provided both capping of contaminated material and recreational opportunities (see Figure 3-3 for a map showing a portion of the trail and the river reach from Cataldo to Rose Lake). Recreation sites continue to be a top priority in the lower basin, with a combination of cleanup, communication, and access restrictions being used to minimize exposure, but these remedies are not aimed at reducing inputs to the rivers or the Lake.

Due to the complexity of remediating contaminated wetlands and floodplains that serve as waterfowl habitat, EPA has begun projects to convert uncontaminated agricultural land to wetlands to provide more desirable, clean habitat for birds in the region (e.g., Schlepp project 2007–2014; 2015 Robinson Creek project; Moreen, 2021b). Although this does not remediate contaminated areas, it does reduce wildlife exposure to contaminants by diverting the birds to cleaner sites that have been excavated and capped and have low risk of recontamination during flooding. Future lower basin remedial activities will include source control, additional agriculture-to-wetland conversion activities, and actual wetland remediation, such as soil treatment to reduce lead bioavailability (Moreen, 2021b).

As in OU-1, OU-2, and the upper basin of OU-3, property remediation has been occurring in the lower basin. Priorities are to control sources in and out of the river, manage exposure risks, and prevent floodplain (re)contamination. Some remedies are identical to those elsewhere, such as excavation and replacement of soil in residential areas (850 properties since 2019; EPA, 2020) along with remedies to protect sites from recontamination.

River, Riverbank, and Floodplain Sources of Contaminants

As previously mentioned, a major challenge in the lower basin is remediation of the large amount of lead contamination in floodplains, river banks, and river bed sediments. Sediment cores collected in transects of the CDA River in 2012 (CH2M Hill, 2015, 2017; Maul Foster and Alongi, 2020) show typical concentrations of thousands of mg/kg of lead in the river sediment, although this does vary, with some stretches of the river showing much lower levels (tens of mg/kg). For a particular transect, levels are usually relatively consistent, but variation

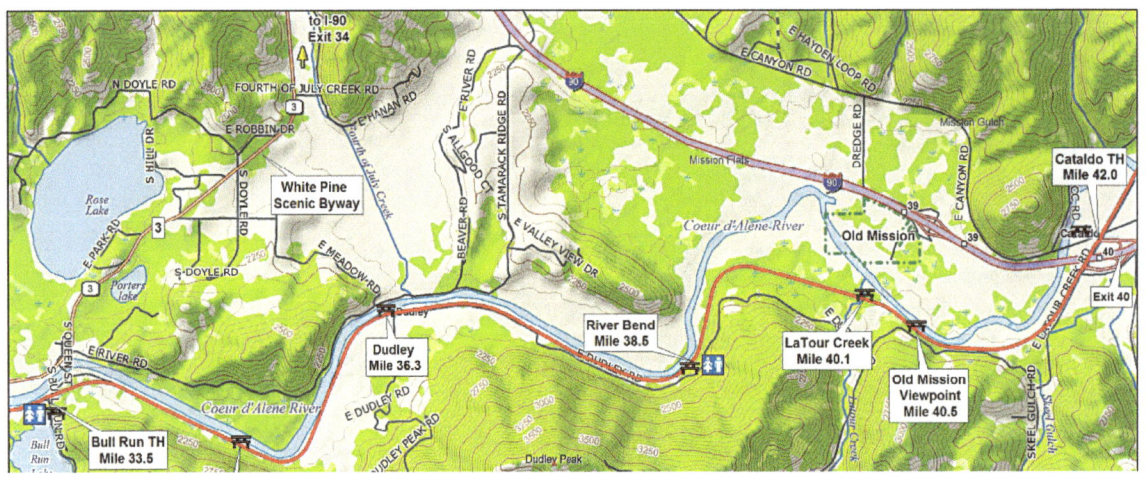
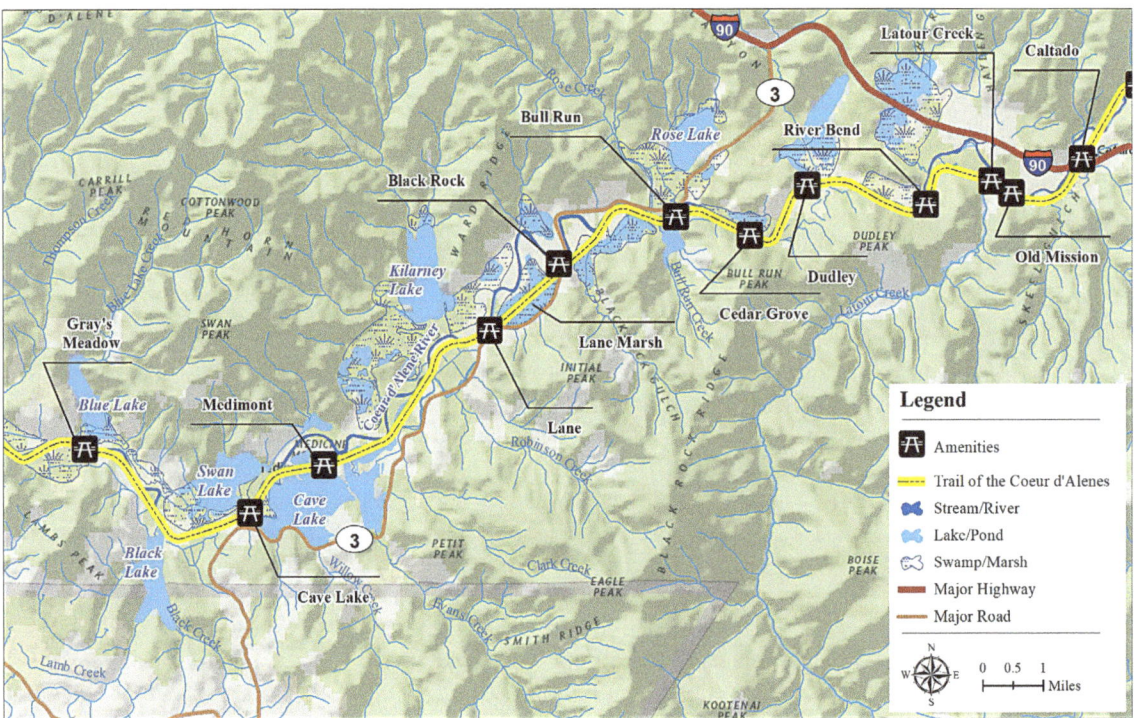

FIGURE 3-3 (A) Map of the stretch of the lower CDA River from Cataldo to Rose Lake, where recreation sites were remediated. The red line is the Trail of the Coeur d'Alenes. SOURCE: https://parksandrecreation.idaho.gov/parks/trail-coeur-d-alenes/maps/. (B) Larger map of the lower basin area along the CDA River, showing Gray's Meadow and Lane Marsh. SOURCE: Generated by the committee using data from IDEQ, CDA Tribe and the USGS National Map and associated datasets.

does occur with depth. For the ~3 m cores collected in 2012, the largest concentrations were found in the deeper portions of the core, potentially indicating deposition of less contaminated sediment from either upstream or the floodplain during more recent times (although the flood plains/banks of the river are also contaminated). In other stretches, the most contaminated material is at the shallow depth. The 2019 cores were only ~0.66 m in depth and showed comparable lead and zinc levels that did not vary with depth. Over a 1-mile stretch, sediment concentrations could vary from tens of mg/kg to > 25,000 mg/kg for both lead and zinc.

These varying levels reflect the complicated bathymetry of the system, with scour and deposition affecting metals concentrations in different stretches as water flows and flooding transports materials. Laboratory erosion experiments generally showed coarser, non-cohesive surface sediment but more cohesive, fine material at depth (CH2M Hill, 2015, 2017; Maul Foster and Alongi, 2020). The depth of the erodible layer varied from only the top 5 cm of the core to 20 cm. Erosion pins that were installed as part of the 2019 sampling and monitoring have the potential to provide valuable information regarding sediment transport and deposition.

Remedies in the lower basin are not without risk of recontamination. For the contaminated river banks and beds, dredging would remove the metals from the system, but this plan also risks resuspension of contaminated material. Capping has a lower risk of release during implementation, but must be robust enough to ensure the contaminated sediments are not re-exposed in the future during flooding or changing system hydrodynamics.

A field experiment at Lane Marsh (see Figure 3-3B) used thin layer capping, a method designed to mimic the deposition of clean sediment in multiple applications to allow vegetation to recover. Results showed that a thickness of 6 inches (15 cm) reduced lead exposure, and vegetation recovered within two years. Active capping using biochar, which is also being tested in the lower basin, could increase the sorptive capacity of the capping material (Knox et al., 2014; Wang et al., 2018). As noted in Lane Marsh, transport of contaminated material to a capped site from an upstream or neighboring untreated area may re-contaminate a treated area, and this could be exacerbated during a flooding event in which contaminated materials are more widely dispersed. It is also necessary to account for the connection between the river and the flood plains in remediation strategies. If remediation were to focus primarily on highly contaminated river bed sediments, the relative contribution of the floodplains may increase as they continue to replenish the bed sediments, unless remediation also isolates those sediments from the river or reduces major flood frequencies.

The 2020 Optimization Report for OU-3 (EPA Region 10, 2020) states that "there is considerable uncertainty in evaluating discharge of metals from sediments both spatially and temporally" and it emphasized that a conceptual site model is critical to implementing remedial activities given the amount of lead present in the floodplains and riverbanks/riverbed that can be transported to the Lake. A conceptual site model that includes detailed hydrodynamics is being developed by EPA (Prestbo, 2021a; EPA, 2020). Any models or analysis of the lower CDA River must give full consideration to the evolving nature of the channel and floodplain system as it continues to respond to the changing rates of sediment input and metals input from both the South Fork and North Fork CDA watersheds.

Future Remedy Priorities

EPA's goal is to adaptively manage remedial activities in the lower basin, with a focus on prevention of sediment (and thus lead) transport from the lower basin of the CDA River to the Lake (EPA, 2020). In 2017–2018, multi-objective decision analysis was used to prioritize and select remedies and projects for the lower basin (Prestbo, 2021b). Overall, wetland restoration and capping scored high in this process. Current projects in the planning phase include Gray's Meadow restoration (Figure 3-3B) to reduce metals, improve water quality, and provide wetland habitat, with construction likely to begin in 2022. In Lane Marsh, research on thin layer capping continues. Pilot testing of biochar amendments to limit lead bioavailability in wetlands is also occurring. Near Harrison, dredging and capping of the riverbanks is the identified remedy, with characterization set to begin in 2022 and construction planned for 2026. The Dudley reach of the CDA River (Figure 3-3B) is highly contaminated and hydrologically challenging (deep channel), such that new technologies are likely to be needed to limit downstream transport of metals. Pilot implementation of dredging/capping scenarios is planned for 2024 or 2025.[3] Waste consolidation areas for these efforts are currently being sited.

Assessment of Land Use Changes Coincident with Remedy Implementation

As mentioned in Chapter 2, the USGS has been measuring metal concentrations and fluxes within the upper and lower CDA basin as part of the Basin Environmental Monitoring Program to evaluate the long-term effects of the Bunker Hill Superfund site remediation. These data are further analyzed by the committee (later in this chapter)

---

[3] https://cumulis.epa.gov/supercpad/SiteProfiles/index.cfm?fuseaction=second.stayup&id=1000195

to better understand the potential effects of remedial actions not only on metals, but also on sediment and phosphorus. As a complement, the committee analyzed land cover information to determine if it could reveal changes indicative of Superfund remedy implementation. This analysis made use of the National Land Cover Database (NLCD),[4] a national dataset with information on land use and land cover for various time periods for all areas of the United States. Each grid cell represents a particular land cover category and was derived from classification algorithms that processed Landsat satellite imagery. There are NLCD datasets available that represent 2001, 2004, 2006, 2008, 2011, 2013, and 2016 and 2019; these datasets all use the same land use descriptions and codes and can be compared directly. In addition, a 1992 NLCD dataset is available, but the land-use classification system was different than those beginning in 2001. Finally, there is a historical dataset, compiled and published by USGS, that provides land use and land cover from the 1970s and 1980s. This dataset is in a different format (polygons versus raster) and uses a categorization scheme different from either the 1992 NLCD data or the 2001–2019 data. To compare land use in OU-1 and OU-2 over time, it was necessary to reclassify the earlier datasets. Once the categories were realigned to the 2001–2019 classification system, the CDA watersheds that encompass OU-1 and OU-2 were isolated and the land use areas were determined using ARCGIS Pro. Comparison of the land cover in these areas is presented comprehensively in Table 3-1 and Figures 3-4A, B, and C for years 1992, 2001, and 2019.

The data in Table 3-1 highlight several key trends in land use in and around OU-1 and OU-2, although caution must be exercised in comparing recent data to pre-2001 data and especially to the historic data for 1970–1985 due to changes in methods and instrumentation. First, barren land represented a major portion of the land use prior to major remediation efforts that started in 1992, which is not surprising since barren land is a common characteristic of floodplains and forested lands severely impacted by mine wastes (Moore and Luoma, 1990). The acreage of barren lands has been reduced dramatically since the 1970–1985 time period, indicating the success of landscape revegetation that was part of the Superfund activities. Second, throughout the 1970–1985 period, forest land was the dominant land use in the area. This is in large part because the areas within these watersheds extend beyond the mining impacted areas in OU-1 and OU-2. However, there has been an increase in the amount of evergreen forest since 2001, which has coincided with a decrease in shrubland and herbaceous land cover. In other words, since 2001 acreage of evergreen forests has increased and acreage of shrubs has decreased, but the total of the two has remained relatively constant. Third, developed lands (the sum of all four developed categories) have been increasing since 1992 from 7.29 to 9.48 km². Finally, acreage of emergent wetlands appears to have increased over the past decade. These trends are consistent with expectations of improved landscape within an area of ongoing remediation.

SOURCES OF PHOSPHORUS TO COEUR D'ALENE LAKE

Beyond metals, the other major input of concern from the CDA basin to CDA Lake is that of nutrients that might increase in lake productivity, which could enhance anoxia in the hypolimnion and result in the possible release of metals from lake bed sediments. Nutrient inputs to CDA Lake are the primary concerns of the 2009 Lake Management Plan (LMP; IDEQ and CDA Tribe, 2009) and have been the subject of investigation by both parties over the past decade. In particular, in 2020 IDEQ and the CDA Tribe produced the report, *Coeur d'Alene Lake Management Program: Total Phosphorus Nutrient Inventory, 2004–2013* (IDEQ and Tribe, 2020), which includes an accounting of phosphorus sources and loads to the Lake, using monitoring data, estimates from modeling exercises, and previous reports to determine how the sources of phosphorus to CDA Lake had changed from 2004 to 2013. That report noted that nitrogen had not yet been subjected to the same analysis, although this was planned for the near future. The nitrogen to phosphorus ratios (such as TN/TP and SRP/DIN)[5] for CDA Lake suggest that this system may be mainly phosphorus-limited, but that during the summer the system may trend toward co-limitation or even nitrogen limitation at times (discussed further in Chapter 5). The 2020 Phosphorus Inventory report was used extensively to inform the section below, which describes the major point and nonpoint sources of phosphorus to CDA Lake.

---

[4] https://www.usgs.gov/centers/eros/science/national-land-cover-database

[5] TN = total nitrogen, TP = total phosphorus, SRP = soluble reactive phosphorus, DIN = dissolved inorganic nitrogen = $NO_3 + NH_4$

TABLE 3-1 Distribution of Land Uses for the Sub-Watershed Areas Including and Surrounding OU-1 and OU-2

	OU-1 and OU-2 Areas by Land Use Based on 2019 Designations (Areas in km²)						
lucode_2019	NLCD Land Cover Class	1970–85	1992	2001	2006	2011	2019
11	Open Water		0.86	0.21	0.11	0.11	0.10
21	Developed, Open Space	0.53	0.15	1.42	1.41	1.41	2.05
22	Developed, Low Intensity	4.34	3.72	3.11	3.11	3.12	2.88
23	Developed, Medium Intensity		0.01	3.08	3.13	3.15	3.70
24	Developed, High Intensity	5.20	3.41	0.38	0.41	0.41	0.85
31	Barren Land	12.63	1.62	4.91	1.90	1.84	1.53
41	Deciduous Forest		0.21	0.14	0.14	0.14	0.53
42	Evergreen Forest	48.63	38.18	36.87	41.35	41.45	46.23
43	Mixed Forest		2.72				0.30
52	Shrub/Scrub	11.01	13.69	30.50	29.47	29.37	22.77
71	Herbaceous		18.54	1.91	1.43	1.46	0.99
81	Hay/Pasture		0.00	0.05	0.05	0.05	0.00
90	Woody Wetlands		0.08	0.07	0.06	0.06	0.50
95	Emergent Herbaceous Wetlands		0.01	0.54	0.07	0.07	0.73
Total		82.35	83.19	83.19	82.65	82.65	83.17

SOURCE: National Land Cover Database.

Phosphorus Point Sources

Point sources to the Lake include wastewater treatment plants (WWTPs) that discharge to the watershed, including the CTP that treats groundwater from the CIA. With regard to municipal wastewater, most of the more than 175,000 people in Kootenai County (see Chapter 1) are serviced by WWTPs of the cities of Coeur d'Alene and Hayden. The WWTPs for both of these communities have state-of-the-art advanced phosphorus removal, and their phosphorus discharge permits are extremely low (i.e., effluent total phosphorus < 100 μg/L relative to typical wastewater concentrations of 5,000–7,000 μg/L). But even more importantly, the effluents from these facilities are discharged to the Spokane River downstream of CDA Lake, so these plants (and the population they service) are not nutrient inputs to the Lake.

Table 3-2 shows the nine WWTPs that either directly discharge to the Lake (Plummer and Harrison WWTPs) or upstream of the Lake into the CDA River (Page, Mullan, and Smelterville WWTPs) and St. Joe River (St. Maries, Santa-Fernwood, Clarkia, and Emida WWTPs). Of these WWTPs, the Plummer facility employs advanced phosphorus removal and has average effluent total phosphorus concentrations of ≈ 80 μg/L. The other facilities are conventional primary/secondary systems and have average effluent total phosphorus concentrations of 1,500–2,000 μg/L. The overall phosphorus export from these facilities averaged 8.3 MT/yr for the 2009–2017 period (IDEQ and CDA Tribe, 2020). Compared to the total phosphorus inputs from the CDA and St. Joe Rivers (about 133 MT/yr; see later analysis), smaller tributaries to the Lake, and direct atmospheric deposition onto the Lake, these WWTP discharges only account for 6 percent of the annual total phosphorus budget.

The phosphorus in conventional municipal wastewater discharges generally has very high bioavailability (≈ 80 percent; Li and Brett, 2012), whereas the particulate phosphorus draining forested catchments that probably dominates CDA Lake inputs has been shown to have low bioavailability (< 20 percent) (Ellis and Stanford, 1988; Ekholm and Krogerus, 2003). When taking phosphorus bioavailability into account, the discharges from these WWTPs could have a more substantial impact on primary production in CDA Lake than 5 percent suggests (i.e., WWTPs account for 23 percent of the bioavailable phosphorus loading to CDA Lake). Advanced nutrient removal processes in WWTPs can achieve 95 percent phosphorus removal using biological treatment (i.e., effluent total

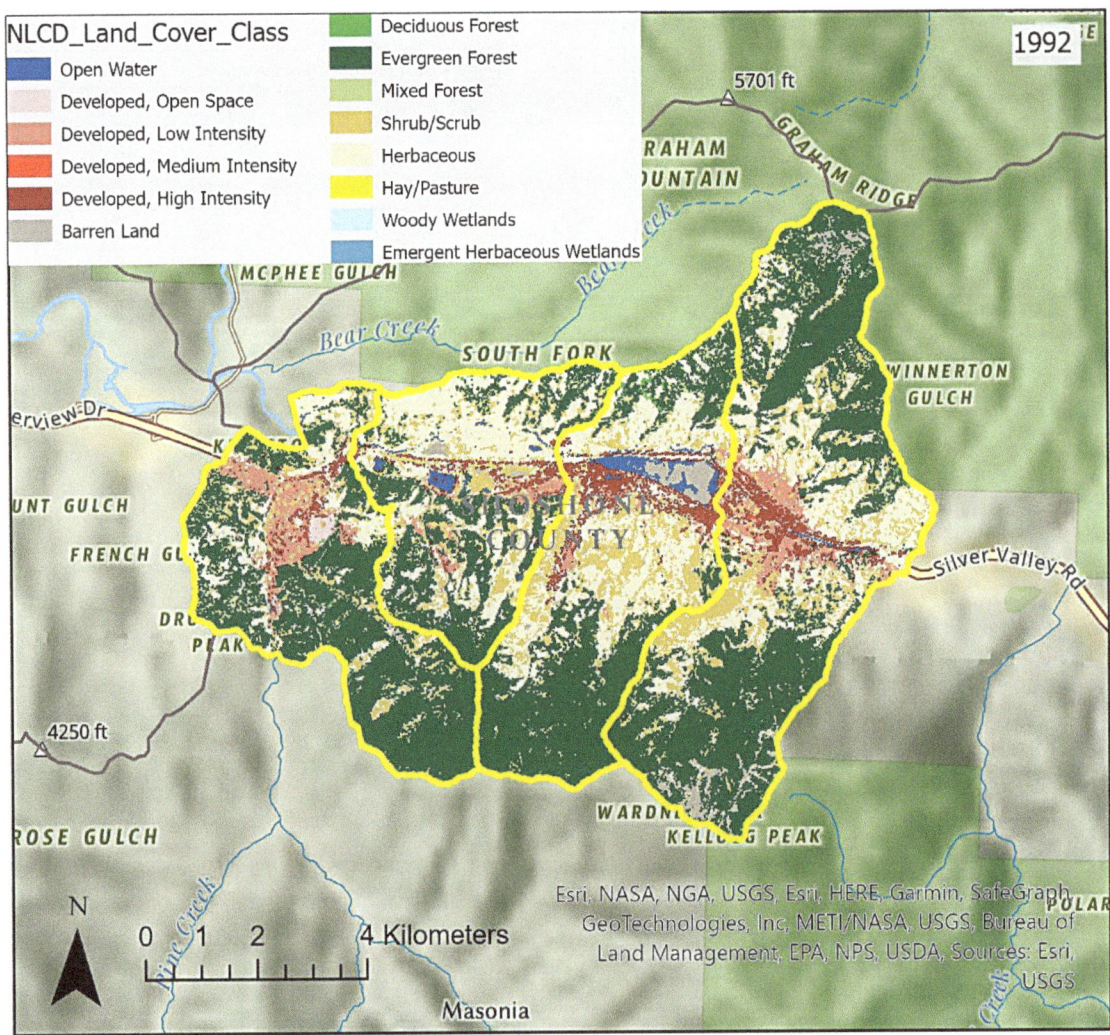

FIGURE 3-4(A) 1992 land use in the Bunker Hill Box. SOURCE: Data plotted by the committee from the National Land Cover Database. https://www.mrlc.gov/data?f%5B0%5D=category%3ALand%20Cover&f%5B1%5D=region%3Aconus.

phosphorus concentrations ≈ 250 µg/L) and 99 percent removal using iron or aluminum based chemical treatment (i.e., effluent total phosphorus concentrations ≈ 50 µg/L) (Li and Brett, 2012, 2015). Given that all of the WWTPs that discharge to the Spokane River downstream of CDA Lake already implement advanced phosphorus removal, consideration of this technology for the WWTPs that discharge upstream of CDA Lake could reduce concerns about these phosphorus inputs into the future.

Phosphorus from the Central Impoundment Area

It has been estimated that the Bunker Hill CIA contributes 5–15 tons[6] of TP/yr to the CDA River (Clark and Mebane, 2014; IDEQ and CDA Tribe, 2020). This facility stores gypsum waste from a former phosphoric acid fertilizer plant. Gypsum is a common soil amendment that is often used to reduce phosphorus runoff from over-fertilized soils

---

[6] All the load values in the Phosphorus Inventory report are in US short tons.

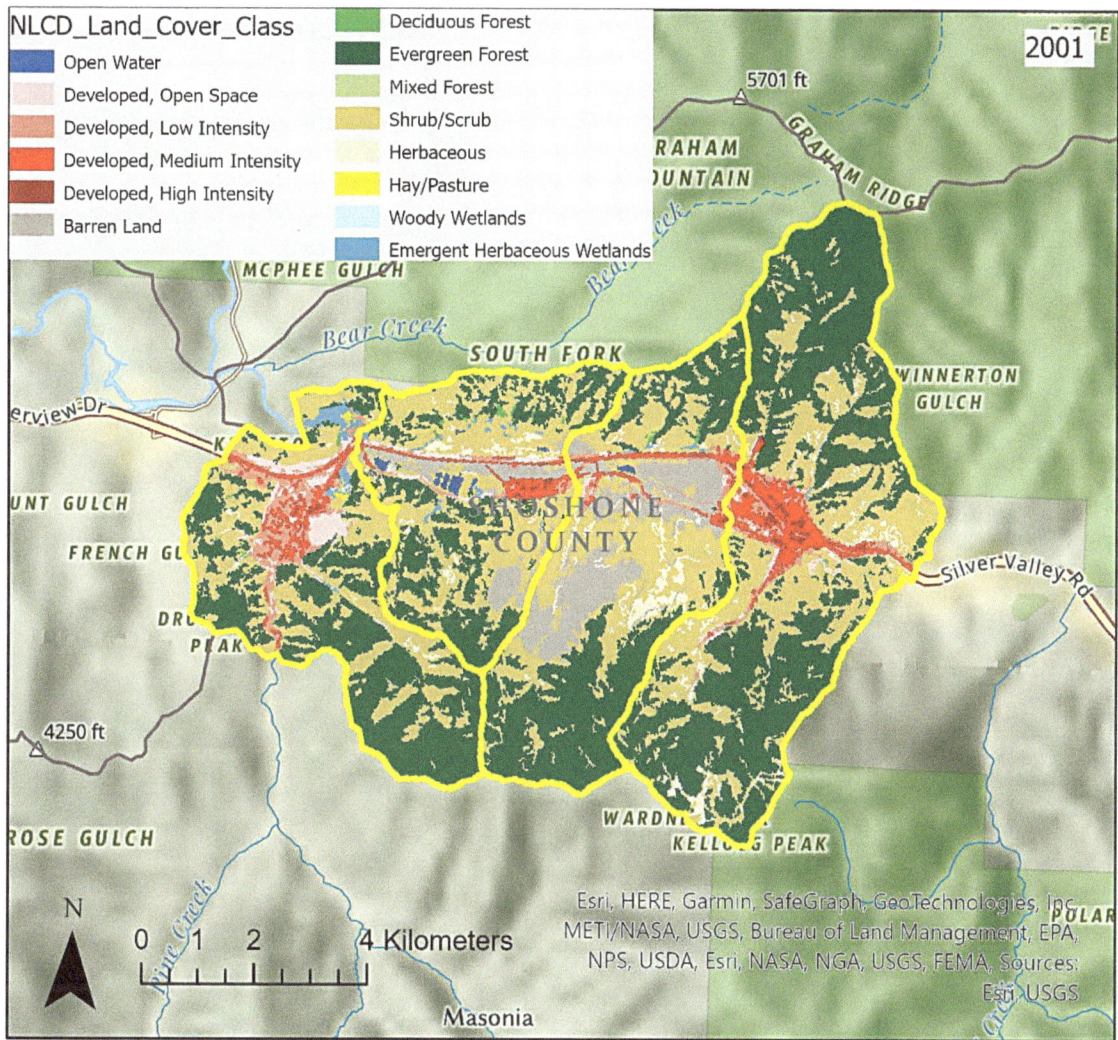

FIGURE 3-4(B) 2001 land use in the Bunker Hill Box. Note that land classifications changed in 2001. SOURCE: Data plotted by the committee from the National Land Cover Database.

(Watts and Torbert, 2009). Because gypsum is calcium sulfate ($CaSO_4 \cdot 2H_2O$), which easily solubilizes ($K_{sp} = 10^{-4.58}$), dissolved calcium concentrations in these deposits would likely be very high, which should result in the formation of calcium phosphate minerals. These have been shown to have widely varying bioavailability depending on which specific mineral complexes form (Li and Brett, 2013). For example, the phosphorus in calcium phosphate ($CaHPO_4$) has been shown to be ≈ 100 percent bioavailable, whereas the phosphorus in apatite [$Ca_5(PO_4)_3(OH,F,Cl)$] had zero bioavailability (Li and Brett, 2013). Other mineral complexes, such as tricalcium phosphate [$Ca_3(PO_4)_2$] probably have intermediate bioavailability depending on the specific environmental conditions (Deubel and Merbach, 2005). The phosphorus released from this facility could have had a substantial impact on productivity in CDA Lake, but this depends on what types of calcium phosphate minerals are formed.

Superfund remediation activities have targeted treatment of the gypsum pond wastes in the CIA. Indeed, EPA has completed construction of a groundwater cut-off wall and groundwater collection system for the CIA that will route the retrieved groundwater through the CTP (see Box 3-1). Although the CTP was designed to capture

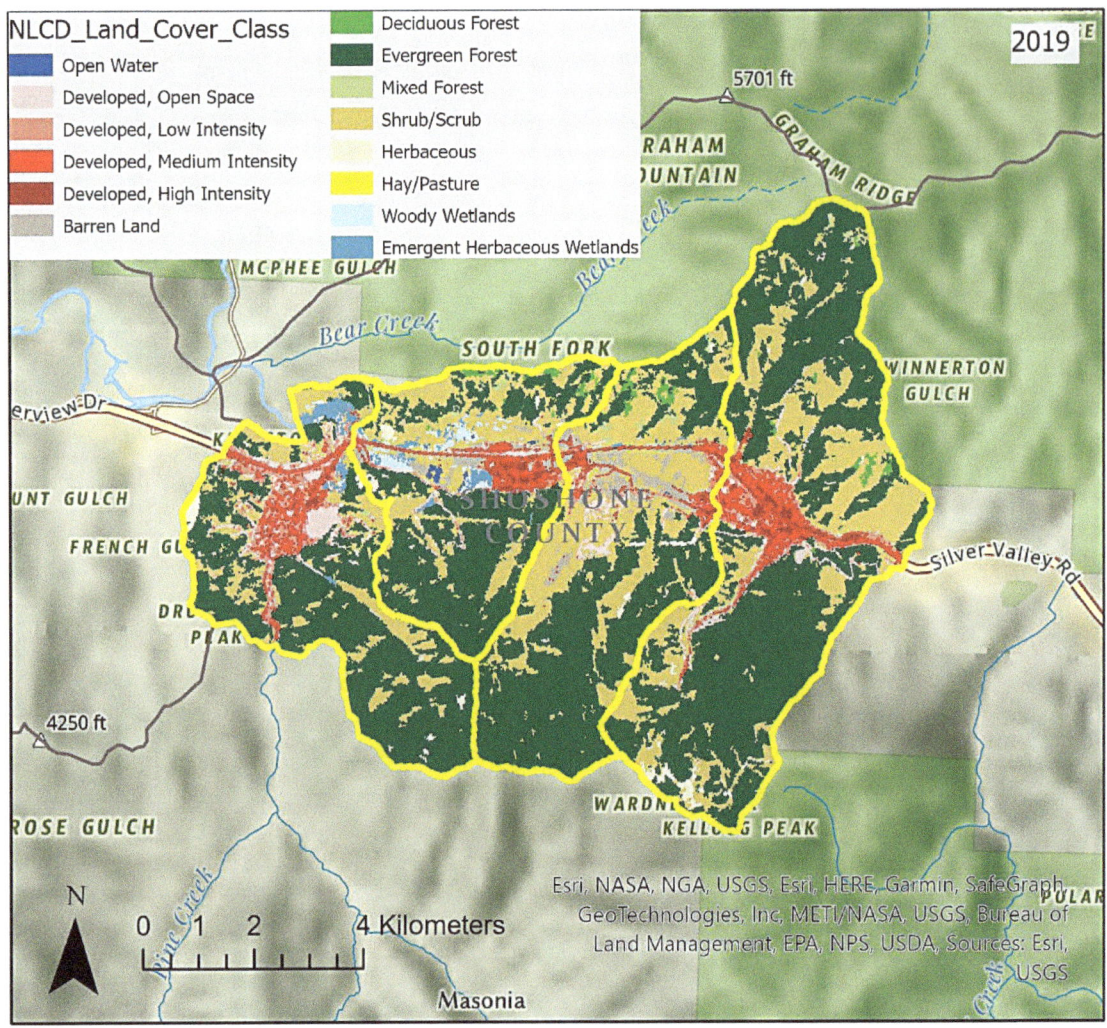

FIGURE 3-4(C) 2019 land use in the Bunker Hill Box. SOURCE: Data plotted by the committee from the National Land Cover Database.

metals being released from the Bunker Hill complex, it is also capturing phosphorus, and the recent upgrades are expected to result in substantial reductions in total phosphorus loads, at least during times of low to moderate river discharge. Indeed, Moreen (2021a,b) presented data suggesting that the upgraded CTP was removing more than 98.4–98.8 percent of total phosphorus with an effluent discharge loading of less than 0.045 kg /day (or about 0.016 MT/yr). (For perspective, the analysis later in this chapter concludes that average fluxes of total phosphorus from the CDA River are currently about 67 MT/yr.)

Nonpoint Sources of Phosphorus

The major nonpoint sources of phosphorus in the CDA basin include stormwater-mediated erosion of soils and sediments, agriculture and forestry, nonmunicipal community wastewater treatment systems and septic systems, and atmospheric deposition, but data are not available to determine the relative importance of these various sources.

TABLE 3-2 Permitted Municipal Wastewater Treatment Plants in the CDA Basin

Name	NPDES SYS ID #	Discharge Location	County	Average load 1991–1992 (kg/yr)[1]	Average load 2009–2017 (kg/yr)[2]	Average annual discharge (MGD)[2]	Average total phosphorus conc (mg/L)[2]	Range of total phosphorus concentration (mg/L)[2]	Range of monthly discharge (MGD)[2]	Range of annual load (kg/yr)[2]
Coeur d'Alene Lake										
Harrison WWTP	ID0021997	Anderson Slough	Kootenai	120	36	0.01	2	–	0 to 0.25	28 to 46
Plummer WWTP[3]	ID0022781	Plummer Ck.	Kootenai	290	25	0.23	0.07	0.0005 to 0.4	0.08 to 1.4	15 to 37
Coeur d'Alene River										
Mullan WWTP	ID0021296	SF CDR	Shoshone	310	260	0.10	1.8	0.3 to 3.7	0.04 to 290	190 to 360
Page WWTP	ID0021300	SF CDR	Shoshone	5,400	5,400	2.0	1.9	0.6 to 3.7	1.1 to 6.9	4,000 to 6,700
Smelterville WWTP	ID0020117	SF CDR	Shoshone	225	140	0.04	2.5	0.13 to 5.7	0.0075 to 0.26	60 to 260
St. Joe River										
Clarkia WWTP[4,5]	ID0025971	St. Maries River	Shoshone	20	34	0.01	2	–	0 to 0.04	16 to 56
Santa-Fernwood WWTP[4]	ID0022845	St. Maries River	Benewah	60	190	0.07	2	–	0 to 0.2	90 to 300
St. Maries WWTP	ID0022799	St. Joe River	Benewah	1,400	2,200	1.3	1.3	0 to 3.9	0.2 to 3.9	1,700 to 2,900
Emida WWTP	ID0028487	Santa Ck.	Benewah	no data	no data	no data	no data	no data	no data	no data
TOTAL (kg/yr)	–	–	–	*7,800*	*8,300*	–	–	–	–	–
TOTAL (tons/yr)	–	–	–	*8.6*	*9.0*	–	–	–	–	–

[1]From Woods and Beckwith (1997).
[2]From average over all years within the period of record to 2 significant figures, unless otherwise noted.
[3]Plummer data are for 2014–2017, after plant upgrades and process testing were completed.
[4]Assumed total phosphorus concentration, based on average from other sites to 1 significant figure.
[5]Average monthly data not available. Estimate based on monthly maximum flow.
SOURCE: Table 12 from IDEQ and CDA Tribe (2020).

An important nonpoint source of phosphorus to CDA Lake is soil erosion that mobilizes particulate phosphorus during storm events. This phosphorus export during erosion is probably anthropogenically enhanced in some places in the watershed (e.g., due to forestry activities and unpaved roads), but the extent to which erosion is enhanced above natural background levels is not currently known. Although some of this phosphorus could be derived from legacy mine tailings that have accumulated in the lower basin, the analysis presented later in this chapter shows that mine-affected watersheds deliver a similar amount of total phosphorus to the Lake as the watershed that is largely free of historic mining activity (total phosphorus loadings for the CDA and St. Joe Rivers are 67 and 66 MT/yr, respectively). Thus, forests, which are the dominant land cover throughout the CDA basin, and hence forestry practices, are probably the largest determinant of phosphorus loss from the watersheds. Agriculture received scant mention in the CDA Lake Phosphorus Inventory report because it is a small percentage of the land use/land cover (0.55 percent from Figures 1-19 and 1-20). Although it is possible that agriculture (other than forestry) plays a small role in the overall export of phosphorus to CDA Lake, there are insufficient data to evaluate its role. The uncertainty about the role of agriculture as a source of nutrients should be addressed in future reports.

The population in the CDA basin that is not serviced by municipal WWTPs is either on septic systems or connected to small, non-municipal community wastewater systems, the latter of which often discharge to a lagoon, drain field, soil adsorption system, or some other type of land application system. The Phosphorus Inventory report catalogued 17 such small community systems in the vicinity (less than 500 ft) of the shores of CDA Lake and concluded that they discharged less than 0.1 tons/yr of TP to CDA Lake. These small community wastewater systems can be found on the CDA Story Map.[7]

Septic systems are generally considered a "low-tech" approach to wastewater treatment. However, the high affinity of phosphate for iron, aluminum, and calcium minerals in soils means that phosphorus export from these systems to surface waters can be lower than for conventional primary/secondary WWTPs, which typically only remove $\approx$ 50 percent of the phosphorus during treatment and usually discharge directly to rivers or lakes. The contribution of on-site septic systems in the immediate vicinity of CDA Lake to overall phosphorus loading is difficult to estimate with available information. The Phosphorus Inventory report concluded that in 2008 there were 3,969 septic systems less than 1 mile from the Lake in Kootenai County, accounting for 1.3–1.5 tons P/yr. This calculation assumed that a three-bedroom home (with three to four people) generates wastewater with a total phosphorus content of 9 mg/L and flow of 250 gal/day (946 L/day), which equates to a load of about 1.04 kg P/person-year. One review indicates that humans discharge an average of 0.61 kg P/person-year associated with their feces and urine (Rose et al., 2015). The Phosphorus Inventory report assumed 90 percent soil retention, which is supported by Robertson et al. (2019). The report acknowledged that there would be population growth in the lakeshore areas beyond 2008; indeed Table 1-4 in Chapter 1 lists 35,599 people in 2021 in the nearshore watersheds. Hence, there is considerable uncertainty about the phosphorus loading from septic sources calculated in the Phosphorus Inventory Report. Furthermore, IDEQ does not keep records on the proportion of the population within the CDA basin connected to WWTPs versus septic systems, although the CDA Story Map shows the region's structures including nearshore properties, population by census blocks, and the boundaries of city limits and small community wastewater systems. A significant proportion of residents in some cities (e.g., St. Maries, South Fork Sewer District) live outside city limits but are served by sewer systems.

Direct atmospheric inputs of phosphorus to CDA Lake are estimated to be 6 MT/yr. This estimate is based only on loading to the surface of CDA Lake, because atmospheric loading to the watershed is accounted for in the tributary estimates. Rather than being based on direct measurements at CDA Lake, these estimates are based on studies done at Flathead Lake by Ellis et al. (2015) and at Fernan Lake by LaCroix (2015), the latter of which is within the watershed and close to CDA Lake, and a global compilation of atmospheric phosphorus deposition estimates (Tipping et al., 2014). The phosphorus deposition estimate used for CDA Lake (i.e., 30 ± 34 kg/km^2·yr) is similar to the average estimate for all of North America (42 ± 39 kg/km^2·yr) from Tipping et al. (2014).

---

[7] https://storymaps.arcgis.com/stories/41e74951a3224d6aad6523b10a9bff09

Superfund and Lake Management Plan Activities on the Landscape That Affect Phosphorus Loading

Over the past three decades there have been a variety of Superfund activities as well as activities done in response to the LMP (outside the purview of Superfund) that may have altered phosphorus loading from the watershed. For example, revegetation of areas denuded by mining waste is often accomplished by hydroseeding to accelerate the establishment of a vegetated land cover. The mulch used for hydroseeding likely has a high concentration of soluble phosphorus, and additional phosphorus in the form of fertilizers may be applied. EPA has estimated that Superfund-related revegetation efforts were associated with total phosphorus inputs of 14 MT in the lower basin (1997–2021), 35 MT in the Box (1997–2005), and 10.8 MT in the upper basin (1997–2021) (personal communication, Ed Moreen, EPA, late 2021). Expressed as a total mass, this is about 60 MT over the 25-year period (1997–2021). It is unlikely, however, that all of this phosphorus was mobilized and transported to the river because the fertilization process is designed to retain the phosphorus on the soil so it can stimulate plant growth. Fertilization for forestry purposes is also a possibility, but the extent and intensity of such activities is not currently known. The analysis of the total phosphorus flux for the CDA River near Harrison (shown in Figure 3-23, later in this chapter) indicates a total of about 1,880 MT for these years (assuming that the 2021 value is the same as the 2020 value). Even if all of the Superfund-applied phosphorus were delivered to the Lake (an extreme assumption) it would constitute only about 3 percent of the total flux from the CDA watershed to the Lake (60 MT/1,880 MT).

There are also numerous restoration activities that occur as part of the LMP and/or are facilitated by the Basin Environmental Improvement Project Commission that might affect phosphorus loading to CDA Lake, but there is no quantification of the phosphorus load reduction from these efforts. Examples include (1) purchase of more than 2,200 acres (890 hectares) by the CDA Tribe for fish and wildlife habitat restoration and enhancement of streams for in-stream habitat; (2) advertisement of low-interest loans for purchase of direct-seed (no till) equipment for farmers; (3) shoreline protection on the St. Joe River (~8.5 miles or 13.7 km) and the CDA River (~11 miles or 18 km) via National Resources Conservation Service and Soil and Water Conservation district programs; (4) stabilization of eroding riverbanks and creeks; and (5) construction of new WWTPs in the city of Plummer, in Heyburn State Park, and for newer developments in and near the city of Coeur d'Alene, as well as upgrades to existing WWTPs (Brunner, 2017). The Restoration Partnership[8] also lists efforts undertaken by a variety of partners (including EPA) to return natural resources to a healthy condition in the CDA basin.

Phosphorus loading data from Zinsser (2020) and presented later in this chapter show extended periods of increased phosphorus fluxes from the CDA River followed by a period of decreased phosphorus fluxes during the 2010s. In addition, the committee's analysis of trends in total phosphorus in rivers not significantly influenced by mining or remediation also show decreases in recent years, suggesting other drivers of total phosphorus change such as land use, wildfire, or climate. It is possible that some remediation activities could lead to total phosphorus release due to pH changes or that total phosphorus could be affected by broader changes such as revegetation associated with the much-improved air quality in the mined area.

Given the LMP's emphasis on reducing phosphorus loadings, a focused effort needs to be carried out on an ongoing basis to record all activities in the CDA Lake watershed that are expected to influence phosphorus loads (positively or negatively). Some type of model (which could be very simple) should be applied so that estimates of the cumulative phosphorus management activities can be evaluated over time and then compared with observed changes in phosphorus loading (as measured at river monitoring stations). Discrepancies between these estimated and observed changes could help to build better models for predicting how management actions (or climate change) may influence phosphorus loadings in the future. Such predictive models are central to an adaptive management for controlling phosphorus loadings going forward.

---

[8] restorationpartnership.org

ANALYSIS OF INPUTS TO COEUR D'ALENE LAKE

The following quantitative analyses investigate inputs of water, lead, zinc, cadmium, sediment, and phosphorus into CDA Lake and throughout the CDA watershed over the past 30 years. Using river data collected by the USGS, the analyses paint a general picture of improving water quality conditions in the rivers and lowering of contaminant loads to CDA Lake.

Water

The water inputs to CDA Lake are dominated by the two major tributaries: the St. Joe River (drainage area 4,788 km^2) and the CDA River (drainage area 3,797 km^2). (Minor inputs to the Lake include direct precipitation on the Lake, which is about 1 percent of the river inputs, and groundwater inputs, which have not been estimated.) The total area of all tributaries to CDA Lake is 9,474 km^2 (Wise, 2021). This means that 51 percent of the total drainage area of the Lake is in the St. Joe River watershed and 40 percent is in the CDA watershed. The remaining 9 percent is in other small, unmonitored watersheds surrounding the Lake.

For purposes of this analysis of inflows to the Lake, long-term discharge records from the two major rivers were used to create a single time series of estimated inflows to the entire Lake. The two records used are the CDA River at Cataldo (USGS gage 12413500) and St. Joe River at Calder (USGS gage 12414500). The period of record used in this section is water years 1987–2020 (October 1, 1986–September 30, 2020). For the analysis described here, the discharge at each of these two streamgages is multiplied by the ratio of the total drainage area of that tributary to the drainage area at the streamgage location (for the St. Joe River at Calder that ratio is 1.8034, for the CDA River at Cataldo it is 1.2135), and the two tributary values are summed. Then, to extrapolate to the ungaged portions of the watershed, these values are multiplied by the ratio of the total watershed area of the Lake to the total watershed areas of these two tributaries (i.e., 1.1036).

Based on the discharge record for water years 1987–2020, the mean daily discharge into the Lake is 209 m^3/s, and the total range of daily values is 15–3,842 m^3/s. There is a strong seasonality to the record, with the highest discharges (based on median monthly values) in May and the lowest in September. The distribution of daily discharge values over the entire period is shown in Figure 3-5. This figure shows that the median discharge for May is 518 m^3/s and the median discharge for September is 35 m^3/s. This pattern comes about as a result of the timing of precipitation (higher values in the winter and spring), the timing of snowmelt, and the influence of high rates of evapotranspiration in the summer and early fall. The cause of the highest discharge events are typically rain-on-snow events. With the warming conditions now taking place in the region (see Chapters 1 and 10), it is

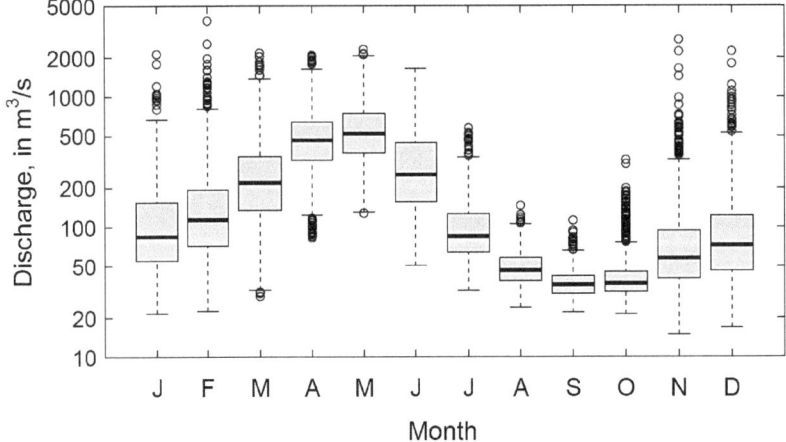

FIGURE 3-5 Boxplot of daily discharge values for the estimated total inflow (all tributaries) to CDA Lake, water years 1987–2020. The boxplots are based on logarithms of the discharge values. SOURCE: Data courtesy of USGS and plotted by the committee.

reasonable to expect that the potential for very high discharges from rain-on-snow events will change and that the peak annual discharges will come earlier in the year.

The dataset can be explored for trends over time. For the annual mean discharge, using a Mann-Kendall test for trend (Mann, 1945), the slope computed is +0.17%/yr but the p-value[9] for the trend is 0.81 (meaning that evidence for the trend is very weak). The slope was computed using a Thiel-Sen slope estimate (Helsel et al., 2020) based on the logarithms of the discharge values. The annual mean values are shown in Figure 3-6A. Trends in low-flow, as characterized by the annual seven-day minimum discharge, were also not significant. They are shown in Figure 3-6B, which indicates a trend slope of –0.21 percent/yr with a p-value of 0.49. Finally, the records were examined for trends in the annual maximum daily discharge. This is shown in Figure 3-6C, which shows a trend slope of +0.13 percent/yr with a p-value of 0.93. There is no significant trend to these measures of discharge over the period 1986–2020. There are slight, but insignificant, indications of increases in average and high flows and decreases in low flows.

One other type of change that is commonly considered in discussions of streamflow trends, particularly in watersheds where there is a significant amount of seasonal snow pack, is a measure of changes in the timing of annual runoff. Here, each water year was evaluated, and the date on which half of the total water-year volume had entered the Lake was determined; these dates range from mid-March to late-May. The relationship between this half-volume date and year is shown in Figure 3-6D. Overall, there is no significant trend in the half-volume date (Mann-Kendall p-value is 0.38), and this nonsignificant trend has a positive slope (the date tends to become later each year). Climate change projections would suggest that this trend may become negative at some point in the future (a tendency toward earlier runoff), but at the present time this is not what the observations are indicating. There does appear to be a shift in the direction of earlier runoff starting in about 2008, but the signal is not conclusive.

Overall, none of the major features of the daily discharge pattern of the river inputs to the Lake has changed over the period 1987–2020 in a manner that can be considered statistically significant. It is plausible that climate change and land use change will result in changes in the future, but at this point such changes are not apparent.

Metals and Sediment

To assess the input of metals from the CDA watershed to the Lake, it is necessary to have a model that estimates contaminant transport at key locations in the river/lake system. The model could be purely statistical or some combination of mechanistic and statistical. What is crucial is that the model be able to take the existing set of concentration and discharge measurements made over the period of record and produce unbiased estimates of contaminant concentration and flux for all days in the period of record. Only by having an estimate of conditions on all days does it become possible to do mass balance calculations, define the spatial and temporal trends, understand the past drivers of the system, and build hypotheses about future concentrations and fluxes in the system.

The committee's analysis uses the statistical method/model known as Weighted Regressions on Time, Discharge, and Season (WRTDS; Hirsch et al., 2010) to make inferences about the history of concentration and flux, on a daily time step, based on the types of records that are typically available in the rivers of the CDA Lake watershed (typically on the order of 250 observations for each of the key contaminants at a given monitoring location over nearly three decades). WRTDS has been used extensively for many river systems in the United States, including in this watershed (see Zinsser, 2020). More sophisticated methods could have been used if the dataset (sampling frequency) had been larger.

The committee's analysis considered four monitoring locations, highlighted on the map in Figure 3-7. Figure 3-7 shows six sites, but two of them (the North Fork of the CDA River at Enaville and the St. Joe River at Ramsdell) have metal fluxes well below 1 percent of the values for the CDA River near Harrison. Hence, these two sites are only considered later in the chapter, in the results on phosphorus. The South Fork of the CDA River at Elizabeth Park lies just upstream of the Box, and the South Fork of the CDA River near Pinehurst lies just downstream of the Box. The CDA River near Harrison is the point at which the CDA River enters CDA Lake and it integrates the inflows from the South and North Forks of the CDA River. The flow at this location is often

---

[9] The p-values reported for this trend test and the subsequent discharge trends discussed are all adjusted for the influence of long-term persistence of discharge at an annual time scale using the zyp package in R. The specific method for adjustment is the Zhang method of pre-whitening (Zhang et al., 2000).

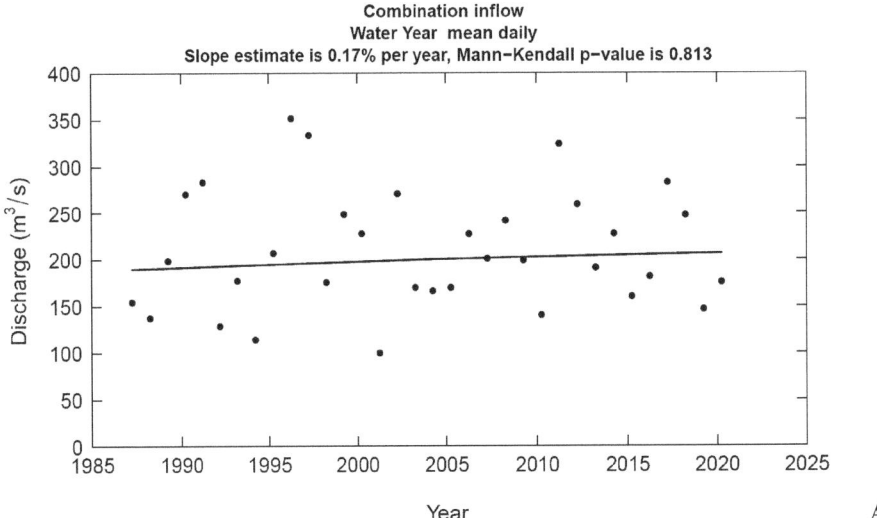

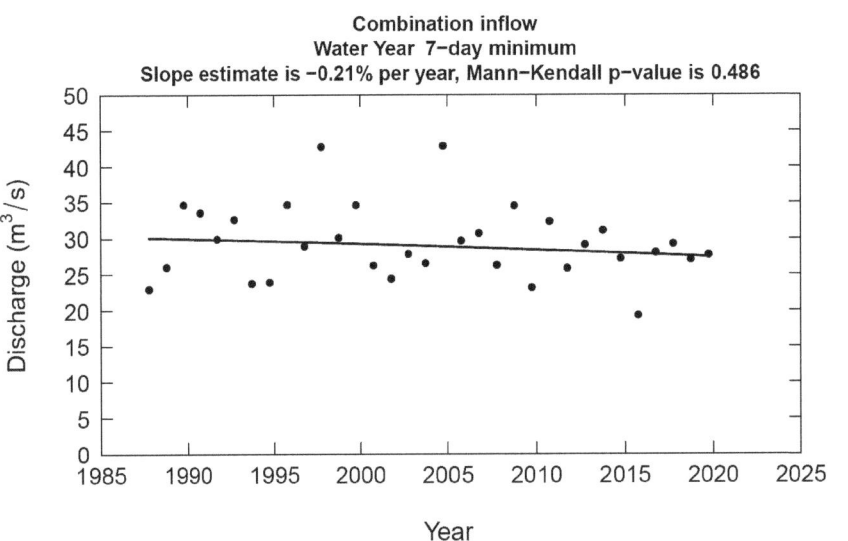

FIGURE 3-6 (A) Plot of **mean** of daily discharge values by water year for the total inflow to CDA Lake along with a loess smooth of these values. (B) Plot of the average seven-day **low flow** values by water year for the total inflow to CDA Lake along with a loess smooth of these values. (C) Plot of the annual **maximum** daily discharge by water year for the total inflow to CDA Lake along with a loess smooth of these values. (D) Graph of day of year when half the water-year volume has entered CDA Lake. For all four statistics evaluated, the trend slopes are not significantly different from zero. SOURCE: Data courtesy of USGS and plotted by the committee.

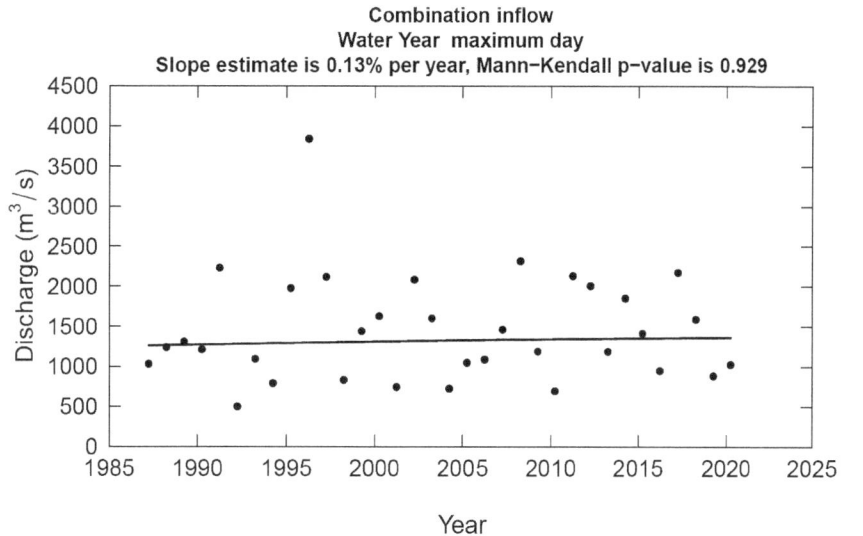

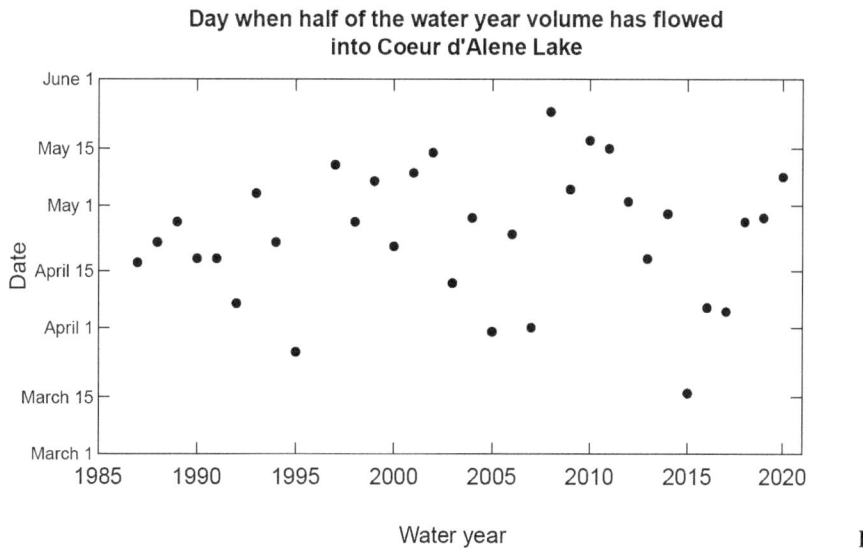

FIGURE 3-6 Continued

influenced by backwater from CDA Lake, which is regulated by the Post Falls Dam. The Spokane River near Post Falls is the regulated outflow from CDA Lake. As the dam only permits surface withdrawals, the material that is discharged would be expected to reflect the conditions in the epilimnion.

River water quality data appear highly chaotic. For example, the record of total lead values measured for the South Fork of the CDA River near Pinehurst is shown in Figure 3-8A (note the logarithmic scale with concentrations ranging over three orders of magnitude). Figure 3-8A suggests a downward concentration trend, but from the data alone, any sort of trend quantification would be highly uncertain. Another way to look at the same data is to consider how total lead varies with river discharge, shown in Figure 3-8B. This figure shows that total lead concentration is weakly related to discharge at low discharges (less than 10 m^3/s), but at higher discharges, total lead concentrations increase substantially with discharge. In addition to these relationships, concentration varies with season, such that for any given discharge, the concentrations tend to be highest in the winter and lowest in the summer. The key to making meaningful interpretations of the water quality data is to build a statistical model that captures these relationships so that the apparently chaotic behavior can be quantified and understood.

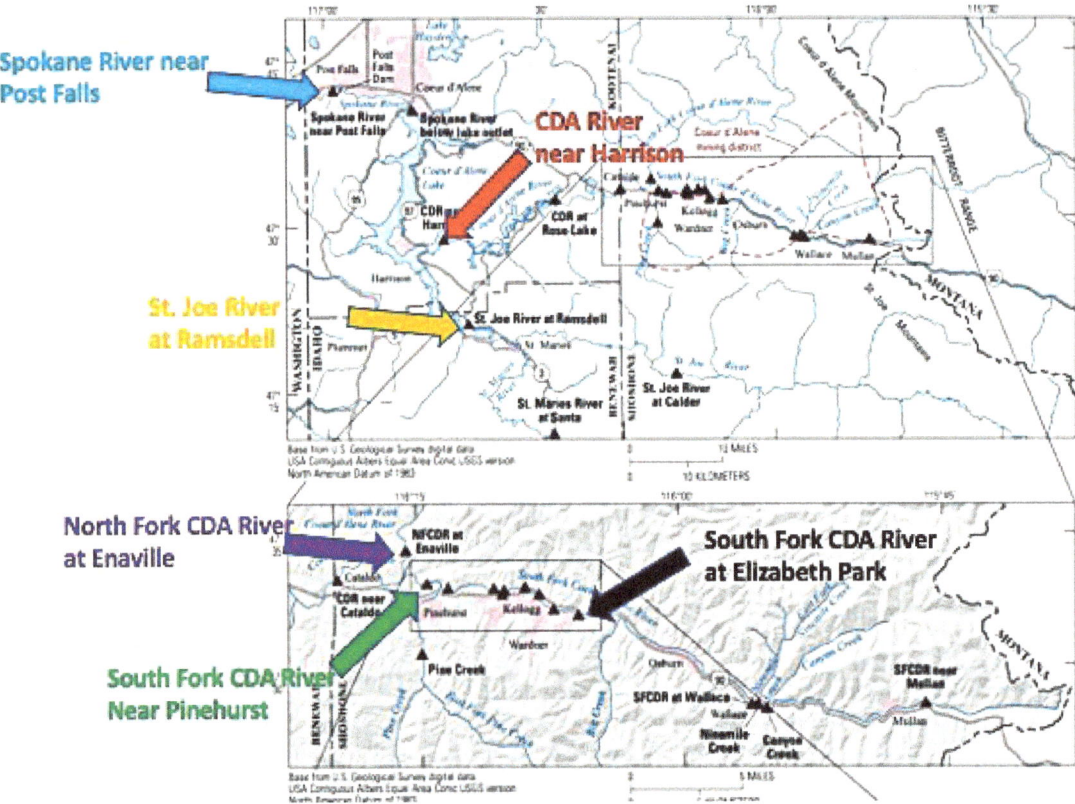

FIGURE 3-7 Map of the CDA River watershed showing the locations of six river gages used in the following analysis of metals and nutrients concentration and flux trends. SOURCE: Zinsser (2020).

The concept of the WRTDS model is to use statistical smoothing to partition the variations in concentration into components that are related to season of the year, watershed hydrologic condition (characterized by the daily mean discharge on the day of sample collection), long-term trend, and a random component (the unexplained portion of the variation). The WRTDS model fitted to these data is shown in Figure 3-8C, which is a contour plot showing the expected value of the total lead concentration for any combination of date and discharge throughout the period of record (regardless of whether that discharge was actually observed on that date). This WRTDS model can then be used to produce several types of outputs, including estimates of the concentration on any given date in the period of record. These daily values can be summarized by taking mean values of these estimates by month or by year. In addition, the model can be used to produce estimates of the expected number of days that concentration might have exceeded some threshold value in any given month or year.

Although the annual averages of concentration are somewhat informative about the long-term trends in lead in the river, these estimates are very strongly influenced by the particular pattern of discharge values that happened in each year. Years such as 2017 or 1996, which had very high discharge, tend to produce high values of mean concentration, and years of low discharge tend to produce lower mean concentrations. Estimates of trends ideally should not be strongly influenced by the specific pattern of wet and dry years. Rather, they should integrate across the seasonally specific frequency distribution of discharge values to produce estimates that are free of the impact of these interannual variations in discharge. The method for doing that is called flow-normalization (see Hirsch and DeCicco, 2015).

The flow-normalized annual mean concentrations are shown by the smooth curve shown in Figure 3-9A, along with the 90 percent confidence intervals around this estimate (based on a bootstrap calculation described by Hirsch et al., 2015). The confidence interval is very wide in the early years, because so much lead was moving in particulate form during high discharge events, and the relationship between discharge and concentration at these

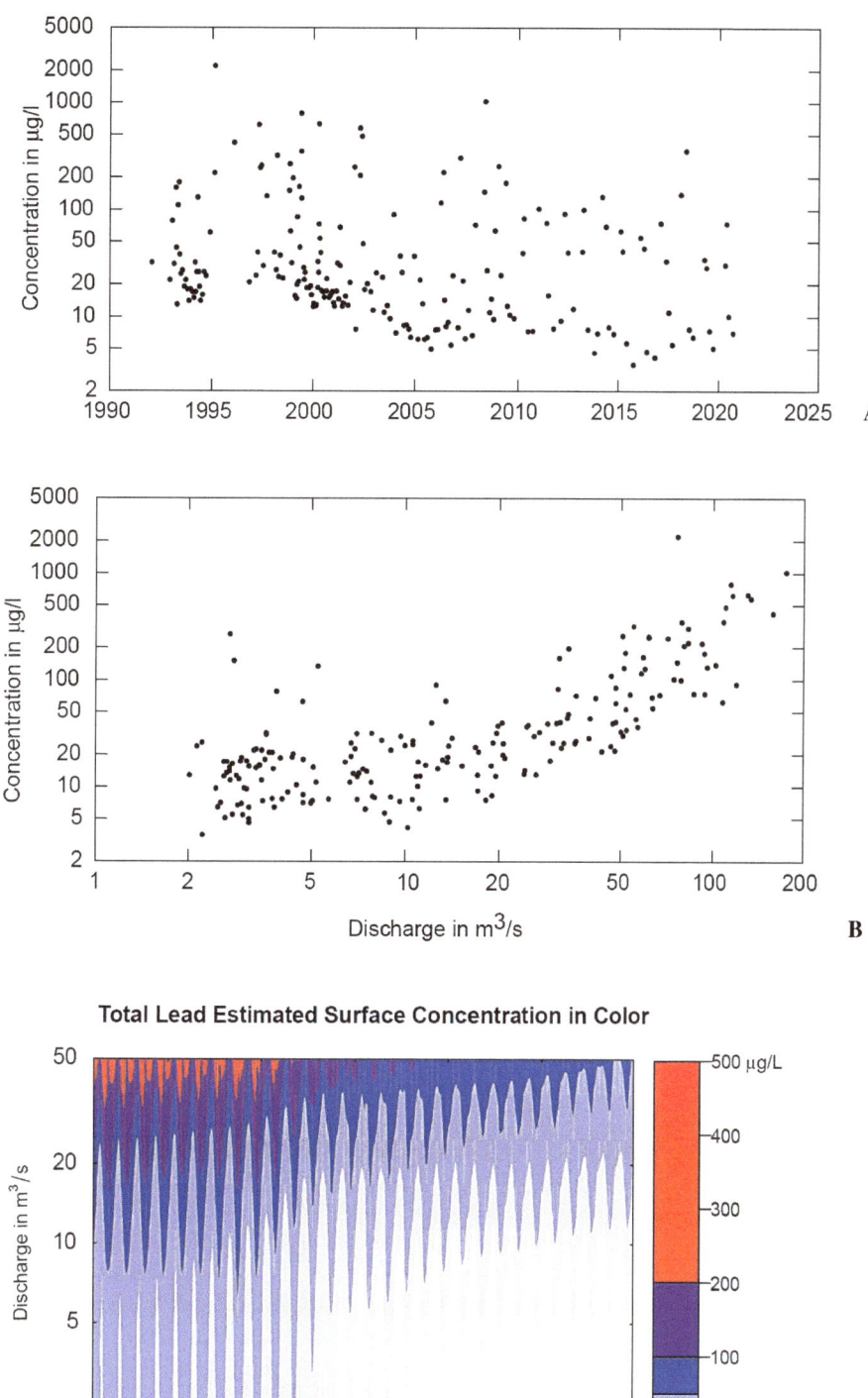

FIGURES 3-8 Concentration of total lead vs. time (A) and versus daily discharge (B) for the South Fork of the CDA River near Pinehurst, Idaho. (C) Contour plot based on WRTDS model of total lead data. The color indicates the expected value of concentration for any combination of date and discharge: white is below 10 µg/L, light gray is 10–20 µg/L, blue gray is 20–50 µg/L, blue is 50–100 µg/L, purple is 100–200 µg/L, and red is 200–500 µg/L. SOURCE: Data courtesy of USGS and analyzed and plotted by the committee.

higher discharges is subject to a high degree of variability. In later years, these high discharge events were much less important contributors to lead in the river; thus, the uncertainty becomes much smaller. The concentration trend at Pinehurst over the period of record, 1992–2020, is a decline in the mean of 54 µg/L (from 71 to 17 µg/L), or a 77 percent decrease.

In thinking about metrics of progress, it makes sense to consider not only the change in mean concentration but also the change in flux (often called "load"). The WRTDS model can also produce estimates of flow-normalized flux in a manner similar to the flow-normalized concentration; the random variable being considered on each day is the flux (the product of concentration and discharge) rather than just the concentration itself. The results for the South Fork of the CDA River at Pinehurst are shown in Figure 3-9B. The metric of change over the full record is a decline of 77 MT/yr (from 98 to 21 MT/yr), or a 79 percent decrease. Trend results for flow-normalized flux are highly relevant to this study because they are building blocks of mass-balance analyses for segments of the river system and the Lake, although the concentration results are useful when thinking about concentrations experienced by the biota (including humans) that interact with the river water on its way to the Lake.

In the sections below, trends in lead, sediment, cadmium, and zinc concentration and flux are assessed using the WRTDS approach for the four monitoring locations shown in Figure 3-7. The text and figures are designed to provide a spatiotemporal analysis of the trends in each of these metals and some interpretation of the results. The section on total lead, which includes Figures 3-8 and 3-9, is the longest because trends on total lead are evaluated to better understand the role of the Superfund remedy in reducing total lead inputs to CDA Lake. See Appendix A for an explanation of the methods.

Throughout the descriptions of trend results for concentrations and flux, "likelihood" terminology is used to describe the degree of statistical uncertainty of the results. The concept is described in Hirsch et al. (2015) and is based on a new perspective of reporting to replace the traditional "p-value" approach for evaluating water quality trends (see McBride et al., 2014; McBride, 2019). The change between the starting year and ending year of the period being analyzed are evaluated using a block bootstrap procedure (see Hirsch et al., 2015), and the result is expressed as the likelihood that the true trend has the same sign as the estimated trend. For example, in Table 3-3, the dark blue shading indicates that the likelihood that the change from the start of the trend period to the end is indeed negative, is greater than 95 percent, and this is called a "highly likely downward trend." The full set of likelihood categories used is shown below:

Likelihood category name	Likelihood results from the bootstrap analysis
Highly likely downward trend	>95% likelihood trend is downward
Likely downward trend	70% to 95% likelihood that trend is downward
Highly uncertain	30% to 70% likelihood that trend is downward
Likely upward trend	70% to 95% likelihood that trend is upward
Highly likely upward trend	>95% likelihood trend is upward

Total Lead

Using the same approach described above for the South Fork of the CDA River near Pinehurst, Idaho, additional evaluations of total lead trends were conducted at the South Fork of the CDA River at Elizabeth Park, Idaho; the CDA River near Harrison, Idaho; and the Spokane River below the Lake outlet. The major results for all four of these sites are shown in Table 3-3 and Figure 3-10.

Throughout the record, the flow-normalized flux at the downstream end of the South Fork (Pinehurst) is always greater than at the upstream site (Elizabeth Park). Furthermore, although fluxes have declined at Elizabeth Park, the rate of decrease, either in mass terms or percentage terms, is less steep than at Pinehurst. This is what one would expect given the high intensity of the remediation effort within the Box. The results show that, if there is to be more decrease in total lead flux from the South Fork CDA watershed, it will need to come almost entirely as a result of future mitigation measures above Elizabeth Park versus any additional remediation done in the Box.

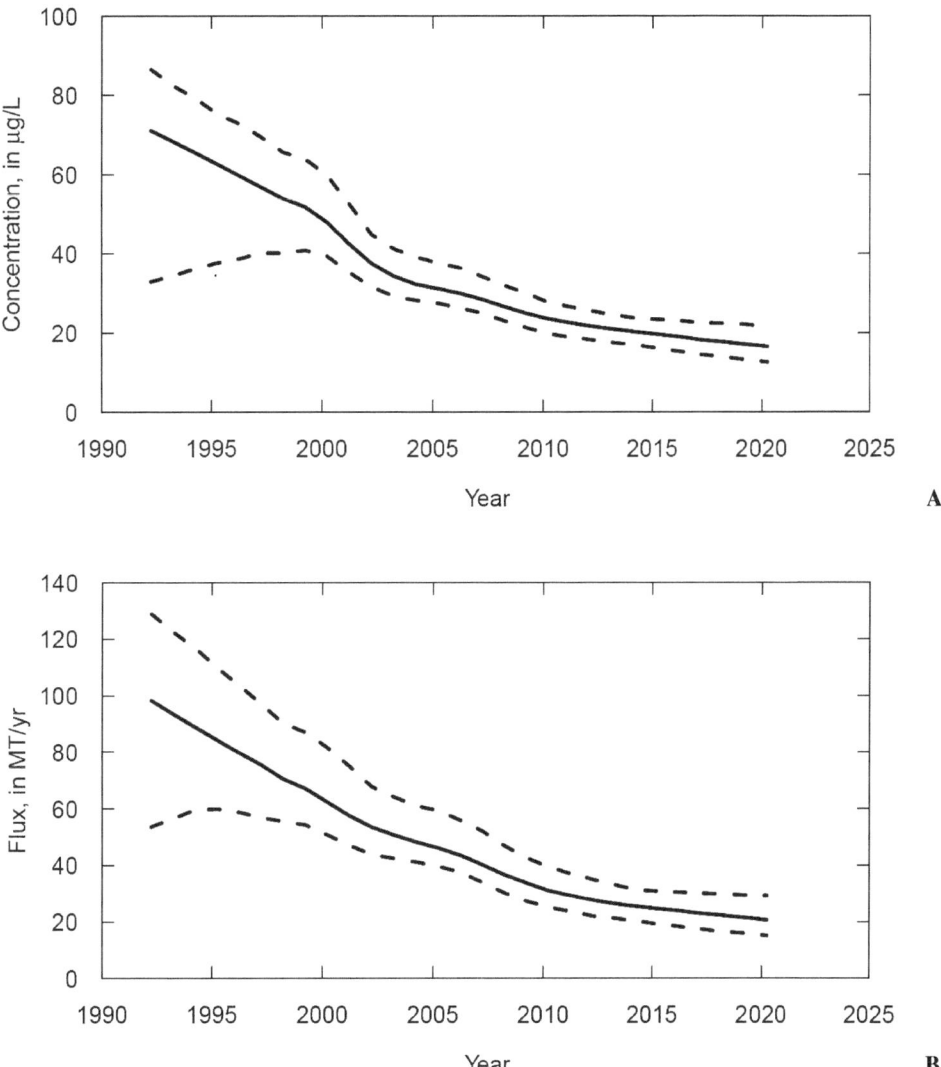

FIGURE 3-9 Graph of annual flow-normalized concentration (A) and flux (B) values for total lead, South Fork of the CDA River near Pinehurst, Idaho. Solid line is based on these annual values (expressed in MT/yr and µg/L). Dashed lines represent the 90 percent confidence interval around the flow-normalized flux and concentration estimates. SOURCE: Data courtesy of USGS and analyzed and plotted by the committee.

Another observation (data not shown) is that the lead transport past Pinehurst is strongly punctuated with periods of high flow events, which transport large amounts of lead. That analysis shows that the net erosion of high-lead sediments in the Box had been strongly suppressed, but in the area above the Box, the tendency for high-flow events to move large amounts of high-lead sediments has continued. This is another indicator that, if future investments are going to be made to control lead transport from the South Fork of the CDA watershed, they would need to focus on erosion in the area upstream of Elizabeth Park.

The flux of total lead out of the lower CDA River (into CDA Lake) is only minimally related to the inputs coming from the South Fork. Throughout the 1990s, the ratio of flow-normalized flux of total lead out of the lower CDA River to the flux into the lower CDA River was about 7:1. This means that most of what was entering the Lake during those years was not derived from the South Fork (or the North Fork, which delivered about one tenth the amount of lead

TABLE 3-3 Trends in Total Lead: Flow-Normalized Annual Mean Concentration (upper table) and Flow-Normalized Annual Flux (lower table)

Site	First year	Percentage Change over period			First year mean	2020 mean
		Start–2020	2000–2020	2010–2020	µg/L	µg/L
SF CDA at Elizabeth Pk	1993	–72%	–55%	–33%	52	15
SF CDA nr Pinehurst	1992	–77%	–65%	–30%	71	17
CDA nr Harrison	1991	–34%	+28%	–32%	84	55
Below Lake Outlet	1991	–88%	–48%	–42%	8	1

Site	First year	Percentage Change over period			First year mean	2020 mean
		Start–2020	2000–2020	2010–2020	MT/year	MT/yr
SF CDA at Elizabeth Pk	1993	–62%	–47%	–29%	43	16
SF CDA nr Pinehurst	1992	–79%	–67%	–34%	98	21
CDA nr Harrison	1991	+27%	+66%	–20%	620	790
Below Lake Outlet	1991	–86%	–54%	–51%	54	7

NOTES: The period of record is different among the four sites, so the first water year is designated in the column labeled "First year." Trends are computed over three time periods, and for each time period the total change, expressed in percent, is displayed. The color code for the boxes is based on the uncertainty about the trend direction. Dark blue indicates a highly likely downward trend (likelihood > 95 percent), light blue indicates a likely downward trend (likelihood between 70 and 95 percent), no shading indicates that the trend direction is highly uncertain (likelihood of a downward trend is between 30 and 70 percent), pink shading indicates a likely upward trend (likelihood of an upward trend between 70 and 95 percent), and red shading (which does not appear in this table but will appear in other similar tables) indicates a highly likely upward trend (likelihood of an upward trend > 95 percent). The last two columns indicate the estimated value of flow-normalized concentration or flux for the first and the last year of the record. SF = South Fork.

as the South Fork) but rather was derived from lead stored in the lower CDA River basin as a legacy of a century of mining activities within the South Fork watershed. In 2000, the situation changed, such that the flow-normalized flux from the lower CDA River basin to the Lake rapidly increased, and this increase lasted most of the decade, even though inputs to the reach were continuing to decline. By approximately 2009, outputs from the lower basin had more than doubled compared to 2000, but inputs to the lower basin had declined by about 45 percent. Thus, the ratio of output to input for the lower CDA River had grown to approximately 27:1 by 2009. Total lead fluxes out of the lower CDA River began decreasing again around 2009 and have been decreasing to the present time.

This general pattern of a decline in total lead flux from the lower CDA basin up until about 2000, then a flux increase between 2000 and 2009, followed by another decline to the present, is also seen for the other constituents investigated in this chapter (see subsequent discussions). The Committee's hypothesis about these patterns is that the 2000–2009 period was a time when the lower river (including the floodplain, bed, and banks) was *readjusting* to major changes in inputs of lead and suspended sediment brought about by the Superfund remediation in the upper basin, such that the decreased input from upstream was approximately balanced by an increase in lead flux generated from within the lower CDA reach. This readjustment involved very significant erosion and downstream transport of legacy sediments, primarily derived from the bed of the lower CDA River, which are rich in lead[10] and are thick (in some cases tens of meters). After about 2010, this period of readjustment ended and declines in total lead flux once again followed a roughly exponential decline. The results suggest that further reduction of South Fork sources will have minor impacts on delivery of total lead to the Lake.

Studies conducted by CH2M used a different method for calculating annual lead flux values (CH2M, 2015) and did not compute a flow-normalized flux at these locations. However, that report also identified the large discrepancy between lead entering the lower CDA River from upstream and the amount discharged to the Lake.

---

[10] Bookstrom et al. (2013) defined lead-enriched as > 1,000 mg/kg dw. Lead concentrations in a 1-mile stretch of river bed sediments can vary from 10 to greater than 25,000 mg Pb/kg.

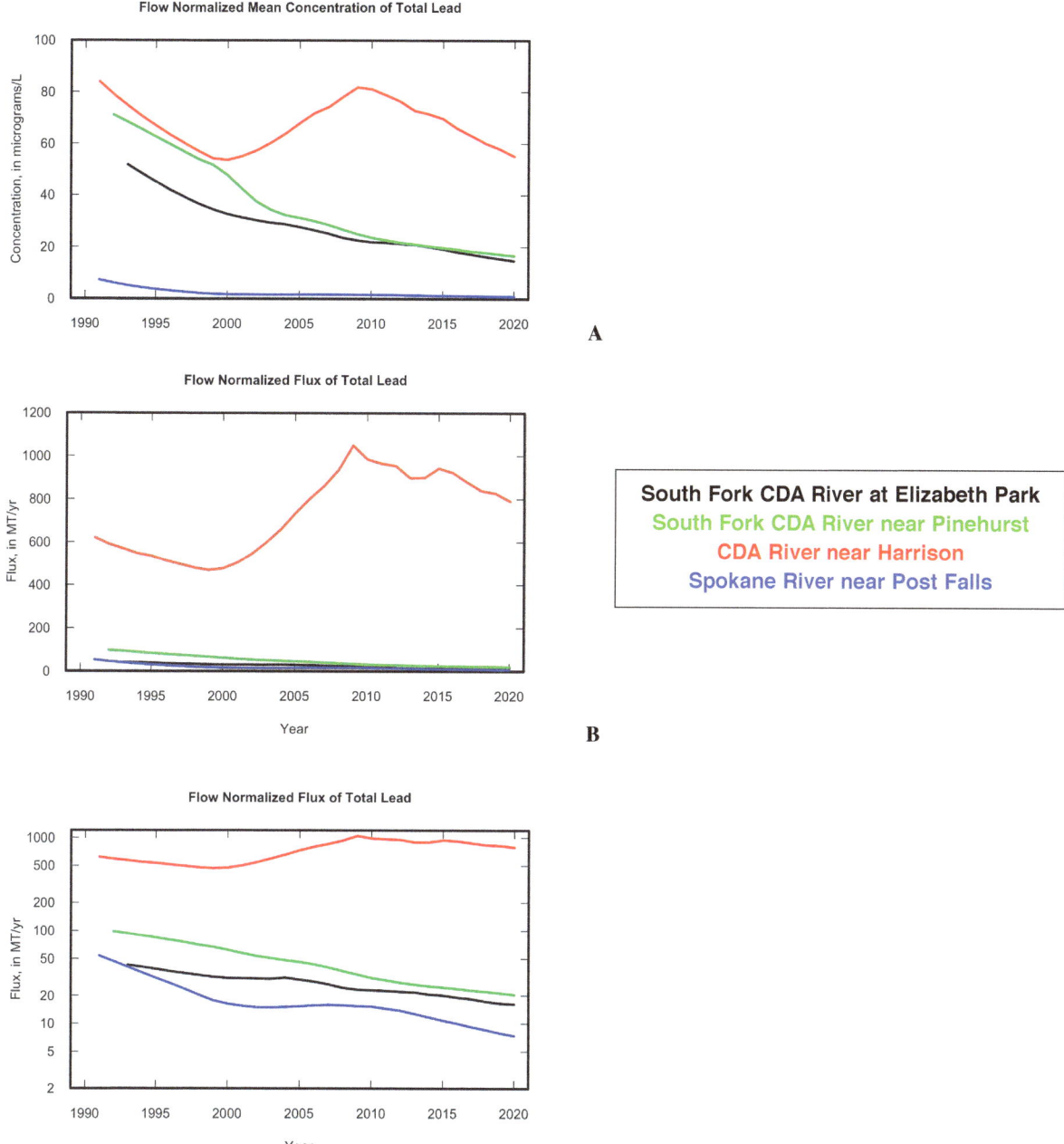

FIGURE 3-10 Flow-normalized annual mean concentration (A) and flux (B) of total lead in µg/L and MT/yr, respectively. (C) is the same as (B) but plotted on a logarithmic scale to show detail at the sites with smaller fluxes. SOURCE: Data courtesy of USGS and analyzed and plotted by the committee.

Considering 1988–2012, they estimated total lead flux from upstream of about 34 MT/yr. Through very detailed studies of processes taking place in the reach, they also estimated bank erosion of about 32 MT/yr and bed erosion of about 250 MT/yr. They also identified an important sink (about 68 MT/yr) in floodplain deposition. Their estimate of the net flux out of the watershed and into the Lake was about 250 MT/yr. For the comparable time period, the WRTDS method estimates a flux to the Lake of 594 MT/yr. The large discrepancy between the two calculations is likely related to differences in approach. First, it does not appear that CH2M considered the statistical issue of retransformation bias associated with the use of regressions based on the logarithms of concentration (see Cohn et al., 1992; Helsel et al., 2020, pp. 256–257). The very high degree of natural variability of this dataset could account for underestimation of as much as 40 percent. Second, the CH2M calculations assumed that the rating curve (the relationship between concentration and discharge) was trend-free over the entire period. Third, their calculations are based on estimating suspended sediment fluxes and then multiplying these fluxes using an estimated mass of lead per unit mass of sediment.

Even though the committee's findings are quantitatively different from the CH2M findings, the most important result is consistent: the vast majority of lead transported to the Lake in recent decades is derived not from contemporary inputs of lead from the mined area, but from erosion of legacy lead-rich sediments in the lower CDA watershed. The Committee's analysis strives to look at the dynamics over time in this imbalance of inputs and outputs, whereas CH2M considers the entire period of 1988–2012 to be static such that year-to-year variations in inputs or outputs are only related to the year-to-year differences in streamflow conditions, rather than being related to evolving river system dynamics that arise in response to the complex history of sediment and lead inputs due to mining and remediation.

The results for the Spokane River below the CDA Lake outlet show very substantial declines in the total lead fluxes out of the Lake—a decrease of 86 percent over the period 1991 to 2020. They also show that Lake outputs in relation to Lake inputs have changed substantially, from outputs being about 7 percent of inputs in 1993 to outputs being less than 1 percent of inputs in 2020. This can be explained by the fact that over time the inputs to the Lake have become more and more dominated by the particulate phase and these particles deposit very readily as they move through the Lake (see later discussion of the dissolved lead).

Figure 3-11 shows the history of total lead flux through the watershed in combination with the history of major remedial activity. What is clear is that the major landscape modification activities in OU-2 correspond in time with the decreases in flux from the South Fork of the CDA River, but the large increase in flux from the lower CDA River follows those activities by five to ten years.

Figure 3-12 provides another way to look at the history of flow-normalized total lead fluxes by tracking the downstream pattern of lead transport in 1993 and 2020. What it clearly shows is that, even in 1993, the South Fork watershed provided an input that was small compared to the output from the lower CDA River and by 2020 the upstream input was an even smaller component of the output of the lower CDA River. It also shows how small the output of the Lake is in comparison to the inputs, indicating that something close to 99 percent of the total lead that enters the Lake is deposited in the bottom sediments of the Lake and never reaches the outlet.

The major take-away messages from Table 3-3 and Figures 3-11 and 3-12 is that the inputs of total lead from the South Fork of the CDA River have been greatly reduced over this nearly three-decade history. Further decreases in total lead from the South Fork basin will have limited consequences for the future of lead inputs to the Lake. Rather, understanding the changes in the lower basin from 2000 to 2010 is crucial to understanding how the delivery of lead from this reach will change going forward and how it will respond to remedial strategies. Box 3-4 provides another analysis, different from the WRTDS approach, of the total lead data that demonstrates the effectiveness of the Superfund remedy in reducing loads of lead to CDA Lake.

Sediment

The fluctuations of the flow-normalized flux of lead described above might be explained by differences in the parts of the legacy materials that have eroded over time. The stratigraphy and chemistry of the bed material of the CDA River is documented extensively in CH2M Hill (2017). The bulk lead concentrations of the sediments vary considerably in the vertical dimension. The oldest materials, which are considered to predate the mining era,

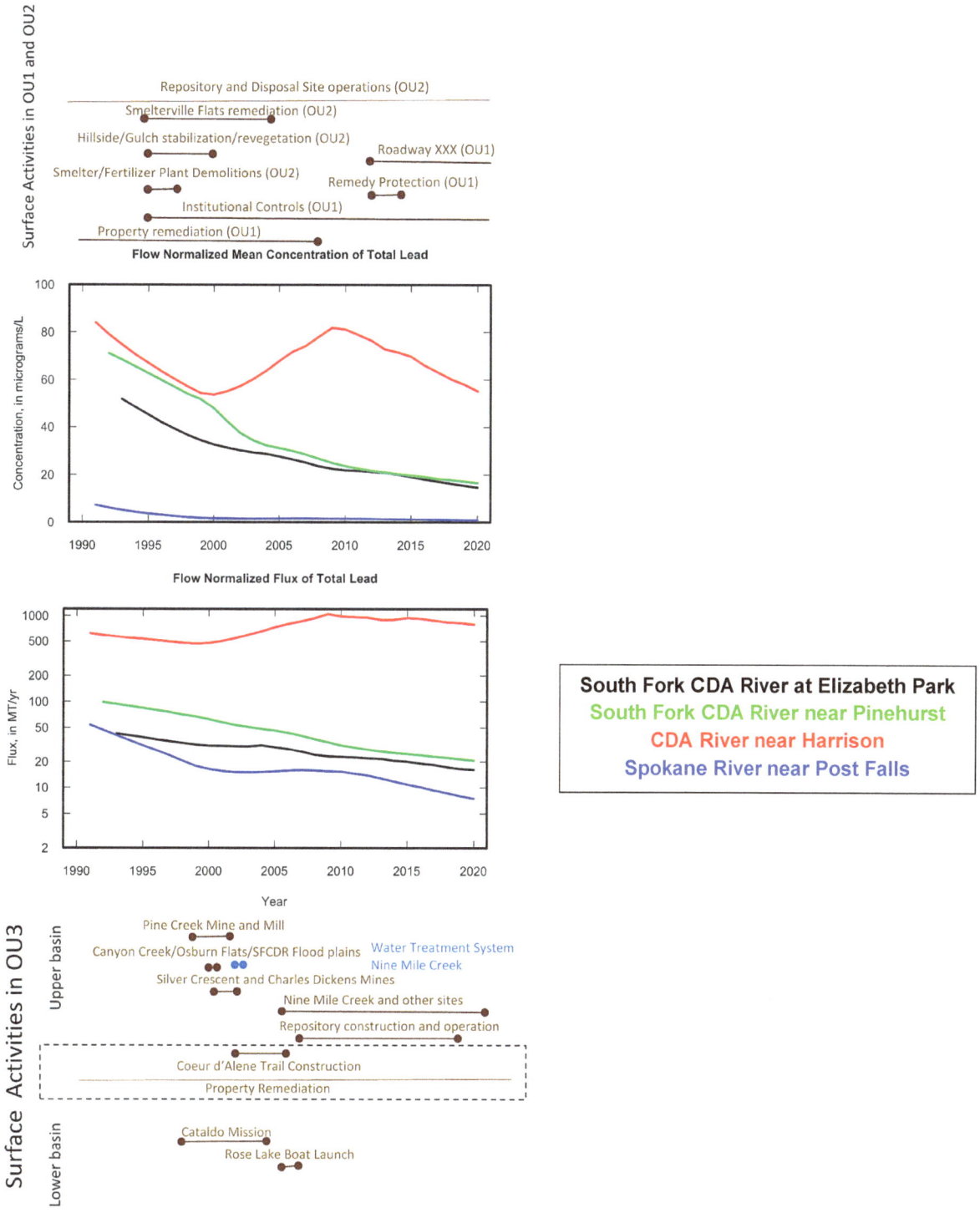

FIGURE 3-11 Total lead concentration and flux through the CDA basin overlaid with Superfund remedial activities in all three operable units. Middle panels correspond to Figure 3-10A and C. SOURCE: Data and information courtesy of USGS and Moreen (2021b) and analyzed and plotted by the committee.

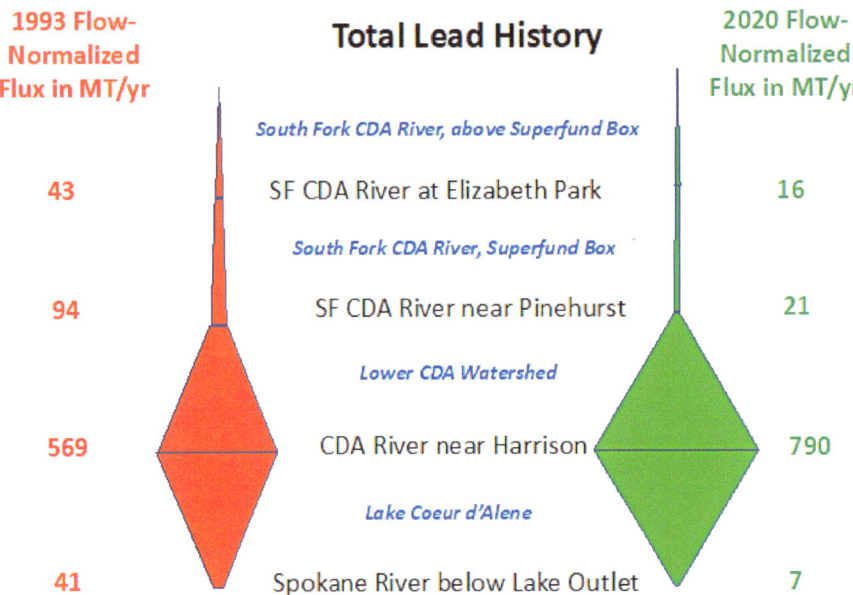

FIGURE 3-12 Schematic graph of downstream fluxes of total lead as of 1993 and 2020. The width of the colored zones of the figure is proportional to the flow-normalized flux at each of the four monitoring locations shown. The names of the four monitoring locations are shown in black, the reach of the river/lake system is identified in the blue italics lettering, and the estimates of flow normalized flux are derived from the WRTDS calculation.

have a lead content of less than 100 mg/kg sediment. The deposits from the early mining period, when milling processes were very crude and large amounts of lead remained in the waste materials discharged downstream, had lead concentrations of ~10,000–20,000 mg/kg, with an observed maximum of 70,000 mg/kg. Sediments from the mid-20th century when ore processing was more effective had lead concentrations of ~5,000–10,000 mg/kg. Finally, near the surface are the modern deposits, representing the periods after the end of mining activity through the Superfund remediation, which have lead concentrations of ~1,000–3,000 mg/kg. The committee analyzed data on the sediments that moved during high discharges (above 250 m^3/s or approximately the 95th percentile of the distribution of daily discharges) over the past two decades. Concentrations of total lead were in the range of 2,000–5,000 mg/kg, consistent with a mixture of the mid-20th century deposits and the modern deposits.

The two relevant data sets are the USGS data on suspended sediment concentration, total lead, and dissolved lead; and the Basin Environmental Monitoring Program, which made direct measurements of lead in several particle size fractions. One hypothesis for the complex patterns of flow-normalized lead flux seen at Harrison emerges from the examination of the suspended sediment records for the South Fork near Pinehurst (1989–2020), the North Fork at Enaville (1989–2020), and the CDA River near Harrison (1999–2020). Sediments were sampled only about six times per year in recent years, and a consistent sampling pattern for all three sites only covers the years 1999–2020. Figure 3-13 shows the flow-normalized flux estimates for sediment at these three sites from 1999 to 2020 and the 90 percent confidence intervals on these estimates.

Even with these low sampling frequencies, it is possible to say that the flow-normalized sediment flux for both the South Fork and the North Fork of the CDA River have declined steadily and substantially over the period 1990–2020. If one looks at the fluxes on a per unit area basis, for the South Fork the decline is from about 40 MT/km^2/yr to about 12 MT/km^2/yr. For the North Fork the decline is from about 29 MT/km^2/yr to about 13 MT/km^2/yr. It is reasonable to conclude that some of this substantial decline in sediment yields from the South Fork watershed is the progress of the Superfund remediation, since stabilization of mine tailings and other waste material was an important remedial objective. However, it is notable that sediment yields also decreased substantially in the North Fork watershed. This may be due to some remediation carried on outside

BOX 3-4
Evidence That Remediation Is Affecting CDA River Concentrations of Lead

To determine whether remediation is having an impact on lead concentrations in the CDA River, the committee investigated the load of total lead in the CDA River at Harrison at times of high discharge, when most of the lead is likely delivered. This analysis was a much simpler approach than the WRTDS approach, with fewer assumptions, but designed to answer a similar question: how is the behavior of the river changing with respect to the transport of total lead? If the relationship between daily discharge and daily total lead concentration on days of high discharge has not changed over time, then it would be a strong indication that the remediation has not had an effect. But, if for any given discharge value (in the range of high discharges) there are lower concentrations over time, then one can say that the remediation is having a beneficial effect.

As shown in Figure 3-4-1, the committee examined all of the data pairs of total lead concentration and discharge, measured in the CDA River near Harrison, in 1999–2020 where discharges are in the top 10 percent of daily discharges in the Harrison record (above 194 m^3/s). (As points of reference, the mean discharge at this site is about 74 m^3/s, the median is about 34 m^3/s, and the maximum daily value over the past 30 years is about 1,900 m^3/s.)

This dataset was restricted to very high flows because these are the flows that are crucial to total lead transport, and the model would get more complex because of the dramatic change in slope that would exist if all of the data were included. Figure 3-4-1 suggests that there is a relationship between discharge and concentration, and it is linear in the logarithms. (Considering the full data set, there is essentially no relationship between discharge and concentration for discharge values below 150 m^3/s.)

The committee built a regression model for total lead concentration versus discharge, ln(C) as a linear function of ln(Q), shown in Figure 3-4-2. The regression results show that total lead concentration is proportional to discharge to the 2.76 power.

If the relationship between discharge and concentration did not change over the period 1999–2020, then one could expect the residuals from the regression to be evenly scattered around zero over time. That is, a value that plots a certain distance above the line (see Figure 3-4-2) could just as easily happen in the early part of the record as in the middle or the late part. The residuals plotted as a function of year are shown in Figure 3-4-3. In Figure 3-4-3, the y axis could be expressed as residuals in percent. As a point of reference, a value of +1 on this graph translates to an observed value that is 170 percent above the regression line, +2 translates to an observed value that is about 640 percent above the regression line, and –1 translates to about 63 percent below the line.

The important feature of Figure 3-4-3 is that there are a lot of high positive residuals around the year 2010, but as one goes from 2010 to 2020, there seems to be an upper bound on the residuals that declines year after year, and in the last two or three years, all of the residuals are negative. If there were no change in the concentration versus discharge relationship over time, then the residuals would be spread symmetrically through the space with a density

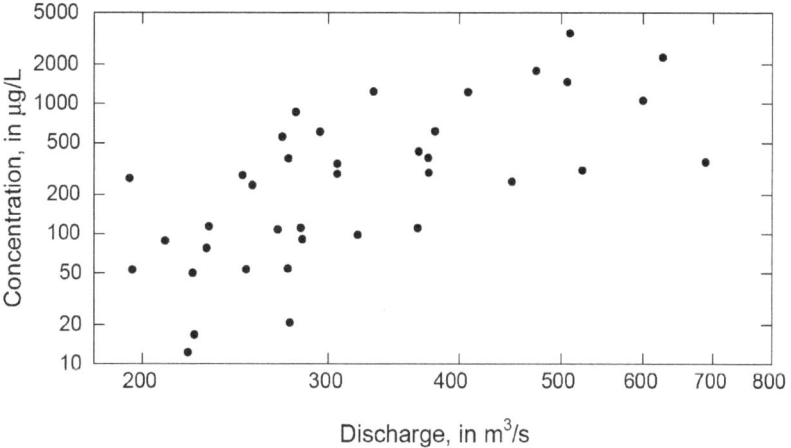

FIGURE 3-4-1 Concentration of total lead as a function of discharge at Harrison, 1999–2020. Note that both axes are on a log scale. SOURCE: Data courtesy of USGS and analyzed and plotted by the committee.

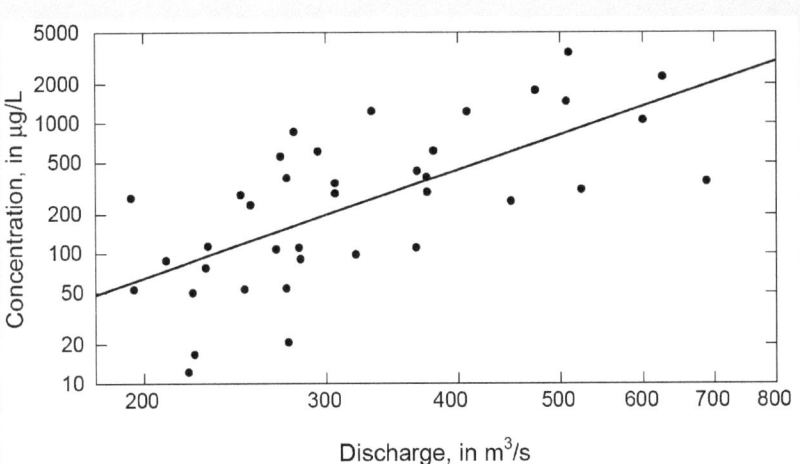

FIGURE 3-4-2 Concentration of total lead as a function of discharge, 1999–2020, regression model. Note that both axes are on a log scale. SOURCE: Data courtesy of USGS and analyzed and plotted by the committee.

that gets lower as one moves away from the zero line, but there should be no relationship between residual values and time. As one moves from 2005 to 2010, the observations tend to fall well above the regression line, and then from 2010 to 2020 the residuals become more and more negative. What this means from a transport perspective is that, for any given value of discharge, the amount of total lead transported at a given Q value (within this range of high Q values) was increasing over time until about 2010 and then it declined gradually over the following ten years.

As discussed previously, the committee hypothesizes that the 1999 to 2010 period was one in which there was a great deal of scour of legacy high-lead sediments in the lower basin, due to changes in sediment delivery from the upper basin that resulted from the remediation process. After 2010, this scour was tending to abate and total lead concentrations came down. Although the decreases observed between 2010 and 2020 could be the result of the increasing stability of the lower basin landscape as a result of the Superfund remedy, there may be other influences that are driving these changes perhaps related to improved forest practices and/or forest health throughout the entire CDA River watershed.

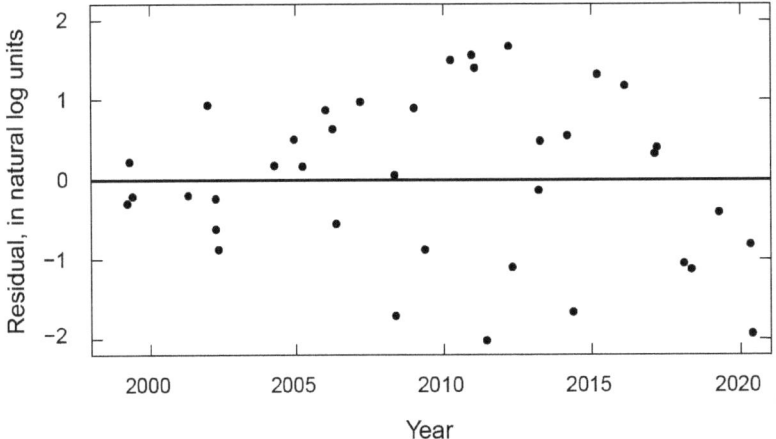

FIGURE 3-4-3 Residuals from the regression of total lead concentration as a function of discharge, Coeur d'Alene River near Harrison, 1999–2020, for discharges above the 90th percentile. SOURCE: Data courtesy of USGS and analyzed and plotted by the committee.

> A regression model can also be built that considers time to be an explanatory variable, and given the curvature seen in the residuals, the obvious first choice is to consider a quadratic trend over time. This is accomplished by adding two more explanatory variables: the first of them is the year minus 2010, and the second is the square of this explanatory variable. This formulation allows the model to include a quadratic trend in the log concentration to log discharge relationship. The inclusion of this quadratic relationship to time results in a modestly significant improvement in the fit of the model (p-value 0.119). This is a good indication that there is a non-monotonic trend in the concentration versus discharge relationship. This trend rises to about 2008, then falls to 2020. This says that the Superfund remedy, and the resulting increased stability of the lower CDA River channel, has been reducing the rate of total lead entering the Lake since about 2008. These conclusions correspond well to the evidence provided by the WRTDS model reported for total lead for the CDA River near Harrison, which is that that flux of total lead has been decreasing in recent years.
>
> A caveat to consider about these results is that the last three years of this record (2018–2020) had only five samples in this high discharge range (they were dry years). As soon as a few new samples have been collected at high discharges, this type of analysis should be run again to determine if this downward trend is continuing.

of the Superfund project, but it is also possible that an important driver of the decline for the North Fork and the South Fork sediment fluxes may be some other activity, such as improved forest management and harvest practices in these watersheds. Regardless of the causes of these trends in sediment delivery to the lower CDA River, from about 2005 to 2015 the flux of sediment out of the lower CDA River far exceeded the input from upstream, meaning a considerable part of the sediment transported out of the lower CDA River came from storage in the lower basin. The findings of CH2M Hill (2017) suggest that most of this sediment was derived from the lead-enriched sediments in the bed of the lower CDA River.

Figure 3-14 shows the flow-normalized mass balance for the lower CDA River. One possible explanation for the increase in total lead flux from the CDA River during the 2000s, followed by declines in the 2010s, is that the channel of the lower CDA River has been adjusting to the declining inputs from upstream (as hypothesized earlier). If the river was carrying a declining amount of sediment from upstream (brown decline in Figure 3-14), increased erosion of the lead-enriched channel bed in the lower basin could occur. Over time, the net effect of this erosion would be a decreasing slope through most of the lower basin's channel length and a gradual decline in the net scour rate (as seen in the final several years in Figure 3-14). There are two implications of this hypothesis that should be considered in projecting future lead transport to the Lake. The first is that as a new equilibrium channel configuration is reached, there may be substantial declines in the rate of scour of lead from the lower CDA River and thus a lower rate of delivery of lead to the Lake. The second is that managing the future inputs to the Lake demands more scientific attention to the changes underway in the lower CDA watershed. Such analysis must consider the transport capacity of the river to be dynamic, responding to the decades-long decreases in sediment delivered from the upstream watershed.

Without an understanding of the changes over the past three decades it will be difficult to project how legacy lead transport will change in the future. In the long run, lead transport will decline over time if the riverbed becomes depleted of lead-rich sediments. However, a crucial fact to consider is the magnitude of the legacy lead that underlies the river. The detailed mapping and stratigraphy study by CH2M Hill (2017) estimates the mass of lead in the lower river bed as between ~5 million and 11 million MT. The analysis of lead transport presented in Table 3-3 indicates that the rates of transport of lead from the CDA River to the Lake as of 2020 are about 790 MT/yr. Using the lower of the two estimates of lead in the bed material, this suggests that, at current rates, the lead would be eroded from the river bed in about 6,000 years. This estimate illustrates that the amount of lead in the lower basin is extremely large in comparison to the rates of delivery to the Lake. The establishment of a new equilibrium condition for the channel, dictated by the dramatic declines in sediment input from both the South Fork and the North Fork, will probably drive the delivery of this legacy lead in the coming decades. If the current rates of delivery of lead to the Lake are considered to be unacceptably high, then the questions that will need to be answered going forward are as follows: How rapidly will the channel of the lower CDA River readjust to the much lower sediment inputs? What will be the pattern of lead flux declines in response to this readjustment?

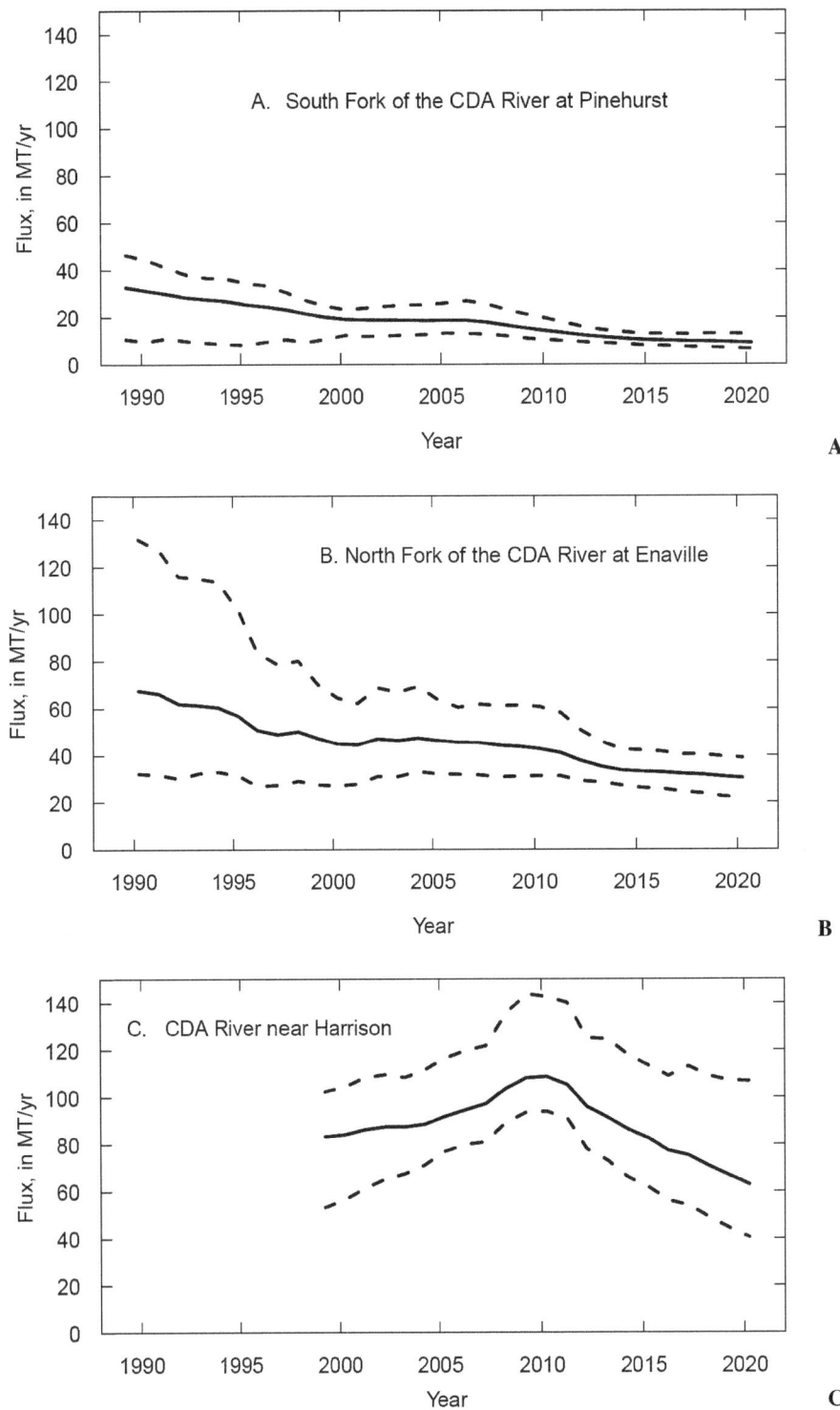

FIGURE 3-13 Flow-normalized flux of suspended sediment. (A) South Fork of the CDA River near Pinehurst (solid line) 1989–2020. (B) North Fork of the CDA River at Enaville (solid line) 1989–2020. (C) CDA River near Harrison (solid line) 1999–2020. Dashed lines are 90 percent confidence intervals. SOURCE: Data courtesy of USGS and analyzed and plotted by the committee.

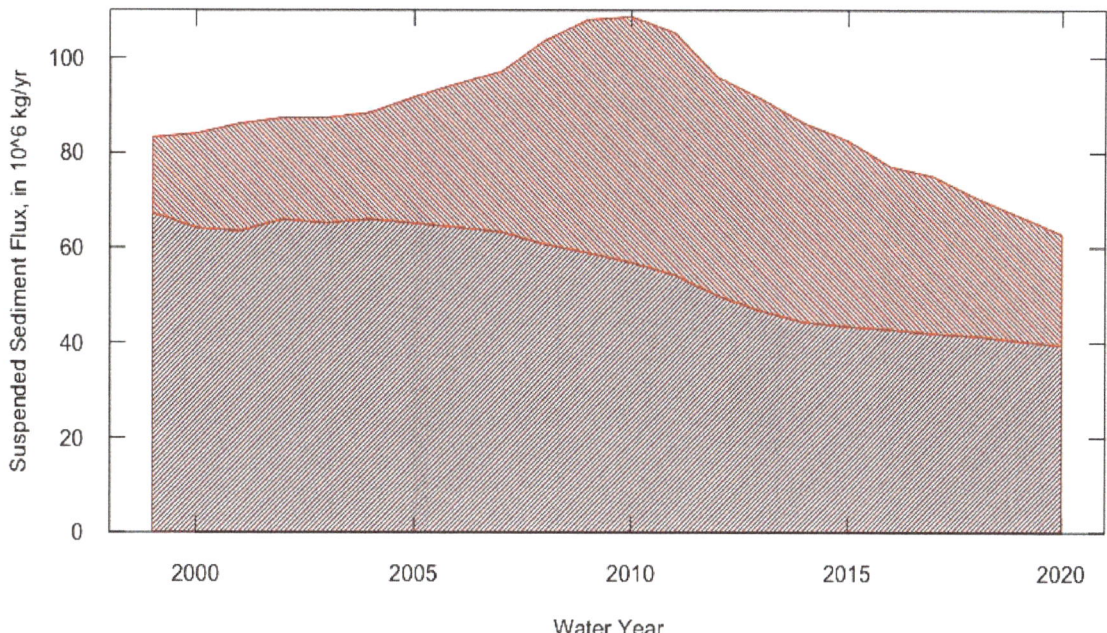

FIGURE 3-14 Estimated sediment balance in the lower CDA River 1999–2020 expressed as flow-normalized fluxes. Brown area is the sum of the inputs from the North Fork and South Fork. Red area represents the net scour of sediment from within the lower CDA watershed. Total shaded area is the flux out of the lower CDA River to the Lake. Units are millions of kg/yr. SOURCE: Data courtesy of USGS and analyzed and plotted by the committee.

If some engineering modifications are proposed, how will they change this trajectory of channel readjustment and lead delivery?

Two kinds of observations will be needed to project how lead delivery to the Lake will evolve into the future. The first are observations of dissolved and total lead as well as suspended sediment into and out of the lower CDA River, at least at the existing monitoring locations of Enaville (North Fork of the CDA River), Pinehurst (South Fork of the CDA River), and Harrison. The goal of this data collection should be to obtain about a dozen high-flow samples every year at each site. This needs to be done in conjunction with ongoing data analysis to determine the total fluxes at each location, change in lead and sediment storage, and average lead content of the sediment transported. This could be enhanced by using insitu turbidity sensors, as turbidity is a statistical surrogate for sediment and lead, but the current level of monitoring is still needed to ensure that the calibration of the statistical relationships remains relevant. Second, there also needs to be a system of regular observations of scour and deposition of bed materials along the 60-km length of the main stem of the CDA River. Modern boat-operated bathymetry technologies (including perhaps autonomous boats) can probably be employed as an efficient system for repeated riverbed change mapping. This would need to be coupled with a data analysis approach that documents the total amount of change and then relates it to the hydrologic conditions that drove the change.

Together, these data can then be used to build and calibrate a reach-scale model that can predict future lead transport to the Lake. In short, the lower CDA River is not simply a "conveyer belt" delivering lead from upstream to the Lake. Rather, it is a dynamic system that will respond to the changing delivery of lead and sediment in a manner that increasingly delivers legacy lead to the Lake or that increases the storage of new sediments that are less contaminated than the legacy materials and thus reduces delivery of lead in the coming decades. An adaptive management strategy will be needed to continuously assess the success of any future control measures (including the no-action option) that may be considered.

Dissolved Lead

Understanding dissolved lead is important because it is potentially more bioavailable in the river and the Lake. In the case of input from the South Fork, the lead flux is dominantly in the suspended fraction (about 5 percent is dissolved) and this has not changed substantially over the past 30 years. WRTDS can be used to estimate the expected number of days for each year when the concentration of dissolved lead for the CDA River near Harrison exceeds relevant aquatic life criteria values—for example, both the EPA acute criteria level of 65 µg/L and the chronic level of 2.5 µg/L (which assume a hardness of 100 mg/L $CaCO_3$—see Chapter 9). These calculations use the WRTDS model and the actual discharge on each day to compute a probability that the concentration would have exceeded the criteria value, and then these are summed over the water year to compute an expected number of days for each year (Figure 3-15). The number of days above the water quality criteria is small for the 65 µg/L threshold but very large for the 2.5 µg/L threshold. In both cases, it is difficult to say that there is a meaningful trend, although the past decade suggests a substantial decrease in the number of days the river may have exceeded these criteria. These are worst-case examples for CDA Lake, since the highest concentrations are found near the mouth of the CDA River and decline with distance as they mix with Lake water.

Finally, comparing results for total and dissolved lead over the history of the datasets, on average the flow-normalized dissolved lead flux for each year is about 3 percent of the flow-normalized total lead flux, declining slightly over time (Table 3-4). Over the past 25 years, the total lead flux to the Lake has been almost entirely in the suspended fraction, suggesting that future remediation strategies should focus on reducing particulate lead.

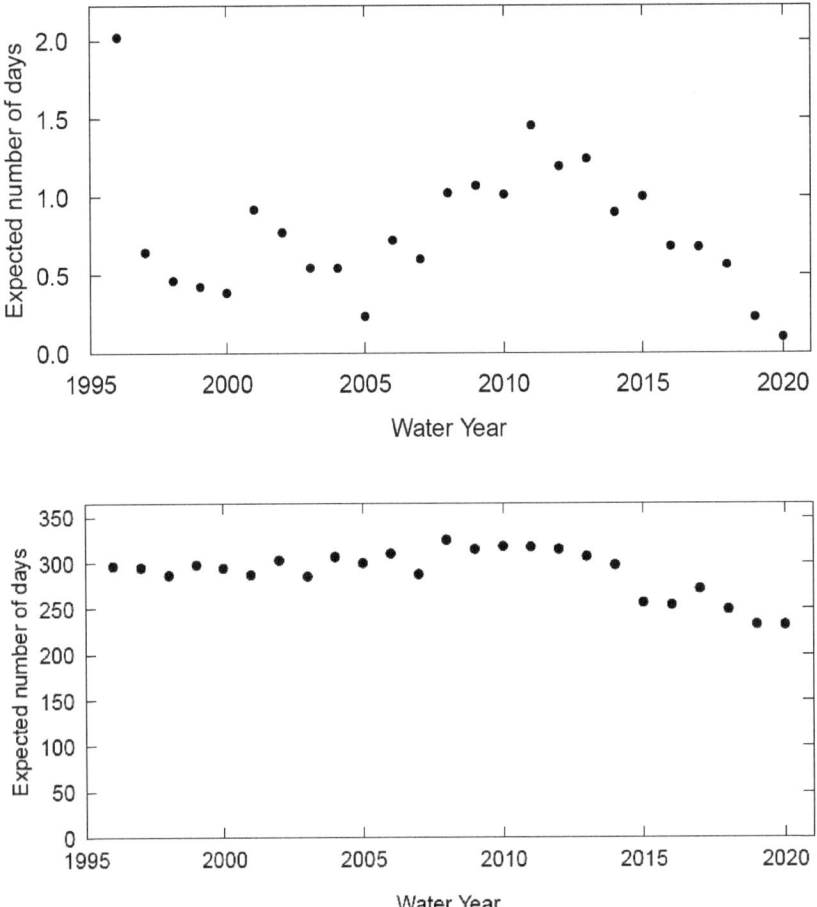

FIGURE 3-15 Expected number of days on which dissolved lead exceeds the acute aquatic life criteria of 65 µg/L (top) or the chronic aquatic life criteria of 2.5 µg/L (bottom); based on the WRTDS model for Harrison, where the CDA River enters CDA Lake. SOURCE: Data courtesy of USGS and analyzed and plotted by the committee.

TABLE 3-4 Trends in Dissolved Lead; Flow-Normalized Annual Mean Concentration (upper table) and Flow-Normalized Annual Flux (lower table)

Site	First year	Percentage change over period			First year mean	2020 mean	Dissolved, as % of total	
		Start–2020	2000–2020	2010–2020	µg/L	µg/L	2000	2020
SF CDA nr Pinehurst	1990	−82%	−70%	−41%	8.0	1.45	10%	9%
CDA nr Harrison	1996	−19%	−23%	−37%	5.6	4.6	11%	8%
Below Lake Outlet	1991	−79%	+25%	+7%	1.9	0.4	19%	45%

Site	First year	Percentage change over period			First year mean	2020 mean	Dissolved, as % of total	
		Start–2020	2000–2020	2010–2020	MT/year	MT/yr	2000	2020
SF CDA nr Pinehurst	1990	−87%	−51%	−21%	7.1	0.9	3%	5%
CDA nr Harrison	1996	−16%	−19%	−35%	21	17	4%	2%
Below Lake Outlet	1991	−78%	+38%	0%	17	4	17%	51%

NOTES: The period of record is different among the three sites, so the first water year is designated in the column labeled "First year." Trends are computed over three time periods, and for each time period the total change, expressed in percent, is displayed. The color code for the boxes is based on the uncertainty about the trend direction. Dark blue indicates a highly likely downward trend (likelihood > 95 percent), light blue indicates a likely downward trend (likelihood between 70 and 95 percent), no shading indicates that the trend direction is highly uncertain (likelihood of a downward trend is between 30 and 70 percent), pink shading indicates a likely upward trend (likelihood of an upward trend between 70 and 95 percent), and red shading (which does not appear in this table but will appear in other similar tables) indicates a highly likely upward trend (likelihood of an upward trend > 95 percent). The next two columns indicate the estimated value of flow-normalized concentration or flux for the first and the last year of the record. The last two columns compare the estimated values of dissolved lead to total lead for the years 2000 and 2020.

Total and Dissolved Cadmium

Cadmium was analyzed using the same approach used in the previous section on lead, but focusing on total cadmium with only brief mention of the dissolved phase. The reason is that there is a strong similarity between the results for total and dissolved cadmium at all sites, and the dissolved cadmium constitutes a very large fraction of the total (unlike the case with lead). The overall behavior of the total cadmium data (summarized in Table 3-5) is much more regular and much less variable than that of the total lead data. The results for dissolved cadmium are summarized in Table 3-6. Comparing the total and dissolved cadmium data, for flow-normalized flux over the 1997–2020 period, the dissolved cadmium makes up about 83 percent of the total cadmium, and this percentage is consistent over the whole record. For flow-normalized concentration, dissolved cadmium makes up about 96 percent of the total cadmium.

Throughout the record, the total cadmium concentrations (Figure 3-16A) tend to be highest at Pinehurst, and the values at Elizabeth Park are nearly as high. With the substantial dilution from the North Fork of the CDA River, the concentrations at the mouth of the CDA River are much lower (as of 2020, about one-fourth of those at Pinehurst). Finally, as a result of both dilution and loss processes (i.e., settling and biological uptake), the levels below the Lake are about 15 percent of the levels at Harrison, and about 4 percent of the levels upstream at Pinehurst. The total cadmium flux results (Figure 3-16B) show, by far, the highest annual flow-normalized values being at Harrison, with the Pinehurst values second highest and about a half the values at Harrison. The flux at Elizabeth Park is generally about half of the value that it is at Pinehurst. Finally, the flux at the outflow (Spokane River) is much smaller than it is at the inflow at Harrison, being only about one-third of the Harrison value. An interesting feature that stands out is that all of the records—except for the CDA River near Harrison—show something like a linear or exponential decline in flux over the period of record. At Harrison there is a plateau between about 2000 and 2010, which is similar to features seen with total lead as well. It is indicative of some type of readjustment happening within this reach around this period of time (see previous discussion on page 88).

TABLE 3-5 Trends in Total Cadmium: Flow-Normalized Annual Mean Concentration (upper table) and Flow-Normalized Annual Flux (lower table)

Site	First year	Percentage change over period			First year mean	2020 mean
		Start–2020	2000–2020	2010–2020	µg/L	µg/L
SF CDA at Elizabeth Pk	1993	–57%	–44%	–21%	8.0	3.4
SF CDA nr Pinehurst	1992	–65%	–51%	–27%	10.8	3.8
CDA nr Harrison	1991	–78%	–49%	–35%	4.3	0.9
Below Lake Outlet	2001	–37%	NF*	–28%	0.22	0.14

Site	First year	Percentage change over period			First year mean	2020 mean
		Start–2020	2000–2020	2010–2020	MT/year	MT/yr
SF CDA at Elizabeth Pk	1993	–54%	–41%	–19%	1.8	0.8
SF CDA nr Pinehurst	1992	–66%	–53%	–28%	3.9	1.3
CDA nr Harrison	1991	–63%	–34%	–33%	8.0	3.0
Below Lake Outlet	2001	–41%	NF*	–28%	1.5	0.9

NOTES: Trends are computed over three time periods, and for each time period the total change, expressed in percent, is displayed. The color code for the boxes is based on the uncertainty about the trend direction. Dark blue shading indicates a highly likely downward trend (likelihood > 95 percent). The last two columns indicate the estimated value of flow-normalized concentration or flux for the first and the last year of the record. NF* indicates "not feasible" because the data set starts in 2001.

TABLE 3-6 Trends in Dissolved Cadmium: Flow-Normalized Annual Mean Concentration (upper table) and Flow-Normalized Annual Flux (lower table)

Site	First year	Percentage change over period			First year mean	2020 mean	Dissolved, as % of total	
		Start–2020	2000–2020	2010–2020	µg/L	µg/L	2000	2020
SF CDA Elizabeth Pk	1997	–49%	–43%	–20%	6.6	3.4	95%	97%
SF CDA nr Pinehurst	1990	–64%	–50%	–27%	10.0	3.6	92%	95%
CDA nr Harrison	1996	–62%	–52%	–38%	1.9	0.7	81%	76%
Below Lake Outlet	2001	–29%	NF*	–21%	0.2	0.1	79%	90%

Site	First year	Percentage change over period			First year mean	2020 mean	Dissolved, as % of total	
		Start–2020	2000–2020	2010–2020	MT/year	MT/yr	2000	2020
SF CDA Elizabeth Pk	1997	–47%	–41%	–18%	1.4	0.7	84%	85%
SF CDA nr Pinehurst	1990	–68%	–50%	–25%	3.6	1.2	83%	88%
CDA nr Harrison	1996	–55%	–43%	–20%	3.4	1.6	60%	52%
Below Lake Outlet	2001	–29%	NF*	–22%	1.1	0.8	78%	93%

NOTE: Trends are computed over three time periods, and for each time period the total change, expressed in percent, is displayed. The color code for the boxes is based on the uncertainty about the trend direction. Dark blue shading indicates a highly likely downward trend (likelihood > 95 percent). The next two columns indicate the estimated value of flow-normalized concentration or flux for the first and the last year of the record. The last two columns compare the estimated values of dissolved cadmium to total cadmium for the years 2000 and 2020. NF* indicates "not feasible" because the data set starts in 2001.

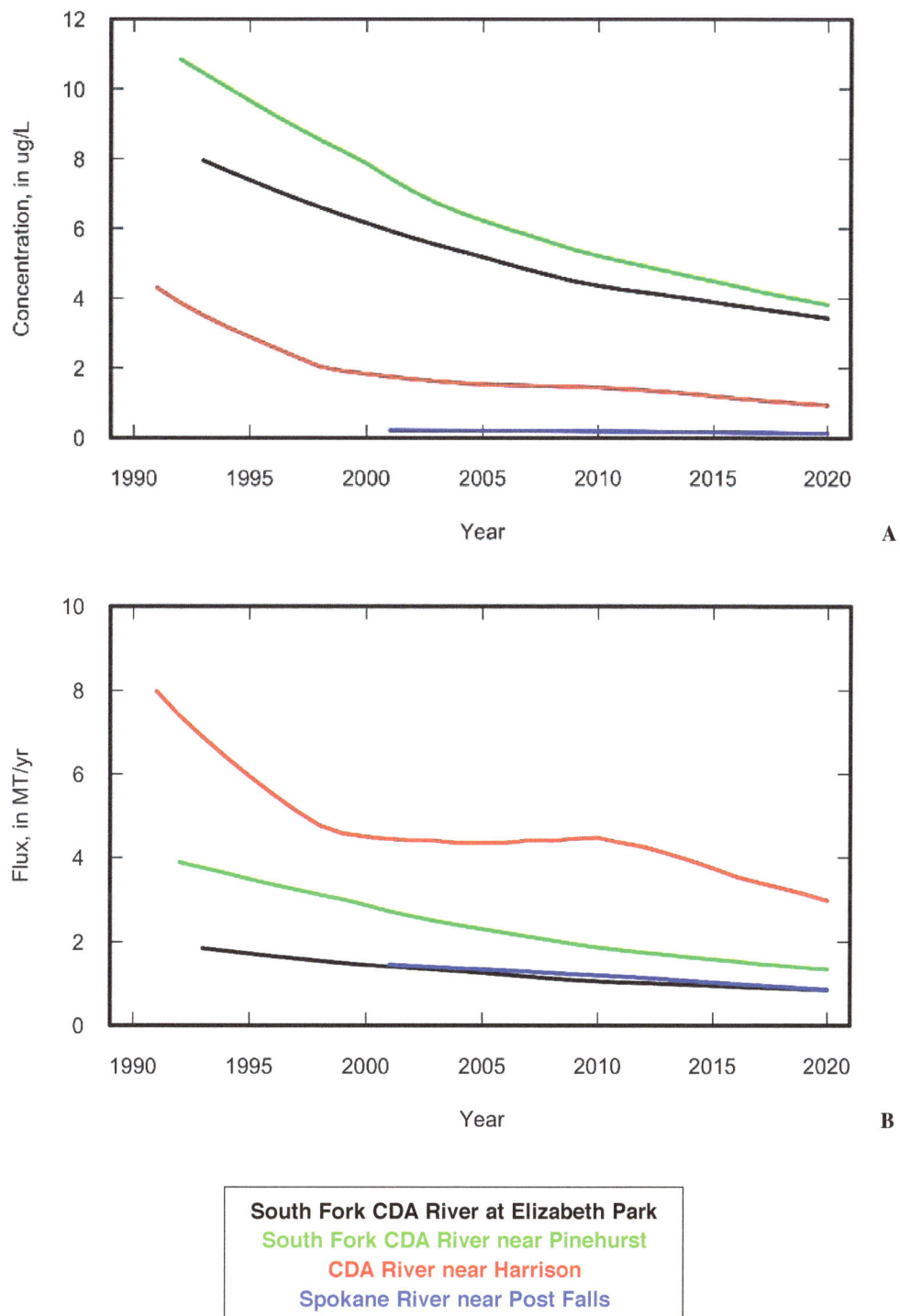

FIGURE 3-16 Estimated flow-normalized annual mean concentration (A) and flux (B) of total cadmium at four sites in the CDA watershed. SOURCE: Data courtesy of USGS and analyzed and plotted by the committee.

Figure 3-17 shows snapshots of the flow-normalized flux for total cadmium through the system in 2001 and 2020. First, in contrast to the total lead history, as of 2020 the inputs of cadmium from the South Fork of the CDA River to the lower CDA River were 43 percent of the outputs of the lower CDA River to the Lake (1.3/3). For lead, the 2020 inputs of lead from the South Fork of the CDA River to the lower CDA River were only about 3 percent of the output of the lower CDA River to the Lake (21/790). This suggests that further reductions in cadmium coming from the South Fork *are* likely to be important to reducing cadmium inputs to the Lake. Second, the Lake is a sink for cadmium, with outputs that are 30 percent of the inputs, but less so than for lead, for which Lake outputs are only about 1 percent of Lake inputs. Both of these differences are primarily a result of the fact that cadmium is transported mostly in the dissolved phase, but lead is transported mostly in particulate form.

Figure 3-18 shows the yield (flux per unit area) of total cadmium at each of the monitoring sites. What is particularly noteworthy is that in the 1990s the yield at Pinehurst was substantially greater than the yield at Elizabeth Park. This means that the incremental area between the two monitoring sites was delivering more total cadmium per unit area than was coming from the watershed area above Elizabeth Park, which is not a surprise since this incremental area was known to be the part of the watershed with the greatest amount of contamination. Moving toward the present, the yields for these two sites have become virtually equal, suggesting that the areal average yield is now no different in the Box than in the watershed upstream of the Box. This is illustrative of the effectiveness of the Superfund remediation within the Box, with respect to cadmium.

In general, all indications from Tables 3-5 and 3-6 are that total cadmium fluxes and concentrations are continuing to trend downward, at least since the year 2010. Compared to the total lead record, the total cadmium record shows more continuous trends and the magnitude of the seasonal and year-to-year fluctuations are lower. The uncertainty of results is much lower than is the case for total lead, which is what one might expect given that so much of the total cadmium is in the dissolved form and its transport is not subject to the highly variable processes of deposition and resuspension that are so important for total lead. Reductions in cadmium fluxes from the South Fork of the CDA River have been very substantial (a decline of 66 percent at Pinehurst since 1992). Further reductions of cadmium inputs to the Lake will probably depend both on further decreases in flux from the South Fork as well as on controlling the losses of cadmium from storage in the lower basin.

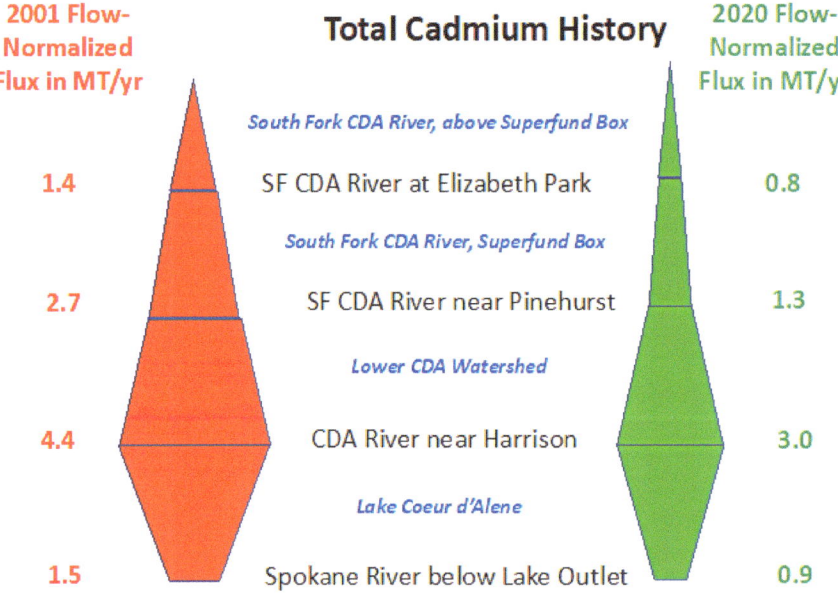

FIGURE 3-17 Schematic graph of downstream fluxes of total cadmium as of 2001 and 2020. The width of the colored zones of the figure is proportional to the flow-normalized flux at each of the four monitoring locations shown. The names of the four monitoring locations are shown in black, the reach of the river/lake system is identified in the blue italics lettering and the estimates of flow-normalized flux are derived from the WRTDS calculation.

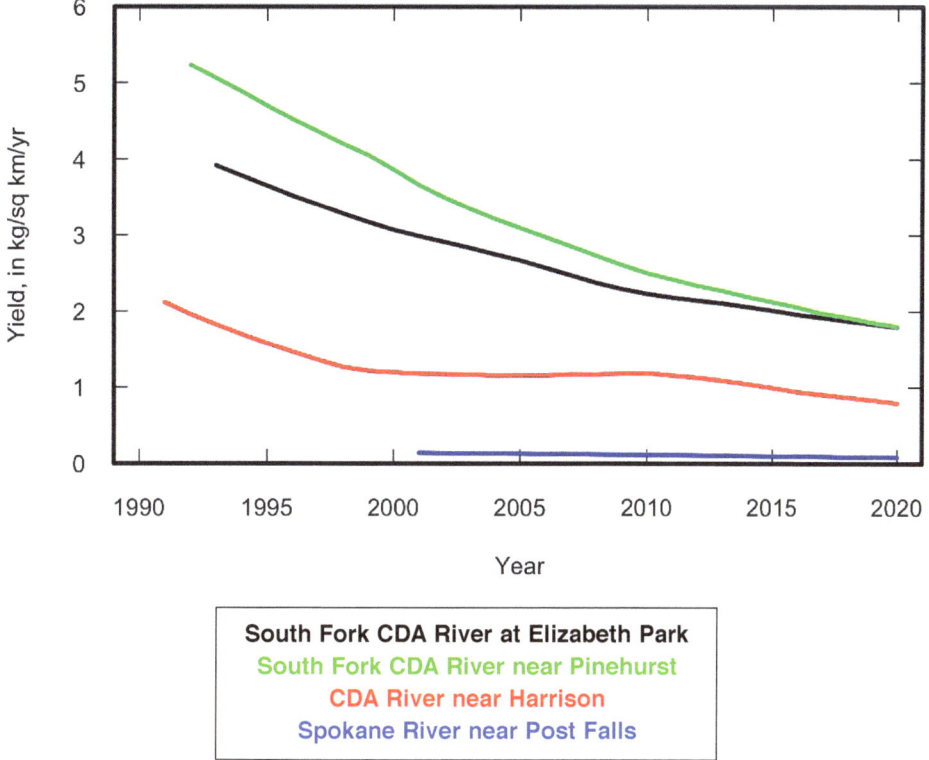

FIGURE 3-18 Estimated flow-normalized annual yield of total cadmium at four sites in the CDA watershed. Not all sites in the legend are in the graphs. SOURCE: Data courtesy of USGS and analyzed and plotted by the committee.

An additional analysis done by the committee (data not shown) is that in the years before about 2005, the total cadmium concentrations were high at both low and high discharge, but after about 2005 the discharge versus concentration relationship became essentially monotonic, with concentrations declining with increasing discharge. This indicates that the primary source of total cadmium is now in the base flow, presumably coming from groundwater.

Looking at the dissolved cadmium concentrations of CDA River inflows to the Lake, in the 1990s the number of days per year that concentrations exceeded the EPA chronic criteria of 0.43 µg/L (assuming 50 mg/L $CaCO_3$ hardness—EPA, 2016) was nearly 365. By 2020, this had decreased to about 350 days per year. For the EPA acute criterion of 0.94 µg/L, between 1996 and 2000 the dissolved cadmium concentration exceeded the criterion about 325 days per year; this has decreased to about 50–100 days of exceedance per year for 2015–2020. These results are further indications of substantial improvement in cadmium discharges to the Lake. They should be thought of as worst-case numbers in the sense that these represent concentration values where the CDA River enters the Lake and do not account for the dilution and losses that may take place within the Lake.

Total and Dissolved Zinc

Total and dissolved zinc were considered at the same four sites where lead was analyzed, with results shown in Tables 3-7 and 3-8. Dissolved zinc is generally more than 80 percent of the total zinc, except for the CDA River near Harrison, and trends at all four sites are generally very similar (in percentage terms).

Figure 3-19 summarizes the flow-normalized total zinc concentration records at these four sites. Throughout the record, the concentrations tend to be highest at Pinehurst; the values at Elizabeth Park are nearly as high. With the substantial dilution from the North Fork of the CDA River, the concentrations at the mouth of the CDA River

TABLE 3-7 Trends in Total Zinc: Flow-Normalized Annual Mean Concentration (upper table) and Flow-Normalized Annual Flux (lower table)

Site	First year	Percentage change over period			First year mean	2020 mean
		Start–2020	2000–2020	2010–2020	$\mu g/L$	$\mu g/L$
SF CDA at Elizabeth Pk	1993	–54%	–38%	–16%	1157	537
SF CDA nr Pinehurst	1993	–64%	–50%	–22%	1688	605
CDA nr Harrison	1992	–60%	–44%	–30%	424	170
Below Lake Outlet	1991	–63%	–43%	–23%	94	35

Site	First year	Percentage change over period			First year mean	2020 mean
		Start–2020	2000–2020	2010–2020	$MT/year$	MT/yr
SF CDA at Elizabeth Pk	1993	–50%	–36%	–14%	262	131
SF CDA nr Pinehurst	1993	–62%	–50%	–23%	545	205
CDA nr Harrison	1992	–45%	–29%	–29%	854	470
Below Lake Outlet	1991	–62%	–46%	–24%	575	219

NOTES: The period of record is different among the sites, so the first water year is designated in the "First year" column. Trends are computed over three time periods and for each time period the total change, expressed in percent, is displayed. The color code for the boxes is based on the uncertainty about the trend direction. Dark blue shading indicates a highly likely downward trend (likelihood > 95 percent). The last two columns indicate the estimated value of flow-normalized concentration or flux for the first and last year of the record.

TABLE 3-8 Trends in Dissolved Zinc: Flow-Normalized Annual Mean Concentration (upper table) and Flow-Normalized Annual Flux (lower table)

Site	First year	Percentage change over period			First year mean	2020 mean	Dissolved, as % of total	
		Start–2020	2000–2020	2010–2020	$\mu g/L$	$\mu g/L$	2000	2020
SF CDA Elizabeth Pk	1997	–46%	–42%	–23%	919	494	97%	92%
SF CDA nr Pinehurst	1990	–67%	–50%	–27%	1732	659	95%	95%
CDA nr Harrison	1996	–63%	–53%	–34%	352	131	91%	77%
Below Lake Outlet	1990	–62%	–44%	–38%	86	32	88%	90%

Site	First year	Percentage change over period			First year mean	2020 mean	Dissolved, as % of total	
		Start–2020	2000–2020	2010–2020	$MT/year$	MT/yr	2000	2020
SF CDA Elizabeth Pk	1997	–45%	–40%	–22%	195	108	88%	82%
SF CDA nr Pinehurst	1990	–67%	–49%	–26%	557	208	88%	90%
CDA nr Harrison	1996	–57%	–47%	–30%	587	205	72%	54%
Below Lake Outlet	1990	–63%	–45%	–29%	559	207	92%	95%

NOTES: The period of record is different among the sites, so the first water year is designated in the "First year" column. Trends are computed over three time periods, and for each time period the total change, expressed in percent, is displayed. The color code for the boxes is based on the uncertainty about the trend direction. Dark blue shading indicates a highly likely downward trend (likelihood > 95 percent). The next two columns indicate the estimated value of flow-normalized concentration or flux for the first and last year of the record. The last two columns compare the estimated values of dissolved zinc to total zinc for the years 2000 and 2020.

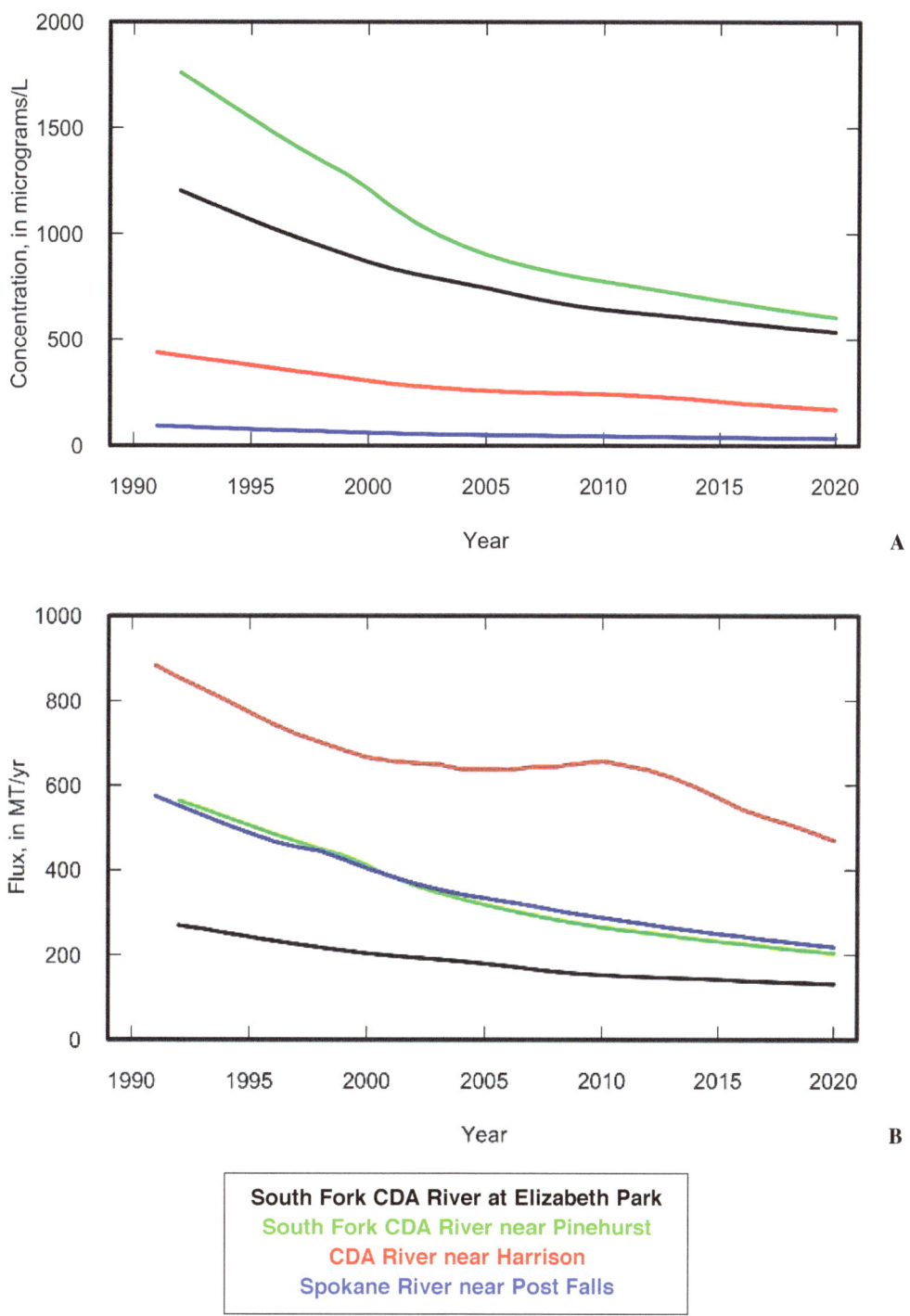

FIGURE 3-19 Flow-normalized concentration (A) and flux (B) of total zinc at four sites. SOURCE: Data courtesy of USGS and analyzed and plotted by the committee.

are much lower (as of 2020 they are about one-fourth the magnitude of those at Pinehurst). Finally, as a result of both dilution and loss processes (i.e., settling and biological uptake), the total zinc concentrations below the Lake are about 20 percent of the levels at Harrison and about 5 percent of the levels upstream at Pinehurst.

The flux results in Figure 3-19B show the highest annual flow-normalized total zinc values being at Harrison, with the Pinehurst values slightly less than half of the values at Harrison. This means that the lower basin is a significant source of total zinc. The flux at Elizabeth Park is generally about half of the value that it is at Pinehurst. Finally, the flux at the outflow is only about one-third of the Harrison value. A substantial portion of the observed decrease over time at Pinehurst is almost certainly the result of continuous improvements in treatment by the CTP. In 1993, 283 MT/yr of total zinc entered the river between Elizabeth Park and Pinehurst, and in 2020, this was reduced to 74 MT/yr. Based on the information in Box 3-1 and recent performance data, up to 75 percent of this amount could be from removal at the CTP (potentially less, depending on what CTP performance was in 1993). An interesting feature of all of the records except for the CDA River near Harrison is the exponential decline in total zinc flux over the period of record. At Harrison, there is a plateau between 2000 and 2010, similar to what was seen with total lead, indicative of some type of readjustment happening within this reach around this time.

Figure 3-20 shows the yield (flux per unit area) of total zinc at each of the monitoring sites. What is particularly noteworthy is that in the 1990s the yield at Pinehurst was well above the yield at Elizabeth Park. This means that the incremental area between the two monitoring sites was delivering more total zinc per unit area than the area above Elizabeth Park (which is not a surprise, since this was known to be the part of the watershed with the greatest amount of contamination). But, moving toward the present, the yields for these two sites become virtually equal, suggesting that the areal average yield is now no different in the Box than in the watershed upstream of the Box.

All indications are, at least since the year 2010, that total zinc fluxes and concentrations are continuing to trend downward. Overall, and compared to the total lead record, the total zinc record is much more limited in its variability. This is to be expected given that so much of the total zinc transport is in the dissolved form and not subject to the highly variable processes of deposition and resuspension that are so important in the case of total lead. Figure 3-21 shows the declines of 45 to 62 percent at all four sites from the early 1990s to 2020 and indicates that the South Fork watershed as well as sources within the lower basin continue to be nearly equal contributors to the fluxes to the Lake. There is a strong similarity to the shapes of the schematic figure for the two time periods, but the latter is about half as wide as the former, with the exception that the reach between Elizabeth Park and Pinehurst narrows the most (because it has been the focus of the most intensive Superfund cleanup activity).

The spatial and temporal patterns of total zinc flux trends are very similar to those for cadmium. The inputs to the lower CDA River in 2020 are 44 percent of the outputs of the lower CDA River (this was 43 percent for cadmium). The total zinc outputs of the Lake are 47 percent of the inputs (this was 30 percent for cadmium). Similar to the

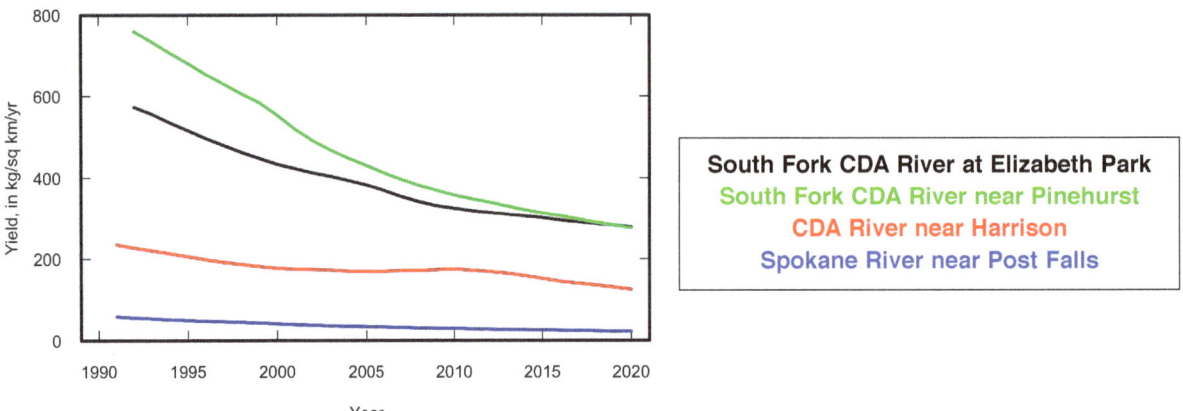

FIGURE 3-20 Flow-normalized yield of total zinc at four sites. SOURCE: Data courtesy of USGS and analyzed and plotted by the committee.

findings for cadmium, it is clear that further reductions in zinc entering the Lake will depend both on reductions in the lower CDA River as well as on reductions in the losses of zinc from the contaminated sediments in the lower basin.

The number of days that dissolved zinc at the inflow to the Lake is above the acute and chronic life criteria of 117 µg/L (corrected to 100 mg/L $CaCO_3$) shows a decline from about 365 days per year in the late 1990s to about 200 to 250 days per year around 2020 (Figure 3-22). Although this is a substantial improvement, it still represents a large number of days above these criteria.

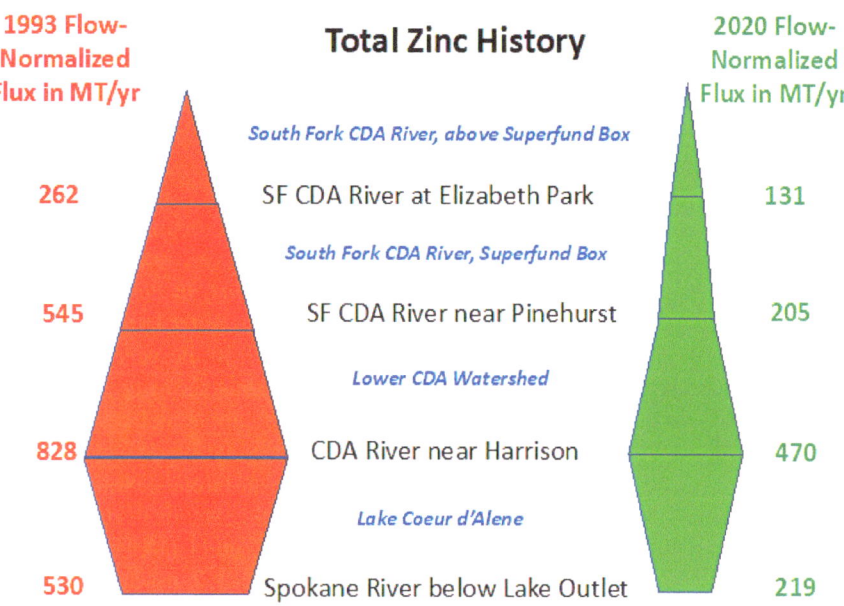

FIGURE 3-21 Schematic graph of downstream fluxes of total zinc as of 1993 and 2020. The width of the colored zones of the figure is proportional to the flow normalized flux at each of the four monitoring locations shown. The names of the four monitoring locations are shown in black, and the reach of the river/lake system is identified in the blue italics lettering. The estimates of flow-normalized flux are derived from the WRTDS calculation described in this chapter.

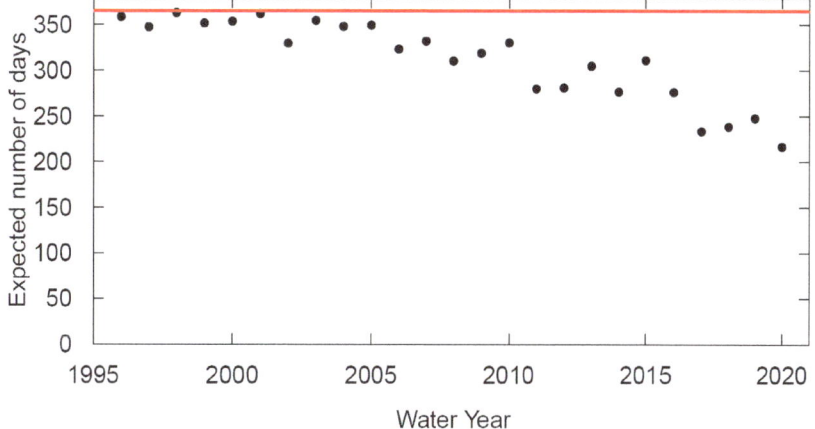

FIGURE 3-22 Expected number of days per year when concentrations of dissolved zinc in the CDA River near Harrison are above the acute and chronic aquatic life criteria of 117 µg/L, based on the WRTDS model and the daily discharge record. The red line represents the maximum value possible, 365 days of exceedances per year. SOURCE: Data courtesy of USGS and analyzed and plotted by the committee.

Phosphorus

Before proceeding into a discussion of the available data about total phosphorus (TP) in the rivers of the CDA watershed and total phosphorus at the outlet of the Lake, it is worth reviewing what has been written on this topic in recent years. In the report, *Coeur d'Alene Lake Management Program: Total Phosphorus Nutrient Inventory, 2004–2013* (IDEQ and CDA Tribe, 2020), the following summary of key trend findings is provided.

> "Comparison of current phosphorus loading with reported values for CY 1991–1992 indicate that phosphorus loading to Coeur d'Alene Lake has increased significantly. Flow-normalized loads from the Coeur d'Alene River and St. Joe River appear to have increased approximately two fold since the early 1990s. The load estimates for both the current and historic data are subject to considerable uncertainty, and thus the magnitude of the observed change is also uncertain. Even so, much of the observed change is likely real. The observed increases in phosphorus loading are consistent with trends in phosphorus concentrations in Coeur d'Alene Lake, which have also increased by approximately two fold since calendar years 1991–1992."

In reading this summary, it is very important to note that the data used in preparing this report only covered the time frame through 2013.

The other publication that discusses river water quality trends is *Trends in Concentrations, Loads, and Sources of Trace Metals and Nutrients in the Spokane River Watershed, Northern Idaho, Water Years 1990–2018* (Zinsser, 2020). The overall summary in that report regarding trends in total phosphorus in the years 2009–2018 is this: Of the nine monitoring sites considered for this period, five were characterized as showing "somewhat likely down" or "likely down" trends in flux, and the other four were characterized as "about as likely as not" to be downward trends. From a concentration perspective, six of the sites showed "somewhat like down" or "likely down" trends, and the remaining three sites were "about as likely as not" to be downward trends. None of the sites showed even moderately convincing evidence of increases over this period. This finding stands in sharp contrast with the statement from the previous report, but is understandable given that the IDEQ and CDA Tribe (2020) study considered a different period (1991–2013) than the Zinsser (2020) report (2009–2018). Zinsser (2020) did consider one site (South Fork of the CDA River near Pinehurst) that had data covering the same period as IDEQ (1991–2013) and it did show an approximate doubling of fluxes between 1990 and 2007, but then a decline from 2007 to 2018. In addition to differences in the period of record evaluated, the statistical method used by Zinsser (2020) was one that allowed for the characterization of non-monotonic trends, which was not a part of the IDEQ method. It should also be recognized that the periods of record covered a relatively narrow window of time within which to separate trends from persistent long-term stochastic variability (Cohn and Lins, 2005).

Against this backdrop of a potentially changing set of conditions, the synthesis presented here employs the same basic method used by Zinsser (2020) but using total phosphorus data through the end of 2020. This synthesis evaluates data at all six sites, which includes the four included in the analysis of metals and sediment but adds two additional sites in watersheds that did not have any Superfund cleanup activity. The results of the WRTDS results for flow-normalized concentration, flux, and yield of total phosphorus are shown in Figure 3-23. It should be noted that a similar exercise was not done for nitrogen because the total nitrogen data set was considered too short for this analysis.

Looking at the yield graph (Figure 3-23C), clearly the highest total phosphorus yield is for the South Fork of the CDA River near Pinehurst. This is likely a result of the fertilizer-production waste that is located in that watershed and perhaps the additions of fertilizer to the landscape for stabilization purposes, particularly during the 2000s. The yield is generally more than twice as high as the next upstream site, the South Fork of the CDA River at Elizabeth Park. The CDA River near Harrison ranks next and is always lower than Pinehurst and the pattern is similar but shifted to the right by about five years. This reflects the dilution effect of the North Fork of the CDA River and it also suggests that the lower CDA River reach may have stored phosphorus from the high input time of 2000–2008 and then released it a few years later.

The St. Joe River ranks next, being the watershed with more agricultural activity and population as compared to the North Fork basin, which has a yield that is about two-thirds of the St. Joe River yield. The total phosphorus yield for the outlet below the Lake is lower than any of the others, and this largely reflects the substantial losses

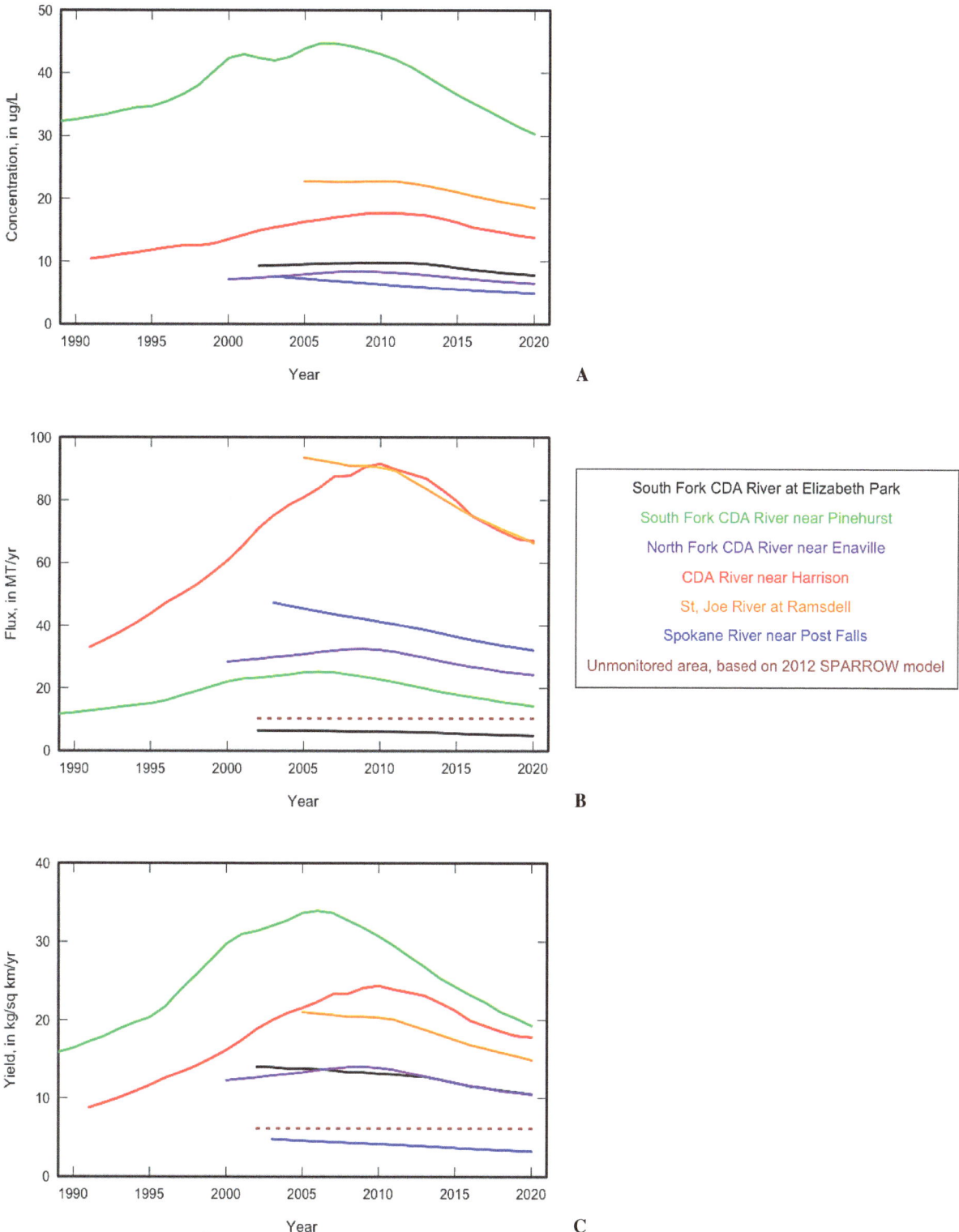

FIGURE 3-23 Trends in total phosphorus concentration (A), flux (B), and yield (C) over time for all six locations in the CDA watershed. SOURCE: Data courtesy of USGS and analyzed and plotted by the committee.

that take place in the Lake. Finally, the unmonitored part of the watershed is shown on the figure as a horizontal dotted line. It is based on the USGS 2012 SPARROW model (Wise, 2021), and there is no information available about possible trends in this unmonitored part of the watershed. It is difficult to argue that this nearshore area should show a lower yield than some of the relatively unmined watersheds (such as the North Fork and the St. Joe) because parts of it have substantial human populations and some parts have agriculture. Better characterization of this unmonitored area should be a high priority (see Chapter 8). However, since it is only about 16 percent of the land area of the total watershed, even a substantial increase in the estimated yield for this portion would have a relatively small effect on the total inputs to the lake. Using the estimates shown here, if the estimated flux of the unmonitored area were doubled, it would result in only a 7 percent increase in the total phosphorus flux to the Lake.

The temporal patterns of total phosphorus concentration, flux, and yield at Pinehurst and Harrison are complex. At Pinehurst, flow-normalized flux increased 114 percent from 1989–2006 and then decreased from 2006–2020 by 43 percent. For both low and high discharges, concentrations of total phosphorus rose dramatically between 1998–2010, with some concentrations higher than 200 µg/L at high discharge. The committee speculates that these very elevated total phosphorus concentrations were related to high rates of erosion of surficial sediments (soil and stream banks and bed) coincident with a period of intense landscape modification as part of the Superfund remedy. Once those processes settled down and vegetation became more established, the rates of erosion and transport of sediments high in total phosphorus would have decreased.

For the CDA River near Harrison, the pattern is similar. The increase in flow-normalized flux from 1991–2010 was +178 percent and then −27 percent from 2010–2020. This pattern might be explained by the same drivers as those above Pinehurst, but at Harrison it might have been extended in time by the re-working of recently deposited materials from the South Fork (which may have diminished between 2011–2020 as the landscape of the lower basin became more stable). These observations of patterns and hypotheses about causative factors were also addressed by Zinsser (2020). The difficulties in explaining the strong reversal of total phosphorus concentrations and fluxes at these two sites between 2005 and 2010 points to the need for a much more robust scientific effort to understand the historic and current drivers of total phosphorus movement in the watershed. Undoubtedly one of the key issues for maintaining high water quality in CDA Lake will be control of total phosphorus inputs, and without a strong scientific understanding of these very large changes over the past three decades, any conclusions about future strategies for phosphorus control would have to be viewed as highly uncertain.

Table 3-9 presents the overall results for total phosphorus concentration and flux for all six of the sites considered here, and it shows downward trends in total phosphorus flux at almost all locations in the most recent ten years.

Looking at the flow-normalized total phosphorus flux values for 2020 in Table 3-9, the lower CDA watershed value is 38.6 MT/yr (14.3 from the South Fork and 24.3 from the North Fork). The output of the lower CDA River is estimated to be 67 MT/yr, which means that as of 2020 the lower basin is a source of about 28.4 MT/yr. Similar to the metals, there is clearly a legacy of phosphorus in the lower basin that is being released over time, at least since about 2010. This source can be expected to continue into the future, even if inputs to the lower basin decrease.

The flow-normalized total phosphorus input to the Lake (excluding the unmonitored areas) is about 133 MT/yr in 2020 (in almost exactly equal amounts from the CDA River and the St. Joe River). The output from the Lake is estimated to be about 32 MT/yr. Hence, losses of phosphorus in the Lake are about 100 MT/yr, or about 25 percent of the phosphorus input to the Lake becomes output. The columns labeled "2010–2020" in Table 3-9 show trends over the most recent decade. For total phosphorus concentration, trends range from −18 to −29 percent, with all but one being considered "highly likely" to be truly downward trends (and the other being categorized as "likely" to be downward). (Box 3-5 argues why this ten-year trend is valid and not a result of the last two years of the record.) For total phosphorus flux, the range of trends runs from −20 to −37 percent, with likelihood categories similar to those for phosphorus concentration.

There is concern among various stakeholder groups in the CDA region that increasing phosphorus inputs to CDA Lake or phosphorus concentrations in the Lake could have serious consequences for the Lake's future water quality, which is not unreasonable. The documents and presentations available to the committee, with the exception of Zinsser (2020), generally suggest that phosphorus inputs to the Lake have been rising in recent years. However, ongoing increases in either phosphorus flux or concentration in the inputs to the Lake are *not* indicated by the data from the major rivers for the past decade.

TABLE 3-9 Summary of Trends in Concentration and Flow of Total Phosphorus from the Subwatersheds of CDA Lake. Upper Table Is for Trends in Flow-Normalized Annual Mean Concentration, Lower Table Is for Trends in Flow-Normalized Annual Flux

Site	First year	Percentage change over period			First year mean	2020 mean
		Start–2020	2000–2020	2010–2020	µg/L	µg/L
SF CDA at Elizabeth Pk	2002	–16%	NF*	–20%	0.009	0.008
SF CDA nr Pinehurst	1990	–7%	–28%	–29%	0.033	0.030
NF CDA at Enaville	2000	–9%	–9%	–22%	0.007	0.007
CDA nr Harrison	1991	+33%	+2%	–22%	0.010	0.014
St. Joe River at Ramsdell	2005	–18%	NF*	–18%	0.023	0.018
Below Lake Outlet	2003	–36%	NF*	–23%	0.008	0.005

Site	First year	Percentage change over period			First year mean	2020 mean
		Start–2020	2000–2020	2010–2020	MT/year	MT/yr
SF CDA at Elizabeth Pk	2002	–25%	NF*	–20%	6.6	5.0
SF CDA nr Pinehurst	1990	+17%	–35%	–37%	12.2	14.3
NF CDA at Enaville	2000	–15%	–15%	–22%	28.4	24.3
CDA nr Harrison	1991	+103%	–10%	–27%	33.0	67.0
St. Joe River at Ramsdell	2005	–29%	NF*	–27%	93.5	66.3
Below Lake Outlet	2003	–32%	NF*	–22%	47.3	32.1

NOTES: The period of record is different among the six sites, so the first water year is designated in the column labeled "First year." Trends are computed over three time periods, and for each time period the total change, expressed in percent, is displayed. The entry "NF*" indicates that calculations for that trend period are not feasible because the record does not extend back to 2000. The color code for the boxes is based on the uncertainty about the trend direction. Dark blue indicates a highly likely downward trend (likelihood > 95 percent), light blue indicates a likely downward trend (likelihood between 70 and 95 percent), no shading indicates that the trend direction is highly uncertain (likelihood of a downward trend is between 30 and 70 percent), pink shading indicates a likely upward trend (likelihood of an upward trend between 70 and 95 percent), and red shading indicates a highly likely upward trend (likelihood of an upward trend > 95 percent). The last two columns indicate the estimated value of flow-normalized concentration or flux for the first and the last year of the record.

Going forward, it will be important to use trend analysis methods that allow for depiction of trends as having changing slopes over time and even be non-monotonic (the CDA River near Harrison provides an excellent example of that). The trend analysis will need to be revisited frequently. For example, the Chesapeake Bay trend analysis of river nutrient flux data is carried out annually at nine major sites and biannually for a set of more than 100 additional sites. These analyses are published roughly a year after the end of the final water year of the period being analyzed, and those reports provide results for the most recent ten years as well as results covering the whole record (typically around 35 years).

All of the sites considered (whether in the mined and Superfund affected areas or not) showed strong indications of a downward trend in total phosphorus concentration and flux over the 2010–2020 period, and the mean concentrations at the end of this period were highest in sites with the greatest intensity of mining and cleanup activity and lowest at the outlet of the Lake. It is also worth noting that a major improvement in the mitigation of phosphorus outflows (the CTP) began in late 2020 and is not yet reflected in the results at Pinehurst (or the sites downstream from Pinehurst). Based on current inputs to the plant, the total release *without* any treatment would be 1.5–2 MT/yr, and the amount actually being released is < 0.02 MT/yr (Box 3-1). Based on past and current treatment effectiveness, the CTP was a minor contributor to the phosphorus loads over the past decade, and any further improvements would not be expected to meaningfully alter the loads to the river or Lake. So, it is difficult to propose any empirical basis for concern about rising total phosphorus levels in the watershed or downstream of the Lake. The one possible exception to that is for the unmonitored tributaries close to the Lake, which constitute

> **BOX 3-5**
> **Total Phosphorus Trends in 2019–2020**
>
> Given the conclusion about rather pervasive downward trends in riverine total phosphorus fluxes in the most recent ten years, it is worth considering the question of whether this apparent downtrend is an artifact of the fact that the last two years of this record were relatively low flow years (2019 and 2020). For several reasons, the Committee believes that these downward trends are meaningful. First, consider the results presented by Zinsser (2020). That report used data up through the end of water year 2018, so these particularly dry years were excluded. Those results are quite similar to those presented here. For water years 2009–2018, the South Fork at Pinehurst and North Fork at Enaville were characterized as having "likely downward trends," the St. Joe River was characterized as having "somewhat likely downward trend," and the result for the CDA River near Harrison indicated downward trends being "about as likely as not" although the estimated trend over the period was in the downward direction.
>
> Second, the WRTDS trend analysis methodology was designed specifically to be resistant to the effects of year-to-year differences in discharge. One of the major challenges of trend analysis for river water quality is that such a large portion of the variability is due to the variability of discharge and any analysis that does not have some means of filtering out this influence will be confounded by this large source of variance, resulting in a low signal-to-noise ratio for the trend detection process. This property of the WRTDS method is illustrated by Rowland et al. (2021).
>
> To further evaluate this issue that the last two years of phosphorus data may have undue influence on the results because they were years of low flow, the committee conducted a separate analysis that used a much simpler approach than is used in WRTDS to verify the trend findings. That approach focused on the observed trends in concentrations for the CDA River and St. Joe River. For each of these sites, the total phosphorus data were segregated into four groups based on the quartile of the discharge distribution on the day of sampling. For the CDA near Harrison, the analysis shows a very clear decline in total phosphorus concentration over the last few years of the record in both the lowest and highest flow quartiles, and essentially no change in the middle two quartiles. The decline in the lower quartile could be related to improvements in control of total phosphorus from groundwater in the Box, and the decline in the highest quartile may be related to improvements in landscape stability. For the St. Joe River data, the concentrations of total phosphorus show declines in the past several years in all discharge quartiles. These results, along with the WRTDS results, strongly suggest that total phosphorus concentrations and fluxes are declining for both rivers and that these declines are not simply an artifact of relatively dry years. That being said, it will be important to re-evaluate the total phosphorus trends at the river sites at such time as there are data available for a relatively high-discharge year.

16 percent of the total watershed land area. The Committee has not been able to obtain any empirical evidence of trends in total phosphorus from this area although it is logical to have some concern that they might be increasing in some areas due to growing population.

To make an argument about the threat to the Lake from future increases in phosphorus from the watershed, it will be necessary to first understand the reason for these observed decreases (in both mined and unmined watersheds) and then develop a scenario under which these may increase. Factors that may have influenced the total phosphorus decreases within the CDA watershed include the following: (1) declines in availability of phosphorus on the landscape following the end of major fertilizer applications (through hydroseeding) during the intense phase of Superfund landscape revegetation activities; (2) better controls of the waste from the fertilizer plant; (3) riverbank erosion control carried out under the LMP that may be limiting internal loading of phosphorus from river banks to river water; (4) improvement in the vigor of vegetation (perhaps due to air quality improvements since the end of the mining era), and/or (5) improved forest practices that may reduce erosion and hence reduce phosphorus loss. It is also possible that regional climate change may have some influence on total phosphorus fluxes but the mechanisms for that are unclear. Understanding these changes, especially outside of the Superfund area, should be an important research topic to be considered in pursuit of the LMP goals. Without a good conceptual model of the reasons for past decreases, it is difficult to have confidence about the impacts of future total phosphorus control strategies. The parties involved in the protection of the watershed and the Lake should focus on trying to understand the dynamics of phosphorus over the past 30 years.

Summary of Trends across Multiple Constituents and Locations

There are several commonalities and several differences between the trend patterns of the four constituents evaluated here (lead, cadmium, zinc, and phosphorus) across the monitoring sites. Table 3-10 considers all four of these constituents at four sites in the watershed arranged from upstream to downstream. It is a narrative, intended to present general patterns rather than focusing on precise numerical results.

It is also useful to look at the how the frequency distributions of concentration vary across the multiple sites and over the duration of the records and to compare the concentrations at sites downstream of the Superfund remediation with those not downstream from the remediation. The approach was to take the WRTDS model for four sites (St. Joe River, CDA River at Harrison, North Fork of the CDA River, and the South Fork of the CDA River) and estimate the concentrations one would expect on each day of the period of record. Box plots were constructed representing all the days in each of two different periods: water years 1993–1996 and 2017–2020. It was not possible to do the 1993–1996 period for any metals for the St. Joe River because data do not exist for that period). Box plots were constructed from these daily concentration estimates rather than being constructed from the raw data because the distribution of the raw data values is strongly influenced by the particular set of days that happened

TABLE 3-10 Summary of the Temporal Patterns of Change for Four Constituents at Four River Monitoring Sites

	South Fork CDA River near Elizabeth Park	South Fork CDA River near Pinehurst	CDA River near Harrison (input to CDA Lake).	Spokane River below Lake outlet
Total Pb	Steady downward trend over 3 decades. In the early years, this site has a flux about half of the Pinehurst flux, but by 2020 it constitutes about ¾ of the Pinehurst flux.	Steady downward trend over 3 decades. Decline is very steep from about 2000 to 2004, but steady decline continues through 2020. Flux is about 2x the Elizabeth Park flux at the start but by 2020 almost all of the flux comes from above Elizabeth Park	Steady decline to 2000. Most of the flux comes from lower CDA. Flux increases abruptly around 2000 and continues up to around 2009. Large amounts coming out of Lower CDA storage. Since 2009 fluxes are headed down, but slowly. Most of the flux coming out of the Lower CDA watershed.	Declines over the entire 3 decades, very steep 1991–2000, then level 2000–2010, then steep again 2010–2020. In early years the flux out of the Lake was about 10% of the input from CDA River. This declined to about 3% of input by 2000, then to 1% of input from 2010 to 2020. Thus, trapping efficiency has been increasing over time.
Total Cd	Steady downward trend over 3 decades.	Steady downward trend over 3 decades.	Downward steeply to about 1998, then plateau from 1998 to 2010. More and more coming out of storage in lower CDA. Then a downward trend again through 2020, ending up around 2x the Pinehurst flux. The Lower CDA watershed continues to be a major source area.	Steady decline over the 2 decades of monitoring. 2020 level is about 1/3 of the input from CDA River.
Total Zn	Steady downward trend over 3 decades.	Steady downward trend over 3 decades.	Downward steeply to about 1998 (flux is 1.5x the Pinehurst flux), then plateau to about 2010, then steeply downward to 2020. By 2000 it is 2x Pinehurst flux. The lower CDA continues to be a major source.	Steady decline over 3 decades. In early 1990s, flux is about 2/3 of the input from the CDA River, but by 2020 it is less than half of the input.
Total P	Small declines throughout entire 2 decades of monitoring record.	Rises gradually from 1989 to 2006 (roughly 2x increase) then downward from 2006 to 2020. Flux in 2020 is just slightly higher than 1990 flux.	Rises steeply from 1991 to 2010 (roughly 3x increase, increase mostly out of Lower CDA storage). Decreases from 2010 to 2020 (27% decline). Overall, an increase of about 100% over the 3 decades. In recent years outputs from the lower CDA have been about 167% of the inputs.	Steady declines over the two decades of monitoring. Flux out of the Lake as of 2020 is about ¼ of the sum of the inputs from the two main tributaries (CDA and St. Joe Rivers).

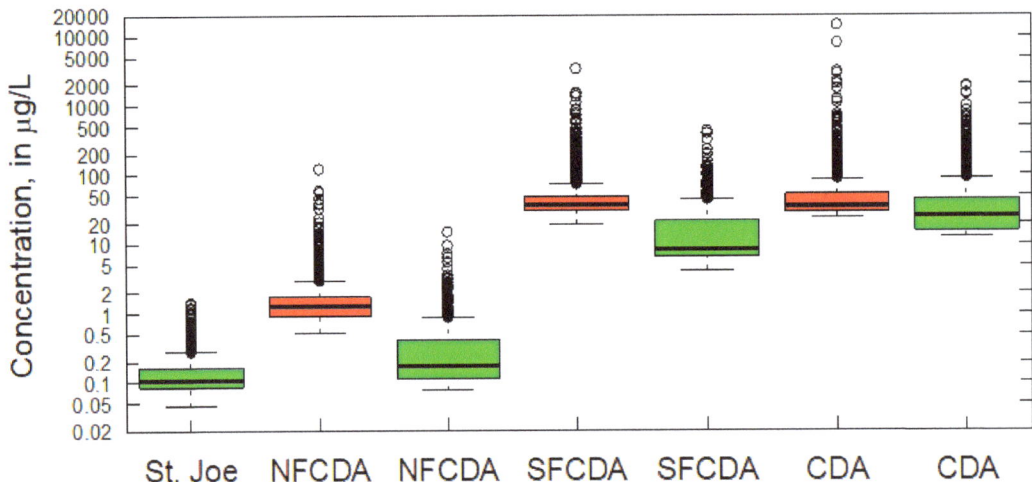

FIGURE 3-24 Boxplots of estimated daily concentration of total lead at four sites (St. Joe River at Ramsdell, North Fork of the CDA River (NFCDA) at Enaville, South Fork of the CDA River (SFCDA) near Pinehurst, and CDA River near Harrison). The estimates are outputs from the WRTDS model for total lead at each site, for two specific time periods (1993–1996 in red and 2017–2020 in green). Note that the y-axis is log scaled. SOURCE: Data courtesy of USGS and analyzed and plotted by the committee.

to be sampled. The goal was to make the comparisons across sites and time periods that are representative of the general behavior of the individual watersheds, showing the broad range of temporal variations that are driven by the variations in discharge and season. The result should also be responsive to trends that may exist between the two time periods being evaluated. The two sets of years that were examined both contain individual years of very high discharge (1996 and 2017). The result of this analysis for total lead is shown in Figure 3-24.

The St. Joe River data provide an upper bound to what might be considered modern regional background levels of total lead. There is likely to be some small amount of lead mining and/or milling that has taken place in this watershed, but it was never sufficiently intense to warrant this watershed being included in the Superfund remedy. The North Fork of the CDA River in the more recent period shows estimated daily lead concentrations that are slightly higher than those of the St. Joe, although they are still much less than those in the CDA River basin (Figure 3-24). Even though the North Fork watershed was not part of the Superfund remedy, it is clear that the concentrations declined between the two periods, presumably because of the decline in mining activity and/or remediation that has happened outside of the Superfund. It would not be surprising if at least some of the soil contamination in the North Fork and St. Joe watersheds originated from smelting activities in the basin of the South Fork. Nonferrous smelters resulted in widespread dispersion of secondary contamination through the first century of mining activity in the western United States. Moore and Luoma (1990) cited soil contamination with arsenic, lead, and cadmium that still affected vegetation and cropland in the late 1980s and covered 300 km^2 surrounding the Anaconda smelter in the Clark Fork mineral extraction complex in Montana. Declining trends in metals contamination would be expected with revegetation and recovery of those soils over time.

Moving to the South Fork of the CDA River in the earlier period, the median concentration is about 30 times higher than that of the North Fork of the CDA River (and about 350 times higher than that of the St. Joe River). The change at the South Fork of the CDA River between the two periods (expressed on the basis of median values) is about 80 percent, which one can attribute to the end of mining and the Superfund remedy. This is an impressive decrease, yet an additional 99 percent reduction from current levels would have to occur for lead levels at the South Fork of the CDA River to reach modern background levels. For the CDA River at Harrison, the reduction in the median between the two periods is about 30 percent; in order for levels here to reach the low levels seen in the St. Joe River, a further 99.6 percent reduction from current levels would be needed.

Figure 3-25 shows a similar analysis for total zinc. It shows that in the more recent period, the St. Joe River had concentrations that were about a factor of two lower than the North Fork. Similar to the lead analysis, there

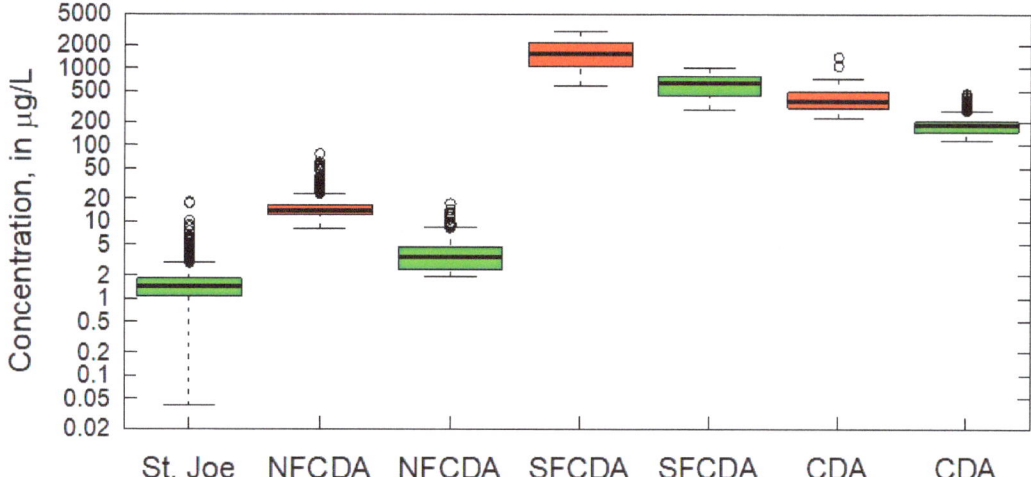

FIGURE 3-25 Boxplots of estimated daily concentration of total zinc at four sites (St. Joe River at Ramsdell, North Fork of the CDA River at Enaville, South Fork of the CDA River near Pinehurst, and CDA River near Harrison). The estimates are outputs from the WRTDS model for total zinc at each site, for two specific time periods (1993–1996 in red and 2017–2020 in green). Note that the y-axis is log scaled. SOURCE: Data courtesy of USGS and analyzed and plotted by the committee.

is a substantial decrease (about 75 percent) in concentrations from the earlier period to the later one, indicating improvements in this much less contaminated watershed, even in the absence of the Superfund remedy. For both the South Fork of the CDA River and the CDA River at Harrison, declines in median total zinc concentrations between the two periods are around 55 percent. It is logical that the reductions should be similar because the declines come from improvements in the South Fork basin, and the CDA River at Harrison results simply reflect the dilution of South Fork concentrations by water from the North Fork of the CDA River. A further reduction of about 99 percent from current levels would be required for the CDA River inputs to the Lake to be similar to a regional background level, over and above the 55 percent reduction that has already taken place.

For cadmium, Figure 3-26 illustrates the distribution of concentrations during the two time periods for the North Fork of the CDA River at Enaville, the South Fork of the CDA River near Pinehurst, and the CDA River

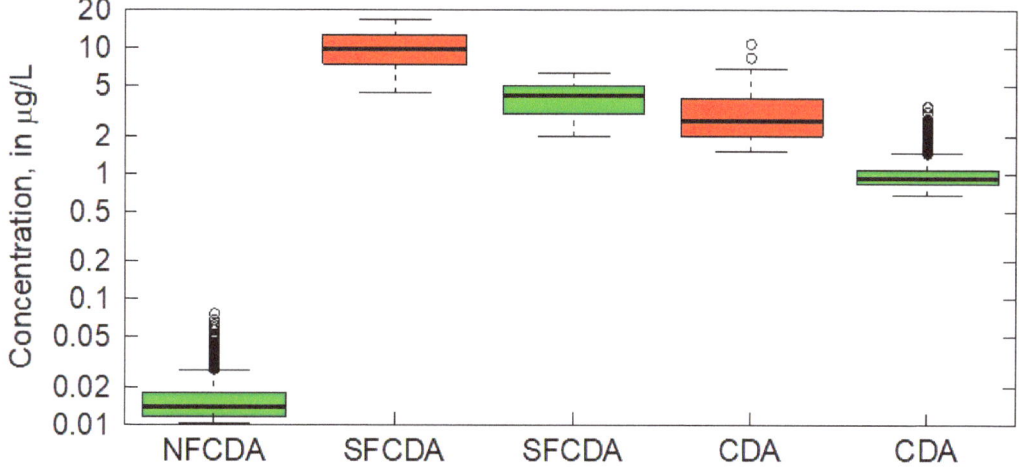

FIGURE 3-26 Boxplots of estimated daily concentration of total cadmium at three sites (North Fork of the CDA River at Enaville, South Fork of the CDA River near Pinehurst and the CDA River near Harrison). The estimates are outputs from the WRTDS model for total cadmium at each site, for two specific time periods (1993–1996 in red and 2017–2020 in green). Note that the y-axis is log scaled. SOURCE: Data courtesy of USGS and analyzed and plotted by the committee.

near Harrison. It was not possible to create a WRTDS model of total cadmium for the St. Joe River because the vast majority of the data (55 out of 59 observations) were less than the reporting limit of 0.03 µg/L.

The declines in total cadmium at Pinehurst and Harrison between the two time periods were both about 60 percent, and the concentrations at the CDA River near Harrison are approximately what would be expected based on the South Fork values modified by dilution with the "clean" North Fork water. Comparing the recent period for the CDA River at Harrison to the background level represented by the North Fork of the CDA River, the concentrations at the CDA River at Harrison would need to decrease by an additional 98.5 percent to be equivalent to background.

What all of these figures for lead, zinc, and cadmium indicate is that the remediation has, in most cases, brought about a sizable reduction in concentrations over the past two and a half decades (the one exception being lead for the CDA River near Harrison). Nevertheless, the metals concentrations remain one to two orders of magnitude above regional background levels.

Comparison of Trends by Location

Another perspective is to view the trends in flow-normalized flux for all four constituents at a single site on one graph, with each of the records rescaled to a common range defining the maximum flow-normalized flux for that constituent to a value of 1. Figure 3-27 shows the results for the South Fork of the CDA River at Elizabeth Park. For this site, the three metals follow a virtually identical temporal pattern. All of them decline from the early 1990s, ending up at about 40 to 50 percent of their maximum value in 2020. The phosphorus record is shorter than the others, but it also declines throughout its record. For phosphorus, the rate of decline appears to be accelerating in the final years as compared to the other constituents. Figure 3-27 suggests that the Superfund remediation in the upper basin above Elizabeth Park has had a very favorable impact on the three metals, yet all of them continue to be far above regional background levels. The reason for the decline in phosphorus is not well understood.

Moving downstream, Figure 3-28 shows the results for the South Fork of the CDA River near Pinehurst, just downstream of the Box. At this site, all three metals have declined in a roughly exponential pattern, with cadmium and zinc having virtually identical time histories and declining slightly more than 60 percent since 1992. The decline in lead has been steeper than for cadmium or zinc, with a total decline of about 80 percent since 1992. There is a much longer record of phosphorus at this site than at the Elizabeth Park site. Phosphorus was trending upwards steadily from the late 1980s to about 2006 and has turned downward since then, falling by about 40 percent from its highest value in 2006. The reasons for this pattern are not clear but they do suggest that the early years of the Superfund remedy may have caused the mobilization of phosphorus from the gypsum-fertilizer wastes in the Box.

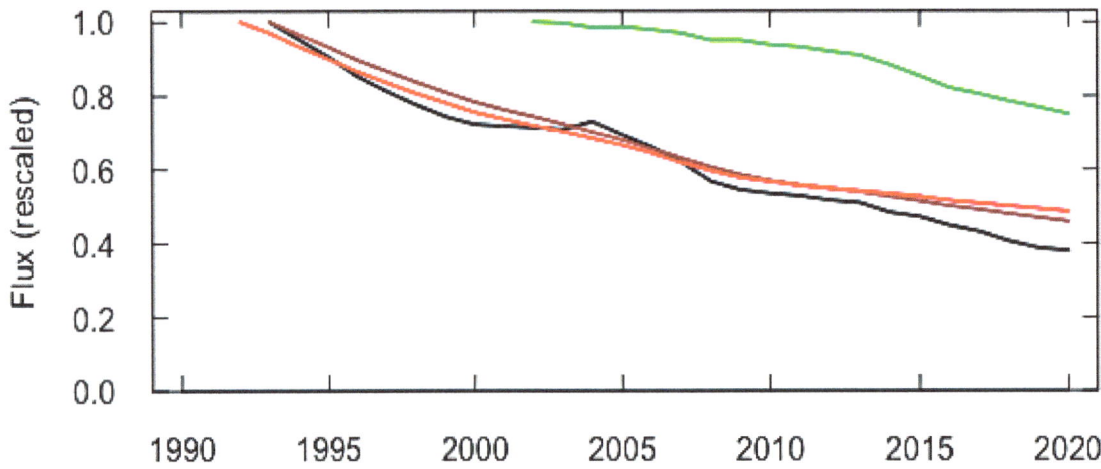

FIGURE 3-27 Relative values of flow-normalized flux for the South Fork of the CDA River at Elizabeth Park over time. Black is total lead, brown is total cadmium, red is total zinc, and green is total phosphorus. SOURCE: Data courtesy of USGS and analyzed and plotted by the committee.

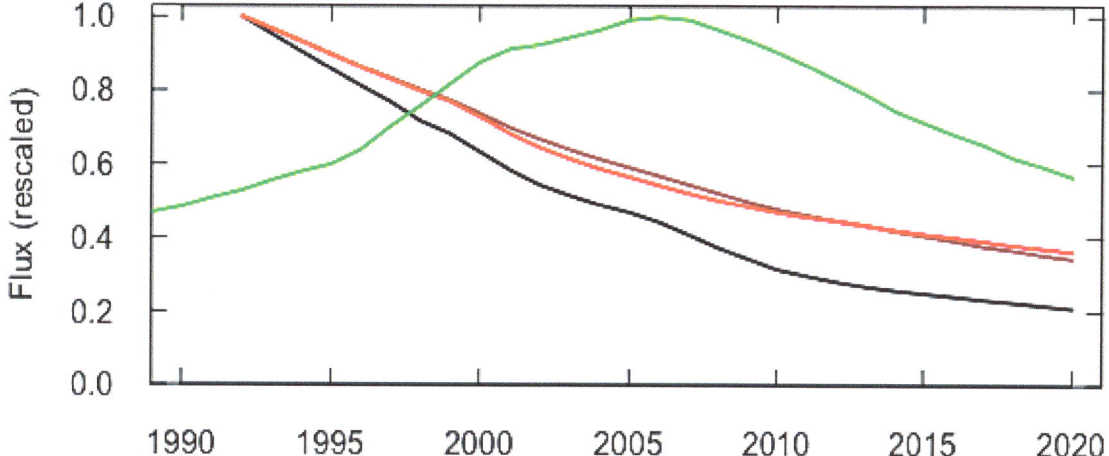

FIGURE 3-28 Relative values of flow-normalized flux for the South Fork of the CDA River near Pinehurst over time. Black is total lead, brown is total cadmium, red is total zinc, and green is total phosphorus. SOURCE: Data courtesy of USGS and analyzed and plotted by the committee.

Phosphorus inputs to this watershed segment may have also increased because of vigorous revegetation efforts, including hydroseeding, which resulted in large additions of phosphorus to the Superfund landscape. Many of these activities were at or near completion by the mid-2000s; thus, this source of phosphorus has greatly diminished since that time. There is no reason to expect it to reverse and trend upward in the future.

Figure 3-29 shows the flow-normalized fluxes for the four constituents at the outflow of the CDA River near Harrison. The temporal patterns here are quite different than those observed at Pinehurst. For zinc and cadmium, the declines are both steep initially, they are nearly constant from about 2000 to 2010, and they return to fairly steep declines after 2010. The initial rate of decline is greater for cadmium than for zinc, but after 2000 they follow nearly identical patterns. The lead record is more complex, with a downward trend from 1990 to 2000 and then a sharp reversal from 2000 to about 2009. This must represent a release from sources within the lower CDA

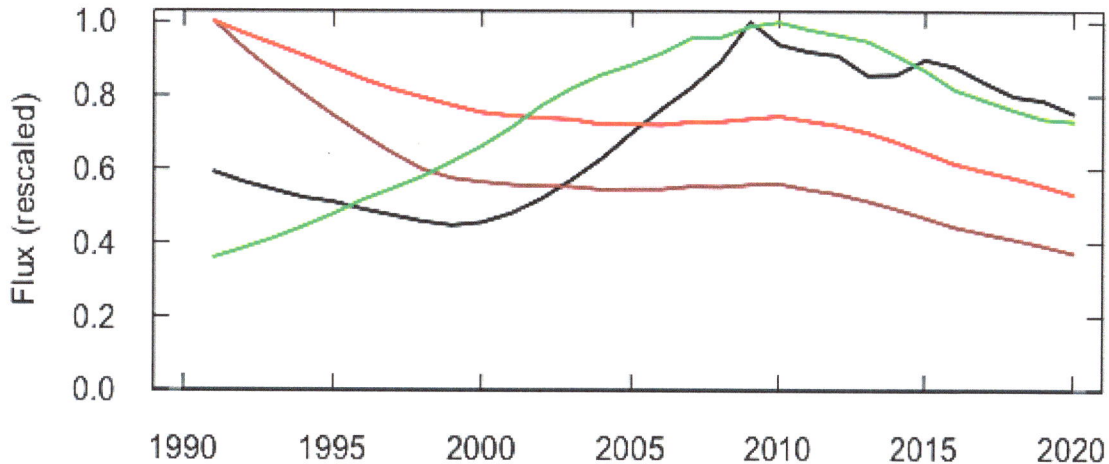

FIGURE 3-29 Relative values of flow-normalized flux for the CDA River near Harrison over time. Black is total lead, brown is total cadmium, red is total zinc, and green is total phosphorus. SOURCE: Data courtesy of USGS and analyzed and plotted by the committee.

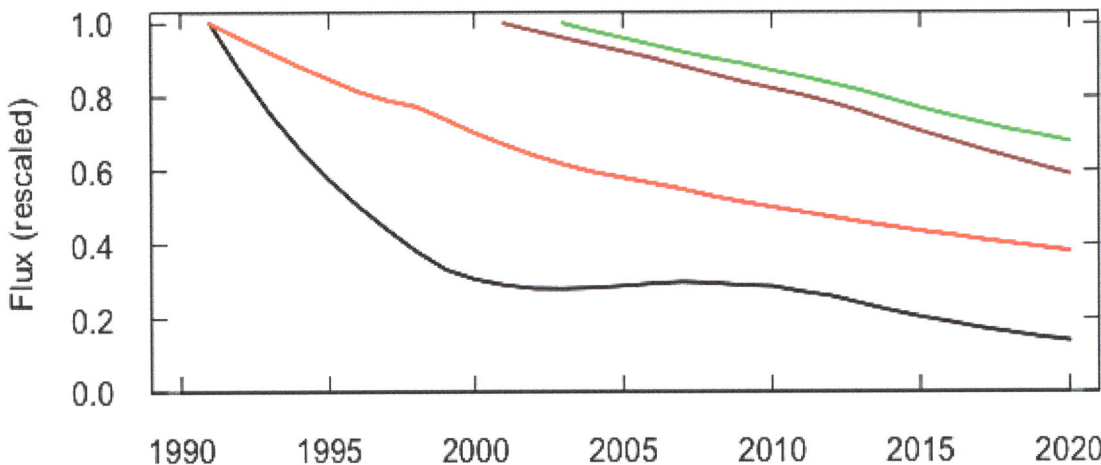

FIGURE 3-30 Relative values of flow-normalized flux for the Spokane River below the Lake outlet over time. Black is total lead, brown is total cadmium, red is total zinc, and green is total phosphorus. SOURCE: Data courtesy of USGS and analyzed and plotted by the committee.

watershed (because one sees no such increase in the inputs *to* the lower CDA watershed). Then, after 2009, the decline is somewhat erratic and fairly modest in slope (slightly more than 20 percent over the 2009–2020 period). Total phosphorus rises from 1990 to 2010 (similar to the inputs from the South Fork) and then turns down from 2010 to 2020 and roughly matches the timing and relative decline of lead. This suggests conditions stabilizing during the most recent decade.

Finally, Figure 3-30 shows the trends in flow-normalized fluxes for the Spokane River below the Lake outlet. In the early years, the relative rate of decline in lead was much greater than for zinc. But, the interpretation of the figure requires care because of the large differences in the record starting dates (1991 for lead and zinc, 2001 for cadmium, and 2003 for phosphorus). To compare their rates of decline in the common period of record (2003–2020), the results for those years are plotted again in Figure 3-31, which shows that the declines for zinc,

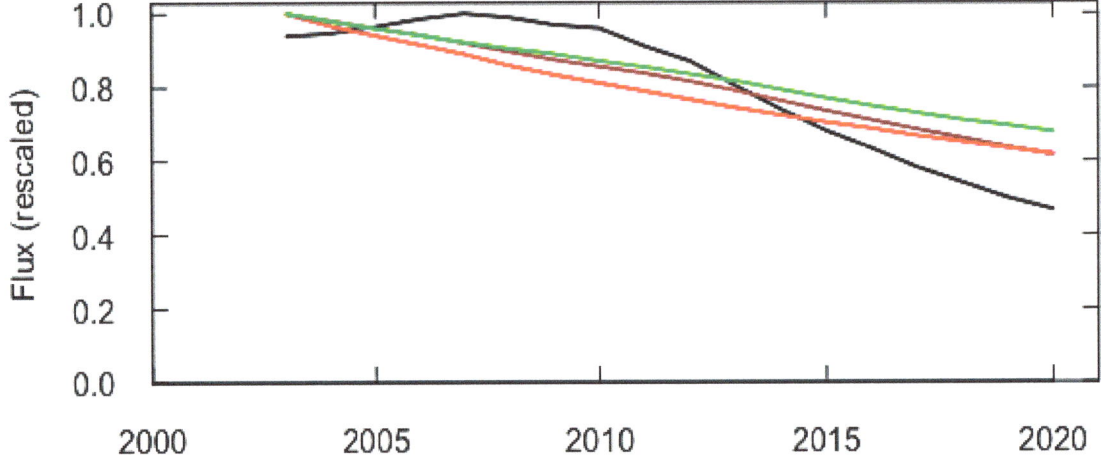

FIGURE 3-31 Relative values of flow-normalized flux for the Spokane River below the Lake outlet for 2003–2020 only. Black is total lead, brown is total cadmium, red is total zinc, and green is total phosphorus. SOURCE: Data courtesy of USGS and analyzed and plotted by the committee.

cadmium, and phosphorus are nearly identical over these 18 years. Using cadmium as an example, the rate of decline in the early years of this period (around 2004) was about 2 percent per year and in the last few years it has been about 3.5 percent per year. The recent decline in zinc has been about 2.7 percent per year, and for phosphorus the recent decline is only about 2.4 percent per year. However, for lead the decline started later than the others but has become relatively much steeper, with declines of about 6.4 percent per year.

CONCLUSIONS AND RECOMMENDATIONS

Inputs of lead, cadmium, and zinc to CDA Lake reflect the century-long legacy of mine waste deposition in the Lake's watershed. The frequent floods that transport wastes downstream, the geochemical reactions within groundwater that mobilize metals, and the minimal dilution that occurs between the source of the primary contamination and the Lake all contribute to ongoing metal deposition in the Lake. Although the Superfund remediation has reduced metal inputs from the upper basin, the lower basin comprises an immense stockpile of metal-enriched particulates poised for transport to CDA Lake. Reducing metal inputs in the future will increasingly depend upon controlling the drivers of inputs from the lower basin as remediation progresses. Based on the committee's analysis of numerous data sets to better understand changes in metal, sediment, and nutrient loading to CDA Lake over the past 30 years, the following detailed conclusions and recommendations are made.

1. **Observed land use changes in the upper basin are consistent with a recovering landscape where remedial activities, revegetation, and improvements to air pollution are ongoing.** The analysis of land use suggests that halting of mining activities and revegetation activities are reducing the extent of barren land and altering the balance of vegetated land cover, although the urban footprint has not changed. The reduction in barren land through remedial activities and the associated stabilization of soils have likely contributed to lowering metal loads from the upper basin to the river and Lake.

2. **Rates of streamflow entering CDA Lake over the past three and a half decades have been remarkably free of trends.** This is not only true for average annual flow rates but also true for annual low flows and annual high flows. Also, there does not appear to be a shift in the timing of runoff. These findings stand in sharp contrast with other parts of the western United States, which have generally seen substantial declines in streamflow and a shift of the center-of-mass of annual runoff to earlier in the year.

3. **Cadmium, lead, and zinc concentrations and loads into the mainstem Coeur d'Alene River from the South Fork have declined over the past 30 years, and Superfund activities have likely contributed to this decline.** For the South Fork of the CDA River at Elizabeth Park, fluxes of the three metals have declined since the early 1990s, with 2020 values being 40 to 50 percent of their maximum. Similarly, at the South Fork of the CDA River near Pinehurst, just downstream of the Box, fluxes of zinc and cadmium have declined slightly more than 60 percent since 1992 while the decline in lead flux has been about 80 percent. Stabilization of the landscape, capping, and sequestration activities have likely been effective at reducing fluxes of particle-associated lead. For zinc, remedial activities in the upper basin and particularly in the Box, including the continuous improvements at the Central Treatment Plant, have helped substantially lowered concentrations and fluxes. Continued remediation efforts and institutional controls are required to maintain the progress that is underway.

4. **Reductions of total lead fluxes from the South Fork of the CDA River were offset by processes in the lower basin that released lead from roughly 2000–2010, such that present-day lead inputs to CDA Lake are still substantial.** That is, lead concentration and flux at Harrison decreased from 1991 to 2000, increased from 2000 to 2010, and decreased from 2010 to 2020, for a net overall increase of 27 percent. Overall, lead flux to the Lake at Harrison was still 1.3 times higher in 2020 compared to the 1990s because of the increase in fluxes between 2000–2010. In 2020, lead fluxes into the lower basin at Pinehurst were only 2.6 percent of lead flux to CDA Lake at Harrison, demonstrating that there are large

reservoirs of metals in the river sediments and floodplains of the lower basin. Furthermore, it is clear that future decreases in lead fluxes into the Lake will be determined much more by evolving storage and release mechanisms in the lower basin than by further efforts at controlling pollution in the South Fork watershed. The committee's analysis of total lead in high-flow discharges at Harrison shows that lead concentrations in these flows have decreased over time, suggesting that remediation is having a beneficial effect. Remediation of the lower basin will require careful planning so as not to remobilize metals and transport them to the Lake.

5. **Sediment transport in the CDA watershed has been going through a period of adjustment, driven by substantial decreases in delivery of sediment from both the North and South Forks of the CDA River to the lower CDA River.** The committee hypothesizes that this has brought about scouring of sediment in the lower basin, leading to an increase in legacy lead delivered to the Lake. The slope of the channel downstream profile should decline over time as the river adjusts to dramatically lower sediment inputs, such that this change in profile should eventually result in a decrease in the rate at which legacy lead- and phosphorus-enriched sediments are scoured from the lower CDA river bed and banks. To better understand the future trajectory of sediment (and hence lead and phosphorus) delivery to the Lake, there will need to be a substantial increase in sediment data collection at the upstream and downstream limits of the lower CDA River, coupled with repeated measurement of channel slope and geometry and predictive modeling of the readjustment of channel transport and morphology. This will also be vital to evaluating any proposed remedial measures dealing with the legacy sediment and lead stored in the lower basin.

6. **There have been downward trends in cadmium and zinc concentrations and fluxes throughout the CDA basin, and fluxes of both metals to the Lake (measured at Harrison) were lower in 2020 than in 1992 (by 63 and 45 percent, respectively).** At the CDA River at Harrison, the cadmium and zinc fluxes leveled off during 2000–2010 but are declining again in the most recent decade. Unlike total lead, as of 2020 the inputs of total cadmium and total zinc from the South Fork of the CDA River to the lower CDA River were 43 and 44 percent (respectively) of the outputs of the lower CDA River to the Lake. This suggests that further reductions in cadmium and zinc coming from the South Fork *are* likely to be important to reducing inputs of these metals to the Lake. Targeted studies and trend data show that the primary sources of cadmium and zinc are now base flow, presumably coming from the groundwater system. The Central Treatment Plant is now a minor source of zinc.

7. **Over the past decade, total phosphorus fluxes and concentrations at monitoring sites in the CDA River, the St. Joe River, and the Spokane River below the lake outlet have all been declining (typically 20 to 30 percent reductions during the 2010–2020 decade).** In the case of the CDA River, this is a reversal of the trend observed over the prior decade. Like lead, total phosphorus flux to the Lake in 2020 was higher (by 2.3 times) than in the early 1990s. Projecting future trends of phosphorus in CDA Lake will require a sustained effort at monitoring and regular data synthesis for phosphorus across the whole watershed (including the 16 percent of the watershed that is not in the CDA or St. Joe River watersheds), with monitoring efforts closely connected to research (including modeling) aimed at understanding the reasons for this current decline. Without a better understanding of phosphorus transport in the whole watershed, there is no basis for projecting future phosphorus transport or the potential for future increases in phosphorus in the Lake.

8. **All of the wastewater treatment plants that discharge to the CDA basin should be considered for advanced phosphorus removal,** similar to those that discharge to the Spokane River downstream of CDA Lake. Phosphorus emanating from wastewater treatment plants is likely to be much more bioavailable than nonpoint source phosphorus inputs from the basin. By improving understanding and taking prudent action, CDA Lake can prove to be an exception to the broader rule of worsening trophic status and increasing algal blooms seen in lakes elsewhere in United States.

REFERENCES

Axtmann, E. V., and S. N. Luoma. 1991. Large-scale distribution of metal contamination in the fine-grained sediments of the Clark Fork River, Montana. *Applied Geochemistry* 6:75–88.

Balistrieri, L., S. Box, A. Bookstrom, R. Hooper, and J. Mahoney. 2002. Impacts of Historical Mining in the Coeur d'Alene River Basin. Chapter 6 *In:* Pathways of Metal Transfer from Mineralized Sources to Bioreceptors: A Synthesis of the Mineral Resources Program's Past Environmental Studies in the Western United States and Future Research Directions. USGS Bulletin 2191. L. S. Balistrieri and L. L. Stillings, eds.

Barton, G. 2002. Dissolved Cadmium, Zinc, and Lead Loads from Ground-Water Seepage into the South Fork Coeur d'Alene River System, Northern Idaho, 1999. USGS Water-Resources Investigations Report 2001-4274. Prepared in cooperation with U.S. Environmental Protection Agency.

Bookstrom, A. A., S. E. Box, J. K. Campbell, K. I. Foster, and B. L. Jackson. 2001. Lead-rich sediments, Coeur d'Alene River Valley, Idaho: area, volume, tonnage, and lead content. U.S. Geological Survey Open-File Report, 01-140.

Bookstrom, A. A., S. E. Box, R. S. Fousek, J. C. Wallis, H. Z. Kayser, and B. L. Jackson. 2013. Baseline, Historic and Background Rates of Deposition of Lead-Rich Sediments on the Floodplain of the Coeur d'Alene River, Idaho. U.S. Geological Survey Open-File Report 2004-1211. http://pubs.usgs.gov/of/2004/1211/.

Box, S. E., A. A. Bookstrom, M. Ikramuddin, and J. Lindsay. 2001. Geochemical analyses of soils and sediments, Coeur d'Alene drainage basin, Idaho: sampling, analytical methods, and results. U. S. Geological Survey Open-File Report 01-139.

Brunner, J. 2017. Lake Management Successes 2010-2017. Idaho DEQ.

CH2M Hill. 2015. Technical Memorandum Addendum E-1—Riverbank Characteristics, Erosion Rates, and Lead Contribution. Architect and Engineering Services Contract No. 68-S7-04-01. December 2015.

CH2M Hill. 2017. Technical Memorandum Addendum E-6—Riverbed Characterization. EPA Contract No. EP-W-06-021. June 2017.

CH2M Hill for EPA. 2013. Central Impoundment Area Groundwater Collection System Design Definition Report. Bellevue, Washington: CH2MHill.

CH2M Hill and URS Corp. 2001. Final Ecological Risk Assessment. Coeur d'Alene Basin Remedial Investigation/Feasibility Study. Prepared by URS and CH2M HILL for EPA. Contract No. 86-W-98-228. Work Assignment No. 027-RI-CO-102Q. 1,820 pp. May 18, 2001.

Clark, G. M. 2003. Occurrence and Transport of Cadmium, Lead, and Zinc in the Spokane River Basin, Idaho and Washington, Water Years 1999–2001. USGS Water-Resources Investigations Report No. 2002-4183.

Clark, G. M., and C. A. Mebane. 2014. Sources, Transport, and Trends for Selected Trace Metals and Nutrients in the Coeur d'Alene and Spokane River Basins, Northern Idaho, 1990–2013. U.S. Geological Survey Scientific Investigations Report 2014-5024.

Cohn, T. A., D. L. Caulder, E. J. Gilroy, L. D. Zynjuk, and R. M. Summers. 1992. The validity of a simple statistical model for estimating fluvial constituent loads: An empirical study involving nutrient loads entering Chesapeake Bay. *Water Resour. Res.* 28(9):2353–2364.

Cohn, T. A., and H. F. Lins. 2005. Natures style: Naturally trendy. *Geophysical Research Letters*. 32(23):5 pp. https://doi.org/10.1029/2005GL024476.

Deubel, A., and W. Merbach. 2005. Influence of Microorganisms on Phosphorus Bioavailability in Soils. *In:* Microorganisms in Soils: Roles in Genesis and Functions. F. Buscot and A. Varma (eds.). Berlin Heidelberg, Germany: Springer-Verlag.

Downs, C. J., and J. Stocks. 1977. *Environmental Impacts of Mining*. Wiley, New York.

Ekholm, P., and K. Krogerus. 2003. Determining algal-available phosphorus of differing origin: routine phosphorus analyses versus algal assays. *Hydrobiologia* 492:29–42.

Ellis, B., and J. Stanford. 1988. Phosphorus bioavailability of fluvial sediments determined by algal assays. *Hydrobiologia* 100:9–18.

Ellis, B. K., J. A. Craft, and J. A. Stanford. 2015. Long-term atmospheric deposition of nitrogen, phosphorus and sulfate in a large oligotrophic lake. *Peer J* 3:e841. https://doi.org/10.7717/peerj.841.

EPA (U.S. Environmental Protection Agency). 2010. OU-2 and OU-3 Removal and Remedial Actions Timeline, Bunker Hill Superfund Site. Focused Feasibility Study, Upper Basin of the Coeur d'Alene River.

EPA. 2012. Upper Basin Interim ROD Amendment—Interim Record of Decision Amendment, Upper Basin of the Coeur d'Alene River, Bunker Hill Mining and Metallurgical Complex Superfund Site, August 2012. https://semspub.epa.gov/work/10/664107.pdf.

EPA. 2013. Optimization Evaluation, Bunker Hill Mining and Metallurgical Complex Superfund Site, Central Treatment Plant (CTP), Kellogg, Shoshone County, Idaho. EPA 542-R-13-004. https://www.epa.gov/sites/production/files/2015-07/documents/bunkerhill_optimizationreport_final_jul2013.pdf

EPA Region 10. 2015. Fourth Five-Year Review Report for Bunker Hill Superfund Site. Shoshone and Kootenai Counties, Idaho. Seattle, Washington: EPA Region 10.

EPA. 2016. Aquatic Life Ambient Water Quality Criteria Cadmium—2016. EPA-820-R-16-002. Washington, DC: EPA Office of Water.

EPA. 2019. Superfund FY 2019. Annual Accomplishments Report. EPA 540R20001. https://semspub.epa.gov/work/HQ/100002479.pdf.

EPA. 2020. Adaptive Management Project Management Plan for the Lower Basin of the Bunker Hill Mining and Metallurgical Complex, Shoshone County, Idaho.

EPA Region 10. 2020. Optimization Review Report Remedial Action Process Optimization Study: Lake Coeur d'Alene Bunker Hill Mining and Metallurgical Site Operable Unit 03 Coeur d'Alene, Kootenai County, Idaho.

EPA Region 10. 2021. Fifth Five-Year Review Report for the Bunker Hill Mining and Metallurgical Complex Superfund Facility (Bunker Hill Superfund Site) Shoshone, Kootenai, and Benewah Counties in Idaho, Spokane County in Washington.

Helgen, S. O., and J. N. Moore. 1995. Natural background determination and impact quantification in trace metal-contaminated river sediments. *Environ. Sci. Technol.* 30:129–135.

Helsel, D. R., R. M. Hirsch, K. R. Ryberg, S. A. Archfield, and E. J. Gilroy. 2020. Statistical methods in water resources. U.S. Geological Survey Techniques and Methods, Book 4, Chapter A3, 458 p. https://doi.org/10.3133/tm4a3.

Hirsch, R. M., D. L. Moyer, and S. A. Archfield. 2010. Weighted regressions on time, discharge, and season (WRTDS), with an application to Chesapeake Bay river inputs 1. *Journal of the American Water Resources Association* 46(5):857–880.

Hirsch, R. M., and L. A. De Cicco. 2015. User guide to Exploration and Graphics for RivEr Trends (EGRET) and dataRetrieval: R Packages for Hydrologic Data (version 2.0, February 2015). U.S. Geological Survey Techniques and Methods Book 4, Chap. A10. doi:10.3133/tm4A10.

Hirsch, R. M., S. A. Archfield, and L. A. De Cicco. 2015. A bootstrap method for estimating uncertainty of water quality trends. *Environmental Modelling & Software* 73:148–166.

Hornberger, M. I., S. N. Luoma, M. J. Johnson, and M. Holyoak. 2009. The influence of remediation in a mine-impacted river: do improvements upstream impact metal trends over large spatial and temporal scales? *Ecological Applications* 19(6):1522–1535. https://doi.org/10.1890/08-1529.1.

Horowitz, A. J., K. A. Elrick, J. A. Robbins, and R. B. Cook. 1993. The effect of mining and related activities on the sediment-trace element geochemistry of Lake Coeur d'Alene, Idaho. Part II, Surface sediments. U.S. Geological Survey Open-File Report 93-656.

Idaho State Historical Society. 2019. Idaho State Historical Society Silver Valley Tour. 44 pp.

IDEQ and Coeur d'Alene Tribe. 2009. Coeur d'Alene Lake Management Plan. State of Idaho Department of Environmental Quality.

IDEQ and CDA Tribe. 2020. Coeur d'Alene Lake Management Program: Total Phosphorus Nutrient Inventory, 2004–2013.

Jenne, E. A., and S. N. Luoma. 1975. Forms of Trace Elements in Soils, Sediments, and Associated Waters: An Overview of Their Determination and Biological Availability. *In:* Biological Implications of Metals in the Environment. Proceedings of the Fifteenth Annual Hanford Life Sciences Symposium at Richland, Washington, September 29–October 1, 1975. H. Drucker and R. E. Wildung, Eds., Published by Technical Information Center Energy Research and Development Administration, Oak Ridge, Tenn., CONF-750929.

Knox, A. S., M. Huntz Paller, and K. L. Dixon. 2014. Evaluation of Active Cap Materials for Metal Retention in Sediments. *Remediation Journal* 24(3):49–69. https://doi.org/10.1002/rem.21394.

LaCroix, T. 2015. A Nutrient Mass Balance of Fernan Lake, Idaho and Directions for Future Research. M.S. Thesis, College of Graduate Studies, University of Idaho, Moscow, ID.

Langman, J. B., J. D. Ali, A. W. Child, F. M. Wilhelm, and J. G. Moberly. 2020. Sulfur species, bonding environment, and metal mobilization in mining-impacted lake sediments: column experiments replicating seasonal anoxia and deposition of algal detritus. *Minerals* 10(10):849.

Li, B., and M. T. Brett. 2012. The impact of alum based advanced nutrient removal processes on phosphorus bioavailability. *Water Research* 46:837–844.

Li, B., and M. T. Brett. 2013. The influence of dissolved phosphorus molecular form on recalcitrance and bioavailability. *Environmental Pollution* 182:37–44.

Li, B., and M. T. Brett. 2015. The relationship between operational and bioavailable phosphorus fractions in effluents from advanced nutrient removal systems. *International Journal of Environmental Science and Technology* 1–12.

Long, K. R. 1998. Production and disposal of mill tailings in the Coeur d'Alene mining region, Shoshone County, Idaho: Preliminary Estimates. U.S. Geological Survey Open-File Report 98-595.

Luoma, S. N., and P. S. Rainbow. 2008. Metal Contamination in Aquatic Environments: Science and Lateral Management. Cambridge: Cambridge University Press.

Mann, H. B. 1945. Nonparametric tests against trend. *Econometrica* 13:245–259.

Maul Foster and Alongi. 2020. 2019 Bathymetric Survey, Sediment Pin Installation, and Riverbed Sediment Sampling.

McBride, G. 2019. Has water quality improved or been maintained? A quantitative assessment procedure. *Journal of Environmental Quality* 48(2):412–420. http://doi.org/10.2134/jeq2018.03.0101.

McBride, G., R. G. Cole, I. Westbrooke, and I. Jowett. 2014. Assessing environmentally significant effects—A better strength-of-evidence than a single P value? *Environmental Monitoring and Assessment* 186(5):2729–2740. https://doi.org/10.1007/s10661-013-3574-8.

Moore, J. N., and S. N. Luoma. 1990. Hazardous wastes from large scale metal extraction: a case study. *Environ. Sci. Technol.* 24:1279–1285.

Moreen, E. 2021a. The groundwater below contaminated floodplains of the CDA valley can be of low pH if the sulfuric acid generated from the wastes is not oxidized. Presentation to the NASEM Committee. May 4, 2021.

Moreen, E. 2021b. Mining Legacy & Remedial Actions Bunker Hill Superfund Site. Presentation to the NASEM Committee. July 19, 2021.

Morra, M. J., M. M. Carter, W. C. Rember, and J. M. Kaste. 2015. Reconstructing the history of mining and remediation in the Coeur d'Alene, Idaho Mining District using lake sediments. *Chemosphere* 134:319–327. https://doi.org/10.1016/j.chemosphere.2015.04.055.

National Research Council (NRC). 2005. Superfund and Mining Megasites: Lessons from the Coeur d'Alene River Basin. Washington, DC: National Academies Press. https://doi.org/10.17226/11359.

Nimick, D. A., and J. N. Moore. 1991. Prediction of water-soluble metal concentrations in fluvially deposited tailings sediments, Upper Clark Fork Valley, Montana, USA. *Appl. Geochem.* 6(6):635–646.

Paulson, A. J. 2001. Biogeochemical removal of Zn and Cd in the Coeur d'Alene River (Idaho, USA), downstream of a mining district. *Science of the Total Environment* 278(1-3):31–44. doi:10.1016/s0048-9697(00)00886-x.

Prestbo, K. 2021a. Hydraulic and Sediment Transport Characteristics Lower Basin Coeur d'Alene River. Presentation to NASEM Committee. EPA. February 2021.

Prestbo, K. 2021b. Presentation to the NASEM Committee. EPA. July 2021.

Robertson, W. D., H. Brodie-brown, D. van Stempvoort, and S. Schiff. 2019. Review of phosphorus attenuation in groundwater plumes at 24 septic system sites. *Science Tot. Environ.* 1–42.

Rose, C., A. Parker, B. Jefferson, and E. Cartmell. 2015. The characterization of feces and urine: A review of the literature to inform advanced treatment technology. *Crit Rev Environ Sci Technol* 45:1827–1879.

Rowland, F. E., C. A. Stow, L. T. Johnson, and R. M. Hirsch. 2021. Lake Erie tributary nutrient trend evaluation: Normalizing concentrations and loads to reduce flow variability. *Ecological Indicators* 125:107601.

Tipping, E., S. Benham, J. F. Boyle, P. Crow, J. Davies, U. Fischer, H. Guyatt, R. Helliwell, L. Jackson-Blake, A. J. Lawlor, D. T. Monteith, E. C. Rowe, and H. Toberman. 2014. Atmospheric deposition of phosphorus to land and freshwater. *Env. Sci. Proc. Impacts* 16:1608–1617.

Wang, M., Y. Zhu, L. Cheng, B. Andserson, X. Zhao, D. Wang, and A. Ding. 2018. Review on utilization of biochar for metal-contaminated soil and sediment remediation. *Journal of Environmental Sciences* 63:156–173. https://doi.org/10.1016/j.jes.2017.08.004.

Watts, D. B., and H. A. Torbert. 2009. Impact of gypsum applied to grass buffer strips on reducing soluble P in surface water runoff. *Journal of Environmental Quality* 38(4):1511–1517.

Wise, D. 2021. Landscape Changes and Phosphorus Delivery to Coeur d'Alene Lake. Presentation to the NASEM Committee, May 4, 2021.

Woods, P. F., and M. A. Beckwith. 1997. Nutrient and trace-element enrichment of Coeur d'Alene Lake, Idaho. USGS Water Supply Paper 2485. https://doi.org/10.3133/wsp2485.

Zhang, X., L. A. Vincent, W. D. Hogg, and A. Niitsoo. 2000. Temperature and Precipitation Trends in Canada during the 20th Century. *Atmosphere-Ocean* 38(3):395–429.

Zinsser, L. M. 2019. Trace metal and nutrient loads from groundwater seepage into the South Fork Coeur d'Alene River near Smelterville, northern Idaho, 2017. U.S. Geological Survey Scientific Investigations Report 2019–5113. https://doi.org/10.3133/sir20195113.

Zinsser, L. M. 2020. Trends in Concentrations, Loads, and Sources of Trace Metals and Nutrients in the Spokane River Watershed, Northern Idaho, Water Years 1990-2018. USGS Scientific Investigations Report 2020-5096. https://doi.org/10.3133/sir20205096.

4

In-Lake Processes: Hydrodynamics

The goal of the next three chapters is to analyze Coeur d'Alene (CDA) Lake water column data collected over the past 30 years in order to assess trends in water quality. In conducting its analyses, the committee set about to elucidate the main water column processes at work affecting water quality in order to make estimates about future conditions, which are elaborated on in Chapter 10. Unlike Chapter 3, which focused on metals and phosphorus, the in-lake analyses consider additional parameters relevant to the condition of the Lake, including physical parameters such as water motions, specific conductance, pH, and temperature (Chapter 4); dissolved oxygen, nutrients, and other parameters indicative of the Lake's productivity, such as chlorophyll a (Chapter 5); and metals, including arsenic (Chapter 6). In addition to illuminating water quality trends, the available in-lake spatial and temporal data are evaluated by the committee to better understand the Lake's potential for eutrophication and, ultimately, metals mobilization from sediments, which is the subject of Chapter 7.

INTRODUCTION TO LAKE PROCESSES

Seasonal and long-term water quality trends within a lake are the result of the interactions among the key physical, chemical, and biological processes and the associated process drivers that alter inputs, outputs and internal dynamics within the lake, as illustrated in Figure 4-1. The main high-level processes relevant within CDA Lake include the lake hydrodynamics, sediment transport, sediment biogeochemistry, and primary productivity (examples of specific processes within those four categories are given below). Each of these processes is influenced by a subset of drivers that can affect the water quality within a lake both spatially and temporally. These drivers include riverine inputs (e.g., metal, nutrient, and sediment loads) associated with river flows and lake inputs, such as the meteorological forcing at the air–water interface. The drivers determine the mixing, chemical loads, and photochemical and biogeochemical dynamics within a lake that lead to changes to water quality parameters, such as pH, phosphorus and nitrogen concentrations, dissolved oxygen, and chlorophyll a.

Figure 4-1 illustrates a number of the key interactions among processes and drivers in CDA Lake and is referred to repeatedly in the next four chapters. This chapter covers relevant hydrodynamic processes, such as the upstream discharges from the St. Joe River and CDA River. These inflows enter CDA Lake at variable depths, depending on both river and lake temperatures (which change seasonally), and they contribute to the general south-to-north advection within the Lake. The CDA River provides the main source of particulate and dissolved metals, while the cleaner St. Joe River serves to dilute metal concentrations in the southern reach of the lake (e.g., at locations such as C5), with the extent of dilution controlled by in-lake hydrodynamics and physical processes

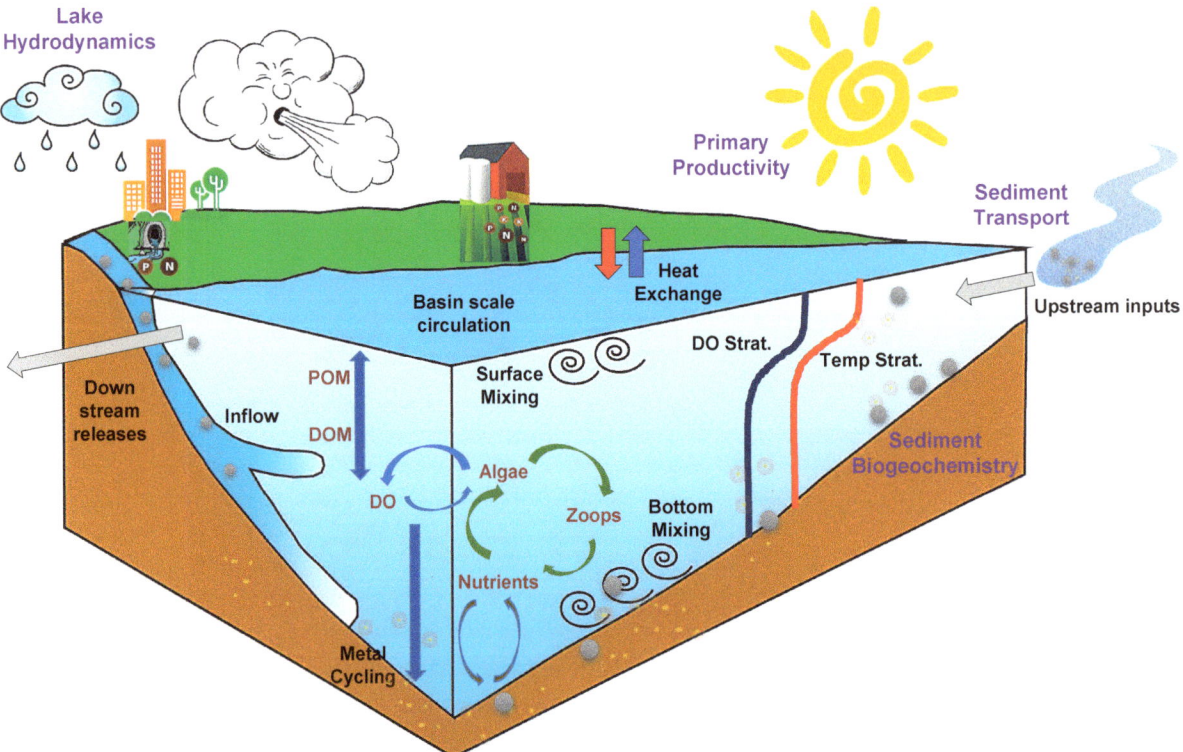

FIGURE 4-1 Schematic of in-lake processes, drivers, and water quality parameters. From right to left are shown a stream input to the lake; temperature (Temp.) profile showing warming at the surface and the sharp thermocline separating the warm epilimnetic water from the cool hypolimnetic water; a dissolved oxygen (DO) profile, with reduced DO in the hypolimnion; surface mixing due to wave breaking and nighttime cooling; bottom mixing due to currents interacting with the rough lake bottom; basin scale horizontal gyres that can develop due to persistent winds; downstream release of water; and a cross-section of a second stream inflow indicating that the inflow will either flow to the bottom or intrude at intermediate depth depending on the relative density of the inflow and the lake. The heat exchange arrows at the surface represent the impacts of meteorology acting at the surface of the lake. NOTE: DOM = dissolved organic matter, POM = particulate organic matter, Zoops = zooplankton.

such as sediment transport. Internal hydrodynamics have redistributed particulate metal contamination through the northern and southern Lake, as evidenced by widespread sediment contamination. The details of these hydrodynamic processes are not well known (as discussed below) but could be critical to understanding recovery of Lake water quality into the future.

Also discussed in this chapter are sediment transport processes that can have a significant impact on water quality within CDA Lake. Turbidity in the water column impacts photosynthesis by reducing the depth of light penetration. Settling of particles including inorganic solids and algae detritus can transport metals and nutrients from the euphotic zone of the water column to bottom waters and the sediment. If remediation reduces the particulate metal load to the Lake, then particles entering the Lake that settle to the bottom can serve as a capping layer over previously deposited sediments.

Discharges from the rivers as well as nonpoint sources can influence phosphorus concentrations in the Lake, which may lead to increases in phytoplankton biomass and production in the Lake as lake temperatures warm and the days get longer in the spring and summer. Dissolved and particulate organic matter in the water column can bind metal ions and alter their concentrations. A yearly occurrence in lakes (including CDA Lake) is thermal stratification, in which colder and denser bottom waters become isolated from warmer surface waters. Thermal stratification allows some dissolved metals to accumulate in bottom waters, impacts sediment transport through reduced velocities, and can lead to lower dissolved oxygen concentrations in the lake bottom waters (hypolimnion).

Such in-lake processes affecting dissolved oxygen, nutrients, and chlorophyll *a* are explored in Chapter 5, while the processes affecting metals are discussed in Chapter 6.

At work in the bottom waters of CDA Lake are sediment biogeochemistry processes that control metal speciation, adsorption/desorption to sediments, and pore water solubility through changes in pH and redox conditions, as well as lake hydrodynamics like mixing that control the inputs and exchanges to the sediment. In addition, reduced pH caused by organic particle decomposition in low dissolved oxygen areas of the water column (Davison, 1993) can lead to metal ion (e.g., zinc) release from settling particles and from bottom sediments into bottom waters, as observed by Kuwabara et al. (2000). These processes are all discussed in Chapter 7.

Figure 4-1 is representative of in-lake processes *during the summer*. The drivers that influence metal concentrations and speciation, thermal stratification, nutrient cycling, and primary and secondary productivity all change seasonally. For example, spring snowmelt leads to higher river flow rates and the associated increase in sediment transport. Particles entering the Lake in the spring are primarily inorganic materials from the watershed, including colloidal/nanosized materials (see, e.g., Davison, 1993). When inflows decrease in late spring and early summer, inorganic materials settle in the Lake or are exported via the Spokane River, the Lake clears, and productivity increases. Particulate material becomes dominated first by phytoplankton and then by detrital organic particles through summer and fall. In some lakes, the settling of organic particles during and after periods of phytoplankton growth can strip some metals from the water column and rerelease them in deeper waters (Balistrieri et al., 2002). Seasonal changes in dissolved oxygen and pH due to photosynthesis can also have significant impacts on metal speciation and the abundances and diversity of biota in the Lake—both at the sediment–water interface and within the water column itself. Higher water column oxygen concentrations in the winter and spring result from the higher oxygen solubility at lower temperatures, but phytoplankton productivity in the summer can further increase dissolved oxygen concentrations in the epilimnion. Throughout this and the two subsequent chapters, the available data are analyzed to assess the impact of seasonality on metal and nutrient concentrations.

LAKE HYDRODYNAMICS

The hydrodynamics of a lake involve the water movements that arise on account of external forcing (such as the surface heat exchange, wind, inflows and outflows) and internal responses to that forcing. Both occur within a physical domain (the system boundaries). In the case of the external forcing, that physical domain is the watershed (described by the topography and land use types), while for a lake it is the lake basin, defined by its bathymetry and the water surface.

The external forcing is impacted by the surrounding topography; the spatial characteristics of the watershed; and the temporal changes in the hydrology, meteorology, and climate. These temporal changes occur over a broad range of scales (minutes through years). Alterations in land use within the watershed can also exert impacts. For example, after land clearing, enhanced erosion can increase the suspended sediment concentration of a stream, and hence its nutrient load and, in extreme cases, its density relative to the lake water. These changes generally occur at longer timescales than variability in meteorology and hydrology.

The responses of the lake to the forcing are numerous and complex, particularly at those times of year when a lake's water column is thermally stratified (and hence density stratified). When a lake is thermally stratified, physical forces bring it to a state in which isotherms (contours of equal temperature) are horizontal, thereby preserving a minimal potential energy state. When external forcing is imposed, such as by a wind blowing over the surface, the isotherms become tilted, and lake motions are initiated to reestablish horizontal isotherms. Such motions are characteristically oscillatory and are often referred to as internal waves. Stratification generally occurs during the warmer times of year (spring, summer, fall), but even winter can exhibit weak stratification that exerts an influence on the hydrodynamics. This is even true for times of the year when a lake or part of it is ice-covered (see Wüest and Lorke, 2003).

Figure 4-2 shows a simplified case of a two-layer stratified lake to illustrate how a lake responds to external forcing. The responses are of two general types: barotropic and baroclinic. Barotropic responses are exemplified by the action of the wind driving large, basin scale, horizontal gyres of the surface waters of the lake. The surface drag force allows momentum from the wind to be transferred to the water, and the gyres are constrained by the

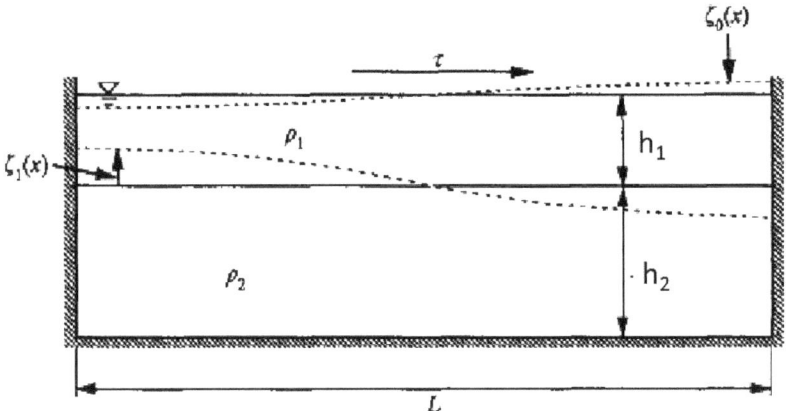

FIGURE 4-2 A cross-section of a hypothetical two-layer stratified lake. The upper layer of depth h_1 and density ρ_1, rests upon the lower layer of depth h_2 and density ρ_2. Wind exerting a shear stress τ on the surface causes surface seiches, resulting in a lake surface tilt with a maximum amplitude ζ_o. In response, an internal wave is set up at the interface resulting in an opposite tilt of the interface that has a magnitude of ζ_i. Generally, $\zeta_o \ll \zeta_i$. SOURCE: Stevens and Imberger (1996).

shape of the lake basin. The wind can also cause sloshing motions of the water within the basin, known as surface seiches. These basin-scale surface waves typically have periods on the order of minutes for lakes the size of CDA Lake, and amplitudes of millimeters (see Figure 4-2, where the surface seiches are shown to have an amplitude of ζ_o). More apparent are wind waves that move across the lake surface. These typically have periods of 2 to 5 seconds and maximum amplitudes of up to 1 to 2 meters. Though these are often the most visible lake motion, they are generally of little importance outside of the littoral zone, where they can break and cause shoreline erosion and resuspend sediments. For lakes the size of CDA Lake, the earth's rotation has no effect on barotropic motions.

Baroclinic motions arise when horizontal pressure gradients are established, due to tilting of a lake's isotherms, as described previously. Water of two different temperatures at the same horizontal level across the lake (if, for example, the lake isotherms were tilted) would constitute a horizontal pressure gradient. This arrangement is gravitationally unstable, and internal lake motions ensue that eliminate such gradients. The motions take the form of internal waves and vertical upwellings (water rising up at the boundary) and downwellings (water forced down near the boundary). In Figure 4-2, the internal wave amplitude is shown as $\zeta_i(x)$. On the left side of the schematic, where the interface is deviated upward, upwelling is occurring, while on the opposite side there is downwelling. Once the wind stops, the internal wave oscillates back and forth until it is damped by viscosity. The effect of the earth's rotation complicates these motions further, and phenomena such as Kelvin waves (edge-trapped waves) can also arise. The baroclinically driven motions occur over long timescales (hours to days), can generate high current velocities (tens of centimeters per second), and often exhibit large vertical amplitudes (meters to tens of meters).

The hydrodynamic responses of a lake to external forcings can be broken down into a number of individual processes that will now be briefly described. As the responses depend to a great degree on whether a lake is stratified or not, the discussion of processes is separated into "summer" and "winter" to denote the strongly stratified (summer) conditions and the well-mixed or weakly stratified conditions (winter).

Summer Hydrodynamics

During summer in temperate latitudes (see Figure 4-1), lakes stratify in temperature due to a net positive heat exchange at the surface (more heat gained than lost). Surface mixing processes due to wind mixing, and convective cooling of the surface result in the formation of a surface mixed layer (epilimnion) separated from the deeper water by a sharp gradient in temperature (the thermocline). The underlying water is referred to as the hypolimnion, and it is often weakly stratified. Stratification is typically a continuum, with the strongest stratification at the thermocline and the weaker stratification below. An important feature of stratification is that it inhibits the vertical exchange of dissolved substances, including nutrients, metals, and dissolved oxygen. The latter is particularly important because in deep portions of a lake below the thermocline, dissolved oxygen is generally being lost due to decomposition and sediment oxygen demand. After sufficient time, this loss can lead to hypoxia (low oxygen) and eventually anoxia (no oxygen)—conditions that favor the release of some nutrients, metals, and metalloids from the sediments.

Inflows from the streams arrive with a temperature that determines their density. In extreme cases, high suspended sediment concentrations can increase that density. When a stream inflow reaches a stratified lake, its density can be either lower than, higher than, or the same as the lake's. Depending on which case it is, the stream will either overflow, underflow, or form an interflow near the depth of the thermocline. Whichever it does, the inflow carries its "load" of constituents and these get inserted at that depth. If the loads are particulate, they will in time settle out. Note that the stream temperature and the lake surface temperature are both changing throughout the day, often with a significant time lag for the air temperature. Thus, the relative density difference between the lake and the inflow can vary throughout the day and throughout the season. The flowrate and the shape of the inflow channel also play a part in enhancing the mixing and dilution that occurs to the inflow as it is entering the lake. The dilution of the particulate and dissolved contaminant concentrations in the river inflow can sometimes be as high as a factor of 10 (Ayala et al., 2014).

Bottom mixing processes are driven by currents moving across the lake bed. Generally, hypolimnetic currents in a lake are weak (on the order of 1 cm/s), resulting in relatively little mixing. However, higher velocities can be generated by baroclinic motions and internal waves (10–50 cm/s), and these can result in mixing. They can also result in sediment resuspension (Roberts et al., 2021) and reoxygenation of the sediments.

Basin-scale gyres (see Figure 4-3) at the surface of a lake are generally the result of the wind blowing across the lake surface. Though playing little role in the vertical mixing process of the lake, they can distribute epilimnetic material very efficiently and could, for example, trap algal blooms in a particular embayment.

In winter, the same range of hydrodynamic processes are at play, but with some important differences. Often the lake temperature is isothermal due to lake cooling processes that keep the water body well mixed. As a result, dissolved and suspended material are generally isotropic (no direction gradients). Weak stratification can sometimes persist. If ice forms, a reverse temperature gradient can be established that will stratify the lake, with the heaviest 4°C water at the lake bottom, and lighter 0°C water immediately below the ice. Under these conditions, dissolved oxygen will again be lost at the bottom (although decomposition slows at the lower temperature), and it is possible for hypoxia to occur at the lake bottom. With the physical ice cap on the lake, surface aeration is cut off, and dissolved oxygen can be depleted at a faster rate than in summer. Typically, inflows are low during winter, so while the same range of processes can all occur, they generally have a small effect on a lake.

Hydrodynamics Data Analysis

There have been relatively few physical limnology and hydrodynamic studies of CDA Lake, beyond the long-term monitoring conducted by state and tribal entities (see Chapter 2). This has led to a paucity of information on the key physical processes at play in the lake, and a near total absence of time-series data that would allow such an understanding to be developed. The two exceptions are (1) a detailed 12-day field measurement program conducted in 2005 and associated three-dimensional (3-D) model development completed in 2007 (Dallimore et al., 2007; Hipsey et al., 2007; Morillo et al., 2008) and (2) the deployment of a continuous profiling station by the CDA Tribe at station C5 during three periods (in 2011, 2015, and 2019).

2005 Experiment and Modeling

The 2005 experiment involved the installation of a thermistor chain and weather station called LDS, repeated transects along the thalweg of the Lake (plus several cross-sections) using a multi-parameter probe (F-Probe), and microstructure profiling at one station (see Figure 4-4). The experiment ran from approximately June 1 to June 11, 2005, and coincided with a time when the St. Joe River at Chatcolet had flow rates falling from 3,590 cfs (112 cms) to 2,350 cfs (67 cms), and the CDA River at Harrison had flow rates falling from 1,860 to 1,260 cfs.

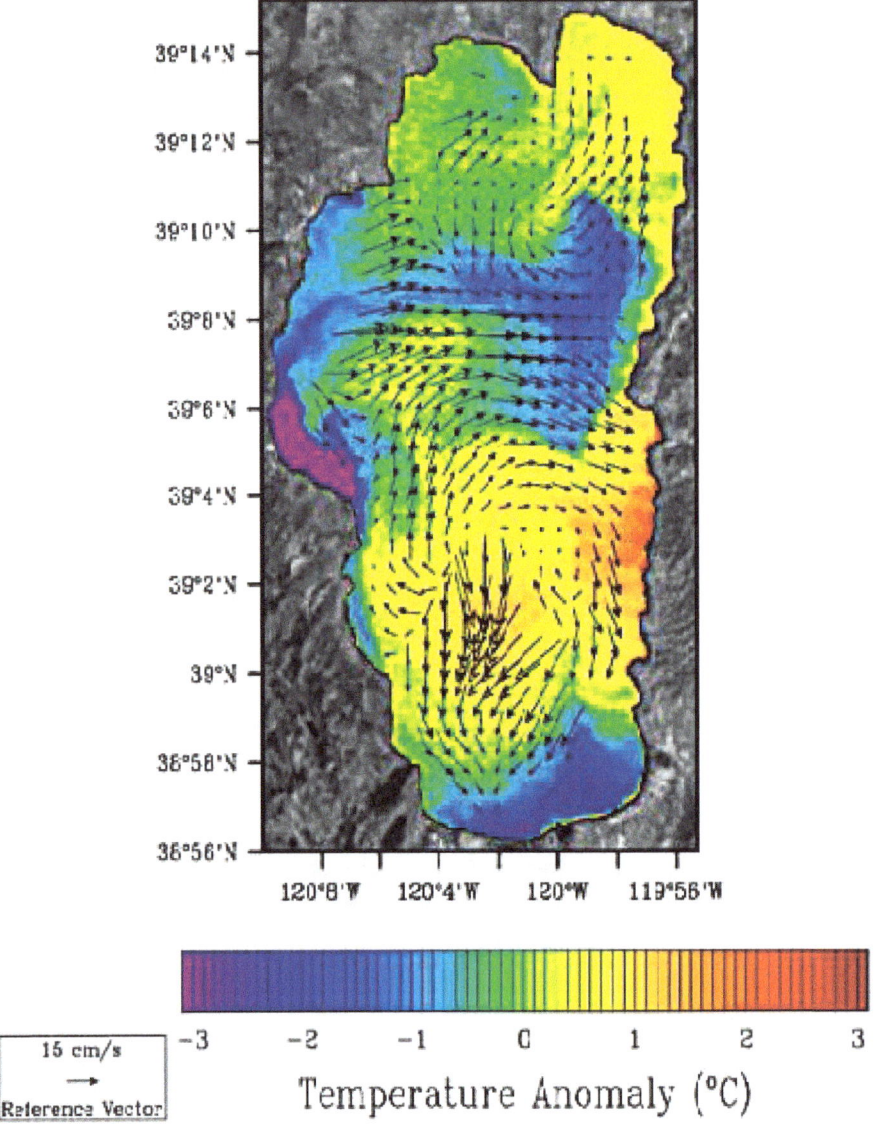

FIGURE 4-3 An example of basin-scale gyres recorded in Lake Tahoe. The circulation cells are indicated by the water velocity arrows and indicate a clockwise gyre in the south and a counterclockwise gyre in the north. The temperature anomaly indicates upwelling of very cold water on the west side (purple) and the formation of a jet as this water gets advected across the lake. Data from ETM+ Band 6 (high gain) temperature anomaly, June 3, 2001, 18:28 UTC. The ETM+ image was interpolated to a 90-m grid using bilinear interpolation. ETM+ refers to an Enhanced Thematic Mapper Plus sensor. SOURCE: Adapted from Steissberg et al. (2005).

These flowrates are mentioned specifically because their magnitudes appear to be an important determinant in the response of the Lake throughout the subsequent summer months.

Some representative temperature transects from the experiment are shown in Figure 4-5A. The mouth of the St. Joe River is to the left on the transects, and the mouth of the CDA River is at station 13. Station 60 (far right) is the main basin of CDA Lake opposite the city of Coeur d'Alene. The transects are composed of a number of F-Probe vertical profiles (locations marked by upside down triangles), and each transect took several hours to complete.

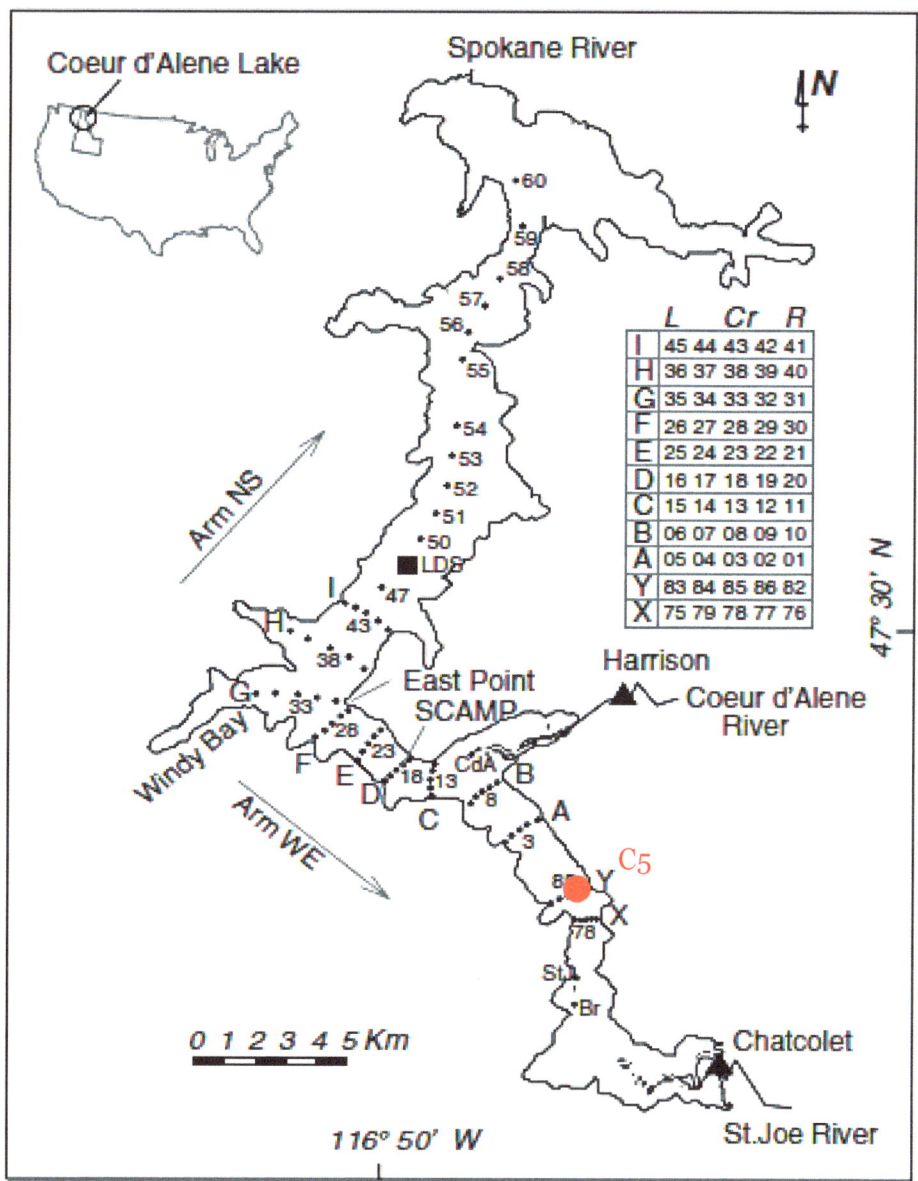

FIGURE 4-4 Map of CDA Lake showing main inflows, outflow, location of the temperature and weather station (LDS), U.S. Geological Survey (USGS) gage sites (Harrison and Chatcolet), and F-Probe profiling stations (dots) with their associated cross transects (X, Y, and A to I). The embedded table shows F-Probe profiling station names at each cross-transect organized from the left to the right shoreline. The microstructure profiler station is indicated as SCAMP. SOURCE: Morillo et al. (2008).

The vertical positions of the river inflows can be inferred from the lowest salinity (lowest specific conductivity) water in lake profiles, as both rivers are relatively low in conductivity compared to the Lake. In Figure 4-5B, which shows the corresponding transects for salinity, the dark blue region of low conductivity at the left on Panel 2 shows that the river water from the St. Joe entered the Lake as an interflow, meaning that because of its density relative to the stratification of the Lake, it intruded below the thermocline but above the lake bottom. The bright red region of high conductivity on the third panel shows an overflow being produced by the CDA River, intruding into the epilimnion above the thermocline. These observations suggest that the river inflows are not likely to play

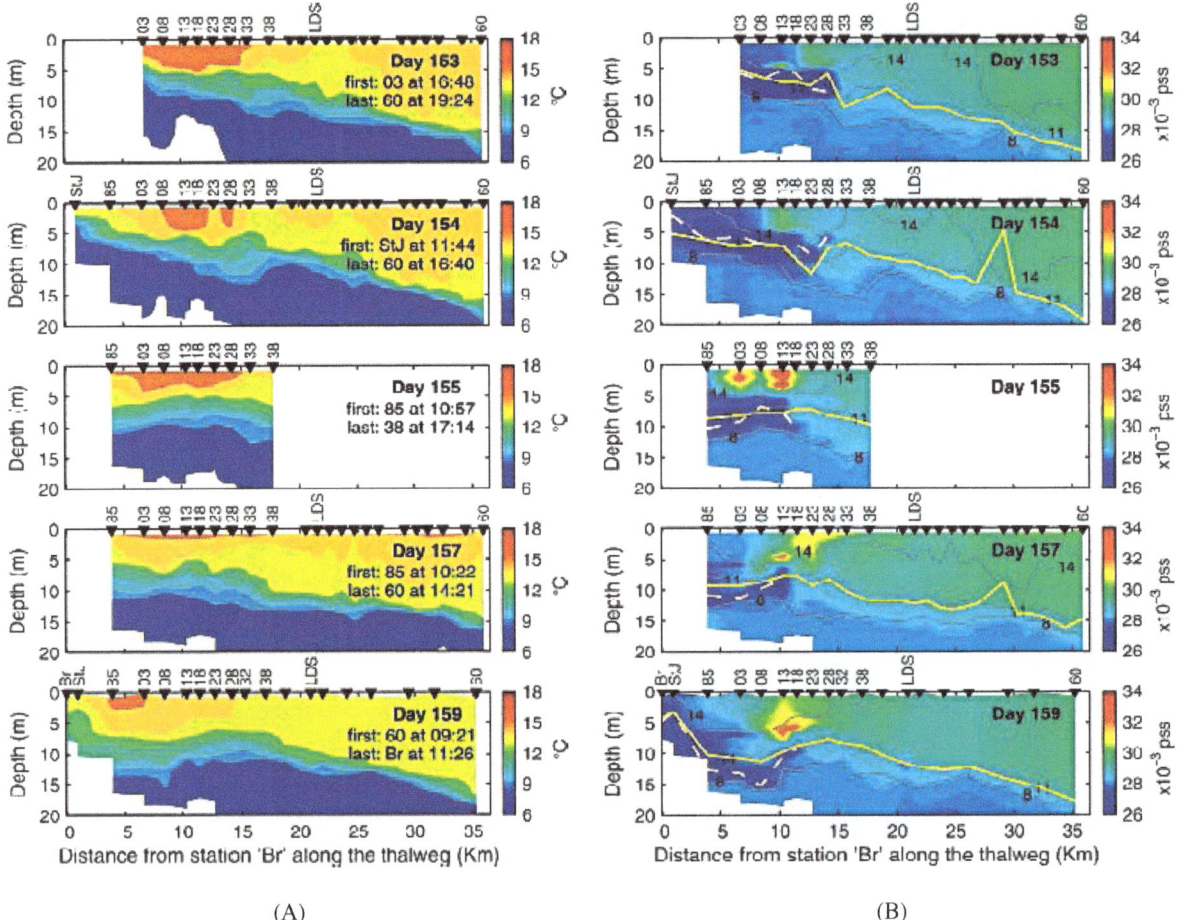

FIGURE 4-5 Representative temperature (A) and salinity (B) transects from the June 2005 experiment. The St. Joe River is to the left and the city of Coeur d'Alene is to the right. White dashed lines show depth of minimum salinity. Solid yellow lines indicate depth of maximum chlorophyll *a*. NOTE: pss = practical salinity scale. SOURCE: Morillo et al. (2008).

a major role in resuspending sediment within the Lake under the conditions shown. The low Froude numbers of the inflows (< 0.21), a metric that expresses the magnitude of the inflow momentum relative to the strength of the vertical stratification, also suggested that the intrusions did not possess sufficient momentum at this time to produce significant mixing, and that any water column mixing would either come from internal wave motions or from external wind-driven processes. Clearly, this may be different under higher flow conditions, although the flows would likely have to increase by a factor of 5 for this to occur.

Internal wave motions (not shown) were evident in the thermistor chain data (collected at 60-second intervals). Spectral analysis (a mathematical technique that allows identification of dominant, periodic motions) did not identify any obvious peaks, possibly because of the short observation period (11 days). There was little to suggest that internal wave (baroclinic) motions would be a major factor in CDA Lake mixing, although the data are very limited.

The ELCOM hydrodynamic model (see subsequent section), which underwent limited calibration and validation using the short data record from this experiment, was run to examine the transport between the lower east–west portion of the Lake, and the upper north–south portion of the Lake. The field data and the model results indicated that the dominant horizontal seiches had the effect of temporarily trapping river water in the lower part of the lake, in the region of the two river inflows. This observation may in part explain subsequent observations of high suspended metals concentrations at station C5 (see Chapter 6). Numerical "tracer" experiments were also conducted with the model, whereby modeled tracers were released from the St. Joe and the CDA Rivers. The model results suggested that the tracers from both rivers quickly became distributed through the epilimnion, as the observations suggest. It should be noted that the model was not run to explore conditions outside those observed during the experimental period or to infer what may be happening later in the summer or during the winter months.

Continuous Profiling at C5

CDA Tribe data from continuous measurements at site C5 were collected over the summer and fall of three different years, and therefore offer insights into the Lake response under three different sets of hydrologic conditions. While some of the observations that were made in the 2005 experiment are confirmed, the data suggest that the magnitude and timing of each year's river flow is a key variable in both the physical limnology and the water quality response. Furthermore, it should be remembered that there is a distinct difference in both the physical and biogeochemical behavior of the northern and southern (south of the mouth of the CDA River) Lake, such that the C5 data are not representative of the northern part of the Lake.

Figure 4-6 presents annual (water year) hydrographs for the St. Joe River for 2010–2020. Although each year is different, a notable characteristic that each year displays is the steep, monotonic falling limb, commencing around June. During high flows earlier in the year, the lower part of CDA Lake where station C5 is located is dominated by the flow from the St. Joe River, and it is only when the flow has fallen low enough that the Lake stratification starts to dominate the hydrodynamics and the water quality, particularly in the hypolimnion.

Station C5 (see Figure 4-4) was the location of the continuous profiling instrument that was operated by the CDA Tribe for several months in 2011, 2015, and 2019. In 2011 and 2015, data on temperature, electrical conductivity, pH, and dissolved oxygen were collected. In 2019, only temperature and electrical conductivity data were collected. The C5 profiler time series of temperature, specific conductivity and dissolved oxygen from 2011 are shown in Figure 4-7. The red bars at the top of the figure indicate when data were available (there are several data gaps that have been filled by interpolation). The flow for the St. Joe River between June 1, 2011, and December 31, 2011, is shown at the bottom.

As can be seen in Figure 4-7, temperature and specific conductance both indicate the presence of vertical oscillations throughout the data.[1] The temperature data indicate that thermal stratification is very surficial until early August, but then increases in intensity, and the depth of the thermocline descends from about 5 m to 8 m before cooling commences in early September. Stratification persists through mid-October. The specific conductance figure (middle of Figure 4-7) shows the clear signal of the St. Joe River intrusion as the blue "tongue." At the beginning of the plot, the dark blue color indicates that the water from the St. Joe is entering at the surface, but as the flow rate reduces over the summer, the intrusion is more dilute (it has entrained lake water) and is located lower in the water column but above the bottom of the Lake. The lowest panel shows dissolved oxygen concentration; low oxygen forms at the bottom of the Lake, but only falls to about 5–6 mg/L. The transition from inflow-dominated conditions to stratification-dominated conditions would appear to commence in early August, when the river flow was in the vicinity of 1,000 cfs.

The year 2015 was very different, particularly for stream inflow (Figure 4-8). At the beginning of the recorded period at C5, stratification was already established, with surface temperatures in excess of 20°C. Unlike 2011, there was no inflow-dominated period evident, as river flows were already below 800 cfs when profiling at C5 commenced. The specific conductivity indicated that the river inflow was present as a distinct underflow (dark blue in Figure 4-8), one that persisted throughout the year until vertical mixing occurred in late November. Most notably, dissolved oxygen became depleted at the bottom of the hypolimnion in mid-July and remained so until full vertical mixing occurred. The shape of the dissolved oxygen distribution suggests that the cause of the dissolved oxygen decline was sediment bed–induced, rather than due to biological oxygen demand from the river inflow. At times, dissolved oxygen levels were below 2 mg/L. It is not known how close to the sediment–water interface the instrument profiled, but it is believed to be approximately 1 m off the bottom.

---

[1] Although the zigzags represent internal waves, the committee does not interpret these waves to mean that there is high-intensity turbulence in the Lake that produces a lot of mixing. Waves are very efficient at transporting energy and produce mixing only when they break.

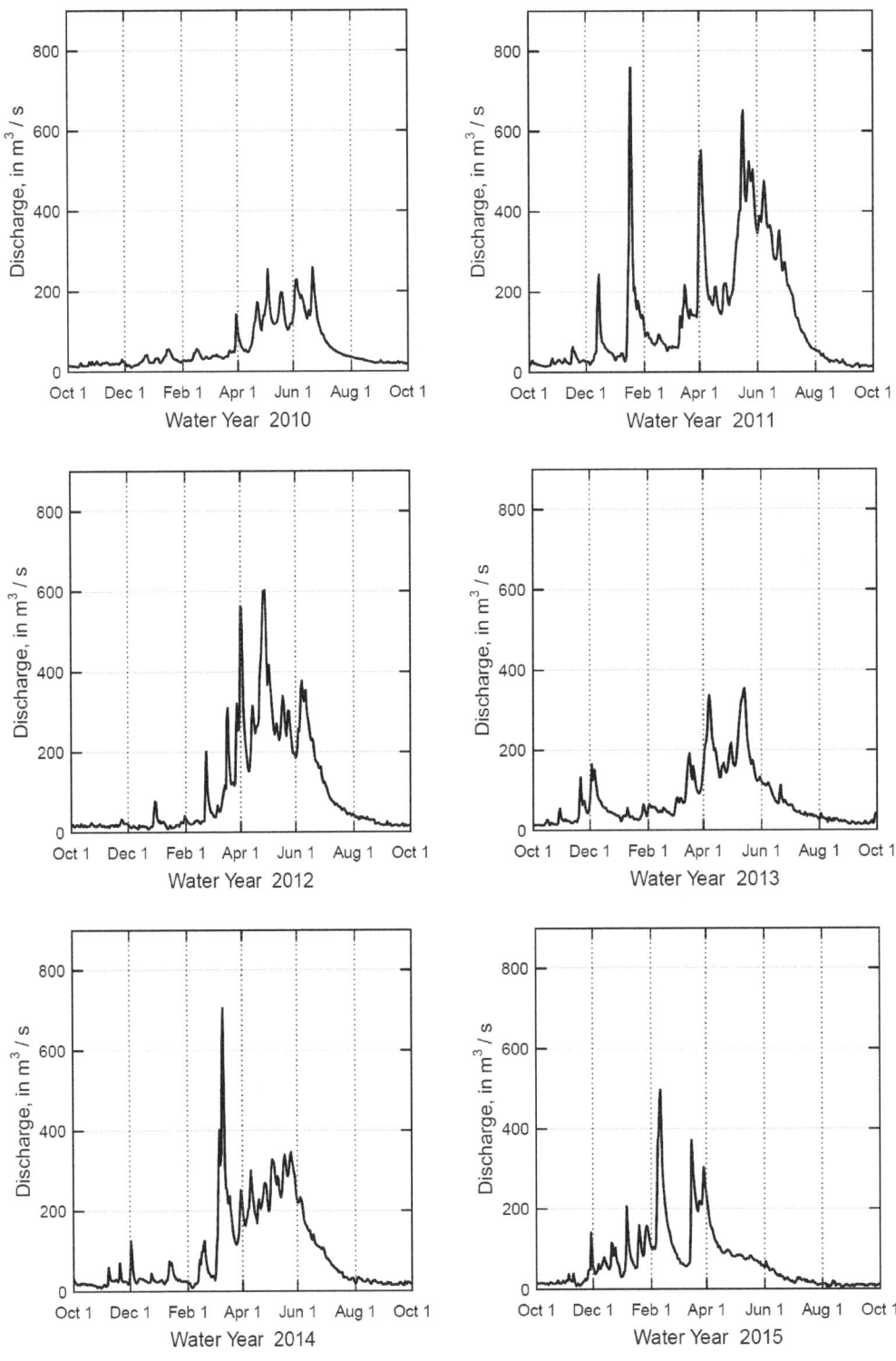

FIGURE 4-6 Annual hydrographs of the St. Joe River. SOURCE: Data courtesy of USGS and graphed by the committee.

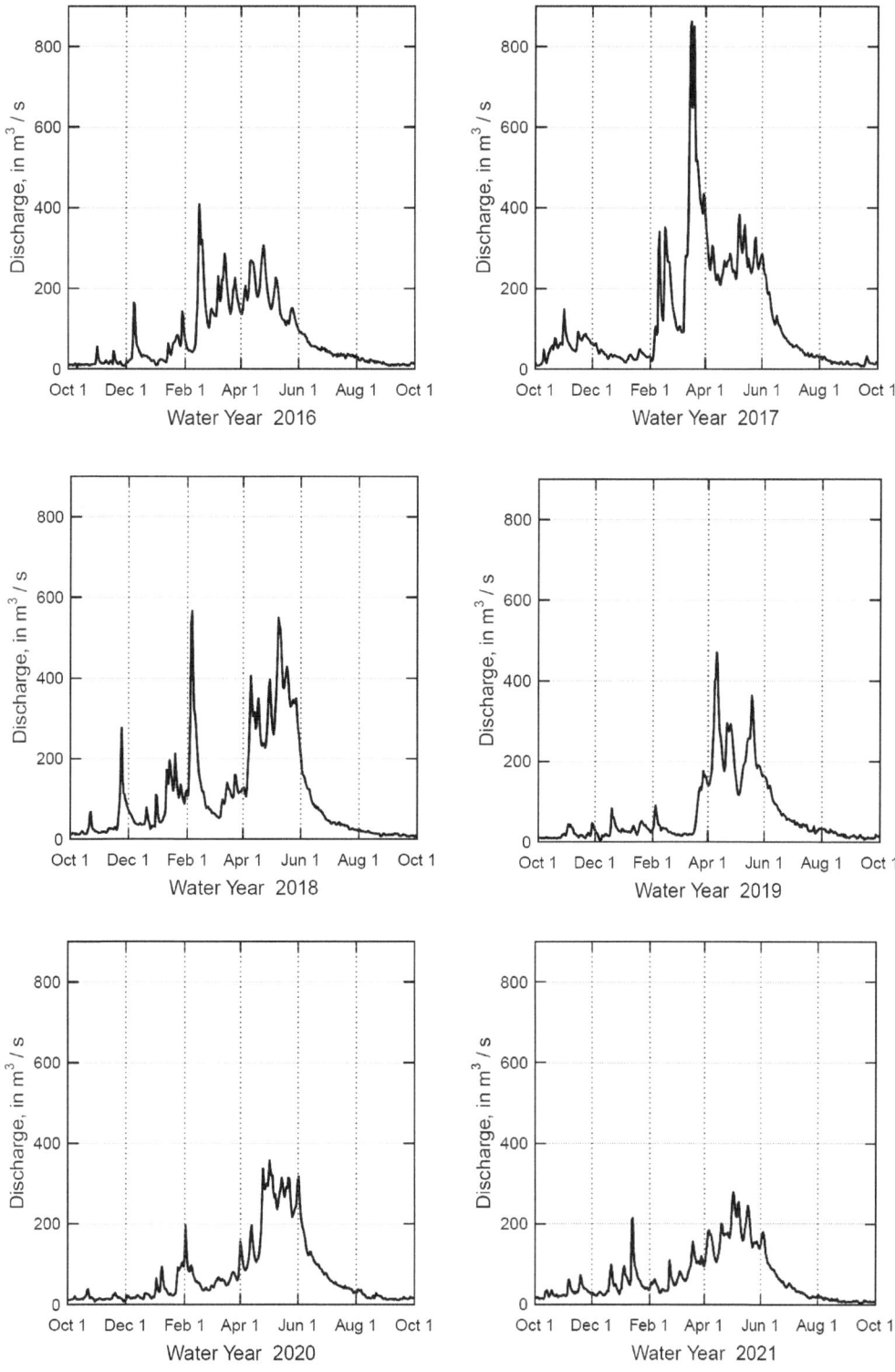

FIGURE 4-6 Continued

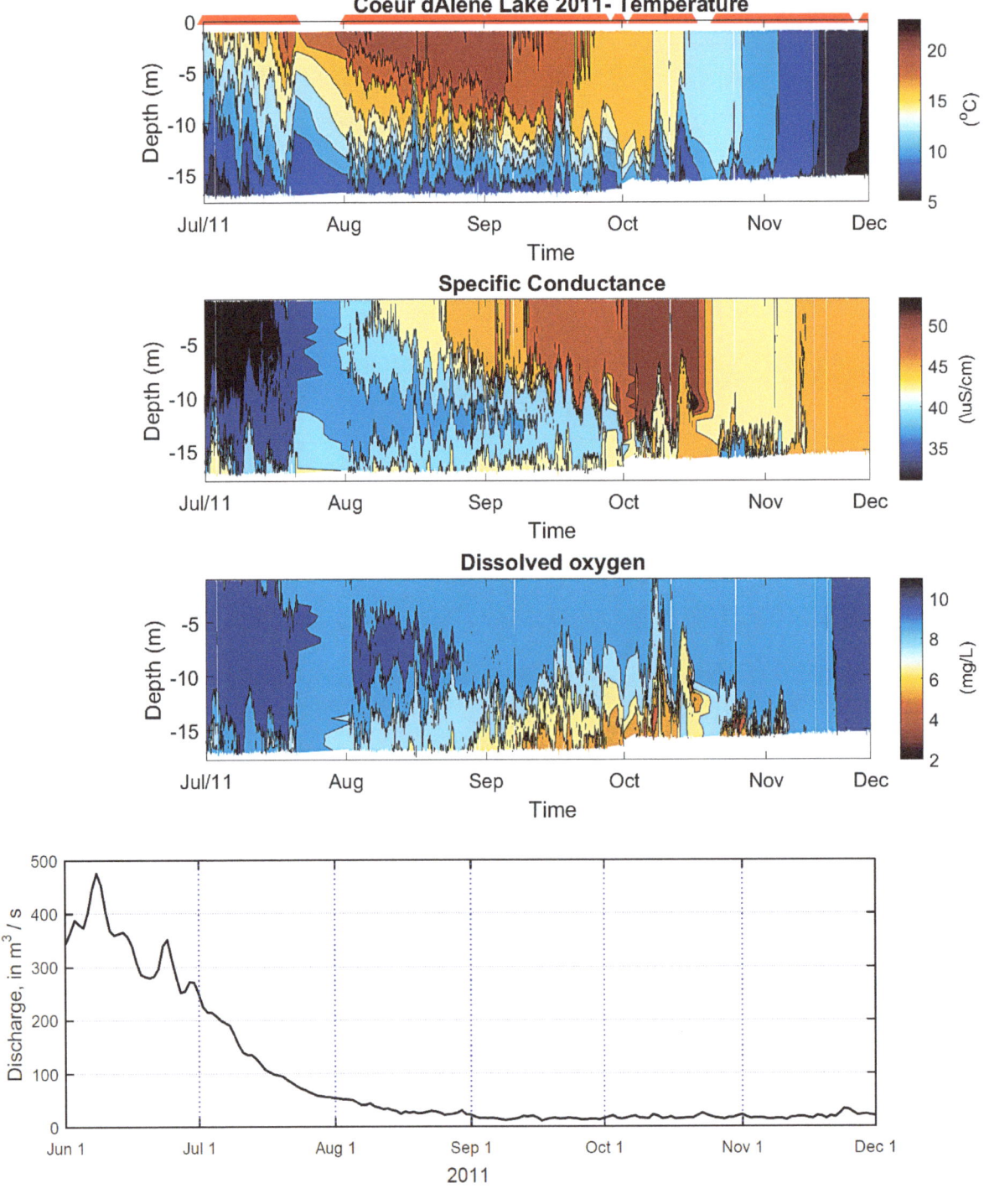

FIGURE 4-7 C5 profiler time series of temperature, specific conductivity, and dissolved oxygen from 2011. Note that for temperature and specific conductivity, redder colors indicate higher values. For dissolved oxygen, redder colors indicate lower values. Bottom panel is corresponding St. Joe River flow in 2011. Units of specific conductance are microsiemens per centimeter. SOURCE: Data plotted by the committee, courtesy of the CDA Tribe.

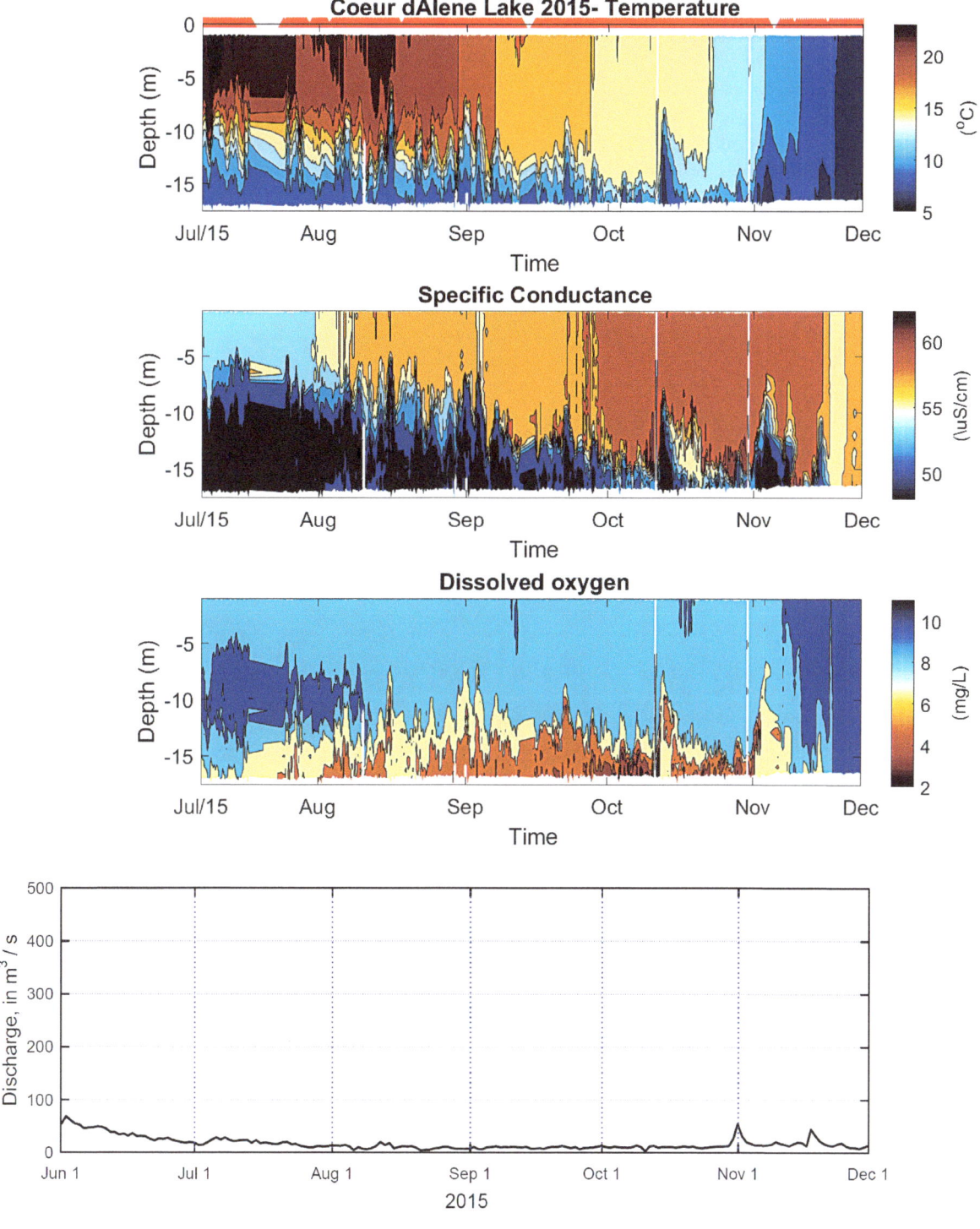

FIGURE 4-8 C5 profiler time series of temperature, specific conductivity, and dissolved oxygen from 2015. Note that for temperature and specific conductivity, redder colors indicate higher values. For dissolved oxygen, redder colors indicate lower values. Bottom panel is corresponding St. Joe River flow. SOURCE: Data plotted by the committee, courtesy of the CDA Tribe.

Conditions for the deployment of the C5 profiler in 2019 were intermediate between the two earlier deployments. The deployment was shorter (three months) and only temperature and specific conductivity were measured (Figure 4-9). Based on the position of the conductivity minimum, the river intrusion occurred as an interflow. Unfortunately, the absence of dissolved oxygen measurements precludes a better understanding of the impacts of the interflow. As was the case in 2015, the river flow fell below 1,000 cfs prior to commencement of profiling, so the transition between river-dominated and buoyancy-dominated conditions could not be observed.

Plotting the St. Joe River inflow against dissolved oxygen and pH at near-bottom (16 m depth) at station C5 reveals a relationship between river flow and water quality. In Figure 4-10, the 2011 profiler data from midnight on

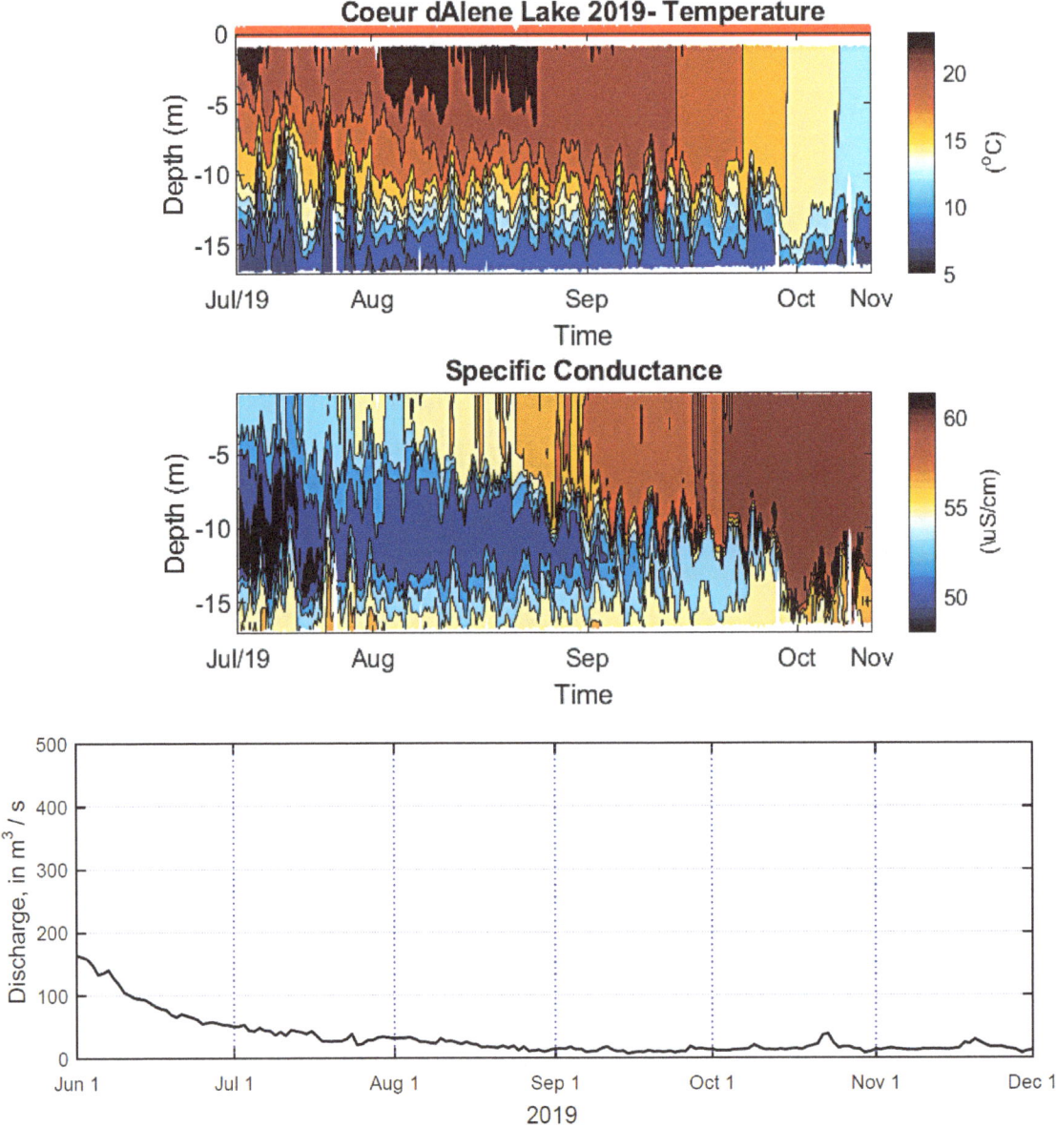

FIGURE 4-9 C5 profiler time series of temperature and specific conductivity from 2019. Note that for temperature and specific conductivity, redder colors indicate higher values. Dissolved oxygen data were not available for 2019. Bottom panel is corresponding St. Joe River flow. SOURCE: Data plotted by the committee, courtesy of the CDA Tribe.

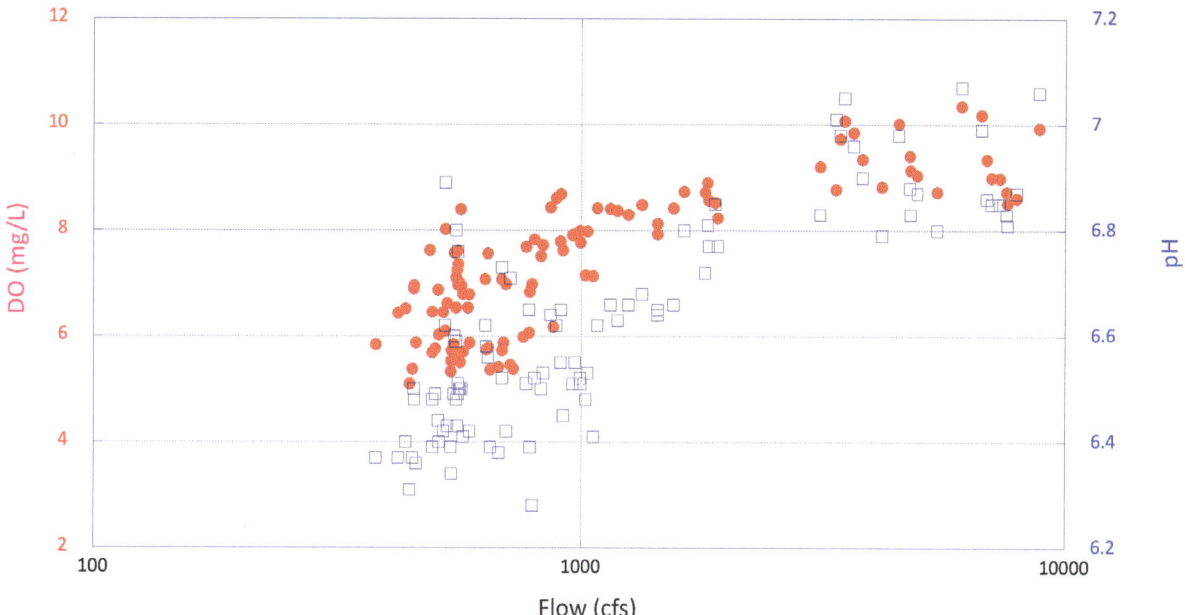

FIGURE 4-10 St. Joe River flow versus dissolved oxygen (red circles) and pH (open squares) at Site C5 (16 m depth) for the period July 1–October 31, 2011. SOURCE: Data plotted by the committee, courtesy of the CDA Tribe.

each day have been plotted (profiling actually took place at two-hour intervals). The year 2011 was a large water flow year, such that the St. Joe River did not fall below 1,000 cfs until early August. Figure 4-10 clearly shows that water quality (dissolved oxygen and pH) does not change systematically when flows are above 1,000 cfs and the system is in the inflow-dominated mode. Below this flow, when the system is transitioning to the buoyancy-dominated mode, both variables decrease in near-linear fashion, with dissolved oxygen falling from approximately 9 mg/L to 5 mg/L and pH falling from approximately 6.9 to 6.5. That is, below 1,000 cfs, the system transitions from river-dominated to the summer stratification period; this may happen in June, July, or August.

In the low-flow year 2015, the St. Joe River was flowing well below 1,000 cfs when the profiler at C5 was installed. Looking at an equivalent plot for 2015 (Figure 4-11), the data all lie along the falling trend-line seen in Figure 4-10, with dissolved oxygen decreasing from 9 mg/L to below 4 mg/L and pH falling in the range of 6.9 to below 6.4.

It is noteworthy that the 2005 hydrodynamic experiment (see previous section), probably the most extensive set of hydrodynamic measurements existing for CDA Lake, was made over a ten-day period when the flow in the St. Joe River was dropping from 3,590 cfs to 2,350 cfs. In other words, the upper portion of the Lake (in the vicinity of C5) where the largest water quality impacts are currently observed was in the midst of the inflow-dominated regime, and the impacts of thermal stratification were likely not occurring and therefore were not captured.

The take-home message from these analyses of hydrodynamic data is that "lake-like" conditions in CDA Lake only occur at low St. Joe River flows (defined as less than 1,000 cfs). Currently at C5, the length of this period is variable, such that hypoxic conditions start to form in some years where 1,000 cfs is reached sooner. If earlier river peaks were to occur under climate change, this period of lake-like conditions could commence sooner, extending the time for oxygen loss. Likewise, if warmer air temperatures in summer due to climate change protracted the stratified season (as it has done in many lakes around the world), then the stratification season is lengthening even further. This would be pushing the lower Lake (at C5) toward anoxia, and may create conditions in the northern Lake for hypoxic episodes in the future. The committee cautions that the C5 data are not representative of the northern Lake; thus, these conclusions may not extend to the northern Lake.

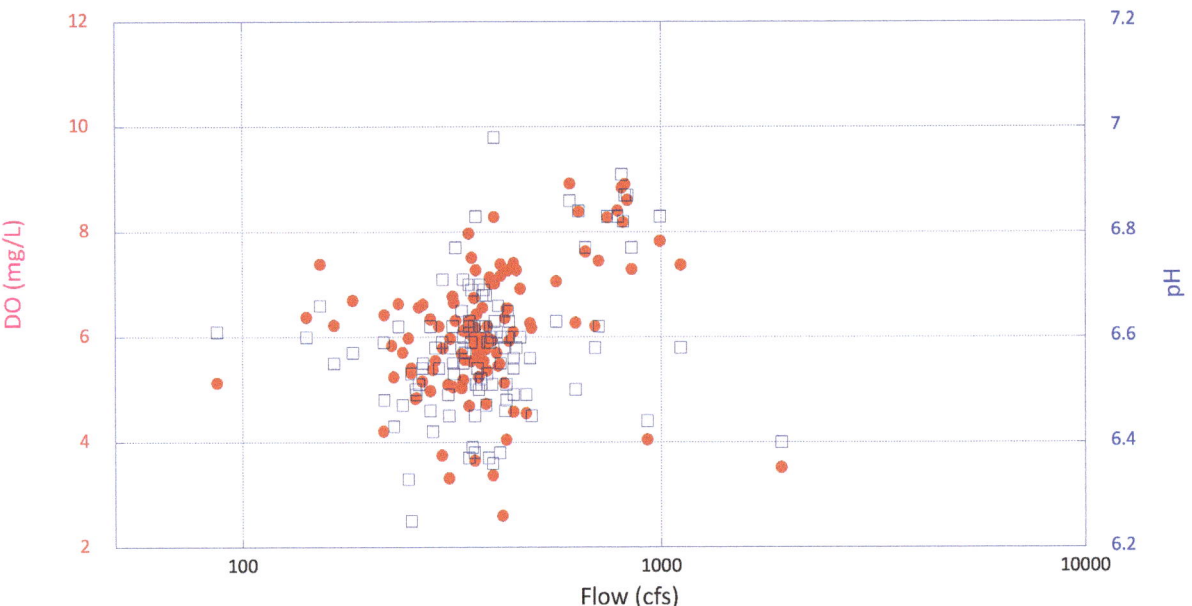

FIGURE 4-11 St. Joe River flow versus dissolved oxygen (red circles) and pH (open squares) for the period July 1–October 31, 2015. Data collection commenced as the flow was approaching 1,000 cfs, but the declining trend in DO and pH as flows decrease is evident. SOURCE: Data plotted by the committee, courtesy of the CDA Tribe.

SEDIMENT TRANSPORT

Sediment transport is a key physical process for the transport and distribution of metals and nutrients attached to sediment particles entering the Lake. The movement of sediments also impacts lake algal productivity and ecosystem dynamics because these sediments can control the availability of nutrients. Lastly, sediment particles in the water column also have an influence on in-lake water quality, for example, by increasing light attenuation and reducing light availability for algal productivity. Because there is no regular collection or analysis of sediment data in CDA Lake, the section below discusses the few special studies that have illuminated sediment transport processes.

Studies of In-Lake Sediments

There are several approaches to understanding sediment transport and morphological change in CDA Lake. These include investigating changes in lake bathymetry over time, analyzing surface and subsurface sediment samples, analyzing sediment core geochronology, and measuring suspended sediments in the water column over time and space. Each of these approaches can shed light on where sediments are moving within the Lake and whether they pass downstream of the Lake. Data on sediment grain size distribution, organic carbon content, and attached nutrients and metals are essential to understanding how suspended and deposited sediments influence the overlying water column chemistry, including algal productivity.

Bathymetry

Changes in lake bathymetry over time can be an indication of sediment erosion and deposition occurring in different areas of a lake. In order to interpret differences between bathymetric datasets, it is important to understand the spatial resolution and the accuracy of the data for each time period to assess the appropriateness of a comparison. Unfortunately, there appears to be limited bathymetric data over time for CDA Lake. Bathymetric surveys were conducted in 1994 by the USGS (Woods and Berenbrock, 1994), as shown in Figure 4-12, and in 2004 on behalf of Avista Corporation as part of its Federal Energy Regulatory Commission (FERC) relicensing process (CDA Tribe, 2004), as shown in Figure 4-13. The 1994 bathymetric survey consisted of only 561 data

IN-LAKE PROCESSES: HYDRODYNAMICS

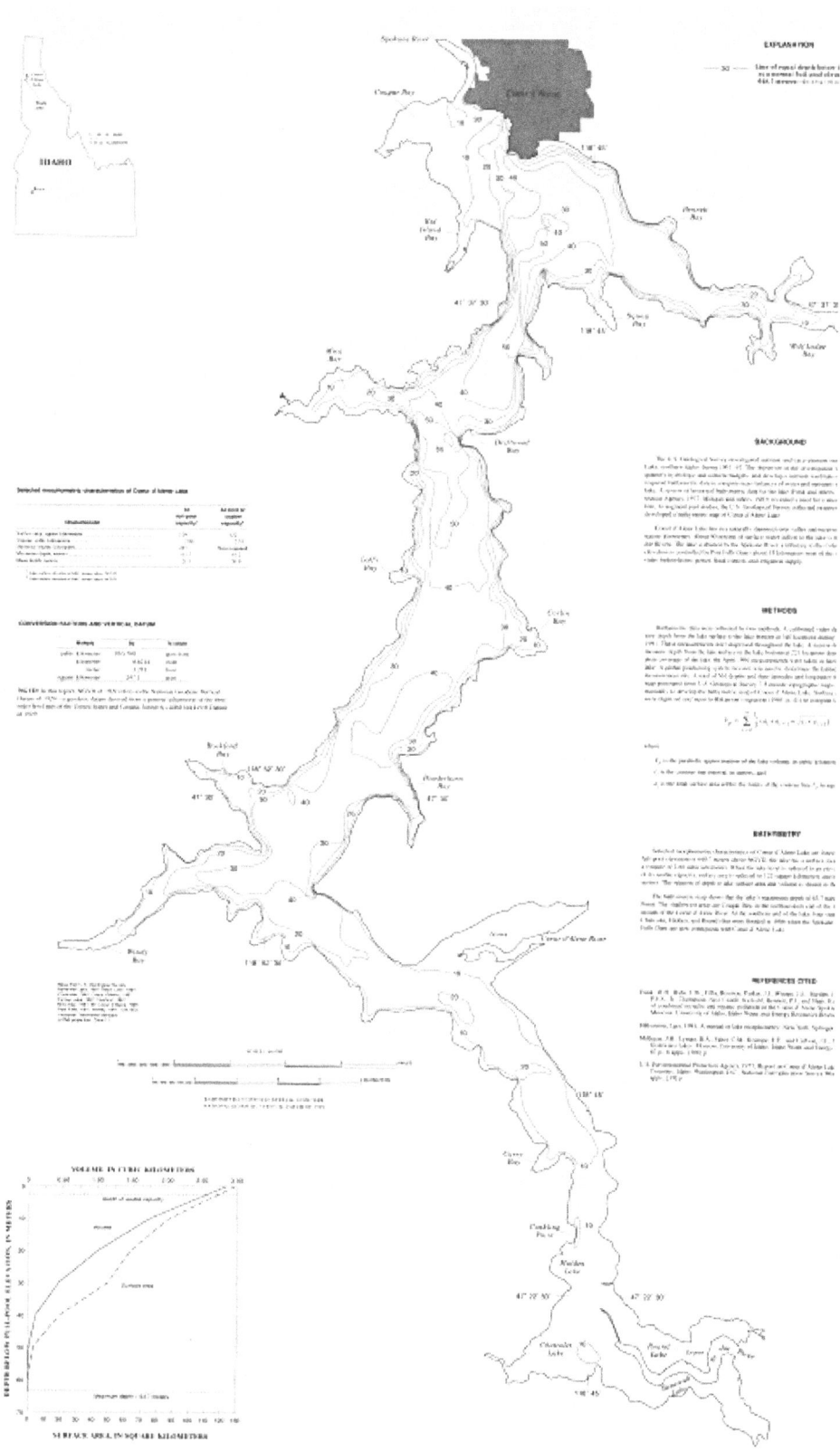

FIGURE 4-12 Bathymetric survey of CDA Lake. SOURCE: Woods and Berenbrock (1994).

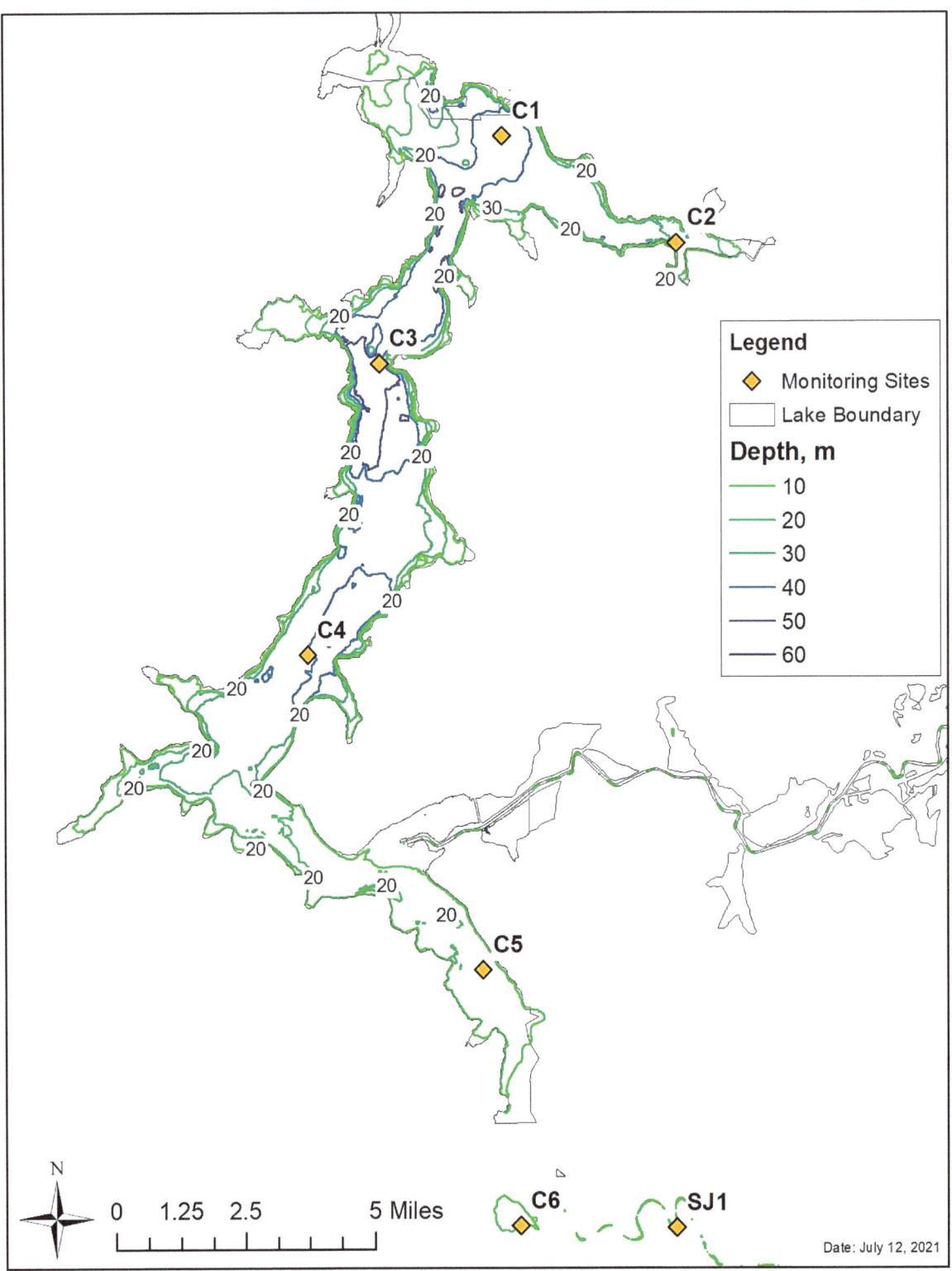

FIGURE 4-13 Bathymetric survey of CDA Lake, reproduced from GIS data. SOURCE: CDA Tribe (2004).

points used to create the map, which limits the ability to interpret the accuracy of the depth contours. The 2004 bathymetric survey consists of thousands of data points based on the geographic information system (GIS) data layer (CDA Tribe, 2004). As a result of this disparity, it is not possible to do a quantitative comparison between the data sets and identify erosion and depositional areas. Qualitatively, there appears to be little difference between the maps, and, given a rough sedimentation rate of 4 cm/decade (Horowitz et al., 1993), it is not surprising that there is little difference over the ten-year time period between surveys. In order to assess if some of the Lake bottom has filled in or eroded over time, high-resolution bathymetric surveys are needed periodically.

The bathymetric maps are useful for a few qualitative interpretations. The general flow direction in the Lake is from south to north from the St. Joe and CDA Rivers to the outlet at Post Falls Dam. Both figures show that where constrictions in the lake bathymetry occur, the lake deepens. The deepest parts of the Lake are along the center, with shallower areas along the sides and in the various embayments. These indicators suggest that the predominance of the flow is along the middle of the Lake, which is expected, but vertical velocity profiles are unavailable. The sides of the Lake and the embayments would be expected to have slower velocities and therefore be more depositional to sediments than the main channel. There are seasonal variations in flow (see Chapter 3), and in suspended sediments entering the Lake. These seasonal variations will impact where sediments entering the Lake will be deposited, whether sediments will travel farther into the Lake, or whether they will pass downstream. In order to better understand sediment transport in the Lake throughout the year, suspended sediment and velocity profiles would need to be collected several times over the year.

Sediment Sampling

Horowitz et al. (1993) collected surface sediment samples in CDA Lake (see open dots in Figure 4-14), the CDA River, and the St. Joe River in 1989 and additional river bank samples in 1991 from the South Fork of the CDA River and the mainstem CDA River. Surface sediments in the Lake from south of Rockford Bay to the mouth of the St. Joe River contained an extremely cohesive, light gray silty clay grain size layer 1.5–3 cm below the surface—representing the ash from the Mount St. Helens eruption in 1981, which implies a recent sedimentation rate of 0.3–0.5 cm/yr. St. Joe River sediments were observed to be substantially coarser than the Lake surface sediments, with fine to medium sand with some silt indicating a clear distinction between sources. The CDA River sediments were observed to be coarser-grained than in the river delta and in the Lake itself but were characterized as finer-grained than material along the banks of the river.

Horowitz et al. (1993) measured the Lake sediment samples for silver (Ag), arsenic (As), copper (Cu), cadmium (Cd), mercury (Hg), lead (Pb), antimony (Sb), and zinc (Zn) and found the chemical distribution patterns in the Lake surficial sediments to be consistent with the CDA River sediments as a major source. Highest concentrations were found adjacent to the CDA River delta, which is consistent with the reduction in water velocity as the river enters the Lake, reducing sediment transport capacity and the sediments dropping out of the water column. However, high trace element concentrations in sediments from the CDA River delta south to Conkling Point and well within Windy, Rockford, Mica, and Wolf Lodge Bays were also observed. Horowitz et al. (1993) suggested there may be secondary source(s) of some of the high trace elements or physical factors other than the predominant flow direction (south to north), such as wind generated wave/currents remobilizing fine grained trace element-rich sediments. In reviewing the data presented in Horowitz et al. (1993, Table 3), the sediment size fraction < 2 μm constituted the higher trace element concentrations farther from the CDA River delta, while the sediment size fraction of 8–16 μm constituted the highest trace element concentrations near the CDA River delta. These observations support the notion that fine-grained materials from the CDA River enter the Lake and get transported throughout the Lake due to high inflows passing through the Lake, wind-driven wave action remobilizing sediments, remobilization from local currents in the Lake, or some combination of the above.

In a complementary study, Horowitz et al. (1995) analyzed subsurface sediments from 12 gravity cores collected in CDA Lake in June 1990 (see black dots in Figure 4-14). The majority of these cores were observed to have a heavily banded upper section and a lower, more homogenous section. The banded sections varied in thickness and number, with the authors stating this was an indication of sediments settling over time and being remobilized. Sediment cores 9 (Windy Bay) and 10 (Wolfe Lodge Bay) had lower metals concentrations in the banded sections

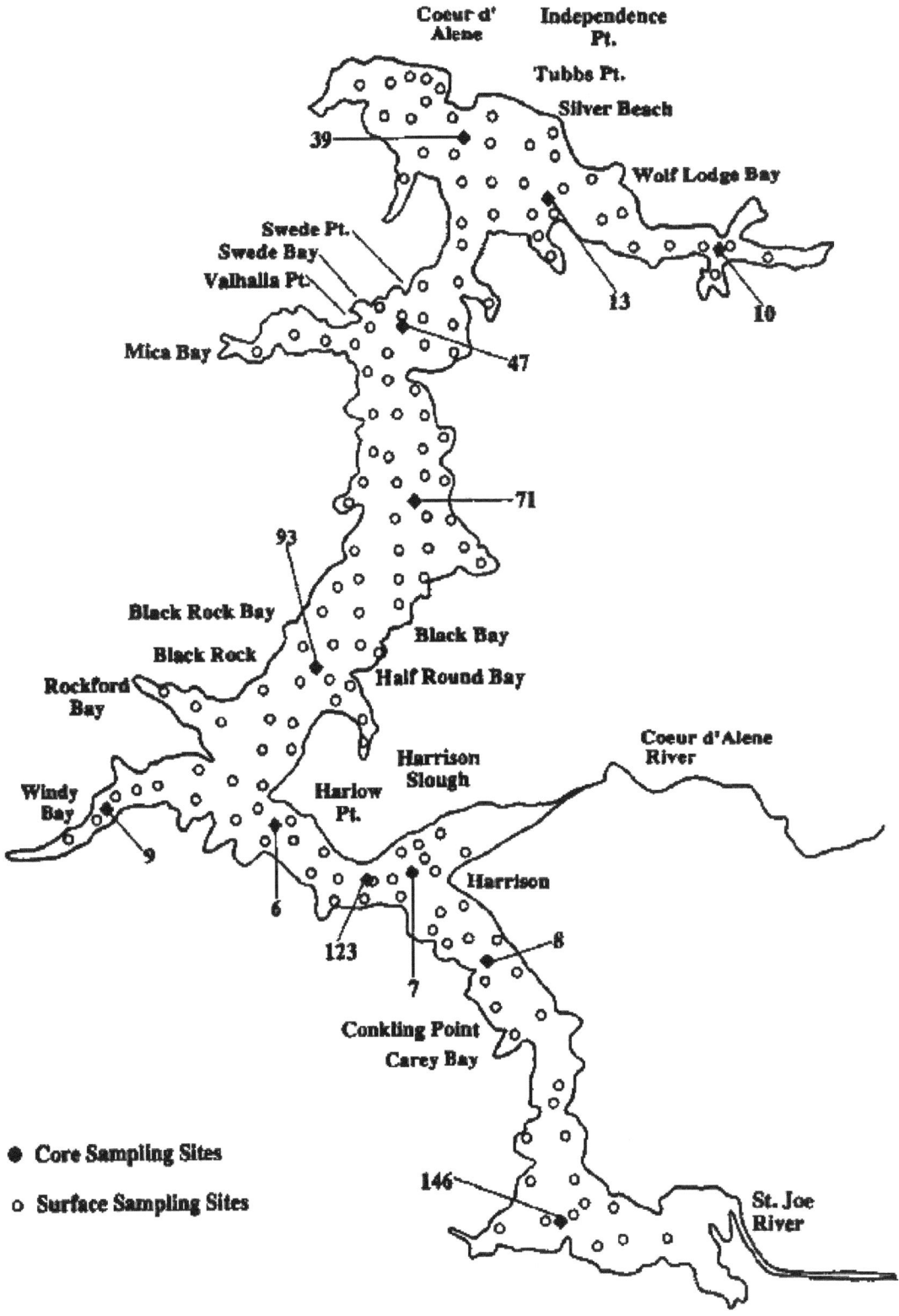

FIGURE 4-14 Map of CDA Lake surface sediment and core sampling sites. SOURCES: Horowitz et al. (1993, 1995).

TABLE 4-1 Sedimentation Rate Estimates for Core 123

Year Range	Sedimentation Rate, cm/yr
1980–1990	2.1
1965–1980	1.7
1959–1965	1.3
1954–1959	1.4

SOURCE: Horowitz et al. (1995).

than in the other cores. This may be due to the dominance of sediment deposition from small local streams and limited transport of sediments from the CDA River to these areas. Lower metals concentrations were observed in the banded section of the core at Site 8 (south of the CDA River), perhaps because St. Joe River flows prevented sediments from the CDA River from depositing (Horowitz et al., 1995).

Trace element data from the sediment cores were used to estimate that the total mass of metal-enriched sediment in CDA Lake is 75 million metric tons of sediments covering 85 percent of the Lake bed, which would indicate significant sediment transport through the Lake over time.

Sediment core 123 was used to conduct a geochronology analysis using cesium-137 activity. Considering a constant rate of deposition between dated points and no compaction, the sedimentation rates in Table 4-1 were estimated. The authors assumed the remainder of the sediment core (58 cm) had a sedimentation rate of 1.35 cm/yr, which means the undated core section represents 43 years and puts the start of the metal-enriched sedimentation in 1911, which is just after the 1880–1890 time period assigned to the onset of mining (Bender, 1991). Bookstrom et al. (2013) further interpreted sediment core 123 from Horowitz et al. (1995) and estimated that the period before 1895 had a sedimentation rate of about 0.1 cm/yr.

Once sediments enter CDA Lake, they tend to spread out based on local flow velocities and wind-driven wave action. This is demonstrated through the in-lake sediment sampling that indicates that trace element concentrations are highest near the mouth of the CDA River and decrease moving away from this location. However, high concentrations can still be found at some of the more secluded bays along the Lake.

Bookstrom et al.'s (2013) analysis of the sediment coring indicates that the net sedimentation rate has gone from 0.1 cm/yr at the turn of the 20th century to 2.1 cm/yr in the 1980s to 1990s, with higher sedimentation rates in the Lake bays than along the Lake axis. The reduced sedimentation along the Lake axis may be due to the Lake bathymetry and general direction of the Lake flow from south to north. The bathymetry surveys from 1994 and 2004 show the Lake has several constrictions along the south-to-north axis along the predominant flow path, and at these constrictions the Lake is deeper, likely maintained by this predominant flow path.

To confirm this conceptual site model of sediment transport in CDA Lake, additional data would be needed on suspended solids concentrations and grain-size distributions in the Lake, from the tributaries into the Lake, and in the flows out of the Lake. In addition, velocity data with depth in several places throughout the Lake—like along the main flow through the Lake, south of the CDA River, and in the bays—are key. Because sediments moving through the Lake carry organic carbon, nutrients, and metals, predicting future changes in Lake water quality will require better understanding of sediment quantities, physical and chemical characteristics of sediments, and how sediments interact biogeochemically across the Lake.

RESERVOIR MODELING OF COEUR D'ALENE LAKE

In lieu of data analysis, model simulations can provide insight into the processes at work in CDA Lake. Two models are discussed below, both of which consider in-lake processes that go beyond hydrodynamics and sediment transport. However, these models are discussed here because the data used to calibrate and validate the models were the C5 profile data discussed above. As argued below, both models would require substantial development before they could be used to simulate and accurately predict in-lake processes.

Aquatic Ecosystem Model 3D

The Aquatic Ecosystem Model (AEM3D) is a coupled hydrodynamics and biogeochemical model used for simulating lakes, reservoirs, estuaries and coastal oceans and is sold and supported by the company HydroNumerics.[2] AEM3D is based on the Estuary, Lake and Coastal Ocean Model–Computational Aquatic Ecosystem DYnamics Model (ELCOM-CAEDYM) developed by the Centre for Water Research in Australia. ELCOM-CAEDYM was used to model CDA Lake in 2007, using data collected in the dedicated field experiment conducted in 2005 and described earlier, along with longer-term seasonal data. In what follows, the model names ELCOM and CAEDYM are used instead of AEM3D, as those were the model names at the time the modeling was performed. AEM3D is presumed to have the same functionality.

ELCOM-CAEDYM can be used to simulate lake processes over timescales of days to years and at spatial scales of meters to kilometers. It includes a library of algorithms that represent many biogeochemical processes influencing water quality under the simulated physical conditions, including primary production; nutrient, carbon, and metal cycling; oxygen dynamics; and the transport and deposition of suspended solids.

The Hydrodynamic Model (ELCOM)

The hydrodynamic model ELCOM (see, e.g., Hodges et al., 2000) solves the unsteady Reynolds-averaged Navier-Stokes equations. ELCOM was run with data from May 20, 2005, to June 9, 2005, with a time-step of 300 seconds. The Coeur d'Alene bathymetry used for the initial simulations consists of a horizontal orthogonal grid with a constant horizontal grid size of 250 m. The vertical grid size is a uniform 1 m for the top 30 m of the water column and increases below this to 4 m at the bottom. The river-forcing data used were water temperature and flow rate data from the USGS gaging stations 12413860 (Coeur d'Alene River near Harrison), 12415140 (St. Joe River near Chatcolet), and 12419000 (Spokane River at Post Falls Dam).

The hydrodynamic model results were compared to high-spatial-resolution thermistor chain data and vertical profiles (described previously) that were acquired over a time period generally coinciding with the modeled period. The simulated temperatures captured some, but not all, of the measured features; no quantitative estimate of the quality of the match was provided. The 250-m model grid was unable to resolve the complex internal wave structure seen in the measured temperature field. Thus, it is uncertain the extent to which the model results can be relied upon to inform about the presence and absence of specific hydrodynamic processes (such as sediment resuspension, internal wave shear, and vertical diffusion). Although the model predicted currents and used these to produce estimates of the transport of modeled tracers, no water velocity data were acquired, and hence the fidelity of these modeled currents is unknown. The fact that the model was run for a relatively short time period (less than three weeks) and during spring also questions its ability to inform about processes occurring under the more strongly stratified and less inflow-dominated summer and fall periods.

A longer-term simulation (February 2003–December 2004) was conducted using straightened bathymetry with a 1,000 m × 250 m grid. Although the simulation results generally captured the onset and breakup of stratification and the overall Lake water balance, there was again no evidence that specific hydrodynamic processes were accurately represented.

The Biogeochemical Model (CAEDYM)

The objective for the application of CAEDYM to CDA Lake was to explore how the Lake may respond to altered loading of zinc and nutrients from the dominant inflows. The CAEDYM simulation includes descriptions of nutrient cycling and phytoplankton dynamics and operates at the same spatial and temporal scale as ELCOM. In addition to the base CAEDYM functionality, extensions were made to allow for more detailed simulation of heavy metals (in particular, zinc) within CDA Lake. The developments included the aqueous speciation of metals in response to changing geochemical conditions, inclusion of benthic fluxes of heavy metals from the sediment to the water column, uptake of metals into the algal biomass (and the associated toxicity effects), recycling of metals incorporated within the algal biomass during algal senescence, and sedimentation of heavy metals accumulated within algal or detrital material.

---

[2] https://www.hydronumerics.com.au/#about

The model was configured to include four phytoplankton groups (cyanophytes, chlorophytes, cryptophytes, and diatoms), based on parameters from a literature review. In addition to the phytoplankton, the simulation included organic matter, nutrients, and relevant inorganic ions. Fish, zooplankton or bacteria were not included in the model setup.

A new aqueous geochemistry sub-model was incorporated to allow for simulation of chemical reactions based on a user-defined set of elements (or components). The module is based on the equilibrium chemistry of aqueous solutions and their interactions with minerals and gases. The sub-model dynamically linked with the model's biological cycles such that any biological activity (e.g., algal nutrient uptake and photosynthesis) could dynamically affect the aqueous speciation. As part of the solution process, pH, ionic strength, and other variables are also calculated.

Model Application to Coeur d'Alene Lake

The ELCOM-CAEDYM model was applied to CDA Lake for 2004, as the best data coverage existed at that time. The results, as reported by Hipsey et al. (2007), did not demonstrate any visual similarity between modeled and measured water quality variables such as dissolved oxygen. No statistical measures of model accuracy were provided, but since field data were available for only six days over the 11-month simulation period, it is doubtful that any such measures would have been meaningful. The predicted values for several of the more conservative cations and anions, such as calcium and chloride, gave some level of agreement with observations, as shown in Figure 4-15. However, predictions of more important variables, such as zinc, showed a questionable match to the data, with no predictive ability for upstream sites (orange and red symbols in Figure 4-15). Although correlations of zinc, sulfate, chloride, and calcium based on all data appeared to provide reasonable values of R^2 (in the range of 0.4–0.6), the zinc and sulfate values, in particular, appeared to show little correlation between measured and modeled concentrations when looked at site by site.

ELCOM-CAEDYM is a potentially useful and powerful tool for both understanding many of the complex hydrodynamic and biogeochemical processes that are taking place within CDA Lake and for testing and informing management actions. However, based on the modeling studies that were performed in 2005–2007, the calibration and validation of both the hydrodynamic model (ELCOM) and the biogeochemical model (CAEDYM) are incomplete. This is in part due to the fact that there were insufficient data of the right types and for a long enough period of time. For the hydrodynamic model, where the results were able to demonstrate the ability to capture some of the complexity in this system, longer sets of time series data for temperature at three to four locations along the length of the Lake are needed. The thermistors should be distributed between the surface and the bottom at each site. A minimum of two years of data would be required to allow a full year of calibration to occur, with a separate year for validation. The collection of full water column velocity profiles would also be important to obtain for different seasons to compare with model simulations. Important processes such as sediment resuspension are largely dependent on bottom current velocities.

For the biogeochemical model, a well calibrated and validated model could play an important part in better understanding some of the knowledge gaps about the CDA system. Some key parameters for which the data could be used in conjunction with temperature data are dissolved oxygen, conductivity, turbidity and pH, all of which can be collected with fast-response probes. Dissolved oxygen is possibly the most important driver for water quality, and having two years of continuous data would assist not only the modeling, but also the fundamental understanding of the biogeochemical processes themselves. The nutrient, phytoplankton, and metals data that need to be collected by direct water sampling would benefit from monthly sampling (over 12 months) along the entire Lake. Combined with the continuous temperature and dissolved oxygen data, this would go a long way to providing precisely what is needed to turn the models into useful tools for both management and science. As described in Chapter 10, some of the largest uncertainties have to do with future climate change impacts, and a reliance on models needs to be a part of addressing those uncertainties.

CE-QUAL-W2 Reservoir Model

Other models have been used for CDA Lake, including the laterally averaged CE-QUAL W2 model that was used by Avista Corporation, which owns and operates the Spokane River Hydroelectric Project. This project consists of the Post Falls, Upper Falls, Monroe Street, Nine Mile, and Long Lake Hydroelectric facilities. Avista used CE-QUAL-W2 as part of the process to renew its FERC license in 2005. Avista's goal was to develop

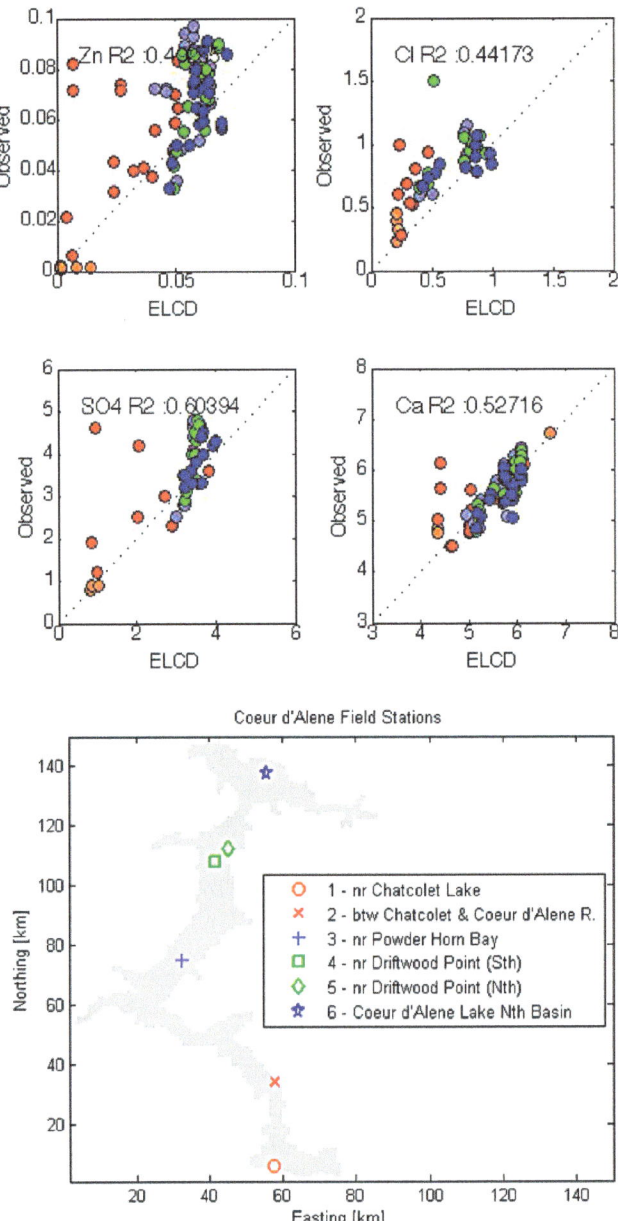

FIGURE 4-15 Scatter comparisons and correlation coefficients for zinc (Zn), sulfate (SO_4), chloride (Cl), and calcium (Ca) based on a comparison of ELCOM-CAEDYM model simulations and 2004 data. Colors correspond to the locations shown in the bottom panel. SOURCE: Hipsey et al. (2007).

CE-QUAL-W2 for CDA Lake to evaluate potential changes in water quality that could result from its lake-level operations (Golder Associates, Inc., 2005). Participants in the relicensing process, including the Washington Department of Ecology and the Idaho Department of Environmental Quality, expressed the need for a better understanding of the effects of project operations on water temperature, dissolved oxygen, pH, aquatic plants and nutrients. These agencies needed sufficient information to establish reasonable assurance that the project would comply with current provisions of the Clean Water Act.

The CE-QUAL-W2 model framework (Cole and Wells, 2016) is a two-dimensional, longitudinal/vertical, hydrodynamic and water quality model. Because the model assumes lateral homogeneity (i.e., it is laterally averaged), it is best suited for relatively long and narrow waterbodies exhibiting longitudinal and vertical water quality gradients. The model has been applied to rivers, lakes, reservoirs, estuaries, and combinations thereof, including entire river basins with multiple reservoirs and river segments. The model framework includes hydrodynamics and aquatic biology and chemistry.

Although CE-QUAL-W2 is a well utilized and robust model framework for understanding water quality in surface water systems, its application in CDA Lake suffers from two key limitations that influence the model's utility for considering the long-term water quality impacts. The first limitation is that the model framework is laterally averaged, which limits the ability of the model to consider 3-D effects in the Lake; medium to large lakes, such as CDA Lake, are inherently 3-D. The second limitation is that the CE-QUAL-W2 model framework does not have a full sediment transport model and it does not track the sediment size fractions that are deposited in a surface water system. This limits the model's ability to accurately simulate sediment transport in the Lake and the nutrients, metals, and other chemicals associated with the sediment particles. Hence, the 3-D model of ELCOM-CAEDYM provides a more robust modeling framework for capturing hydrodynamics, sediment transport, and aquatic biogeochemistry in CDA Lake.

CONCLUSIONS AND RECOMMENDATIONS

This chapter analyzed the limited hydrodynamic data from CDA Lake, along with special studies on hydrodynamics and sediment transport in CDA Lake. The following detailed conclusions and recommendations are made.

1. The limited data available on specific conductivity in the river inflows and CDA Lake showed that water from the St. Joe River entered CDA Lake as an interflow, meaning that, because of its density relative to the stratification of the Lake, it intrudes below the thermocline but above the Lake bottom. The CDA River inflow, on the other hand, was classified as an overflow, intruding into the epilimnion above the thermocline. Hence, **the two major river inflows are not likely to play a major role in resuspending sediment within CDA Lake.** Sediment resuspension could, however, occur in the littoral zones of the lake due to other factors. The identification of those areas most subject to sediment resuspension, together with the relevant water quality impacts, can only be ascertained with a nearshore monitoring program, something that currently does not exist (see Chapter 8).

2. **The committee's analysis of conductivity data found that a St. Joe River inflow of 1,000 cfs is a threshold below which thermal stratification commences and internal lake processes will dominate over riverine influences.** At river inflows above 1,000 cfs, which occur during winter and early spring, river discharge controls water quality (dissolved oxygen and pH) in CDA Lake, and the Lake behaves like a run-of-the-river system with little opportunity for biogeochemical processes to become established. At St. Joe River inflows below 1,000 cfs, which generally occur in June or later, internal dynamics and thermal stratification become important for CDA Lake, especially at site C5. It is during this period that water column processes such as nutrient uptake, phytoplankton proliferation, and decomposition can happen.

3. **More data are needed to characterize thermal stratification, a key variable that controls hypolimnetic dissolved oxygen concentration and pH and ultimately sediment–water interactions.** Stratification impacts all internal processes, including chemical and ecological processes. High-resolution time series temperature data are needed, preferably from four or five thermistor chains installed from south to north along the Lake axis. Instruments should be positioned from top to bottom along each chain. Ideally, profile measurements of dissolved oxygen, turbidity, conductivity, and pH should be collected at those sites. If the 1,000 cfs threshold for St. Joe inflow is reached earlier because of climate change, stratification will start sooner and one will observe a longer reduction of dissolved oxygen and lower pH (see Chapter 10).

4. **There is a lack of data and information on sediment transport in the Lake, including the sediment particle size distribution at various locations.** These are key drivers for the metal concentrations in lake bed sediments and mixing of sediments from various sources. This information can elucidate where sediments from the CDA River end up in the Lake and when and how sediments get transported downstream of the Lake. New data

collection should include the following data in the water column: vertical velocity profiles, thermistor vertical arrays at several locations, dye studies during low-flow and high-flow events, and particle size distribution data along with total suspended solids data.

5. **There is a critical lack of deterministic model usage for the Lake—for both heuristic and predictive purposes—despite having invested in the first stages of developing such a model (ELCOM-CAEDYM).** Development of a powerful 3-D hydrodynamic and water quality model began more than 15 years ago, but little has been done to use the model to better understand key processes within the Lake, the evolution of changes within the Lake, and the likely trajectory of the Lake under future climate changes. All parties have a shared interest to properly calibrate and validate the model with the goals of integrating the various processes that are at play in a complex lake. An initial thrust would be to understand dissolved oxygen dynamics and mass balances within the Lake. Such 3-D models can illuminate what is unknown about the system and what additional data collection could reduce these uncertainties.

REFERENCES

Ayala, A. I., A. Cortes, W. E. Fleenor, and F. J. Rueda. 2014. Seasonal scale modeling of river inflows in stratified reservoirs: Structural vs. parametric uncertainty in inflow mixing. *Environmental Modelling & Software* 60:84–98.

Balistrieri, L., S. Box, A. Bookstrom, R. Hooper, and J. Mahoney. 2002. Impacts of Historical Mining in the Coeur d'Alene River Basin. Chapter 6 *In:* Pathways of Metal Transfer from Mineralized Sources to Bioreceptors: A Synthesis of the Mineral Resources Program's Past Environmental Studies in the Western United States and Future Research Directions. USGS Bulletin 2191. L. S. Balistrieri and L. L. Stillings, eds.

Bender, S. 1991. Investigation of the chemical composition and distribution of mining wastes in Killarney Lake, Coeur d'Alene Area, northern Idaho. MS Thesis, University of Idaho, Moscow, Idaho, 98 pp.

Bookstrom, A. A., S. E. Box, R. S. Fousek, J. C. Wallis, H. Z. Kayser, and B. L. Jackson. 2013. Baseline, Historic and Background Rates of Deposition of Lead-Rich Sediments on the Floodplain of the Coeur d'Alene River, Idaho. U.S. Geological Survey Open-File Report 2004-1211. http://pubs.usgs.gov/of/2004/1211/.

Coeur d'Alene Tribe. 2004. Bathymetry for the Coeur d'Alene River Basin, Idaho. Obtained June 2021 from the Coeur d'Alene Tribe.

Cole, T., and S. Wells. 2016. CE-QUAL-W2: A Two-Dimensional, Laterally Averaged, Hydrodynamic and Water Quality Model, Version 4.0 User Manual. Department of Civil and Environmental Engineering Portland State University. 847 pp.

Dallimore, C. J., M. R. Hipsey, R. Alexander, and S. Morillo. 2007. Simulation Model to Evaluate Coeur d'Alene Lake's Response to Watershed Remediation: Volume 1: Hydrodynamic modeling using ELCOM. Center for Water Research.

Davison, W. 1993. Iron and manganese in lakes. *Earth Science Reviews* 34(2):119–163.

Golder Associates, Inc. 2005. Water Quality Technical Assessment, Lake Coeur d'Alene Temperature, Nutrients, Aquatic Plants, Dissolved Oxygen and pH, Part 2: Modeling Report. Submitted to Avista Utilities and Spokane River Relicensing Water Resources Work Group, Golder Associates, Inc., Redmond, WA.

Hipsey, R. H., R. Alexander, and C. J. Dallimore. 2007. Simulation Model to Evaluate Coeur d'Alene Lake's Response to Watershed Remediation Volume 2: Water quality modeling using ELCOM-CAEDYM. Center for Water Research.

Hodges, B. R., J. Imberger, A. Saggio, and K. B. Winters. 2000. Modeling basin-scale internal waves in a stratified lake. *Limnol. Oceanogr.* 45(7):1603–1620.

Horowitz, A. J., K. A. Elrick, and R. B. Cook. 1993. Effect of mining and related activities on the sediment trace element geochemistry of Lake Coeur d'Alene, Idaho, USA. Part I: surface sediments. *Hydrol. Process.* 7:403–423.

Horowitz, A. J., K. A. Elrick, J. A. Robbins, and R. B. Cook. 1995. Effect of mining related activities on the sediment trace element geochemistry of Lake Coeur d'Alene, Idaho. Part II—subsurface sediments. *Hydrol. Process.* 9:35–54. https://doi.org/10.1002/hyp.3360090105.

Kuwabara, J. S., W. M. Berelson, L. S. Balistrieri, P. F. Woods, B. R. Topping, D. J. Steding, and D. P. Krabbenhoft. 2000. Benthic flux of metals and nutrients into the water column of Lake Coeur d'Alene, Idaho: Report of an August 1999 pilot study. U.S. Geological Survey Water-Resources Investigations Report 2000-4132. https://doi.org/10.3133/wri004132.

Morillo, S., J. Imberger, J. P. Antenucci, and P. F. Woods. 2008. Influence of wind and lake morphometry on the interaction between two rivers entering a stratified lake. *J. Hydraul. Eng.* 134(11):1579–1589.

Roberts, D., G. Egan, A. Forrest, J. Largier, F. Bombardelli, B. Laval, S. Monismith, and S. G. Schladow. 2021. The setup and relaxation of spring upwelling in a deep, rotationally influenced lake. *Limnol. Oceanogr.* 66:1168–1199.

Steissberg, T. E., S. J. Hook, and S. G. Schladow. 2005. Measuring surface currents in lakes with high spatial resolution thermal infrared imagery. *Geophysical Research Letters* 32:L11402. doi:10.1029/2005GL022912.

Stevens, C., and J. Imberger. 1996. The initial response of a stratified lake to a surface shear stress. *J. Fluid. Mech.* V312:39–66.

Woods, P. F., and C. Berenbrock. 1994. Bathymetric Map of Coeur d'Alene Lake, Idaho. U.S. Geological Survey Water Resources Investigations Report 94-4119.

Wüest, A., and A. Lorke. 2003. Small-scale hydrodynamics in lakes. *Annu. Rev. Fluid Mech.* 35:373–412.

5

In-Lake Processes: Dissolved Oxygen and Nutrients

This chapter continues the committee's analysis of trends in water quality in Coeur d'Alene (CDA) Lake by focusing on dissolved oxygen, the nutrients phosphorus and nitrogen, and parameters indicative of the Lake's productivity such as chlorophyll a. The chapter also specifically addresses the bullet in the statement of task that asks whether reduced levels of zinc entering the Lake as a result of the upgrade to the Central Treatment Plant and other upstream activities are removing an important control on algal growth.

EUTROPHICATION, PRODUCTIVITY, AND OXYGEN DEPLETION

Like other ecosystems, lake ecosystems are driven by the inputs and cycling of essential chemical elements that are often supplied at insufficient rates to support growth. These are "limiting nutrients." Among the biologically important elements, nitrogen (N) and phosphorus (P) are most commonly identified as limiting nutrients in lakes (Elser et al., 2007), although there remains discussion (Conley et al., 2009; Schelske, 2009; Schindler and Hecky, 2009) about the timescales over which nitrogen can be limiting due to potential compensation by nitrogen fixation (see below). Overall, elevated nutrient inputs ("loading") accelerate the slow natural processes by which lakes increase in primary productivity (in terms of internal carbon fixation by phytoplankton, periphyton, and macrophytes) over time, leading to the generally undesirable outcome of anthropogenic eutrophication. A eutrophied lake has diminished water clarity due to accumulation of phytoplankton biomass in the water column. Depending on conditions, the phytoplankton community of a eutrophic lake can become dominated by undesirable species such as cyanobacteria, which can produce chemical compounds that are toxic to aquatic life as well as to pets and humans. Eutrophication can also lead to anoxia.

Decreases in Dissolved Oxygen

Elevated production of phytoplankton biomass due to increased nutrient loading can often lead to depletion of oxygen in the bottom waters of a lake, which occurs when organic matter produced by nutrient-driven phytoplankton sinks into the bottom layers and is decomposed by bacteria that consume oxygen. Respiration in benthic sediments also consumes oxygen and leads to its depletion in bottom waters. The term "hypoxia" is used to describe situations in which oxygen is depleted below saturation levels, where saturation refers to the maximum amount of dissolved oxygen that water can hold as a function of temperature and pressure. Hypoxia is usually defined as a depletion that is detrimental to organisms of interest. Dissolved oxygen concentrations that are considered hypoxic are in the range of 2–4 mg/L (Diaz and Rosenberg, 2008; Howell and Simpson, 1994; Paerl et al., 2006); obviously, the exact concentration below which dissolved oxygen has negative effects varies depending on the organism of interest, the environment, and the duration of exposure (e.g., Hrycik et al., 2017; Tellier et al., 2022). In practice, hypoxia is often defined in terms of measurable consequences reflected in the ecosystem, such as the oxygen concentration at which fisheries start to collapse (Renaud, 1986) or a particular biological function becomes impaired (Diaz and Rosenberg, 1995).

The term "anoxia" refers to situations when oxygen is reduced to very low levels (undetectable or < 1 mg/L; Nürnberg, 2002). At these low dissolved oxygen levels, not only is aquatic life (fish, mollusks) threatened but chemical conditions at the sediment–water interface can be altered, releasing nutrients back into the water column and creating a highly undesirable self-amplifying loop of eutrophication. Specifically, when oxygen levels at the sediment–water interface decline low enough (to < 1 mg/L) to allow redox potential to fall below ~200 mV (Mortimer, 1942, 1971), oxidized forms of iron ("ferric") are reduced to their "ferrous" form by iron-reducing microbes and are solubilized. This dissolution of iron releases the PO_4-P that was bound to the oxidized ferric compounds. Release of other metals from the sediments back to the water column can also be enhanced due to development of anoxia (see, e.g., Atkinson et al., 2007). These processes are of high relevance to CDA Lake given concerns about how future eutrophication might affect the legacy of metals stored in the Lake sediments.

Nitrogen Cycling

Nitrogen in lakes can be found in various chemical forms, both organic (i.e., associated with a carbon-based molecule) and inorganic. Organic nitrogen is found in living things (bacteria, phytoplankton, zooplankton, fish) where it is an essential component of proteins, nucleic acids, and other molecules. Organic nitrogen can also be found in nonliving forms (including detritus particles) or dissolved molecules derived from and processed by microorganisms. The dominant inorganic forms of nitrogen are nitrate (NO_3) and ammonia/ammonium (NH_3/NH_4) with nitrite (NO_2), usually at low concentrations. Also present in the water is dinitrogen gas (N_2).

Nitrogen has a relatively complex cycle driven largely by specialized microorganisms that use nitrogen not only to build their biomass (e.g., to make protein or nucleic acids) but also as a source or a receiver of electrons during metabolic processes. The cycle can be thought to begin via "nitrogen fixation," in which certain bacteria (of special importance in lakes) and cyanobacteria convert N_2 gas into biologically useful forms via large expenditures of energy. Once in organic form, this nitrogen can be released into inorganic form as NH_3/NH_4 in a process called "ammonification," when organisms are eaten or their biomass is decomposed. In turn, ammonia can be chemically converted to nitrate (via nitrite) by certain bacteria in a process called "nitrification." Both nitrate and ammonia can return to the cycle by being re-assimilated by phytoplankton or heterotrophic microbes for use in biomass production. Finally, nitrate can be used as an electron acceptor for respiration when oxygen is depleted, resulting in production of N_2 gas. This process is called "denitrification" and largely completes the cycle. (Note that the greenhouse gas nitrous oxide [N_2O] can be formed during denitrification and nitrification. There are other processes of nitrogen cycling that can occur, but for the sake of relative simplicity, they are not described here.)

Fixation of N_2 from the atmosphere is not the only way that nitrogen can enter a lake's water column. Transfer of water from the watershed or other regions of the lake can carry nitrogen in various forms into the lake. Most nitrogen loaded to lakes and rivers in undisturbed catchments originates from nitrogen fixation by symbiotic bacteria–plant root associations in the surrounding forests and grasslands. In disturbed catchments, nitrogen from agricultural runoff or point source discharges is often the dominant source. Nitrogen, especially as NH_3/NH_4, can be released from sediments and reenter the water column. Nitrogen can also enter a lake from the atmosphere, in wet deposition as NO_3 or NH_3/NH_4, in gaseous transfer as NH_3/NH_4, or in particulate form as organic nitrogen (pollen, eroded soil). Finally, denitrification is not the only vector by which nitrogen can leave the water column. It can also be assimilated by periphyton, settle to the lake bottom attached to particles, or be discharged from the lake via its outflow.

Phosphorus Cycling

Like nitrogen, phosphorus in lakes can be found in various organic and inorganic forms. Whether in free dissolved form or in organic molecules, phosphorus is generally bound to four oxygen atoms as phosphate (PO_4). Organic phosphorus in living things includes nucleic acids, adenosine triphosphate (ATP), and membrane lipids. As with nitrogen, organic phosphorus can be found in nonliving detritus or as dissolved molecules derived from and processed by microorganisms. By far the dominant inorganic form is phosphate, but under many conditions, concentrations of phosphate are too low for reliable detection. Inorganic forms of phosphorus are also associated with inorganic particles, such as dust, glacial flour, clays, or eroded soil, often in mineral complexes with calcium, iron, and aluminum.

Compared to nitrogen, the cycling of phosphorus is simple. Phosphorus enters a lake via river and stream inflows and atmospheric deposition, although the direct biological availability of that phosphorus can vary and depends on its chemical form. Phosphorus can also enter a lake water column via processes of "internal loading" or release of phosphorus from sediments—a process that is greatly amplified when oxygen concentrations at the sediment–water interface are depleted to less than 1 mg/L. Once in the lake, cycling is dominated by assimilation of PO_4 by phytoplankton and microbes; release of dissolved forms of phosphorus; mineralization of organic phosphorus back to PO_4 by heterotrophic microbes, zooplankton, and fish; and re-assimilation. Under phosphorus-limited conditions, these internal cycling processes can be extremely fast (on the order of minutes), and phosphorus atoms cycle many times within water column biota before being lost via sedimentation or discharge from the lake.

Interactions among Elements, Including Metals

Biogeochemical cycles of nitrogen and phosphorus do not operate independently of each other. For example, nitrogen and phosphorus are used to produce phytoplankton and bacterial biomass within a relatively narrow range of stoichiometric N:P ratios. Suspended organic matter in lakes has an average N:P ratio of ~20–30:1 (atomic) (Sterner et al., 2008), with variation driven by the role of nitrogen or phosphorus limitation in constraining growth (Sterner and Elser, 2002). The cycling of nitrogen and phosphorus are not independent of other elements, especially various metals that play important biological and geochemical roles. For example, the trace elements iron (Fe) and molybdenum (Mo) are essential for biological nitrogen fixation as well as other processes. Metals also play a role in modulating nutrient availability, as when oxidized (ferric) forms of iron form insoluble ferric hydroxides under aerobic conditions, co-precipitating phosphate from the water column. In turn, depletion of oxygen in sediments and at the sediment–water interface can reduce iron to its soluble ferrous form, releasing the co-precipitated phosphate and contributing to internal nutrient loading (see Chapter 7 for details).

ANALYSES OF DISSOLVED OXYGEN TRENDS

The committee analyzed dissolved oxygen data collected in CDA Lake by the U.S. Geological Survey (USGS), the Idaho Department of Environmental Quality (IDEQ), and the CDA Tribe over the past 30 years. The dissolved oxygen vertical profile datasets were first examined for evidence of trends over time at the four major sampling stations in the Lake (C1, C4, C5 and C6—see Figure 2-1). The committee also conducted a time series analysis for evidence of any long-term seasonal trends in dissolved oxygen concentrations in CDA Lake. Finally, the committee performed a Mann-Kendall test of trends in the monthly-averaged dissolved oxygen data in bottom waters at all four sites.

To conduct these analyses, it was necessary to standardize the profile data with respect to sample collection methodologies, sample depth, and temporal frequency because the available data were quite heterogeneous with regard to these dimensions. For example, data collected during sonde upcasts were discarded, the profile data were interpolated and binned by 1-meter and monthly increments, missing maximum depth data were estimated from available adjacent depths, and diurnal fluctuations in dissolved oxygen were confirmed to be small[1] (see Appendix B for greater detail).

---

[1] The committee hypothesized that diurnal variations in dissolved oxygen in CDA Lake would be small because of surface water mixing and because of the absence of light penetration in the hypolimnion, where anoxia is of concern. This was confirmed by analyzing buoy data on dissolved oxygen at C5 on eight separate days during July–August 2011 and 2015. For both the topmost (1 m) and bottommost (17 m) samples at C5, the average standard deviations of dissolved oxygen data were 0.074 mg/L and 0.34 mg/L, respectively.

Observations of Dissolved Oxygen Data

The first analysis compared the dissolved oxygen percentiles for the CDA Lake sampling stations C1–C6 based on all of the raw profile data collected from 1991 to 2021. As shown in Figure 5-1, there was little evidence of low oxygen concentrations at the C1, C2, C3, and C4 stations. There was some evidence of low oxygen concentrations at the C5 station and clear evidence of depressed dissolved oxygen at the C6 station. However, because these percentile distributions are based solely on extant data, and these data were not always collected in a systematic manner with regard to time and depth, this is only a preliminary analysis. For example, in the earlier years, dissolved oxygen depth profiles were often collected at a finer spatial resolution (e.g., 1 or 2 m) in the upper water column and at a coarse resolution (e.g., 5 m) in the hypolimnion. Because the deeper depths were sampled with a coarser resolution, the frequency of low dissolved oxygen concentration in these percentile estimates was probably somewhat underestimated.

The second analysis characterized the within-year and depth variability at the C1, C4 and C6 stations. These stations were the focus of these analyses because C1 and C4 were the deeper stations that had the most data (compared to C2 and C3) and the C6 station had the clearest evidence of seasonal dissolved oxygen depletion. For station C1, it is evident that the main mode of variability for the intra-annual data is time (Figure 5-2). Although not shown, very similar results were obtained for C4. Higher oxygen concentrations are observed throughout the water column during the winter and spring, and lower concentrations are observed during the summer and fall, when CDA Lake is warmer at the surface. For example, both C1 and C4 dissolved oxygen concentrations averaged 11.2 mg/L from December to May and 8.7 mg/L from July to November. There is also a slight tendency for lower dissolved oxygen concentrations in the hypolimnion in July–November at both stations. For example, oxygen concentrations averaged 9.0 mg/L in the upper 20 m of the water column and 8.2 mg/L at depths of ≥ 31 m during this time.

Finally, at both stations there is evidence suggesting a deep chlorophyll maximum, and associated higher dissolved oxygen concentrations, as evidenced by the spike in dissolved oxygen concentrations in the metalimnion (8–14 m) during mid-July. However, the chlorophyll data were not granular enough to enable the committee to determine which stations and/or sampling dates had a deep chlorophyll maximum present.

In contrast to C1 and C4, the vertical dissolved oxygen profiles for C6 showed much clearer evidence for hypoxic and even anoxic conditions during the summer stratified period (Figure 5-3). From July to mid-September,

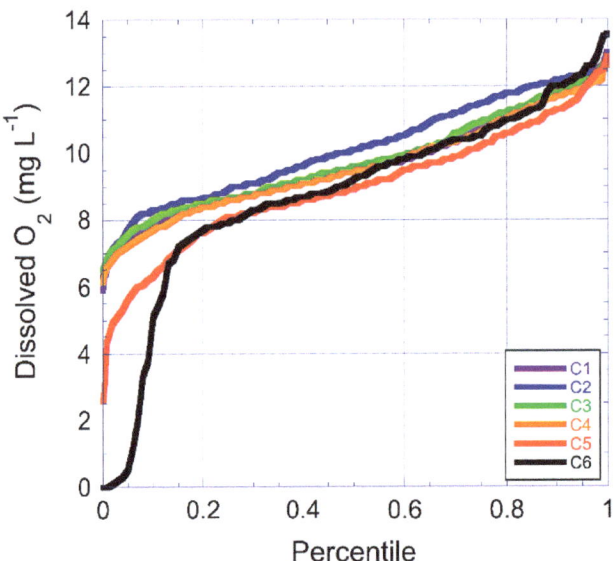

FIGURE 5-1 The percentile distribution for all of the observed dissolved oxygen concentrations data, 1991–2021. SOURCE: Data plotted by the committee, courtesy of USGS, IDEQ, and the CDA Tribe.

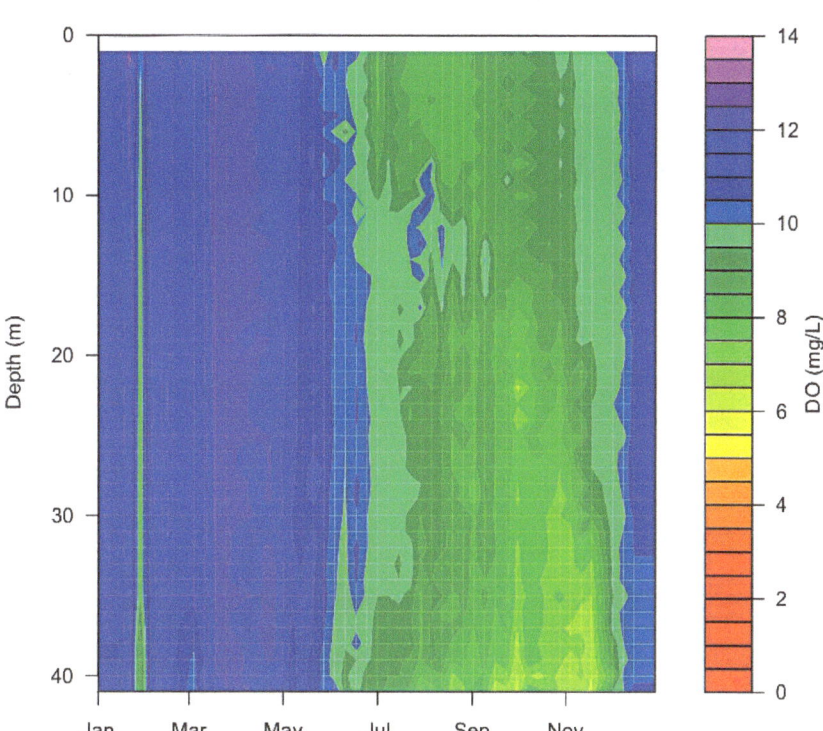

FIGURE 5-2 Intra-annual and depth variation in dissolved oxygen concentrations at the C1 sampling station. This plot is based on an analysis of all profile data collected from this station for all years (1991–2021) but only considered by the Julian date within the year. Thus, this plot represents average conditions at this station by depth and date averaged across all available data. SOURCE: Data plotted by the committee, courtesy of USGS and IDEQ.

dissolved oxygen concentrations only averaged 1.7 mg/L, at depths of 8–11 m, whereas dissolved oxygen concentrations averaged 8.6 mg/L in the upper 5 m of the water column during this same time.

The committee also examined the buoy data collected by the CDA Tribe at station C5 (data not shown). These data were available at a very high temporal resolution (i.e., four or six profiles per day) for most of the period from July to November for the years 2011 and 2015. Buoy data are also available for 2019, but oxygen data are not available for that year. Based on a comparison of the oxygen percentiles for 2011 and 2015, it is evident that on average the two years had similar dissolved oxygen concentrations, but the lowest concentrations were lower during 2015. For example, the bottom quartile of data had an average of 1.5 mg/L lower dissolved oxygen concentrations during 2015. Because 2015 had a tendency for lower dissolved oxygen concentrations, the committee also analyzed the buoy data collected by the CDA Tribe at C5 for the entirety of the temporal and depth distribution. This analysis showed that dissolved oxygen concentrations were lowest below a depth of 15 m during the second half of September and all of October. During this time and at these depths, dissolved oxygen concentrations averaged 5.0 mg/L.

As shown in Tables 5-1 through 5-4, the month-by-year matrix of observed dissolved oxygen concentrations was incomplete for the C1, C4, C5, and C6 sampling stations.[2] For example, the data availability at C1 for the months of November to February was only 24 percent, which increased to 79 percent during the months of July

---

[2] These four paragraphs, and text throughout the report, were edited after report release to reflect corrections to Table 5-3, the previous version of which had data inadvertently shifted by two months.

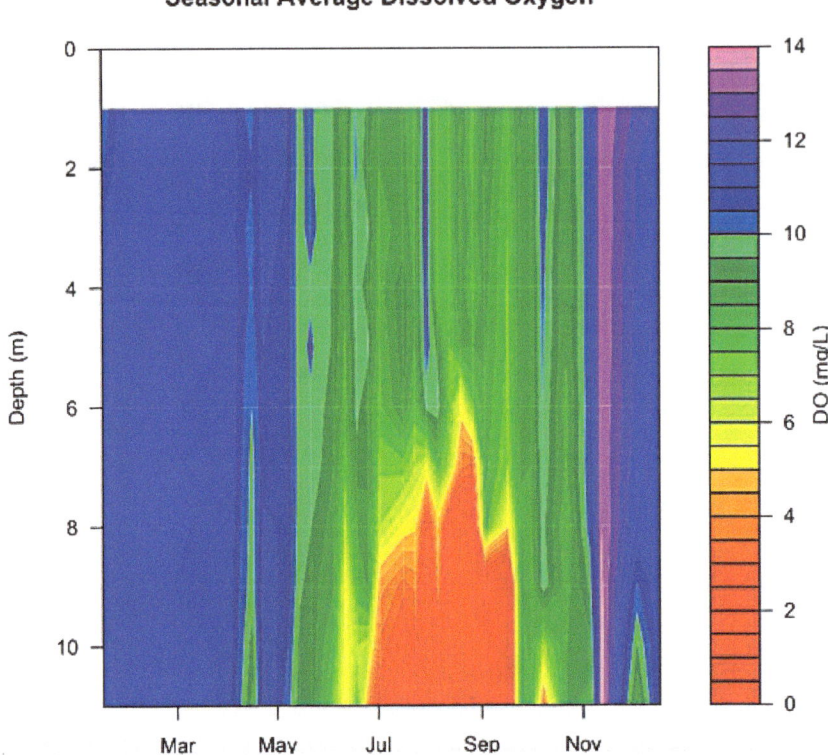

FIGURE 5-3 Intra-annual and depth variation in dissolved oxygen concentrations at the C6 station. This plot is based on an analysis of all profile data collected from this station for all years (1991–2021), but only considered by the Julian date within the year. Thus, this plot represents average conditions at this station by depth and date averaged across all available data.
SOURCE: Data plotted by the committee, courtesy of USGS and the CDA Tribe.

to October. Additionally, the data availability during late summer and early fall has been good since 1995, and data availability in the spring and early summer has improved substantially since 2004.

The average monthly dissolved oxygen concentrations for the C1 station showed clear evidence of modest monthly and depth patterns in the dissolved oxygen concentrations (Table 5-1). A comparison of the monthly average oxygen concentrations in the upper 10 m of the C1 station to the bottom 10 m (i.e., 31–40 m) showed that the entire water column at this station had higher oxygen concentrations during the winter and spring and lower concentrations during the summer and fall (Figure 5-4). Surface and bottom dissolved oxygen concentrations were similar from December to August, and peaked at ≈ 12.0 mg/L in March and April. In October and November, bottom oxygen concentrations averaged 1.9 mg/L less than surface concentrations.

Similar to the C1 station, the long-term monthly mean oxygen concentrations for the bottom of station C4 (30–39 m) showed strong seasonality (as expected) (Table 5-2). Additionally, the average dissolved oxygen concentration in the surface and bottom waters of station C4 were similar during the months of December through August, and peaked at about 11.6 mg/L from January through April (Figure 5-5). During October and November, dissolved oxygen concentrations were about 1.7 mg/L lower on the bottom of station C4 compared to the surface.

Station C5 is considerably shallower than stations C1 and C4, and its seasonal dissolved oxygen dynamics are also substantially different from the deeper stations in several regards (Table 5-3 and Figure 5-6). In particular, the seasonal oxygen depletion evident in the bottom depths (i.e., 14–17 m) was more pronounced and subsided one month earlier in the year at this station. From November to June, monthly dissolved oxygen concentrations at

TABLE 5-1 Long-Term Monthly Averaged Dissolved Oxygen Concentrations (mg/L) for the Bottom 10 m (i.e., 31–40 m) at the C1 Sampling Station

C1	Jan	Feb	Mar	Apr	May	Jun	Jul	Aug	Sept	Oct	Nov	Dec
1991	11.34		11.70	12.00	11.07	9.61	9.52	9.01	8.06	7.47	7.57	10.81
1992	11.89		10.98	10.51	9.74	9.62	8.49	8.06	8.19	7.01	7.34	10.25
1993												
1994												
1995								7.29	8.49	8.46		
1996							8.82	7.67	8.22	7.86		
1997					11.21		10.15	9.06	8.43	7.94		
1998							9.05	8.36	7.20	6.65		
1999							8.35	8.36	7.55	6.84		
2000							8.93	8.61	6.90	6.74		
2001							9.38	8.67	7.05	6.73		
2002								7.83	7.41	6.73		
2003										6.98		9.77
2004		11.28		11.33	10.48	9.96	8.95	8.43		6.73		9.15
2005	9.11			12.09	10.74	9.63	9.14		8.21	7.83		9.83
2006		11.57	12.20		11.31	9.99	9.52	8.70				
2007						9.58	9.07	8.04	7.83	7.23		9.47
2008			12.12	11.99	11.21	10.06	9.26	7.34		7.14		
2009				11.61	10.21	9.45	7.89			7.69		
2010			10.87	11.13	10.59	9.45	8.19	7.47	6.48			9.46
2011		10.83		11.69	10.96	9.77	9.36	9.06		7.41	10.30	
2012			10.56	12.89	11.71	10.31	9.11	8.65	7.68	6.86	6.80	
2013				12.25	10.91	9.94	9.04	7.82		7.42		9.99
2014			11.36	12.57	10.61	9.63	9.18	8.07	7.71			10.10
2015		11.47	11.87		10.79	9.80	9.28	8.52	7.82	7.14	7.14	
2016				11.92	10.75	9.84	9.19	8.59	8.06	7.12	7.08	
2017			12.11	11.48	10.59	9.70	9.42	9.02	8.20	7.67		10.20
2018		11.80	12.48	12.03	11.39	10.73	9.99	9.25	8.97	8.33	7.80	
2019			12.50	12.52	11.13	10.63	9.68	9.21	8.04	7.88		9.68
2020			11.96	11.71	10.79	10.11	9.20	8.63		6.82		9.69
2021	11.0	11.1	12.0	11.9								
Monthly mean	10.8	11.3	11.7	11.9	10.8	9.9	9.1	8.4	7.8	7.3	7.7	9.9
Monthly SD	1.2	0.3	0.6	0.6	0.5	0.4	0.5	0.6	0.6	0.5	1.2	0.4

Note: Red shading denotes less favorable conditions (e.g., low dissolved oxygen), while blue shading denotes more favorable conditions (e.g., high dissolved oxygen). The coloring, created in Excel, is not consistent across Tables 5-1 to 5-4.

TABLE 5-2 Long-Term Monthly Averaged Dissolved Oxygen Concentrations (mg/L) for the Bottom 10 m (i.e., 30–39 m) at the C4 Sampling Station

C4	Jan	Feb	Mar	Apr	May	Jun	Jul	Aug	Sept	Oct	Nov	Dec
1991	10.85		11.21	11.45	11.45	10.05	9.53	9.27	8.70	8.11	7.95	10.98
1992	12.12		11.10	11.11	10.08	9.69	8.82	8.27	8.05	7.28	6.28	10.71
1993												
1994												
1995								7.54	8.36	8.05		
1996							9.12	9.05	8.44	8.17		
1997					10.98		10.40	9.42	8.83	8.18		
1998							9.01	8.82	7.73	6.76		
1999							8.65	8.63	7.90	6.84		
2000							9.07	7.29	6.41			
2001							8.95	7.65	7.21	7.22		
2002								7.98	7.61	8.96		
2003										7.16		10.15
2004		11.24		10.83	9.75	9.64	8.98	8.20		6.69		8.99
2005	9.66			11.42	10.69	9.45	9.33	8.25		7.75		10.13
2006		11.35	12.05		11.16	10.04	9.46	9.05				
2007						9.99	9.38		8.06	7.36		9.59
2008			11.66		10.94	10.40	9.62	7.53		7.42		
2009			11.05		10.39	9.44	8.86			7.18		
2010			10.67	11.42	10.61	9.69	8.26	7.77	6.71			9.83
2011		10.86		11.08	10.44	10.26		9.40		8.04		10.62
2012			10.69	11.91	11.55		9.41	8.76	7.68	7.19	6.79	
2013				11.90	10.64	10.00	8.88	8.13		7.81		10.20
2014			11.83	11.70	11.07	10.06	9.47	8.73	8.12			10.13
2015		11.72	11.67		10.97	10.06	9.19	8.88	8.06	7.25	7.09	
2016				11.60	10.71	9.63	9.27	8.75	8.11	7.15	7.33	
2017			11.30	10.92	10.82	9.05	9.50	9.24	8.61	7.85		10.40
2018		11.80	12.09	11.87	11.32	10.64	9.99	9.49	9.27	8.36	7.93	10.37
2019			12.59	12.08	11.00	10.34	9.71	9.47	8.47	7.68		10.23
2020			12.00	11.51	10.87	10.08	9.22	8.73	7.32	6.67	9.74	11.19
2021	11.43	11.98	11.99									
Monthly mean	11.0	11.5	11.6	11.5	10.08	9.9	9.3	8.6	8.0	7.5	7.6	10.2
Monthly SD	1.0	0.4	0.6	0.4	0.4	0.4	0.5	0.6	0.6	0.6	1.1	0.6

Note: Red shading denotes less favorable conditions (e.g., low dissolved oxygen), while blue shading denotes more favorable conditions (e.g., high dissolved oxygen). The coloring, created in Excel, is not consistent across Tables 5-1 to 5-4.

TABLE 5-3 Long-Term Monthly Averaged Dissolved Oxygen Concentrations (mg/L) for the Bottom 4 m (i.e., 14–17 m) at the C5 Sampling Station[a]

C5	Jan	Feb	Mar	Apr	May	Jun	Jul	Aug	Sept	Oct	Nov	Dec
1991			11.2	10.7	10.1	9.6	8.3	6.9	5.2	3.8	9.7	11.2
1992	11.0		10.4	10.0	9.7	8.3	6.0	4.8	6.6	4.7	9.5	11.4
1993												
1994												
1995								5.1	4.8	9.3		
1996							7.2	6.0	4.8	4.8		
1997					10.4		8.4	6.7	6.6	9.1		
1998							7.0	6.0	3.6	3.2		
1999												
2000												
2001												
2002												
2003										5.9		10.4
2004				10.6	8.9	9.9	6.3	4.8		4.3		10.2
2005				10.7	9.3	8.7	7.0	4.6		7.9		9.6
2006		11.0	10.7		9.9	8.2	7.0	4.8				
2007						7.5	6.6	3.8	4.2			11.1
2008			10.8	9.9	10.4	9.0	7.8	6.3		5.8		10.0
2009				11.1	10.7	9.7	7.9	5.8		5.2		10.3
2010			11.3	11.1	10.4	8.6	6.8	6.4	5.2		10.1	
2011		10.9		10.6			8.9	8.1	5.8	5.4	10.1	
2012			11.9		10.6	9.6	8.2	6.1	5.1	6.3	10.7	
2013				10.7	11.3	10.6	8.6	6.8	5.8	8.1	10.9	
2014			12.9	11.6	11.6	9.6	9.4	7.4	6.2	6.2	10.9	
2015		12.8	12.4		10.9	9.3	7.4	6.4	5.8	5.4	11.0	
2016				11.5	10.5	8.6	8.5	6.2	5.2	6.5	11.3	
2017				11.0	11.1	10.5	9.1	7.2	5.8	7.7	11.5	
2018		12.5	12.9	11.3	11.1	10.6	8.6	6.6	5.9	7.7	8.9	
2019	11.9		12.3	11.0	9.7		8.3	6.7	5.7	6.6	11.7	11.3
2020	12.8			11.2	10.8	9.5	6.9	5.5	5.9	11.4	12.6	
2021												
Monthly mean	11.9	11.8	11.6	10.9	10.5	9.4	7.9	6.1	5.4	6.2	10.6	10.8
Monthly SD	0.9	1.0	0.9	0.6	0.7	0.9	1.0	1.0	0.8	1.7	0.9	0.9

Note: Red shading denotes less favorable conditions (e.g., low dissolved oxygen), while blue shading denotes more favorable conditions (e.g., high dissolved oxygen). The coloring, created in Excel, is not consistent across Tables 5-1 to 5-4.
[a]Table 5-3 was replaced after report release to correct a formatting error in which the data had been inadvertently shifted by two months.

TABLE 5-4 Long-Term Monthly Averaged Dissolved Oxygen Concentrations (mg/L) for the Bottom 3 m (i.e., 8–11 m) at the C6 Sampling Station

C6	Jan	Feb	Mar	April	May	June	July	Aug	Sept	Oct	Nov	Dec
1991				9.1	9.5	8.2	5.1	1.7	1.3	7.1	9.8	
1992				11.2	9.1	5.3	0.4	1.5		8.1	12.2	13.6
1993												
1994												
1995												
1996												
1997												
1998												
1999												
2000												
2001												
2002												
2003										8.1		
2004				10.0	9.4	6.5	3.7	0.4		8.5	10.6	
2005				11.2	9.5	6.7	0.5	2.8		8.4		
2006					10.7	8.1	3.2	0.3				
2007						6.0	1.4	0.18	1.7			12.4
2008			11.0	10.2	9.5	9.9	5.7	0.36		9.2		9.4
2009				10.9	10.2	7.3	3.5	0.18		8.4		11.3
2010				10.6	9.8	2.8		0.20	3.4		10.7	
2011				11.1			8.2	3.68	0.7	7.8	10.8	
2012			12.0		10.7	9.4	4.9	0.24	1.5	10.0	11.2	
2013				11.0	10.8	7.4	1.6	0.07	0.3	10.5	11.9	
2014			12.9	12.7	11.6	9.5	6.1	0.26	2.1	8.8	12.0	
2015		12.7	12.0		8.0	2.9	0.2	0.08	8.1	8.9	11.3	
2016				12.5	10.5	6.3	0.6	0.00	3.2	9.5	11.8	
2017				11.7	11.4	8.5	3.2	0.09	1.3	10.4	12.0	
2018		12.8	13.0	12.2	10.3	8.2	0.7	0.04	2.7	10.1	10.8	
2019				11.5	10.3	6.6	1.8	0.00	1.3	9.7	12.2	
2020				11.6	10.2	5.5	0.48	2.4	9.6	12.3		
2021												
Mean		12.7	12.2	11.2	10.2	7.6	3.1	0.66	2.7	9.2	11.5	11.0
SD		0.0	0.8	1.0	1.0	1.9	2.3	1.0	2.4	1.2	0.9	1.5

Note: Red shading denotes less favorable conditions (e.g., low dissolved oxygen [DO]), while blue shading denotes more favorable conditions (e.g., high DO). The coloring, created in Excel, is not consistent across Tables 5-1 to 5-4.

14–17 m averaged ≈ 0.6 mg/L lower than the corresponding concentrations in the upper 5 m of the water column, and during August through October the bottom concentrations averaged ≈ 3.1 mg/L less than the surface concentrations. The deep water dissolved oxygen concentrations at C5 rebounded in November when these concentrations averaged 10.6 mg/L, which was ≈ 4.4 mg/L higher than the preceding month. In contrast, the bottom dissolved oxygen concentrations at C1 and C4 rebounded one month later in December.

Compared to stations C1, C4 and C5, the data availability for station C6 is more limited (Table 5-4). This site has detailed oxygen profile data from 1991–1992, and then again from 2004 to the present. This site also has the most severe seasonal dissolved oxygen depletion in its bottom waters, with concentrations routinely averaging well below 1 mg/L at 8–11 m depth during the month of August, and concentrations only averaging 2.9 mg/L in July and September. During these months, the bottom dissolved oxygen concentrations also averaged 7.0 mg/L lower than the surface concentrations (Figure 5-7). Similar to station C5, dissolved oxygen concentrations rebounded rapidly in the fall, albeit in October (instead of November as was the case for C5, or December as for C1 and C4). The mean monthly bottom dissolved oxygen concentrations were also much more variable at this station, with large outliers within months and dramatic changes in concentrations from one month to the next. Given its unique hydrodynamics (discussed in Chapter 1), oxygen concentration trends at the C6 station are of unclear relevance to the main body of the Lake as represented by the C1–C5 sampling stations.

As expected, there are obvious seasonal patterns in the monthly dissolved oxygen concentrations at both the surface and bottom of CDA Lake. The differences in the surface and bottom dissolved oxygen concentrations were usually quite small at the deeper profile sampling stations (i.e., C1 and C4). The bottom water dissolved oxygen concentrations at the C6 station were very low during the late summer and of concern for fisheries habitat and for redox reactions at the sediment–water interface, which would cause both nutrients and dissolved metals to be released from the Lake sediments.

In general, the available dissolved oxygen data for CDA Lake are quite patchy, but data coverage has improved markedly since 2004. Winter data are particularly sparse, but this time of the year is of less concern since dissolved oxygen concentrations are consistently high during these months. To improve future time series analyses, it would be best to sample every other winter month on a consistent basis.[3]

The analyses reported herein used a coarser sampling resolution for the bottom samples than is ideal because the most important information is the oxygen concentration immediately above the sediment–water interface. However, there was some variability in the maximum depth sampled from one vertical profile to another, especially during the earliest years of this time series. Some of this variability may be due to intra-annual and interannual changes in the surface level of CDA Lake,[4] which would in turn affect that maximum depth from which oxygen data can be collected. When inferring redox conditions at the sediment–water interface for these profiles, it would be beneficial to know the actual bottom depth associated with each profile so that the bottom dissolved oxygen data can be presented relative to their distance from the sediment–water interface.

In early 2022, the committee was shown data suggesting increasing trends in anoxia at C6 and SJ1 (Chess, 2022). Figure 5-8 shows the number of days each year that the hypolimnion was anoxic (defined as dissolved oxygen concentrations < 1 mg/L) at the C6 and SJ1 sampling sites for the 2011–2021 time period. Whether this plot actually indicates a declining oxygen trend is confounded by the fact that 2011 was an anomalously high year for hypolimnetic oxygen concentrations at the C6 sampling site. Table 5-4 shows the 8–11 m dissolved oxygen concentration at C6 averaged 3.7 mg/L in August of 2011, but 0.7 mg/L in August for all years at this sampling station (n = 19). The year 2011 was also anomalous because it was an unusually high flow year for CDA Lake. In addition, the hypolimnetic anoxia analysis shown in Figure 5-8 started in 2011 (instead of 2004 when consistent profiling at this station started) because a switch to twice monthly profiling occurred in 2011. To test whether the significant trend for days of hypolimnetic anoxia at C6 was simply due to starting this time series at an anomalously high year, the committee showed that had this time series been started in 2012 instead, the trend in increasing days of hypolimnetic anoxia at C6 would not be significant.

---

[3] This paragraph was edited after report release to remove duplication with the section on the Mann-Kendall DO trend analysis.

[4] The committee assumes that the level of CDA Lake varies by ca. ± 1 m within a year and probably less from one year to another. These data presumably exist for the outflow of the Lake, since that controls the flow of water into the Spokane River.

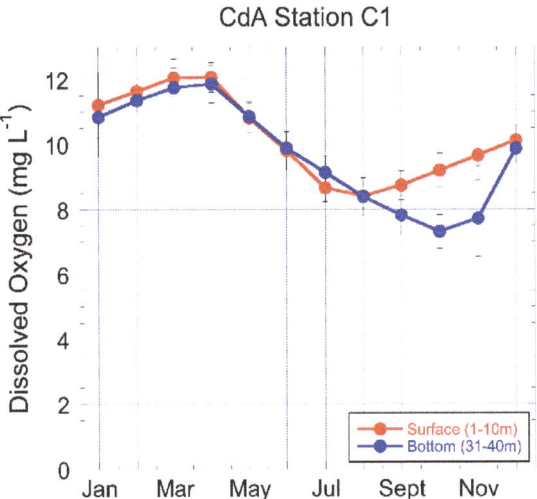

FIGURE 5-4 Monthly average surface (1–10 m) and bottom (31–40 m) dissolved oxygen concentrations at the C1 vertical profile sampling station. The monthly means are for 1991–2021 data, and the error bars represent ± 1 SD. SOURCE: Data plotted by the committee, courtesy of USGS and IDEQ.

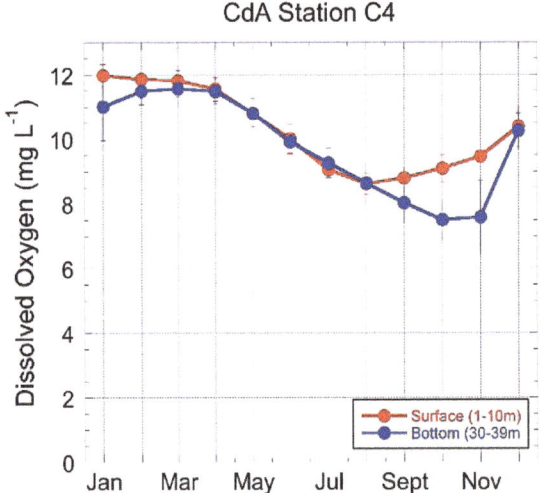

FIGURE 5-5 Monthly average surface (1–10 m) and bottom (30–39 m) dissolved oxygen concentrations at the C4 vertical profile sampling station. The error bars in this plot are ± 1 SD. SOURCE: Data plotted by the committee, courtesy of the USGS and IDEQ.

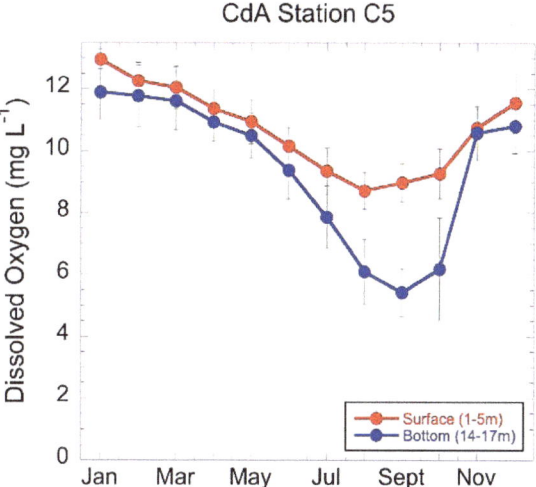

FIGURE 5-6 Monthly average surface (1–5 m) and bottom (14–17 m) dissolved oxygen concentrations at the C5 vertical profile sampling station. The error bars in this plot are ± 1 SD. SOURCE: Data plotted by the committee, courtesy of USGS and the CDA Tribe. Note: Figure 5-6 was replaced after report release to reflect corrections to Table 5-3, the previous version of which had data inadvertently shifted by two months.

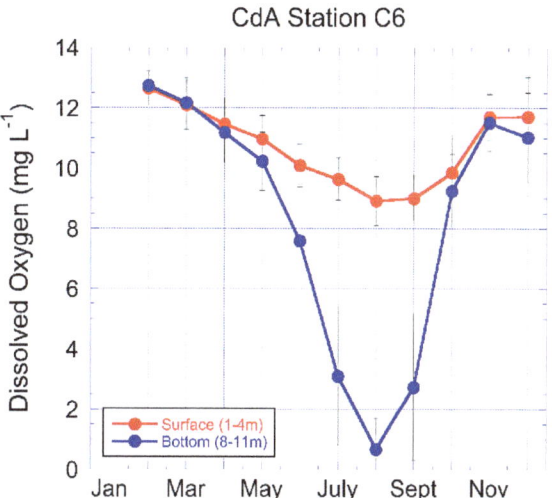

FIGURE 5-7 Monthly average surface (1–4 m) and bottom (8–11 m) dissolved oxygen concentrations at the C6 vertical profile sampling station. The error bars in this plot are ± 1 SD. SOURCE: Data plotted by the committee, courtesy of the USGS and the CDA Tribe.

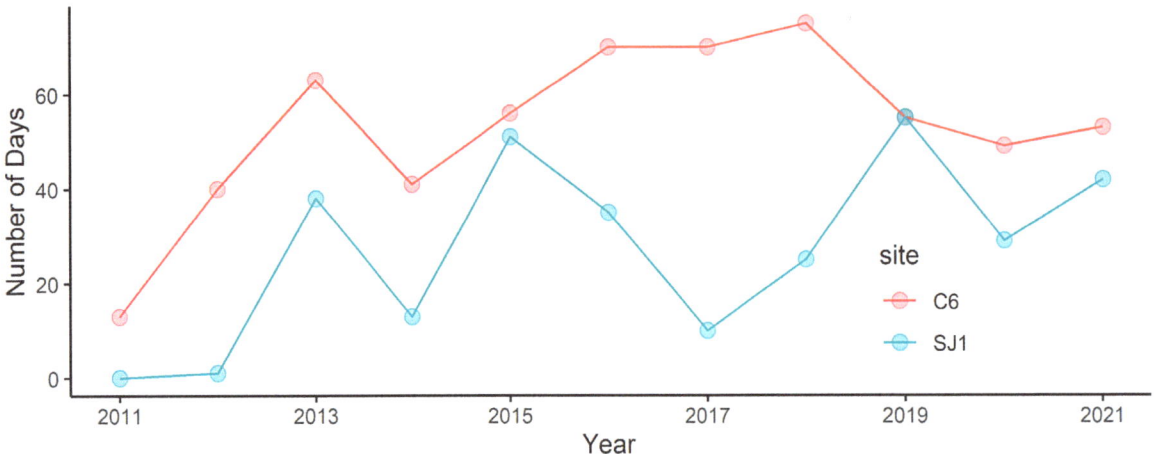

FIGURE 5-8 The number of days the hypolimnion was anoxic at C6 and SJ1. SOURCE: Chess (2022).

Mann-Kendall Dissolved Oxygen Trend Analysis

The committee conducted a formal statistical analysis of all of the bottom-water monthly average data presented in Tables 5-1 through 5-4. The analysis is structured in a manner such that trends over all months are evaluated, but results for each of the individual months can also be considered. The goal of this analysis was to determine if there is an ongoing trend in the direction of anoxia, particularly summertime anoxia, in the Lake. The datasets included in this analysis cover calendar years 1995–2020 plus the months January–March of 2021 at C1 and C4. There are data for 1991–1992 at some of the sites, followed by a gap of several years. Given that there are some concerns about the comparability of the 1991–1992 data to the later years, the committee deleted the 1991–1992 data from this analysis, although an alternative analysis was run with these data included and the results did not change substantially (results not shown).

For each of the four sites, the data were evaluated by doing individual Mann-Kendall trend tests for the time series for each of the 12 months (provided that there were more than three observations in that month during the total record). The significance level was evaluated and categorized as follows. For $p < 0.01$, the trends were considered highly significant; for $0.10 > p > 0.01$, they were considered moderately significant; and for $p > 0.1$, they were considered not significantly different from zero. The Mann-Kendall test was selected for three reasons. It is robust against the influence of a few data points that are widely separated in time from the main body of the data. It is robust against strong asymmetry of the data. It can be combined readily into an overall test of trends across all months (the Seasonal Kendall test described by Hirsch et al., 1982) with a minimum set of assumptions (i.e., there is no need to specify a model of seasonality), but it is resistant to the seasonal differences in the distributions.

For each of the months, the slope of the trend was evaluated using the Theil-Sen slope estimator (Sen, 1968; Theil, 1950), which is closely related to the Mann-Kendall test (such that the sign of the Mann-Kendall test statistic is never the opposite of the Theil-Sen slope) and is highly resistant to the effect of outliers in the data. The results of all of the Mann-Kendall trend tests, for all months and each of the four sites, are shown in Table 5-5. The Seasonal Kendall test (and associated Seasonal Theil-Sen slope) sums the results of the individual monthly results and evaluates the statistical significance of this sum; it also calculates a trend slope across all months. These Seasonal Kendall results are given in the last column of Table 5-5.

The following are some observations about these results. First, for sites C1 and C4, all 22 of the monthly trends that could be evaluated (regardless of significance) are positive and none are negative.[5] Five of those 22 were moderately significant upward trends and none were highly significant. However, when aggregated over the months using the Seasonal Kendall test, both C1 and C4 show highly significant upward trends in dissolved oxygen.

---

[5] Text here and throughout the report was modified after report release to more precisely denote where the trends were observed.

TABLE 5-5 Trends in Monthly Mean Dissolved Oxygen in Bottom Water at C1, C4, C5, and C6, 1995–2020[a]

Site	Jan	Feb	Mar	Apr	May	Jun	Jul	Aug	Sep	Oct	Nov	Dec	All
C1	*	+0.02	+0.04	+0.02	+0.00	+0.03	+0.02	+0.03	+0.02	+0.01	+0.00	0.02	+0.02
C4	*	+0.04	+0.04	+0.04	+0.02	+0.02	+0.02	+0.02	+0.02	+0.01	+0.00	+0.29	+0.06	+0.02
C5	*	+0.16	+0.20	+0.09	+0.08	+0.10	+0.10	+0.07	+0.04	+0.04	+0.14	+0.12	+0.09
C6	*	*	+0.18	+.012	+0.09	+0.05	−0.03	−0.02	+0.05	+0.10	+0.11	*	+0.04

NOTES: The values shown are the estimated slope of the trend in mg/L/yr. Months marked with an asterisk are those with three or fewer monthly mean values. Data are from 1995 to 2020 except for January, February, and March at C1 and C4, which extend to 2021. The color code relates to the significance level of the observed trend. Dark blue indicates a highly significant upward trend ($p < 0.01$), light blue indicates a moderately significant upward trend ($0.10 > p > 0.01$), no shading indicates results are not significantly different from zero ($p > 0.10$), and pink indicates a moderately significant downward trend ($0.10 > p > 0.01$). There were no highly significant downward trends ($p < 0.01$). The last column, labeled "All," is the result of a Seasonal Kendall test on the data from all months. The color coding uses the same assignments as are used for the monthly results.
[a]Table 5-5 was edited after report release to reflect corrections to Table 5-3, the previous version of which had data inadvertently shifted by two months.

At C5, the trends across the 11 months that could be tested were all positive. Of these, four were highly significant and three were moderately significant. The overall result using the Seasonal Kendall test had a highly significant upward trend, which was much larger in magnitude than the trends at C1 or C4.

The data for C6 were much more limited than the data for the other sites. Only nine months could be evaluated. Of these, seven had positive slopes and two had negative slopes (July and August). Of these negative slopes, July was not statistically significant and August was moderately significant. The overall result using the Seasonal Kendall test shows a moderately significant upward trend in dissolved oxygen. The downward trends indicated for July and August are important because they come at the time of year with the lowest dissolved oxygen levels.

In summary, it is difficult to make a general case that there has been deterioration in dissolved oxygen conditions in CDA Lake; rather, the data suggest the contrary conclusion. Of course, these data do not include the lateral lakes, which are known to have hypoxic or anoxic conditions at times. Unfortunately, there are no long-term dissolved oxygen observations in the lateral lakes that would support a similar trend analysis.

These results, coupled with the trend results for phosphorus and nitrogen in the Lake (see below) and entering the Lake (see Chapter 3), indicate that low dissolved oxygen is not expected to become a problem in the main body of the Lake if current trends continue. It will be important to continue to collect dissolved oxygen data, and analyze it every two years, in order to detect changing conditions.

IN-LAKE NUTRIENT ANALYSES

The committee analyzed both phosphorus and nitrogen data from CDA Lake to determine if there are any trends over time that might indicate worsening water quality. The committee plotted nutrient data from five locations in CDA Lake, it created loess smooths of the surface water and bottom water data sets, and it conducted Mann-Kendall tests of trends in the monthly averaged nutrient data at all five sites.

Trends in Total Phosphorus

For total phosphorus (TP), there are regular long-term data available for five sites (C1, C4, C5, C6, and SJ1). For C1 and C4, the data used here cover 2004–2020, and for C5, C6, and SJ1, the data cover 2007–2020. The data at all five sites show strong seasonal patterns, but these patterns are different for the different sites. The data for surface water are displayed in Figure 5-9. Only surface water samples were considered because their nutrient values can be expected to be important drivers of photosynthesis due to light availability.

At both C1 and C4, the highest total phosphorus values in the six years shown are in 2017, which is the year with the highest inflow. In general, the values decline from the first sample of the year at least through the end of the summer at both sites and all years—a pattern that is very similar to the total lead pattern at these sites (see Chapter 6). Both total lead and total phosphorus have a large component carried in the particulate fraction and are thus both responding to the pattern of inputs from the CDA River. Both of these sites show a strong decline

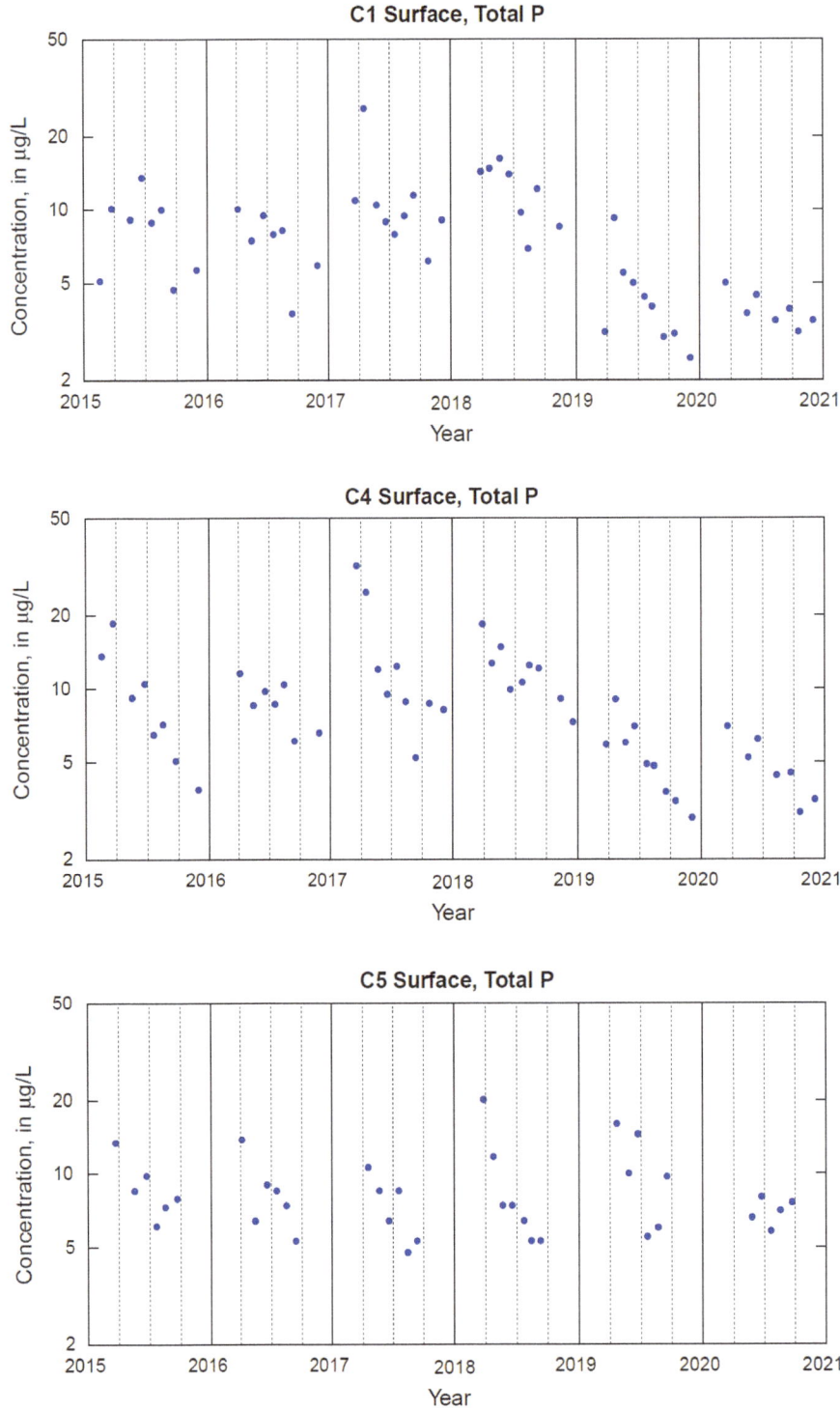

FIGURE 5-9 Concentration of total phosphorus in μg/L, over the years 2015–2020 at five sampling locations: C1, C4, C5, C6, and SJ1. All panels plotted with the same vertical and horizontal scales. Solid vertical lines mark the start of the calendar year, and the dashed lines separate the year into four quarters. SOURCE: Data plotted by the committee, courtesy of IDEQ and the CDA Tribe.

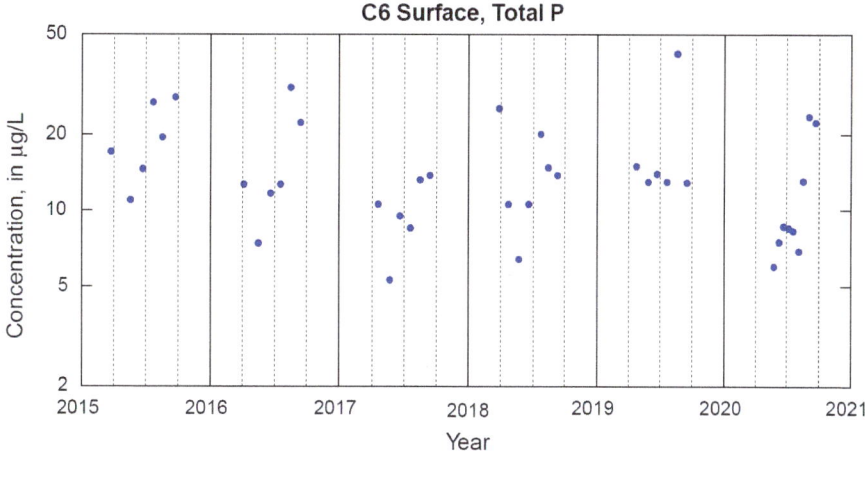

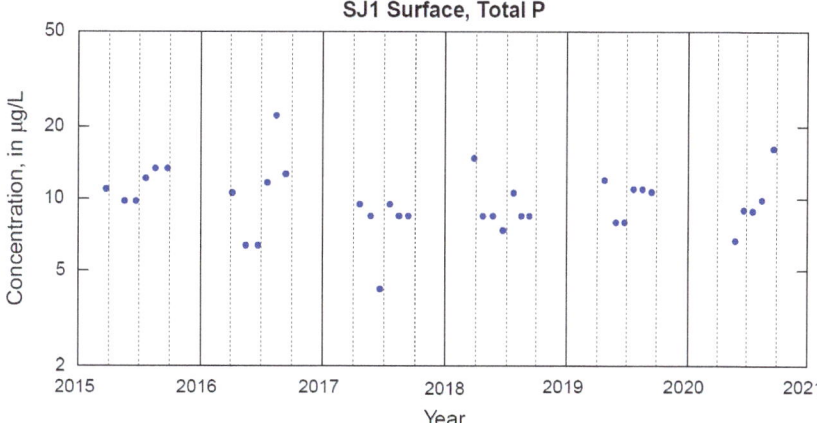

FIGURE 5-9 Continued

after 2017, with the total phosphorus values in 2019 and 2020 having much lower concentration than the prior years. The seasonal pattern at C5 is similar to C1 and C4. The maximum value in 2017 is not as large at C5, but all of the years show declines over the course of the spring and summer.

It is reasonable to ask the following question: might those low values in 2019 and 2020 simply be an artifact of two relatively dry years? Evaluating the inflow records for the January–May period of every year for the CDA River, one sees that the January–May period of 2019 was the third lowest in the past 26 years of record, and January–May of 2020 was the eighth lowest in the past 26 years. Thus, these two years were low flow years but not extremely low years. Furthermore, the total phosphorus concentrations in the CDA River observed in 2019 and 2020 were very low when compared to the other years that are also in the lowest third of the records for the past 26 years. Thus, the committee would argue that these two low concentration years are not simply an artifact of these being low flow years. Nevertheless, seeing this decline in concentrations at the end of the period of record makes it critical that frequent updates of analysis of these records be done (say every other year) to see if this strong departure from previous conditions persists.

The data suggest that river inputs drive the high total phosphorus concentrations of each year at each of these three sites, followed by some combination of biological uptake and particle settling driving the declines from the early peaks. The absence of a sharp decline at C5 for 2019 and 2020 is likely related to the differences in the history of total phosphorus flux for the CDA River versus the St. Joe River. Indeed, the flux of total phosphorus for the CDA River for the years 2018, 2019, and 2020 was 36, 15, and 17 percent, respectively, of the same flux in 2017. In contrast, the St. Joe River fluxes for 2018, 2019, and 2020 were 70, 32, and 30 percent, respectively, of the 2017 value. The discharges for these years were very nearly the same across the two watersheds. This suggests that the impacts of the mining remediation could have been driving the declines in the CDA River and hence driving the declines at C1 and C4. This is further evidence that the Superfund remedy is having a beneficial impact on the phosphorus levels in the northern parts of the Lake. At sites C6 and SJ1 there are no regular seasonal patterns in the total phosphorus concentrations.

Figure 5-10 shows a set of plots for the entire total phosphorus record at all five sites, distinguishing between surface water (or photic zone) and deeper water. The Seasonal Kendall trend results are summarized in Table 5-6. The ten-year trends (or full period of record trends—whichever is longer) are estimated using the Theil-Sen slope estimator. In the cases of sites C6 and SJ1, the change and significance levels are estimated separately for surface water and for bottom water because their trends are distinctly different.

The results for sites C1 and C4 are very similar. The estimated change over the past decade is a *decrease* of 15 percent at C4 and 25 percent at C1 (the p value is about 0.22). Thus, the evidence favors the conclusion that concentrations have declined over this period, but the statistical support for a ten-year decline is weak. Both C1 and C4 show a steady rise in total phosphorus concentrations from the start of the record until about 2016 (Seasonal Kendall trend calculations for C1 or C4 show a significant upward trend during this period). However, this was followed by a decline from 2016 to the end of 2020. The flow-normalized flux record for the CDA River near Harrison (see Chapter 3) also shows a pattern of rising values followed by falling values, but the timing of the reversal in that record comes at an earlier time (around 2010). Finally, surface water and bottom water concentrations are very similar throughout the C1 and C4 records.

The increase in phosphorus concentrations at C1 and C4 up until 2016 might explain conclusions (e.g., IDEQ and CDA Tribe, 2020) that phosphorus concentrations in the Lake have been increasing. But the newer data change the sign of the decadal trend, and unless there is yet another reversal in the record, decadal trends will soon indicate a highly significant downward trend in total phosphorus concentration. Ten years is a short window in which to interpret meaningful trends in something as complicated as total phosphorus, so continued monitoring and timely syntheses are essential to separating stochastic variability from actual trends.

In contrast to C1 and C4, the trend at C5 is steadily downward from the start of the record in 2007 to the end in 2020 (a decrease of 33 percent, with $p < 0.001$). This is in keeping with the observed trend in flux for the St. Joe River (see Chapter 3). The concentrations at C5 are typically slightly higher than those at C1 or C4, and this is also consistent with the somewhat higher total phosphorus concentrations for the St. Joe River versus the CDA River (see Chapter 3). All three of these sites (C1, C4 and C5) have median total phosphorus concentrations that are well within the range of what can be considered oligotrophic (< 12 µg/L).

Farther south in the Lake, sites C6 and SJ1 have more complex temporal patterns of total phosphorus and substantial differences between surface water and bottom water (with bottom water having median total phosphorus concentrations around 30 µg/L). There is much more variability in the record at these two sites compared to C1, C4, and C5, and trends in the deeper water at these two sites are highly uncertain. The surface water at both sites appears to follow the general trend of inputs from the St. Joe River, but the bottom water records at these two sites are much less related to river inputs and show signs that they are responding to internal sources of phosphorus in the Lake bed. Bottom waters at both these locations are anoxic in late summer and fall, a likely cause of the mobilization of phosphorus (see Chapter 7).

The conclusion drawn from the trends analysis is that total phosphorus conditions in CDA Lake are rapidly changing, such that frequent reanalysis (as often as every two years) is crucial. The evidence as of the end of 2020 does not support the hypothesis that total phosphorus concentrations in the Lake are trending upward. For surface water samples at all five sites, the most recent trends in total phosphorus are downward, especially from 2016 to 2020 (see Figure 5-10). The substantial differences in total phosphorus concentrations between surface water and bottom water at the two sites in the southernmost parts of the Lake (C6 and SJ1) indicate that internal processes are significant in the latter areas.

The recent declines in total phosphorus in the Lake at C1, C4, and C5 appear to be linked to the declines in total phosphorus flux for the two main tributaries of the Lake. The possible drivers of these changes (discussed in Chapter 3) are likely the diminution of mining and the effectiveness of the Superfund remedy in terms of soil loss,

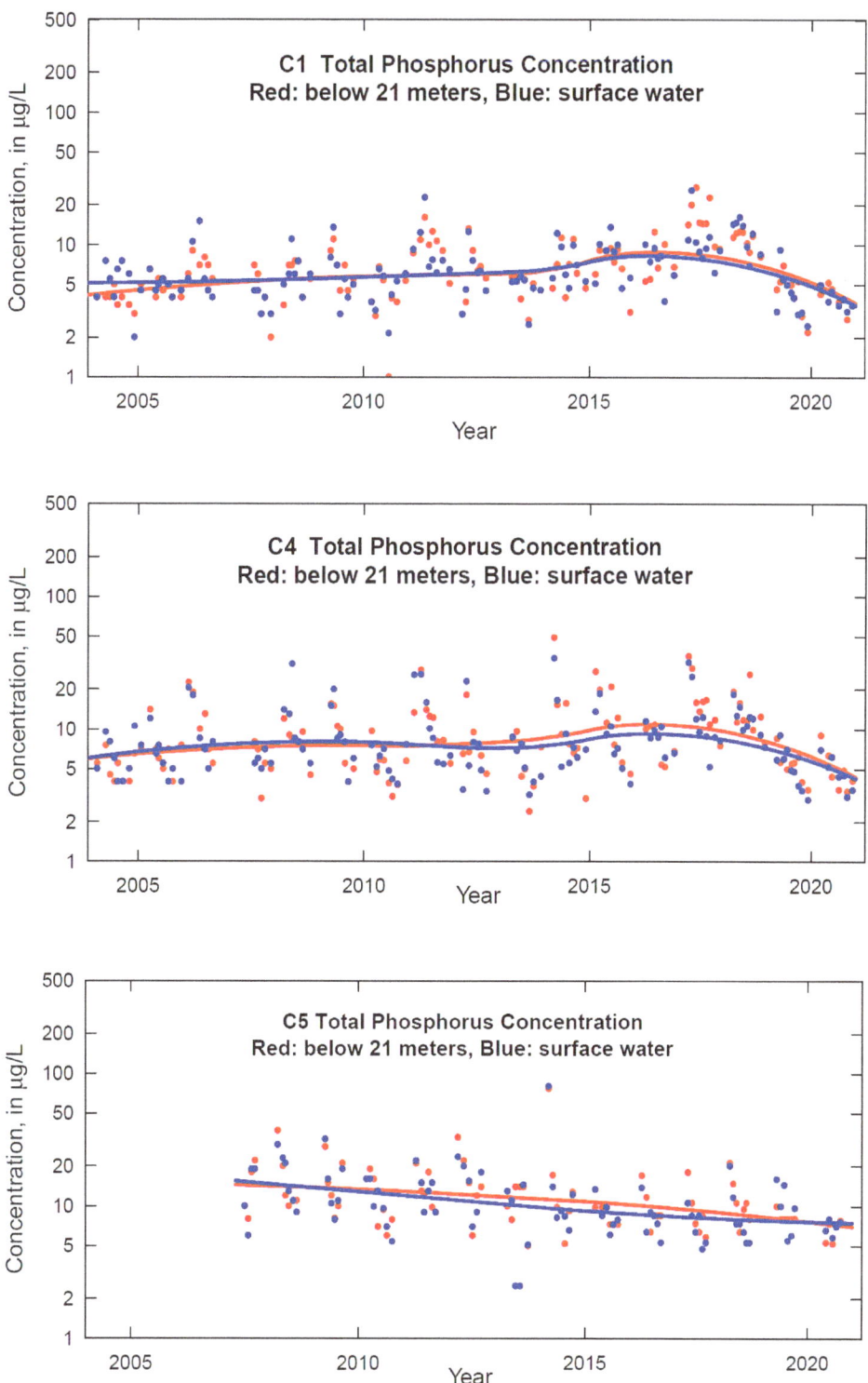

FIGURE 5-10 Total phosphorus concentration versus time at five sampling locations: C1, C4, C5, C6, and SJ1, with red symbols representing bottom water samples (below 21 meters) and blue symbols representing surface water samples. The red and blue curves are loess smooths of the bottom and surface water datasets, respectively. Note that the vertical axes are identical across all of the panels. SOURCE: Data courtesy of IDEQ and the CDA Tribe and analyzed and plotted by the committee.

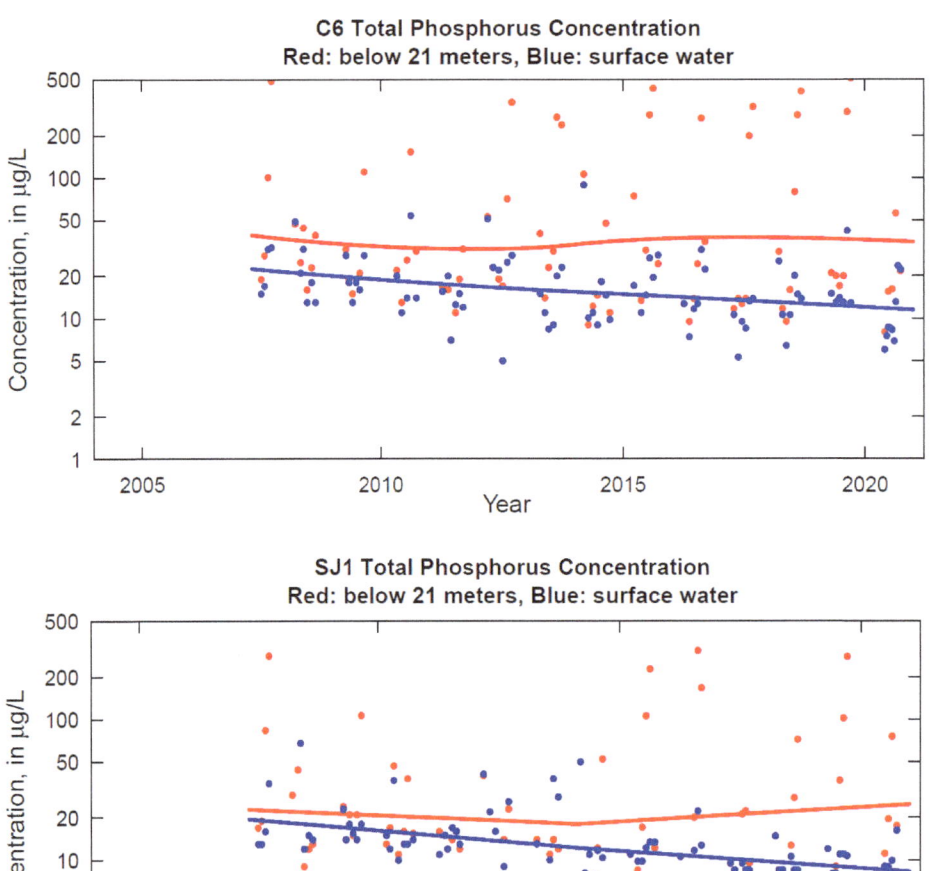

FIGURE 5-10 Continued

TABLE 5-6 Summary of Trends in Total Phosphorus Concentration at Five Monitoring Sites in CDA Lake

Site	Change over past 10 years	2020 median bottom water	2020 median surface water
C1	−25%	3.6 μg/L	3.5 μg/L
C4	−15%	4.3 μg/L	4.2 μg/L
C5	−33%	6.9 μg/L	7.7 μg/L
C6 Surface Water	−29%		12 μg/L
C6 Bottom Water	+9%	35 μg/L	
SJ1 Surface Water	−38%		8.3 μg/L
SJ1 Bottom Water	+49%	25 μg/L	

NOTES: Amounts of change and their significance are determined by the Theil-Sen slope estimator and the Seasonal Kendall test. Dark blue shading indicates a highly significant downward trend ($p < 0.01$), light blue shading indicates trends that are moderately significant ($0.01 < p < 0.1$), and the unshaded trends are not statistically significant ($p > 0.1$). Median values in 2020 are determined using the loess smoothing shown in Figure 5-10.

improved forest practices, and other efforts carried out under the Lake Management Plan to reduce streambank erosion in the watersheds of the CDA and St. Joe Rivers. The ability to project future changes in phosphorus inputs and concentrations in the Lake will depend on research aimed at better understanding the causative mechanisms of these recent trends. Future trends in total phosphorus will depend on a combination of factors, including continuing improvements in landscape stability in formerly mined parts of the watershed and in the channel of the CDA River, changes in rain intensity and river discharges, and the implementation of additional phosphorus control measures under the Lake Management Plan. Forecasting how these changes will influence the delivery of phosphorus to the Lake will depend on the creation of predictive models that take these processes into consideration and utilize the history of the river and Lake total phosphorus records to evaluate the ability of the models to simulate trends.

Trends in Total Nitrogen

Analyses of seasonality and trend for total nitrogen (TN) were run at all five CDA Lake sites, although the records at C5, C6, and SJ1 are all shorter than ten years. The data from the past six years are displayed in Figure 5-11.

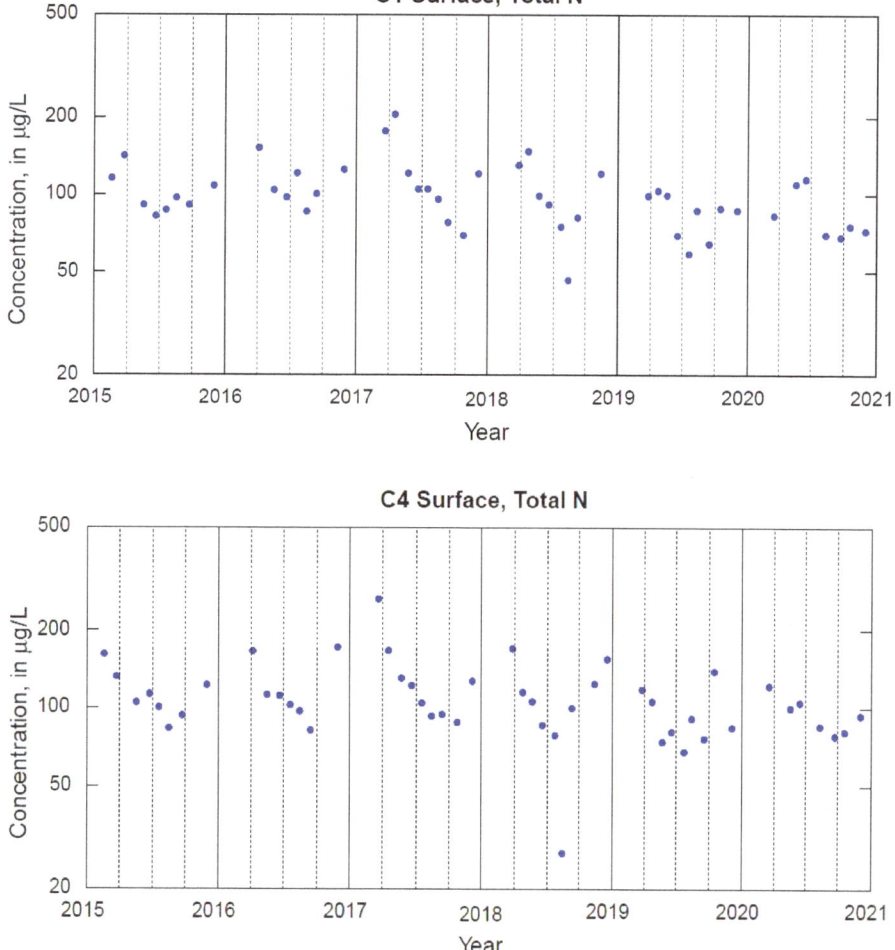

FIGURE 5-11 Concentrations of total nitrogen, in µg/L, over the years 2015–2020 at five sampling locations: C1, C4, C5, C6, and SJ1. All panels plotted with the same vertical and horizontal scales. Solid vertical lines mark the start of the calendar year and the dashed lines separate each year into four quarters. SOURCE: Data plotted by the committee, courtesy of IDEQ and the CDA Tribe.

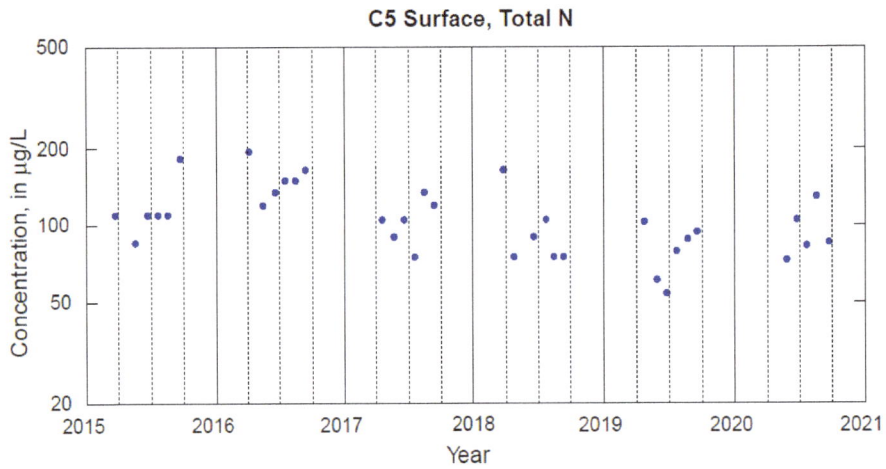
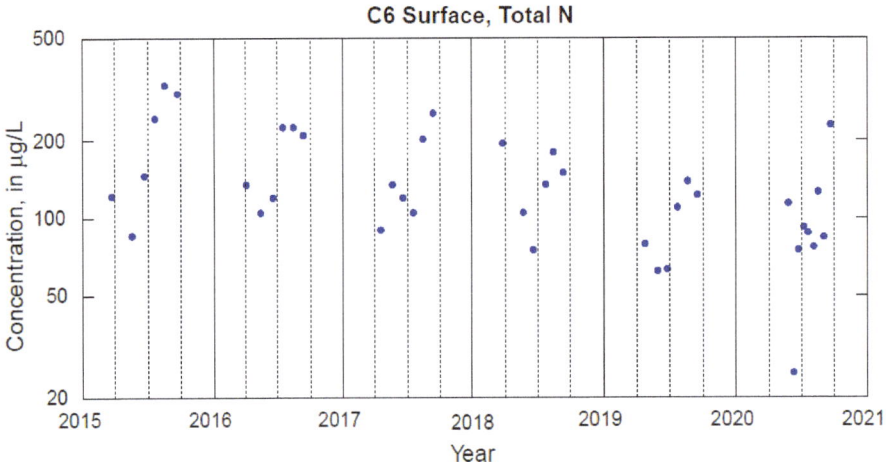
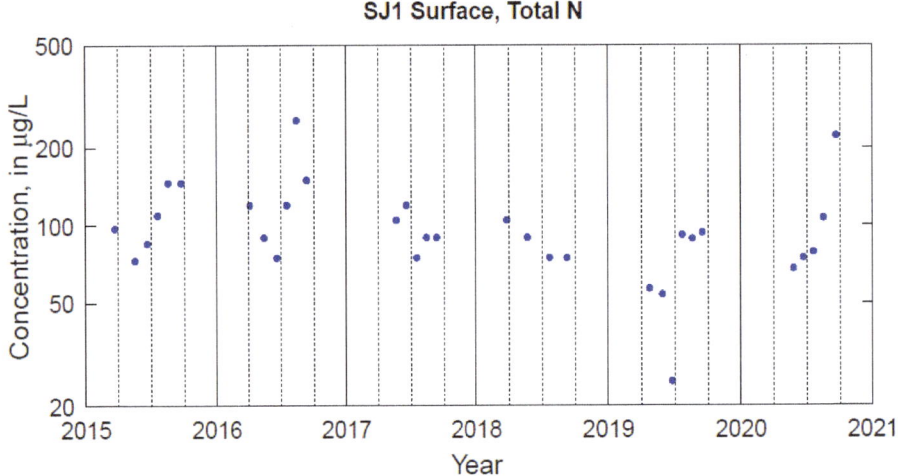

FIGURE 5-11 Continued

The seasonal patterns differ among the sites. At C1 and C4, the patterns are very similar to the total phosphorus record at these sites. The highest values come with the high inflows of the late winter and spring, with concentrations declining through the remainder of the year. The highest values in this period are associated with the year of highest CDA River inflow, 2017. The total nitrogen pattern at site C5 is more complex. The highest values are generally at the first sampled month in the year (such as March or April), but then concentrations are much lower in the summer and appear to rise up to the last sample in the fall. At sites C6 and SJ1 the total nitrogen concentrations tend to climb throughout the entire sampling season and not be so highly related to inflow.

Figure 5-12 shows the trends at all of the sites. Trends in total nitrogen shown in Table 5-7 are quite different across the five sites. These results are consistent with the trends in flow-normalized nitrogen fluxes for the CDA River and St. Joe River reported by Zinsser (2020).

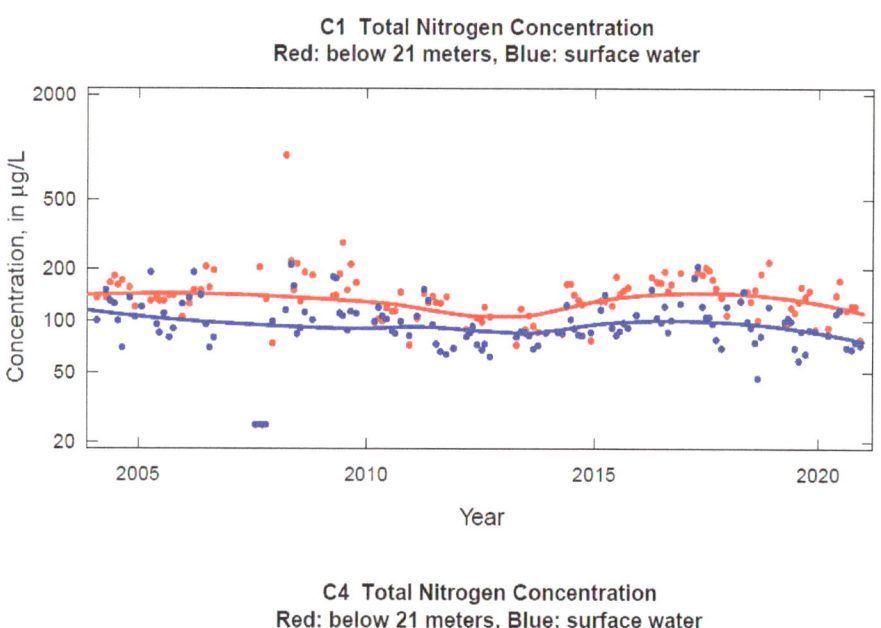

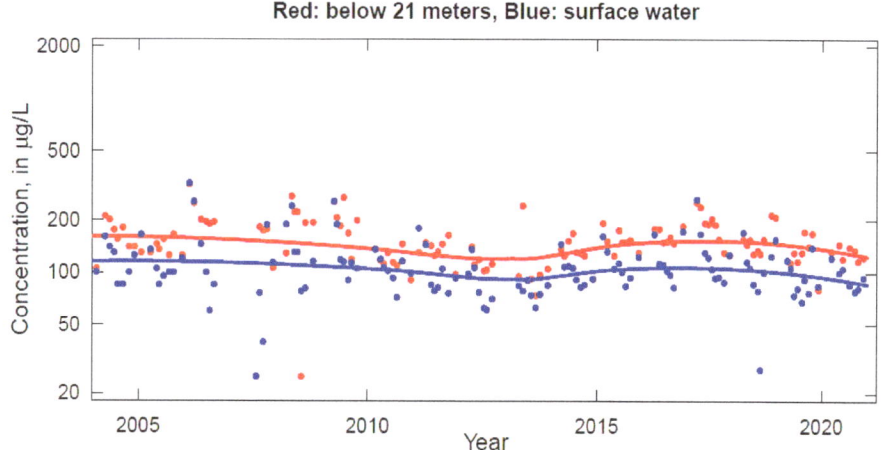

FIGURE 5-12 Total nitrogen concentration versus time at five sampling locations: C1, C4, C5, C6, and SJ1, with red symbols representing water samples below 21 meters and blue symbols representing surface water samples. The red and blue curves are loess smooths of the bottom and surface water data sets, respectively. Note that the vertical axes are identical across all of the panels. SOURCE: Data courtesy of IDEQ and the CDA Tribe and analyzed and plotted by the committee.

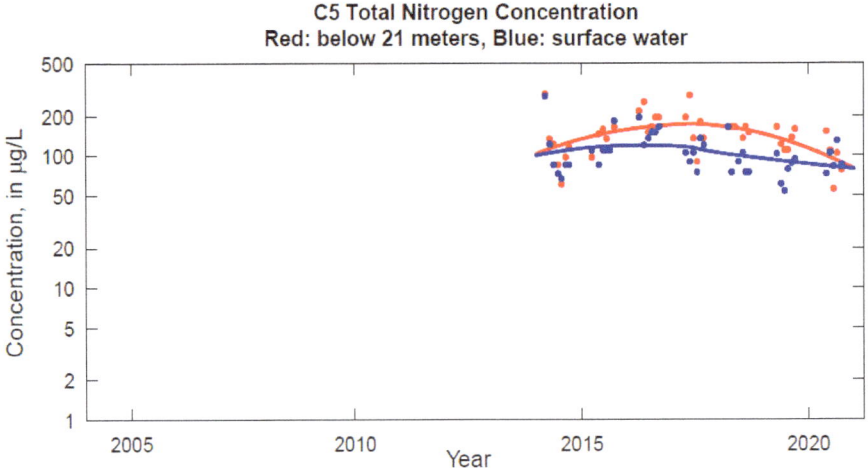

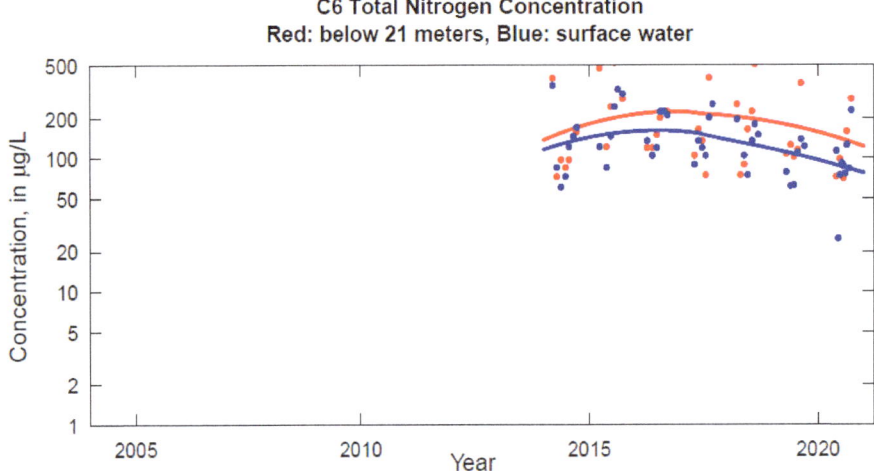

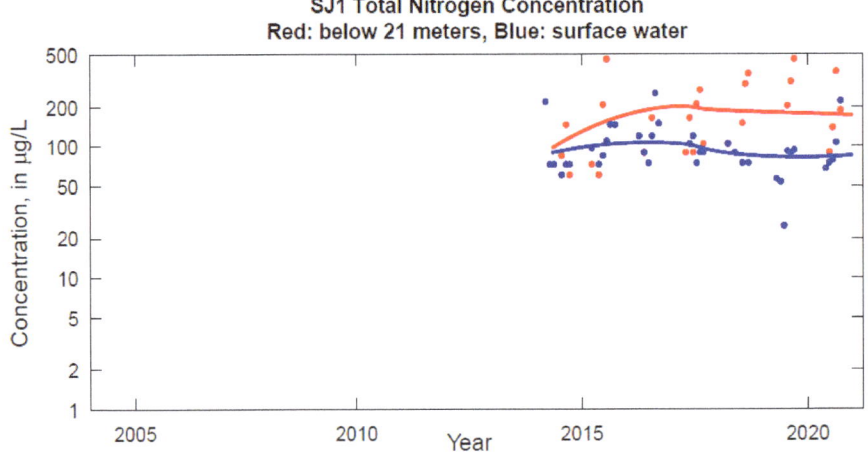

FIGURE 5-12 Continued

TABLE 5-7 Summary of Trends in Total Nitrogen at Five Monitoring Sites in CDA Lake

Site	Change over past 10 years	2020 median bottom water	2020 median surface water
C1	+20%	111 µg/L	76 µg/L
C4	+16%	124 µg/L	86 µg/L
C5	−18%*	79 µg/L	79 µg/L
C6	−45%*	122 µg/L	78 µg/L
SJ1	+15%*	172 µg/L	85 µg/L

NOTES: Amounts of change and their significance are determined by the Theil-Sen slope estimator and the Seasonal Kendall test. The tests include data from all months and from both surface water and bottom water samples. Dark blue shading indicates that each is a highly significant downward trend ($p < 0.01$), pink shading indicates upward trends that are moderately significant ($0.01 < p < 0.1$), and the unshaded trends are not statistically significant ($p > 0.1$). Median values in 2020 are determined using the loess smoothing shown in Figure 5-12. * indicates that the record is seven years rather than ten years.

Changes in atmospheric deposition may also be an important driver of changing nitrogen concentrations in water. The role of atmospheric deposition in this watershed is a topic that has not been studied, but should be part of future evaluations of nitrogen dynamics of the lake and watershed. Across all of the stations evaluated, the bottom waters have higher concentrations (by amounts like 25 percent) than the surface waters. Although phosphorus concentrations have been the primary focus in discussions of possible eutrophication of CDA Lake, nitrogen plays a role as well. All of the available Lake data sets indicate a general improvement in conditions in recent years, but an understanding of the causes is important to being able to evaluate likely future trends. Gaining a better understanding both of the nitrogen cycle, and of nitrogen trends in the Lake and rivers is an important topic that needs to be included in the research agenda supporting adaptive management of CDA Lake trophic status.

IN-LAKE PRODUCTIVITY ANALYSES

To complement the preceding analyses of dissolved oxygen and nutrients, the committee analyzed additional measures of lake productivity—trophic status and chlorophyll a. It also tested the hypothesis that zinc in the water column may be suppressing productivity, in order to better understand whether the Superfund remedy might be removing an important control of eutrophication in CDA Lake.

Trophic Status of CDA Lake with Respect to Nutrients and Chlorophyll a

Data from the 2017 National Lakes Assessment were used to assess CDA Lake with respect to trophic status. The National Lakes Assessment is a regionally stratified random sample of lakes (larger than 1 ha and deeper than 1 m) and thus provides an accurate picture of summer lake conditions in the continental United States. Data for total nitrogen, total phosphorus, and chlorophyll (chl) for more than 1,110 unique study sites were available in the 2017 database (EPA, 2021). The data can also be partitioned by region or state. In Tables 5-8 to 5-10, these data are compared to those obtained for CDA Lake, which were monthly epilimnetic or euphotic averages for summer (June–September) for both IDEQ and tribal data at stations C1, C4, C5, C6, and SJ1. Note that the CDA

TABLE 5-8 Nitrogen Data from U.S. Lakes, Northwest Lakes, and CDA Lake

Total N, mg/L	mean	median	1%	10%
U.S. lakes (n = 1,203)	1.12	0.66	0.058	0.22
NW lakes (n = 152) (WA, OR, ID, MT)	0.77	0.45	0.018	0.083
CDA Lake (n = 86, all data)	0.110	0.090	0.057	0.070

SOURCES: Top two rows: 2017 National Lakes Assessment data, https://www.epa.gov/national-aquatic-resource-surveys/nla. CDA Lake data courtesy of IDEQ and the CDA Tribe.

TABLE 5-9 Phosphorus Data from U.S. Lakes, Northwest Lakes, and CDA Lake

Total P, µg/L	mean	median	1%	10%
U.S. lakes (n = 1,198)	98.5	32.7	2.95	8.86
NW lakes (n = 148) (WA, OR, ID, MT)	78.5	24.3	0.80	6.45
CDA Lake (n = 142; all data)	13.1	11.1	3.50	6.24

SOURCES: Top two rows: 2017 National Lakes Assessment data, https://www.epa.gov/national-aquatic-resource-surveys/nla. CDA Lake data courtesy of IDEQ and the CDA Tribe.

Tribe's dataset had total nitrogen records recorded as "< 50 µg/l"; in these cases, the value was set to 50 µg/L. In the tribal and IDEQ data sets, chlorophyll values coded as "< X" were set to X. Thus, average total nitrogen and chlorophyll values are likely overestimates.

The committee also compared total nitrogen:total phosphorus (TN:TP) ratios in CDA Lake in relation to those in the National Lakes Assessment database. Averaged across stations C1, C4, C5, C6, and SJ1 since 2007, for which paired observations of TN and TP are available, average TN:TP ratios are 40.2 (atomic ± 1.3 s.e.). This value is somewhat higher than the median value of 30.8 (atomic) for lakes in the NLA. A transition to onset of nitrogen limitation occurs when TN:TP ratios are lower than a threshold of 31:4 (atomic) (Downing and McCauley, 1992). Thus, nitrogen limitation of phytoplankton growth seems possible in CDA Lake at sites and times when the TN:TP ratio is low. In general, TN:TP ratios were somewhat lower in the southern part of the Lake (C5, C6, SJ1), implying that nitrogen limitation of phytoplankton growth might be more frequent there than in the northern part of the Lake (C1, C4).

According to the trophic status parameters of Carlson (1977), CDA Lake's median values for total phosphorus (11.1 µg/L) and chlorophyll (1.80 µg/L) concentrations place the Lake in the oligotrophic (< 12 µg P/L, < 2.6 µg chl/L) category.

Chlorophyll Trend Analysis

The committee evaluated the chlorophyll data collected in surface water at sites C1, C4, C5 and C6. For sites C1 and C4, the dataset runs from July 2002 through October 2020. (Data from 1991 and 1992 are not considered because of the long time gap prior to the main body of the dataset.) About 6 percent of these values are censored, meaning that the samples were recorded as "<1 µg/L," and were treated as being equal to half their reporting limit. For sites C5 and C6, the dataset runs from July 2007 through December 2020. Of these data, about 15 percent are censored and were handled in the same manner as the censored data from C1 and C4. Site SJ1 was not considered because more than 50 percent of the data are censored.

For each of the four sites, the data were evaluated by doing individual Mann-Kendall trend tests for the time series for each of ten "seasons." The "seasons" are generally the calendar months, except that January and February data are considered to be one season, and November and December are considered to be one season (because these months had relatively few observations). In the case of C5 and C6, there were insufficient data to conduct the trend analysis on the January–February data.

There was a total of 38 cases, each a combination of season and site, which could be evaluated for trends (4 sites × 10 seasons – C5/C6 Jan/Feb cases). For 9 of the 38 cases that could be evaluated, positive trends were

TABLE 5-10 Chlorophyll Data from U.S. Lakes, Northwest Lakes, and CDA Lake

Chlorophyll, µg/L	mean	median	1%	10%
U.S. lakes (n = 1,208)	24.3	8.63	0.32	1.56
NW lakes (n = 159) (WA, OR, ID, MT)	12.8	2.80	0.20	0.42
CDA Lake (n = 140; all data)	1.84	1.80	0.74	1.09

SOURCES: Top two rows: 2017 National Lakes Assessment data, https://www.epa.gov/national-aquatic-resource-surveys/nla. CDA Lake data courtesy of IDEQ and the CDA Tribe.

TABLE 5-11 Trends in Chlorophyll by Month in Surface Water at C1, C4, C5, and C6, 2002–2020

Site	Jan-Feb	Mar	Apr	May	Jun	Jul	Aug	Sep	Oct	Nov-Dec	All
C1	+0.08	−0.10	+0.03	−0.05	−0.09	−0.02	−0.00	−0.02	−0.05	+0.04	−0.03
C4	−0.10	−0.17	−0.01	−0.06	−0.08	−0.03	−0.00	−0.07	−0.04	+0.02	−0.04
C5	*	−0.14	−0.12	+0.00	+0.05	−0.08	+0.00	+0.09	−0.06	+0.07	−0.00
C6	*	−0.52	−0.08	+0.02	−0.01	−0.10	−0.04	−0.05	−0.05	−0.08	−0.05

NOTES: The values shown are the estimated slope of the trend in μg/L/yr. Months marked with an asterisk are those with three or fewer observations. The color code relates to the significance level of the observed trend. Light blue indicates a moderately significant downward trend ($0.10 > p > 0.01$), and no shading indicates that results are not significantly different from zero ($p > 0.10$). The last column, labeled "All," is the result of a Seasonal Kendall test on the data from all months.

observed, but none of them had even moderate levels of significance ($p < 0.1$). The remaining 29 cases had negative trends; five of them were of moderate significance ($0.01 < p < 0.1$) but none were highly significant ($p < 0.01$).

In addition to those analyses, a Seasonal Kendall test was run for each of the four sites by combining data across all seasons at the site. For sites C1, C4, and C6 the results showed moderately significant decreases (p-values of 0.043, 0.012, and 0.097, respectively). For C5, the sign of the trend was also negative, although it was far from being significant. For each of the months, the slope of the trend was evaluated using the Theil-Sen slope estimator.

The results of all of the Mann-Kendall trend tests for all months and the results of the Seasonal Kendall test for each of the four sites are shown in Table 5-11.

The committee makes the following observations about the results. For sites C1 and C4, three of the monthly trends that could be evaluated (regardless of significance) were positive while 17 were negative. Three of the negative values were moderately significant ($0.01 < p < 0.1$). When aggregated over all of the months using the Seasonal Kendall test (last column), both sites show moderately significant downward trends in chlorophyll.

At C5, the trends in five of the nine months that could be tested were positive, but none of these were even moderately significant. The other four months had negative trends. One of these (the trend in April) was a moderately significant downward trend. The overall trend as determined by the Seasonal Kendall test was slightly negative, but not even close to being significant. At C6, one month (May) had a positive slope (not significant), the other eight months that could be evaluated had a negative slope, and only one of these was significant (April). The overall result for C6 using the Seasonal Kendall test shows a moderately significant downward trend in chlorophyll.

In summary, it is very difficult to make a general case that there is a deteriorating (i.e., increasing) chlorophyll *a* level in the Lake over the most recent 20 years. In fact, the data suggest the contrary conclusion. Of course, these data do not include the lateral lakes, which are known to sometimes have high chlorophyll concentrations. Unfortunately, there are no long-term chlorophyll observations in the lateral lakes that would support a similar trend analysis. Given these results, coupled with the generally declining trends for phosphorus in the Lake and entering the Lake, there has been no tendency toward eutrophication of CDA Lake in the past 20 years. These results, coupled with the lack of increasing trends for phosphorus in the Lake and entering the Lake, show there has been no tendency toward more eutrophication of CDA Lake in the last 20 years. Eutrophication is not expected to become a problem *if current trends continue*.

The Question of Zinc Suppression in CDA Lake

Zinc is essential at low concentrations to most living things (used as an enzyme cofactor), but it is toxic at high concentrations, when it can bind nonspecifically with biologically important molecules and interfere with their function (Gonçalves et al., 2018). Thus, if zinc concentrations reach levels toxic to phytoplankton, zinc can conceivably offset the eutrophying effects of nitrogen and phosphorus loading (Kuwabara et al., 2007). Indeed, undesirable phytoplankton blooms are often, though temporarily, addressed by additions of toxic metals, especially copper (McKnight et al., 1983). Zinc and other metals can also be toxic to biota higher in the water column food web, such as zooplankton and fish.

An important question that the committee was specifically asked to address is the possibility that eutrophication in CDA Lake might occur if zinc concentrations decline to levels sufficient to release lake phytoplankton from inhibition by current high levels of potentially toxic zinc. In theory, this could set up a positive feedback in which increased algal production would lead to depletion of oxygen in the Lake's hypolimnion, leading to release of sediment phosphorus and further amplifying eutrophication. Such concerns are plausible given the high zinc concentrations in some parts of the Lake, along with early laboratory experiments showing that high concentrations of free zinc ion reduced the growth of some phytoplankton taxa isolated from CDA Lake and inhibited their response to phosphorus addition (Woods and Beckwith, 1997; Kuwabara et al., 2007). These responses are concordant with previous experiments with cultured algae, as well as with natural phytoplankton in Lake Ontario (Wong et al., 1978; Wong and Chau, 1989). The committee recognized that it is notoriously challenging to conclusively test or refute such hypotheses with field observations alone. But sufficient weight of evidence from the field can be invaluable in identifying tendencies in the direction predicted by the hypotheses. This element of the statement of task could have important implications for future management, but it has been 15 years since the hypothesis was proposed. The committee felt an analysis, even if limited given constraints imposed by the existing data, would be appropriate.

To evaluate the likelihood of the zinc inhibition hypothesis, the committee performed three analyses with the available field data on chlorophyll that were gathered since the original hypothesis was reported:

1. It compared CDA Lake's chlorophyll values relative to its total phosphorus concentrations and compared these to global relationships. If high zinc concentrations in CDA Lake inhibit phytoplankton biomass, then one would expect CDA Lake chlorophyll values to be lower than those observed in other lakes with similar total phosphorus values. *Furthermore, one would expect that deviation to be most pronounced in parts of the Lake where zinc levels are highest.*
2. It performed multiple regression analyses to predict chlorophyll as a function of total phosphorus and zinc concentrations for CDA Lake. If zinc is important in reducing chlorophyll concentrations, one would expect negative associations of chlorophyll with zinc once total phosphorus was accounted for in a regression model.
3. Since chlorophyll concentrations are likely driven by multiple factors beyond just total phosphorus and zinc, the committee assessed the shape of the relationship between chlorophyll concentration and zinc concentration in different parts of the Lake in order to determine if potential threshold or envelope effects describe how zinc affects chlorophyll–total phosphorus relationships.

A notable limitation of the field data is that all three of these approaches use overall chlorophyll concentrations as an index of phytoplankton performance or proliferation in response to zinc levels in the Lake. Chlorophyll concentration is an aggregate of chlorophyll found in cells across the range of taxonomically and physiologically diverse taxa present. Thus, bulk chlorophyll has the potential to obscure responses of zinc-sensitive taxa that may be ecologically important. Only detailed taxonomic assessments of the phytoplankton community can provide indications of potential zinc impacts on different taxa. However, as with most other lakes, such data for CDA Lake are limited in temporal and spatial scope.

Expected Chlorophyll Concentrations in CDA Lake

The committee sought to assess how CDA Lake conforms to known relationships between nutrient supply (as indexed by total phosphorus concentration) and chlorophyll concentration. Figure 5-13 plots data from a survey of published data (Havens and Nürnberg, 2009) for 389 lakes from North America, Europe, Asia, and New Zealand (all shown in orange) showing the well-known positive log–log relationship between these two parameters. Summer (June, July, August, September) monthly mean values of each parameter for each the years 2007–2020 for CDA Lake are plotted separately for IDEQ (blue) and tribal data sources (black and white). CDA Lake data lie in the low-intermediate range for both parameters relative to the global lake sample. The two parameters are positively correlated, as seen in previous studies. However, the relationship has a shallower slope, and the CDA Lake data have notably lower chlorophyll concentration for any given total phosphorus concentration, especially at higher

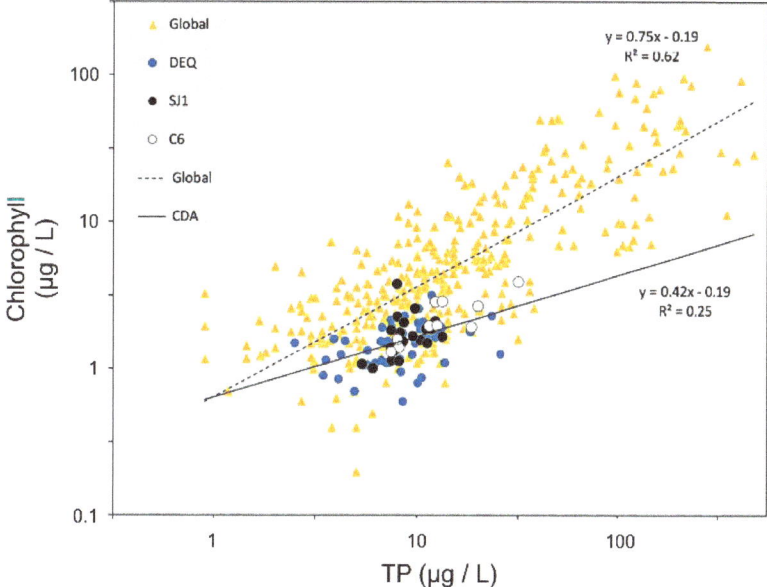

FIGURE 5-13 Relationship between log (chlorophyll concentration, µg/L) and log (total phosphorus concentration, µg/L) for a global set of 370 lakes reported by Havens and Nürnberg (2009) (yellow triangles) and for 82 CDA sites (circles: blue = northern sites sampled by IDEQ; black, white = southern sites sampled by the CDA tribe). Lines fitted to each dataset are shown. SOURCE: Data courtesy of EPA (2021), IDEQ, and the CDA tribe and plotted by the committee.

total phosphorus levels. This reduction in observed chlorophyll concentrations for any given level of total phosphorus is consistent with the hypothesis that high concentrations of zinc (and other metals) depress phytoplankton biomass, moderating the eutrophication response that accompanies nutrient loading.

The committee estimated CDA Lake's expected chlorophyll concentration if the system behaved according to the global relationship reported by Havens and Nürnberg (2009). Using the median of total phosphorus data for CDA Lake (9.53 µg/L; tribal and IDEQ data, summer only), the predicted value from the global chl–TP relationship is 3.50 µg chl/L. In contrast, the observed median chlorophyll value in CDA Lake is 1.55 µg chl/L (summer). From these data, one can infer that, on average, chlorophyll levels in CDA Lake are ~2.25 times lower than expected based on the Lake's total phosphorus concentration. Indeed, if the Lake had the predicted chlorophyll concentration, it would be classified as mesotrophic, not oligotrophic. However, this discrepancy is not uniform across the Lake, as indicated in Figure 5-13. That is, because the slope of the chl–TP relationship is shallower than the one for the global lake data, the discrepancy is larger for samples with higher total phosphorus concentrations.

As seen in Figure 5-13, samples from the southern end of the Lake have generally higher total phosphorus concentrations and chlorophyll concentrations than those at the northern end of the Lake. Thus, in this analysis, southern sites have a greater discrepancy between observed total phosphorus concentration and the chlorophyll concentration expected from the global chl–TP relationship. This holds when considering individual stations within the southern part of the Lake. For example, site C6 is the furthest south and zinc concentrations are < 5 µg/L throughout the recorded data (lowest in the Lake). However, examination of Figure 5-13 shows that it is not distinct in terms of the chl–TP relationship for the Lake as a whole. That is, C6 also shows a lower chlorophyll concentration given its total phosphorus level relative to the global lake dataset. Thus, this lack of a within-lake, zinc-dependent pattern in the chl–TP relationship is not consistent with the hypothesis that high zinc concentrations inhibit phytoplankton biomass in CDA Lake.[6]

---

[6] Our conclusion (lack of support for Zn suppression) from looking at C6 might be tempered by the occurrence of anoxia (and thus unusually low chl and high TP) on some of the dates for C6.

Multiple Regression Analysis

To explore these possibilities further, the committee developed multiple regression analyses of chlorophyll concentrations in CDA Lake as a function of total phosphorus and zinc using datasets provided for sites C1, C4, C5, and C6. First, for each of the hundreds of individual months that might be present in the records (near monthly for most of 1991, 1992, and 2002–2020), the committee extracted all the surface water data ("epi" or "photic zone," always less than 20 m) collected at a given site for a given month. For each month at each site, the median value of total phosphorus, total zinc, and chlorophyll *a* was computed. Cases where there was a "less than" value were coded to be half the reporting level (there were not very many of these except at C6, where most of the zinc values were "less than" values). For purposes of the regression analysis, only sites and months in which all three values were available were used; this comprised a dataset of 458 valid cases. Based on the frequency distributions of each of the variables, all data were natural log-transformed.

The most general model considered looked at all 458 cases (without regard to site or time of year). The form of the model was as follows:

$$\ln(\text{Chl}) = \beta_o + \beta_1 \cdot \ln(\text{TP}) + \beta_2 \cdot \ln(\text{TZn}) + \varepsilon$$

The expectation based on the zinc inhibition scenario is that β_1 should be positive (more total phosphorus leads to more chlorophyll) and that β_2 should be negative (more total zinc leads to less chlorophyll). However, when this model is fit, it has a very low R^2 value (0.05), but it is statistically significant overall (because the sample size is relatively large). The coefficient on ln(TP) is positive (+0.25), as expected, and it is highly significant. However, the coefficient for ln(TZn) is positive (+0.02), not negative, and is not significant. Hence, there is no evidence of zinc inhibition from this analysis.

To consider this hypothesis more deeply, the committee also assessed subsets of the data (e.g., combinations of particular sites or months). In doing so, various outcomes, some supporting the zinc inhibition hypothesis—for example, an analysis of all sites in October yielded a statistically significant positive association for ln(TP) and a statistically significant negative association for ln(TZn)—and some contradicting it, were found. However, such efforts represent a form of "p hacking" (searching among a multitude of correlations for one that fits a preconceived notion) and is not recommended practice. Thus, this regression approach did not prove fruitful in shedding light on the zinc inhibition effect.

Shape of Zinc versus Chlorophyll Relationships

As one final assessment of a possible influence of zinc on chlorophyll concentrations, the structure of the data distribution was considered following Schmidt et al. (2012). This is based on the idea that ecological responses (e.g., chl *a*) to a well-characterized inhibitory factor (e.g., zinc) in the presence of other limiting factors often result in a wedge-like pattern across the gradient of that limiting factor (Terrell et al., 1996; Schmidt et al., 2012). Interpretations of such relationships are vulnerable to coincidental occurrences of events and confounding variables, of course (e.g., Cloern, 2022).

Figure 5-14 shows the relationship between total zinc and chlorophyll for all available data for the photic zone (i.e., data from C1, C4, C5, C6, and the lateral lakes, Swan and Thompson). For the overall dataset, there is a statistically significant ($p < 0.0001$) but very weak *negative* ($r = -0.18$) relationship between chlorophyll and total zinc (Figure 5-14), consistent with zinc inhibition. However, for stations in the main body of the Lake (C1, C4, C5), there is a statistically significant ($p < 0.02$) but very weak *positive* ($r = 0.13$) relationship (data not shown).

High chlorophyll concentrations (> 5 µg/L) in locations with zinc concentrations less than 10 µg/L are evident in the data distribution. However, the only environments with chlorophyll concentrations > 5 µg/L and zinc < 10 µg/L are C6 and some lateral lakes. These, coincidentally, are the most shallow and nutrient-rich locations in the lake for which relevant data are available. At C5, on the other hand, dissolved zinc was also less than 10 µg/L in April–June in 2010, 2011, and 2018 (see Chapter 6), but these were the years with the *lowest* chlorophyll *a* at C5 (< 1 µg/L). The co-occurrence of low zinc and high chlorophyll *a* at C6, therefore, could represent a coincidence tied to the type of environment, rather than evidence that release of zinc inhibition results in high chlorophyll *a*.

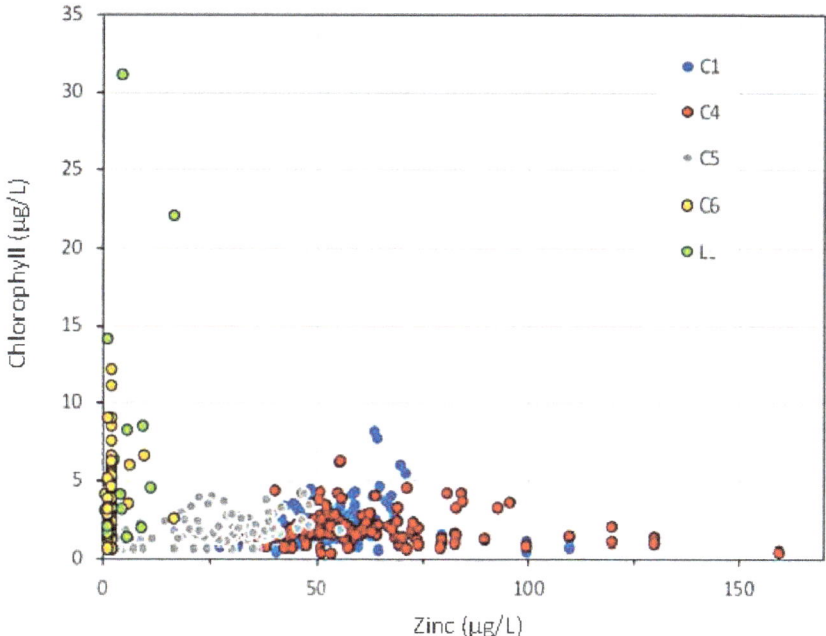

FIGURE 5-14 Photic zone chlorophyll *a* concentrations and total zinc concentrations at C1, C4, C5, and C6 in CDA Lake and in the lateral lakes (data from different sampling stations are indicated with different symbols; LL = lateral lakes). SOURCE: Data courtesy of IDEQ and the CDA Tribe and analyzed and plotted by the Committee.

In those years when zinc concentrations in the euphotic zone were greater than ~60 µg/L (all before 2004), chlorophyll *a* concentrations were always below 5 µg/L and declined as zinc concentrations increased. This could represent inhibition of chlorophyll *a* by zinc, but the threshold for such effects was ~60 µg/L, a concentration higher than currently found in surface waters.

Overall Conclusions Regarding Zinc–Total Phosphorus–Chlorophyll Relations

It has been plausibly hypothesized, based on laboratory experiments, that high zinc concentrations (> 10 µg/L; Wong et al., 1978; Wong and Chau, 1989) inhibit phytoplankton growth in CDA Lake specifically by suppressing the response of some phytoplankton species to phosphorus (Woods and Beckwith, 1997; Kuwabara et al., 2007). This is an important concern because mitigation of zinc loading via the Superfund remedy might release phytoplankton from that inhibition, accelerating eutrophication. Indeed, chlorophyll *a* concentrations in CDA Lake are lower than predicted from total phosphorus concentrations compared to other lakes globally (Figure 5-13), an observation that could be supportive of possible zinc inhibition of phytoplankton growth response to phosphorus supplies. However, from the multiple regression analysis, it is clear overall that zinc inhibition of chlorophyll in the Lake is not consistently detectable in a statistical sense in data from multiple sites across multiple years. Nor is it consistent with the zinc inhibition hypothesis that locations in the Lake with low zinc concentrations also fall below the global chlorophyll–total phosphorus relationship.

Several considerations might explain why the experimental suggestions of negative effects of zinc on phytoplankton response to phosphorus may not result in detectable and consistent relationships among chlorophyll, total phosphorus, and zinc in the euphotic zone.

1. Chlorophyll responds to multiple variables in CDA Lake (e.g., nutrient concentrations including possible nitrogen limitation, temperature, turbidity, residence time, zooplankton grazing, etc.). To drive chlorophyll concentration in a consistent manner, effects of zinc would have to overwhelm these influences. These confounding variables

are not present in the laboratory studies of Woods and Beckwith (1997) and Kuwabara et al. (2007). For example, nitrogen limitation of phytoplankton may also contribute to the shallow slope observed for the chl–TP relationship (Figure 5-13) as nitrogen limitation would constrain chlorophyll biomass below that expected for a given supply of phosphorus.
2. Even if zinc inhibited growth of some green algal taxa or other taxa, compensation in terms of increased production of chlorophyll by diatoms and other less-sensitive species would not be surprising, nor would this result be completely inconsistent with experimental results. Determination and analysis of taxa-specific distributions in the lake could be used to directly test the hypotheses of Kuwabara et al. (2006). That study found less of the zinc-sensitive taxa (the chlorophyte *Chlorella* spp.) at C5 (where zinc concentrations were higher) than at C6 and SJ1 (where zinc concentrations were < 5 μg/L). In contrast, the zinc-tolerant diatom (*Asterionella* spp.) was more abundant at C5 and less abundant than *Chlorella* spp. at C6/SJ1. A preliminary analysis by the committee of *Chlorella* sp. and *Asterionella* sp. from the most recent surveys found the diatom *A. formosa* to be more abundant than the chlorophyte *Chlorella* sp. at both C5 and C6; *Chlorella* sp. were only found at these locations in 2007 through 2009.[7]
3. Zinc concentrations in the photic zone (or epilimnion) are spatially and temporally variable, driven by multiple factors (see Chapter 6). Those influences differ with season and between the northern and southern Lake (see discussion above) and could confound any simple relationships among chlorophyll, total phosphorus, and zinc.

In short, zinc inhibition of some taxa of phytoplankton at the concentrations observed at some places and at some times in CDA Lake cannot be excluded. However, the weight of evidence available to the committee does not support an overall influence of zinc on the Lake's trophic status at present. Nor is there evidence that zinc is currently inhibiting chlorophyll to the extent that reducing zinc inputs to the Lake would cause an increase in chlorophyll concentrations sufficient to cause eutrophication and subsequent anoxia in bottom waters. The reason chlorophyll concentrations in CDA Lake are low relative to phosphorus concentrations remains unclear.

These analyses highlight the difficulties of inferring the impacts of possible zinc toxicity on eutrophication in CDA Lake from either single species experiments or observational data alone. The nature of the Zn–chl relationship in the Lake is dependent on what dataset is selected, and interpretation of observed dependences is complicated by the intervention of various confounding factors. More confidently establishing the nature of this relationship would require controlled field experimentation in which zinc concentrations are enriched in low-zinc zones of the Lake (a feasible experiment). An experiment in which zinc concentrations are selectively lowered in high-zinc zones of the Lake seems less feasible at present. However, this experiment could occur "naturally" over decades if remediation continues to reduce zinc inputs at the present pace. Field experiments testing for zinc effects on particular taxa of CDA Lake phytoplankton across seasons of interest could be especially important given that overall chlorophyll concentrations might mask ecologically important impacts. Given the physiological plausibility of zinc inhibition of phytoplankton growth, continued monitoring and investment in both passive and active adaptive management will be critical tools in the future.

CONCLUSIONS AND RECOMMENDATIONS

This chapter analyzed water column data from CDA Lake over the past 30 years to reveal trends in dissolved oxygen, nutrients, and lake productivity. Based on the committee's analysis of available datasets to better understand changes in dissolved oxygen and nutrient concentrations in CDA Lake, the following detailed conclusions and recommendations are made. It should be noted that these conclusions apply to the main body of the lake, where long-term monitoring data were collected, and not to nearshore or littoral areas, for which data are absent.

1. The evidence that dissolved oxygen concentrations are worsening in the bottom waters of the C1, C4, C5 and C6 stations is equivocal at best. Rather, there is an increasing dissolved oxygen concentration trend particularly

---

[7] This paragraph was updated after report release to clarify the committee's preliminary analysis and to better explain the dates when *Chlorella* data were available.

evident at C5.[8] The analyses reported herein used a coarser sampling resolution for the bottom samples than is ideal because the most important information is the oxygen concentration immediately above the sediment–water interface. It would be beneficial to know the actual bottom depth associated with each profile so that the bottom dissolved oxygen data can be presented relative to their distance from the sediment–water interface. **Coupled with the recent trends in phosphorus concentrations in the Lake (see below) and phosphorus loading to the Lake (see Chapter 3), low dissolved oxygen is not a current problem in the main body of the lake, nor it is expected to become a problem if current trends continue.** However, if climate change were to strengthen thermal stratification substantially in the future, dissolved oxygen concentrations at the sediment–water interface would likely decrease, reversing current trends.

2. **For the most recent decade, the data indicate declines in total phosphorus concentration in CDA Lake, although this decline was not statistically significant in the northern lake.** Declines in total phosphorus in the Lake are consistent with the declines in total phosphorus from the two major rivers entering the Lake. For the northern portions of the Lake, these trends are a reversal of prior increasing trends similar to the reversal of the trend in CDA River total phosphorus inputs. In the southern parts of the Lake at C5, the trends have been steadily downward for more than a decade, similar to the steady downward trend in St. Joe River total phosphorus inputs. The only exception to these patterns is that for deeper water at C6 and SJ1 the total phosphorus trends are upward, suggesting that internal sources of phosphorus from the Lake bed may be important in these areas. The ability to project future changes in phosphorus inputs and concentrations in the Lake will depend on research aimed at better understanding the causative mechanisms of these recent trends in phosphorus delivery from the watershed to the Lake.

3. **Trends in total nitrogen are more complex than those for total phosphorus.** Gaining a better understanding of the nitrogen cycle (such as assessing possible nitrogen fixation in the water column spatially and temporally or identifying hot spots for denitrification), and analyzing chemical speciation of and trends in nitrogen concentration in the Lake and rivers, is an important topic that needs to be included in the research agenda supporting adaptive management of the Lake's trophic state.

4. **CDA Lake is characterized by low total phosphorus and chlorophyll concentrations. According to these widely used metrics of lake trophic status, the Lake can be classified as oligotrophic.** Indeed, median values of total phosphorus and chlorophyll are seven times and five times lower, respectively, in CDA Lake than median values in lakes in the National Lakes Assessment.

5. **The available field evidence does not support the concept that the current high zinc concentrations in CDA Lake suppress chlorophyll *a*.** CDA Lake supports lower amounts of chlorophyll *a* per unit phosphorus than generally observed for lakes worldwide. Although this is consistent with zinc suppression of phytoplankton biomass, beyond this observation there is little evidence from more detailed analysis of field data that the current high zinc concentrations suppress chlorophyll *a*. For example, chlorophyll *a* levels are disproportionately lower than expected in the southern part of the Lake in both zinc-enriched and zinc-poor locations. Consistent with this, multiple regression analysis relating chlorophyll to total phosphorus and total zinc finds a strongly positive and statistically significant association for total phosphorus but a highly nonsignificant association for total zinc. Thus, zinc suppression of overall phytoplankton biomass is not consistently supported, although zinc suppression of individual species of phytoplankton cannot be discounted. Further research involving field experimentation would help develop greater confidence in predicting the response of chlorophyll concentrations and particular taxa of phytoplankton to potential reductions in legacy metal contamination in the Lake.

6. The possibility exists that nutrient enrichment could take place in shallow water areas at the edges of CDA Lake, particularly in areas where land use change could bring about increased inputs of nutrients. These areas could,

---

[8] This conclusion was edited after report release to reflect only the dissolved oxygen trend analyses shown in the report, which were for bottom waters.

in the future, become areas of localized growth of submerged aquatic vegetation or other forms of organisms that could lead to localized hypoxia when they die off. **Attention to monitoring water-column nutrients, chlorophyll, and oxygen in embayments and near-shore environments should be considered, even though the main body of the Lake continues to be oligotrophic and not likely to experience dissolved oxygen depletion.**

REFERENCES

Atkinson, C.A., D. F. Jolley, and S. L. Simpson. 2007. Effect of overlying water pH, dissolved oxygen, salinity and sediment disturbances on metal release and sequestration from metal contaminated marine sediments. *Chemosphere* 69:1428–1437.

Carlson, R. E. 1977. A trophic state index for lakes1. *Limnol. Oceanogr.* 22(2):361–369.

Chess, D. 2022. Limnology Discussion. Presentation to the NASEM Committee. January 28, 2022.

Cloern, J. E. 2021. Use Care When Interpreting Correlations: The Ammonium Example in the San Francisco Estuary. San Francisco Estuary and Watershed Science 19(4): article 1. https://escholarship.org/uc/jmie_sfews/19/4. https://doi.org/10.15447/sfews.2021v19iss4art1.

Conley, D. J., H. W. Paerl, R. H. Howarth, D. F. Boesch, S. P. Seitzinger, K. E. Havens, C. Lancelot, and G. E. Likens. 2009. Ecology. Controlling eutrophication: nitrogen and phosphorus. *Science 323*(5917):1014–1015.

Diaz, R. J., and R. Rosenberg. 1995. Marine benthic hypoxia: a review of its ecological effects and the behavioural responses of benthic macrofauna. *Oceanogr Mar Biol Annu Rev* 33:245–303.

Diaz, R. J., and R. Rosenberg. 2008. Spreading dead zones and consequences for marine ecosystems. *Science* 321:926–929.

Downing, J. A., and E. McCauley. 1992. The nitrogen:phosphorus relationship in lakes. *Limnology and Oceanography* 37:936–945. doi: 10.4319/lo.1992.37.5.0936.

Elser, J. J., M. E. S. Bracken, E. E. Cleland, D. S. Gruner, W. S. Harpole, H. Hillebrand, J. T. Ngai, E. W. Seabloom, J. B. Shurin, and J. E. Smith. 2007. Global analysis of nitrogen and phosphorus limitation of primary producers in freshwater, marine and terrestrial ecosystems. *Ecology Letters* 10(12):1135–1142.

EPA (U.S. Environmental Protection Agency). 2021. National Aquatic Resource Surveys. National Lakes Assessment 2017 (data and metadata files). https://www.epa.gov/national-aquatic-resource-surveys/data-national-aquatic-resource-surveys.

Gonçalves, S., M. Kahlert, S. F. P. Almeida, and E. Figueira. 2018. A freshwater diatom challenged by Zn: Biochemical, physiological and metabolomic responses of *Tabellaria flocculosa* (Roth) Kützing. *Environmental Pollution* 238:959–971.

Havens, K. E., and G. K. Nürnberg. 2009. The phosphorus-chlorophyll relationship in lakes: potential influences of color and mixing regime. *Lake and Reservoir Management* 20(3):188–196. https://doi.org/10.1080/07438140409354243.

Hirsch, R. M., J. R. Slack, and R. A. Smith. 1982. Techniques of trend analysis for monthly water quality data. *Water Resources Research* 18(1):107–121. https://doi.org/10.1029/WR018i001p00107.

Howell, P., and D. Simpson. 1994. Abundance of marine resources in relation to dissolved oxygen in Long Island sound. *Estuaries* 17:394–402.

Hrycik, A. R., L .Z. Almeida, and T. O. Höök. 2017. Sub-lethal effects on fish provide insight into a biologically-relevant threshold of hypoxia. *Oikos* 126:307–317.

IDEQ and CDA Tribe. 2020. Coeur d'Alene Lake Management Program: Total Phosphorus Nutrient Inventory, 2004–2013.

Kuwabara, J. S., Topping, B. R., Woods, P. F., and J. L. Carter. 2007. Free zinc ion and dissolved orthophosphate effects on phytoplankton from Coeur d'Alene Lake, Idaho. *Environ. Sci. Technol.* 41(8):2811–2817. https://doi.org/10.1021/es0629231.

McKnight, D. M., S. W. Chisholm, and D. R. F. Harleman. 1983. CuSO4 treatment of nuisance algal blooms in drinking water reservoirs. *Environmental Management* 7(4):311–320. https://doi.org/10.1007/bf01866913.

Mortimer C. H. 1971. Chemical exchanges between sediments and water in Great Lakes – speculations on probable regulatory mechanisms. *Limnol. Oceanogr.* 16:387–404.

Mortimer, C. H. 1942. The exchange of dissolved substances between mud and water in lakes. *Journal of Ecology* 30(1): 147–201. https://doi.org/10.2307/2256691.

Nürnberg, G. K. 2002. Quantification of oxygen depletion in lakes and reservoirs with the hypoxic factor. *Lake and Reservoir Management* 18(4):299–306. https://doi.org/10.1080/07438140209353936.

Paerl, H. W., L. M. Valdes, B. L. Peierls, J. E. Adolf, and L. W. Harding Jr. 2006. Anthropogenic and climatic influences on the eutrophication of large estuarine ecosystems. *Limnol. Oceanogr.* 51:448–462.

Renaud, M. L. 1986. Detecting and avoiding oxygen deficient sea water by brown shrimp, *Penaeus aztecus* (Ives), and white shrimp *Penaeus setiferus* (Linnaeus). *Journal of Experimental Marine Biology and Ecology* 98(3):283–292. https://doi.org/10.1016/0022-0981(86)90218-2.

Schelske, C. L. 2009. Eutrophication: focus on phosphorus [Review of *Eutrophication: focus on phosphorus*]. *Science* 324(5928):722; author reply 724–725.

Schindler, D. W., and R. E. Hecky. 2009. Eutrophication: more nitrogen data needed [Review of *Eutrophication: more nitrogen data needed*]. *Science* 324(5928):721–722; author reply 724–725.

Schmidt, T. S., W. H. Clements, and B. S. Cade. 2012. Estimating risks to aquatic life using quantile regression. *Freshwater Science* 31(3):709–723. https://doi.org/10.1899/11-133.1.

Sen, P. K. 1968. Estimates of the regression coefficient based on Kendall's tau. *Journal of the American Statistical Association* 63(324):1379–1389. https://doi.org/10.1080/01621459.1968.10480934.

Sterner, R. W., and J. J. Elser. 2002. Ecological stoichiometry: the biology of elements from molecules to the biosphere. Princeton University Press.

Sterner, R. W., T. Andersen, J. J. Elser, D. O. Hessen, J. Hood, E. McCauley, and J. Urabe. 2008. Scale-dependent carbon:nitrogen:phosphorus seston stoichiometry in marine and freshwaters. *Limnol. Oceanogr.* 53(3):1169–1180.

Tellier, J. M., N. I. Kalejs, B. S. Leonhardt, D. Cannon, T. O. Höök, and P. D. Collingsworth. 2022. Widespread prevalence of hypoxia and the classification of hypoxic conditions in the Laurentian Great Lakes. *Journal of Great Lakes Research* 48(1):13–23. https://doi.org/10.1016/j.jglr.2021.11.004.

Terrell, J. W., B. S. Cade, J. Carpenter, and J. M. Thompson. 1996. Modeling stream fish habitat limitations from wedge-shaped patterns of variation in standing stock. *Transactions of the American Fisheries Society* 125:104–117.

Theil, H. 1950. A rank-invariant method of linear and polynomial regression analysis: I, II, and III. *Proceedings of the Royal Netherlands Academy of Sciences* 53:386–392, 521–525, and 1397–1412. https://link.springer.com/chapter/10.1007/978-94-011-2546-8_20.

Woods, P. F., and M. A. Beckwith. 1997. Nutrient and trace-element enrichment of Coeur d'Alene Lake, Idaho. USGS Water Supply Paper 2485. https://doi.org/10.3133/wsp2485.

Wong, P. T. S., and Y. K. Chau. 1989. Zinc toxicity to freshwater algae. NWRI Contribution #89-75.

Wong, P. T. S, Y. K Chau, and P. L Luxon. 1978. Toxicity of a mixture of metals on freshwater algae. *Journal of the Fisheries Research Board of Canada* 35(4):479–81.

Zinsser, L. M. 2020. Trends in Concentrations, Loads, and Sources of Trace Metals and Nutrients in the Spokane River Watershed, Northern Idaho, Water Years 1990–2018. USGS Scientific Investigations Report 2020-5096. https://doi.org/10.3133/sir20205096.

6

In-Lake Processes: Metals

This chapter continues the committee's analysis of status and trends in water quality in the Couer d'Alene (CDA) Lake by focusing on the heavy metals of concern emanating from the Superfund site, particularly lead (Pb), zinc (Zn), and cadmium (Cd). In addition to revealing trends on metals concentrations in the Lake over the past 30 years, the analyses consider aspects of the time series data that are indicative of various processes occurring in the water column. Understanding the status of metals concentrations in the Lake today is an important context for interpreting trends into the future.

IN-LAKE PROCESSES RELEVANT TO METALS

Status and trends of metals in lakes are controlled by interactions among a variety of complex processes, including hydrology, internal hydrodynamics, thermal stratification, vertical movement of particles, biological productivity, and biogeochemical reactions in sediments and the water column (Davison, 1993; Nriagu et al., 1995). These processes control the distribution of metals between dissolved and particulate forms; the degree of biological uptake and release of metals by phytoplankton; the distribution of dissolved and particulate metals between surface and bottom waters; and metal(loid) trapping, rerelease, and fate in the sediments. In turn, the concentrations of metals in various compartments influence biological and ecological processes at multiple levels of biological organization and, ultimately, the kinds and productivity of microbes, plants, and animals that occupy lakes.

Figure 6-1A shows many of these processes along with the forms that metals can take in lake water, from dissolved to particulate-bound. Depending upon water and sediment photobiogeochemistry, the particulate metals can occur in multiple forms, such as in association with different ligands (e.g., iron and manganese oxides), with living organisms, or with organic detritus (Figure 6-1B). As mentioned in Chapter 5, zinc is an essential metal for life and is taken up more readily than lead by phytoplankton and other organisms (see, e.g., Chen et al., 2000). As a result, during a phytoplankton bloom zinc can be depleted from the water column by biouptake (conversion to phytoplankton and detrital particulate forms) more so than lead, and zinc is rereleased more readily in deeper waters as these organisms settle to the bottom of the lake and die. Data and studies directly addressing such processes in CDA Lake are limited. For example, only limited data are available on dissolved metal speciation (e.g., Smith et al., 2015), colloidal metal (Langman et al., 2020), or the forms of particulate metals (e.g., Balistrieri et al., 2003)—all of which affect metal uptake by phytoplankton. Rather, most metal data from the water column are determined as either "total" metal or metal that has passed through a 0.45-μm filter (and classified as "dissolved;" Chess et al., 2012). Therefore, one must infer from time series and studies in other lakes how different

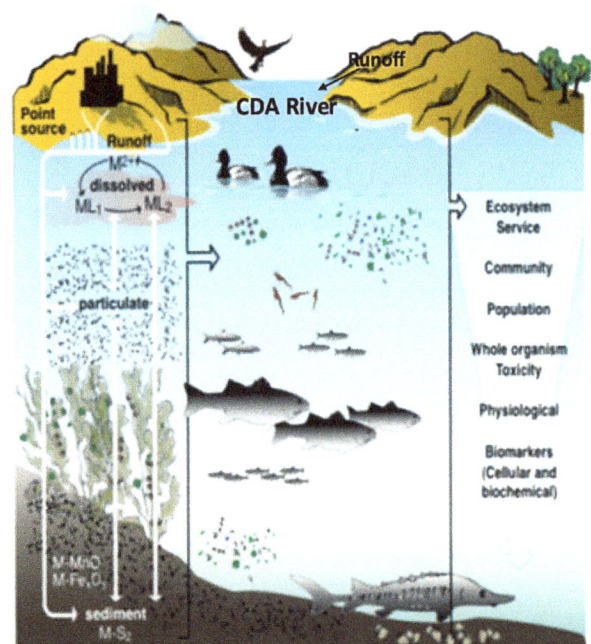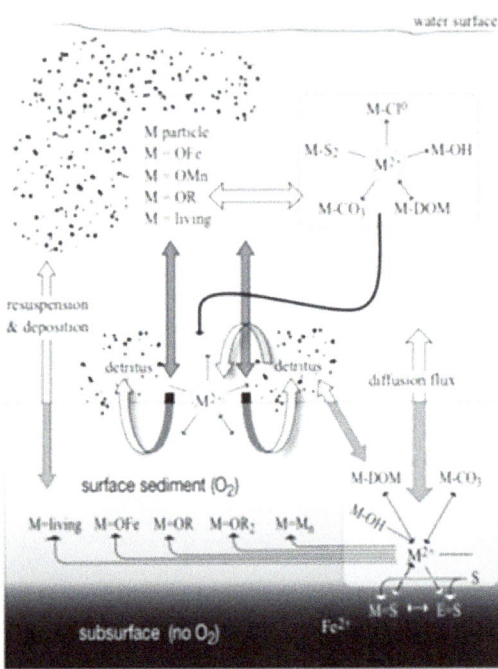

FIGURE 6-1 Simplified conceptualizations of some of the processes affecting the fate and effects of metal(loid)s in lakes. M is metal$_x$ (e.g., Pb, Zn, Cd, As). M-S$_2$ are examples of dissolved metal complexes with chloride (Cl^0), hydroxides (OH), carbonate (CO$_3$), or dissolved organic material (M-DOM). M particles are shorthand for common forms of particulate metals: hydroxides of iron, OFe; manganese, OMn; organic material, OM; or living material, living. SOURCE: Luoma and Rainbow (2008).

processes influence trends in metal concentrations in CDA Lake (Balistrieri et al., 2002). A more comprehensive understanding of at least some processes in Figure 6-1 will be critical to evaluating metal trends and their drivers into the future.

The available metals data in CDA Lake are sufficient to consider whether a metal exists predominantly in the dissolved or particulate phase. This tendency is metal-specific and a function of particle type and concentration, pH, ionic strength, redox conditions, and temperature (Balistrieri, 1998). The relative affinity of metals for iron oxide–dominated particles in an oxidized water column follows the ranking Pb >> arsenic (As) >> Zn ~ Cd (see Chapter 7; Balistrieri et al., 2002). Thus, a given concentration of particulate zinc equilibrates with higher concentrations of dissolved zinc than occurs for lead at the same pH.

The committee found that statistically strong, significant relationships occur among dissolved and total lead and zinc where data are adequate to establish the relationship in CDA Lake (Table 6-1). Because these correlations are ubiquitous spatially and among metals, they suggest that equilibration or exchange between dissolved and particulate metals strongly influences dissolved metal concentrations in CDA Lake.

It is important to understand that "total" metal in the water column does not represent all of the metal associated with particulate material because of methodological constraints.[1] Thus, actual percentages of dissolved to total metal are lower than those found in Table 6-1. Another way to get at the true distribution of metals in the Lake is to consider, as geochemists do, the distribution between total particulate metal and dissolved metal using an operational partitioning coefficient. This is the ratio between total particulate metal concentrations in suspended

---

[1] As explained in Chapter 8, there are three ways of measuring metal: dissolved (filtered through a 0.45-μm filter), total (unfiltered and held at pH 2), and total particulate (metal concentration from a completely decomposed sample). CDA Lake data are reported as "dissolved" or "total" (whatever is desorbed when the unfiltered water sample is stored at pH 2).

TABLE 6-1 Percent of Total Metals in the Dissolved Form in Photic Waters in the CDA River, and in C4, C1, and C5 in CDA Lake for 2015–2020

	Pb	Zn
CDA River	2%	60%
C4	16%	87%
C1	13.6%	98%
C5	24.6%	94%

NOTES: The entries above for the Lake are the slope of the statistically significant correlation between dissolved metal as a function of total metal for the years 2015–2020 at a given location. The entries for the CDA River are the ratio of flow-normalized flux of dissolved lead to the flow-normalized flux of total lead over these years.

material[2] compared to dissolved concentrations in the water column. For the purposes of illustration, the committee calculated very general partitioning coefficients by comparing the zinc and lead concentrations on decomposed particles filtered from the water column of the Lake near Harrison in June 2005 by Kuwabara et al. (2006) to mean dissolved lead concentrations between June and September 2005. The calculated partition coefficient for lead was 6,873 (mg Pb/kg divided by mg Pb/L). Reversing the ratio showed that dissolved lead concentrations were 0.014 percent of total particulate lead concentrations in that example. The calculated partitioning coefficient for zinc was 86, such that dissolved zinc concentrations were 1.1 percent of total particulate zinc concentrations. These operational coefficients illustrate that **extremely large reservoirs of metal are present in the particulate form (e.g., in the bed sediments) that could replenish dissolved concentrations in the water column if conditions were conducive to metal release from the particles.**

***

The following analysis of metals in CDA Lake starts with zinc, then cadmium, and finally lead. For each of these metals, there are observations about how the metal behaves, in particular its relationship to season, depth, location, and lake inflows, followed by an analysis of trends in metal concentration at each station where there are sufficient data. The trend analysis is explored in two ways. One is purely graphical—it is locally weighted scatterplot smoothing. The second is the use of a formal statistical hypothesis test for trend, the Seasonal Kendall test. Both of these methods are described in detail in Appendix B.

ANALYSIS OF TRENDS IN DISSOLVED ZINC

Of the metals examined by the committee, zinc cycling in the water column of CDA Lake is perhaps the most complicated. For each of the lake locations considered below, the raw data are shown, both in graphical form from the past five years and in tabular form for the entire dataset. This is followed by a graph of monthly averaged dissolved zinc concentration data from the photic and bottom zones to reveal seasonal patterns, and finally by a formal trends analysis of dissolved zinc concentration in photic and bottom waters over the past 26 years.

Dissolved Zinc Concentrations at C4

The C4 data shown in Figure 6-2 provide a good example of the patterns of dissolved zinc concentration in CDA Lake from 2015 to 2021. The maximum zinc concentration is always observed in the first month of the

---

[2] Total particulate metals in suspended material (total particulate metal) are the concentrations per mass of particulate determined in a completely decomposed sample (e.g., Balistrieri et al., 2002; Kuwabara et al., 2006). It is challenging to collect enough suspended material for such measurements, so total suspended particulate metal and partitioning coefficients from the water column are not commonly determined.

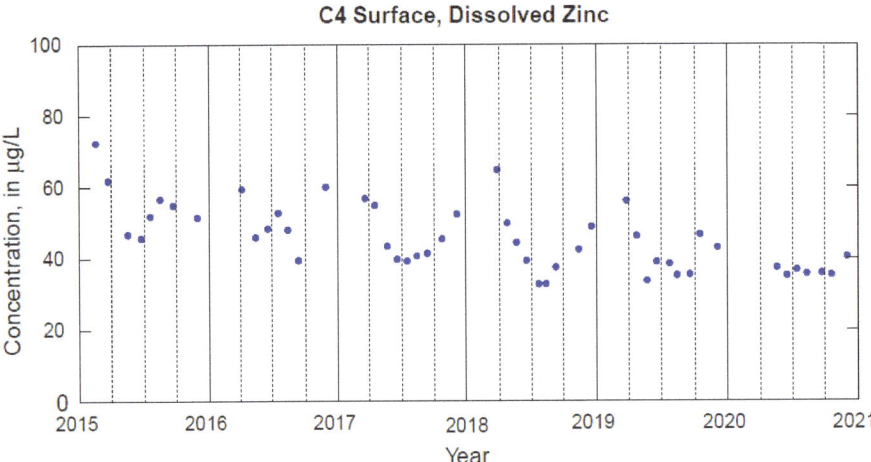

FIGURE 6-2 Time series of dissolved zinc concentrations in the photic zone at C4 from 2015 to 2020. Solid vertical lines mark the start of the calendar year, and the dashed lines separate each year into four quarters. SOURCE: Data courtesy of Idaho Department of Environmental Quality (IDEQ) and plotted by the committee.

calendar year that samples were collected (after each winter data gap). The high concentrations in March reflect zinc inputs from the watershed and, historically, were correlated with discharge from the CDA River (CH2M Hill and URS Corp., 2001). A second consistent aspect of zinc at C4 is the decline of concentrations in the photic zone between the first sampling of the year and September.

Table 6-2 shows the concentration of dissolved zinc in the photic zone of C4, by month and year from 2004 to 2020. Declines over time are evident in the months in which there was extensive sampling. For example, dissolved zinc concentrations in the euphotic zone in 2004 were 15–34 µg/L higher than in 2020 during the same month. Table 6-2 shows that photic zone zinc concentrations are beginning, in the most recent years, to approach the Lake Management Plan target of 36 µg/L during summer stratification. However, exceedances of the 36 µg/L target still occur when the water column is mixed every spring and every fall.

The conceptual model for zinc cycling in stratified deep water lakes involves biogenic stripping (scavenging) of zinc from photic waters and recycling of zinc into bottom waters (Balistrieri et al., 2002). Figure 6-3 shows the seasonal cycling of zinc at site C4. Each year in late spring (mid-April through June), zinc in the photic zone declines rapidly at C4. This period includes the falling limb of the hydrograph (declining inputs), the beginning of stratification, and the clearing of the water column (settling of particles). The decline in photic zone zinc is partly due to trapping—settling of particles from the water column to the bottom of the Lake. Indeed, based on the analysis presented in Chapter 3 (Figure 3-21), about 53 percent of the total zinc coming into the Lake is ultimately trapped.[3] A reason for the decline in photic zone zinc is that phytoplankton begin growing as waters warm, and clear and zinc is an essential element taken up by phytoplankton (Chen et al., 2000). If phytoplankton uptake exceeds zinc inputs during the growing period, the net result is lower concentrations of dissolved zinc in the photic zone via conversion of dissolved zinc to particulate zinc (living and detrital organic particles). Zinc is then rereleased below the thermocline as settling organic particles decompose or are consumed (see, e.g., Lee and Fisher, 1992; Sigg, 1985). The timing of the decline of dissolved zinc in photic waters and increase in bottom waters at C4 shown in Figure 6-3 is consistent with this model.

The bottom waters at C4 follow a seasonal pattern as well, but it is different than the pattern for the photic zone (Figure 6-3). Bottom water concentrations are more constant over the course of the year and tend to decline

---

[3] Percent trapped is the percentage difference in flow-normalized flux between the mouth of the CDA River and the Spokane River below the Lake outlet. (470 − 219 = 251, then divided by 470 = 53%).

TABLE 6-2 Concentrations of Dissolved Zinc in the C4 Photic Zone by Month, in ug/L

Year	Jan	Feb	Mar	Apr	May	Jun	Jul	Aug	Sep	Oct	Nov	Dec
2004		71.8		68.8	64.0	59.0	56.4	50.9		58.0		73.2
2005	65.3			74.6	55.5	44.5	57.7	54.0		51.4		58.4
2006		75.4	92.4		59.3	57.0	56.6	73.6				
2007							55.3	48.9	40.5	51.8		71.2
2008			90.0		72.0	48.0	41.4	41.4		63.1		
2009				65.8	48.1	36.7	57.1	53.5		50.8		
2010			67.8	63.8	61.3	56.7	57.2	55.5	54.0			64.8
2011		67.0		62.9	47.5	35.5	39.6	45.5		47.7		61.2
2012			78.0	75.2	57.2	54.4	45.1	48.9	49.6			
2013				67.2	44.9	52.0	53.5	67.3		55.1		61.3
2014			69.3	58.8	46.2	44.6	43.8	44.8	54.3			52.2
2015		72.5	62.0		46.8	45.7	52.0	56.8	55.0		51.6	
2016				59.5	46.0	48.4	52.9	48.0	39.5		60.2	
2017			56.9	55.0	43.5	39.8	39.3	40.7	41.4	45.5		52.5
2018			65.0	49.9	44.5	39.4	32.8	32.8	37.5		42.5	48.9
2019			56.2	46.2	33.7	39.0	38.5	35.3	35.5	46.7		43.0
2020					37.4	35.2	36.8	35.6	35.9	35.3		40.5

NOTE: A similar table is not shown for C1. The colors follow a scale running from red for the highest values, to brown, then yellow, then green, and finally aqua for the lowest values. SOURCE: Data provided by IDEQ.

in the fall when the Lake destratifies. Zinc concentrations in the photic zone increase at that time through mixing with the enriched bottom water. Averaged over the year, the differences in concentration between bottom water and photic zone water as of 2020 are 12.8 µg/L, with bottom waters having higher concentrations of dissolved zinc (as much as 25 µg/L higher in bottom waters than in surface waters toward the end of the period of stratification). Figure 6-3 shows only the last six years of the record so that the interpretation is not overly confounded by the influence of the interannual trends present in the data.

A pH sag (pH < 7) also begins in deeper waters in CDA Lake in late spring, and pH < 7 is common in many years in bottom waters at C4. Surface water pH in most years is > 7.0 in the photic zone, but drops to as low as pH 6.1 on the bottom as the year progresses. Reduced pH is caused by organic particle decomposition where waters are isolated from oxygen inputs and carbonic acid is produced (Davison, 1993). When pH drops below 7, zinc release from settling particles could be enhanced (see Chapter 7), as could release of zinc from bottom sediments into bottom waters (Kuwabara et al., 2000). The relative importance to bottom waters of zinc release from particle settling compared to zinc release from sediments cannot be determined from existing data. The important point is that zinc release from particulate materials (either suspended or in surface sediments) into the water column in CDA Lake does not require anoxia. Release can occur under oxic conditions, especially if pH declines (also see Chapter 7).

Formal Trend Analysis of Dissolved Zinc

The trend analysis conducted by the committee shows declines in dissolved zinc concentration at C4 over time in both surface and bottom water: by 41 percent over the past 17 years, and by 33 percent over the past ten years (Figure 6-4). The implication is that the Lake is responding to the remediation (reduction) of zinc inputs discussed in Chapter 3. The rate of decline of flow-normalized dissolved zinc flux *entering* the Lake over the same 17-year

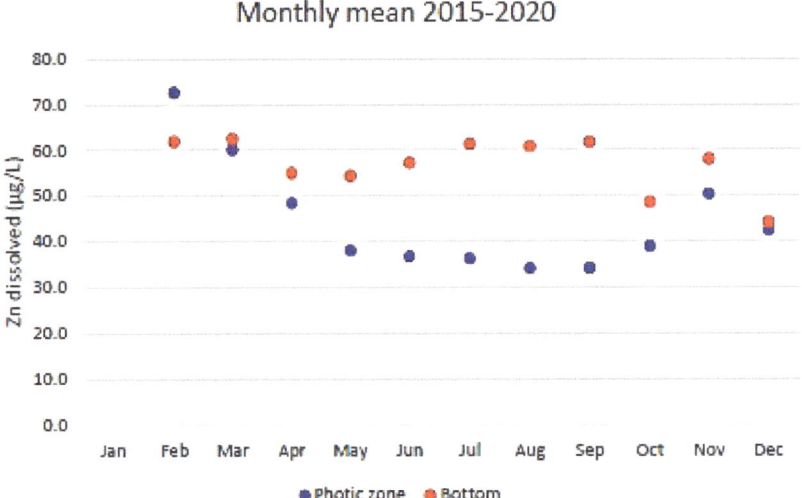

FIGURE 6-3 The geometric mean of dissolved concentrations of zinc in surface (purple) and bottom (red) waters each month of the calendar year over the period 2015–2020 at C4. SOURCE: Data courtesy of IDEQ and analyzed and graphed by the committee.

period is −35 percent. The close correspondence between the rate of decline of the Lake concentrations and the river inputs suggests that the water column of the Lake is responding rapidly to the declining inputs of zinc from the CDA River. Balistrieri et al. (2002) suggested that benthic flux would dominate zinc concentration later in the year, during periods of low river discharge, while influx from the watershed would control zinc concentrations in the Lake during periods of elevated river discharge. The time series data are not an adequate basis upon which to define these processes quantitatively. But if there are seasonal differences in fluxes, they do not detectably slow the overall response of the Lake to the longer-term decline in inputs.

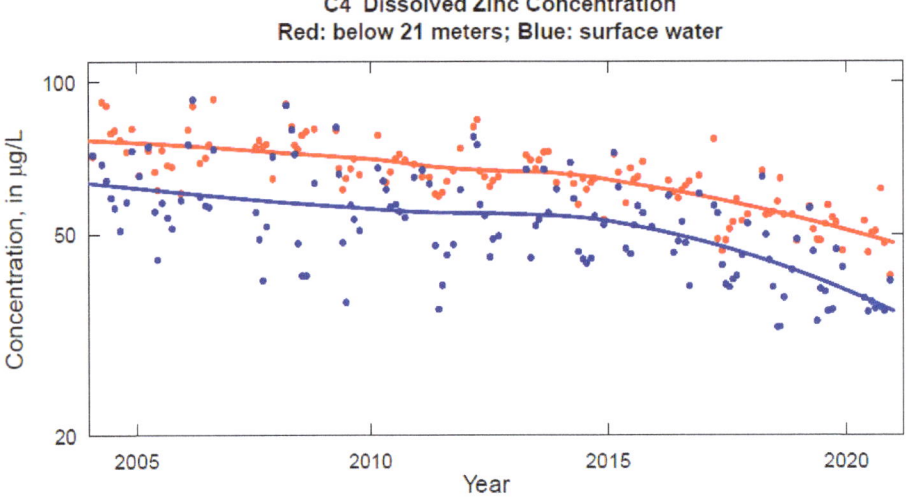

FIGURE 6-4 Dissolved zinc data at station C4, 2004–2020. Surface water data are in blue, and bottom water data are in red. Smoothed estimates of the trends are shown in the red and blue lines. SOURCE: Data courtesy of IDEQ and analyzed and graphed by the committee.

IN-LAKE PROCESSES: METALS   195

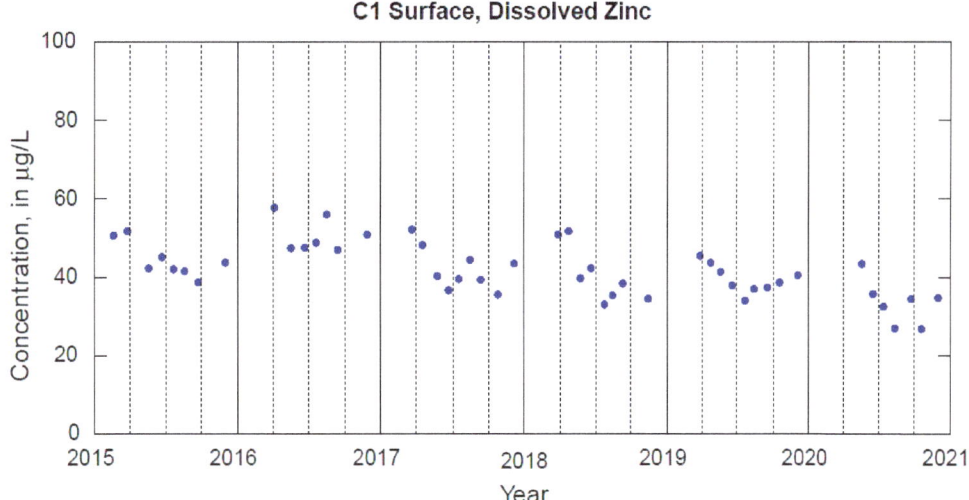

FIGURE 6-5 Time series of dissolved zinc concentrations in the photic zone at C1 from 2015 to 2020. Solid vertical lines mark the start of the calendar year, and the dashed lines separate each year into four quarters. SOURCE: Data courtesy of IDEQ and plotted by the committee.

Dissolved Zinc Concentrations at C1

Data for dissolved zinc in the photic zone at C1 are available for 2004 to 2020, but only data from 2015 to 2020 are shown in Figure 6-5. As seen in Figure 6-5, monthly geometric mean concentrations of dissolved zinc in the photic zone at C1 were highest in February through April, but the spring peak is dampened at C1 compared to C4, by 10–20 µg/L (compare Figures 6-2 and 6-5). The amplitude of the seasonal cycle is somewhat smaller at C1 compared to C4, but the temporal dynamics of the cycle are similar.

Dissolved zinc at C1 is about 8 percent less than at C4 (averaged over 2015–2020, for all months, and averaging photic and bottom waters). Concentrations in the C1 photic zone are similar to those in the C1 bottom waters in February–April, reflecting a vertically mixed lake (Figure 6-6). Concentrations of dissolved zinc in the C1

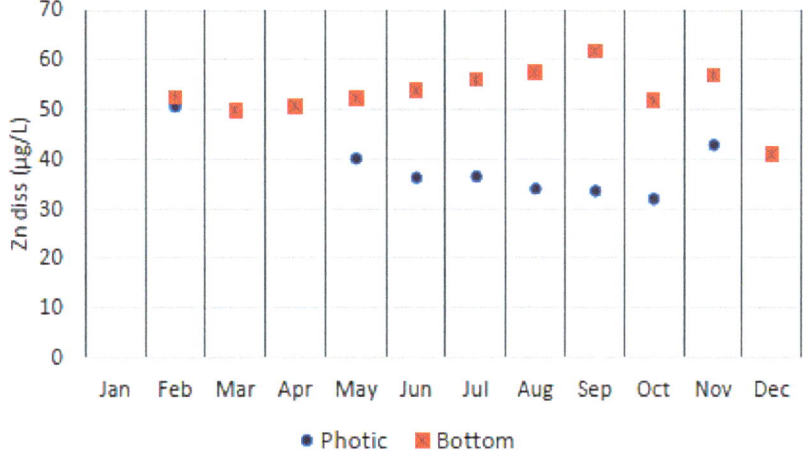

FIGURE 6-6 The geometric mean of dissolved concentrations of zinc in surface (blue circles) and bottom (red squares) waters each month of the calendar year over the period 2015–2020 at C1. SOURCE: Data courtesy of IDEQ and analyzed and graphed by the committee.

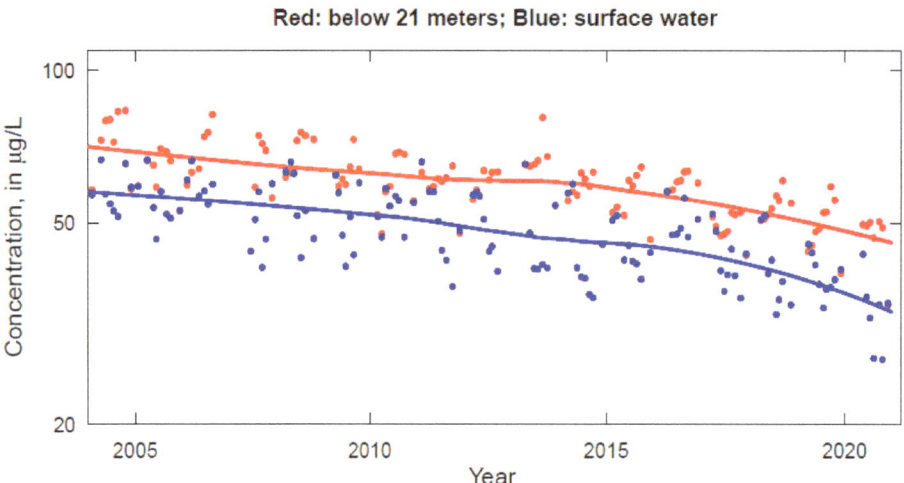

FIGURE 6-7 Dissolved zinc data at station C1 from 2004 to 2020. Surface water data are shown in blue, and bottom water data are shown in red. Smoothed estimates of the trends are shown by the red and blue lines. SOURCE: Data courtesy of IDEQ and analyzed and graphed by the committee.

photic zone begin to decline, and bottom water concentrations begin to increase, in May, such that by September bottom water concentrations exceed photic zone concentrations by 20 µg/L. This difference narrows as the Lake mixes in October through December.

Concentrations of dissolved zinc at C1 in surface waters were similar to C4 in 2015–2020 in summer/fall. Indeed, the differences in zinc concentrations between the two sites in May through September are much smaller (average is only 1.3 µg/L) than the differences over the entire year (average is 3.8 µg/L). Thus, after inputs of zinc from the CDA River begin to slow in late spring, photic zone concentrations of zinc appear to become nearly uniform across the northern Lake, perhaps as a result of widespread mixing. Photic zone concentrations of dissolved zinc at C1 met the Lake Management Plan target of 36 µg/L from June through October from 2017 to 2020 (data not shown), falling into the window between the Lake Management Plan target and the benchmark for ecological disturbance (Chapter 9).

Formal Trend Analysis of Dissolved Zinc

The committee's formal trend analysis for dissolved zinc concentrations at C1 shows highly significant declines over time in both surface and bottom water: by 38 percent over the past 17 years and by 30 percent over the past ten years (2.2–3.0 percent decline per year respectively; Figure 6-7). The average difference between mean concentrations in the photic zone and bottom water as of 2020 was about 12.3 µg/L, which is similar to the difference at C4 (12.8 µg/L).

In summary, the temporal and depth patterns of dissolved zinc concentrations at C1 and C4 have tracked each other for the past 17 years, with concentrations averaging about 8 percent lower at C1 than at C4.

Dissolved Zinc Concentrations at C5

Seasonal dynamics of dissolved zinc in the southern Lake (site C5) show important differences from the northern Lake (C4 and C1). The three panels of Figure 6-8 use data from 2015 to 2020 to illustrate (1) the difference between the annual cycles of zinc in the photic zone and bottom water at C5, (2) the differences between zinc in the photic zone at C4 and C5, and (3) the differences between zinc in the bottom waters of C4 and C5.

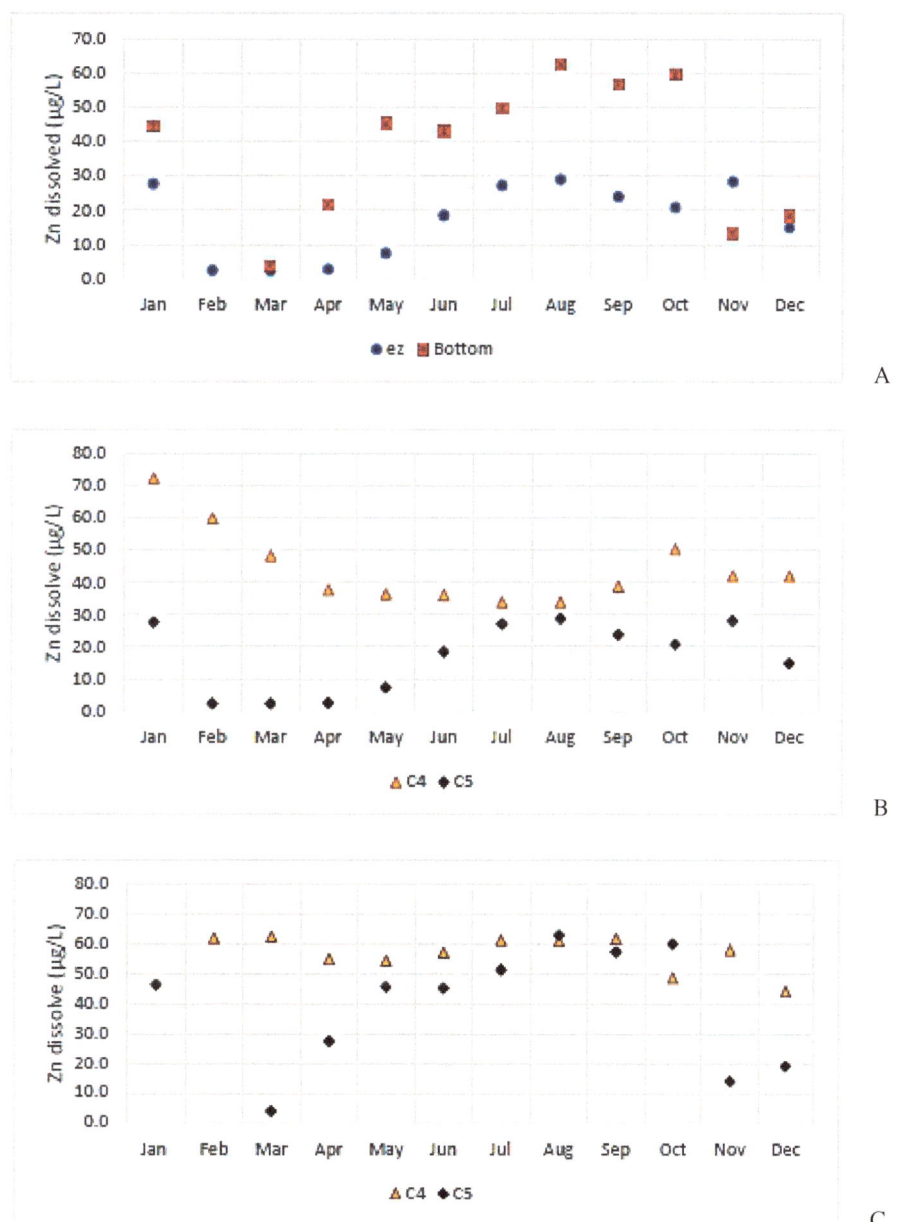

FIGURE 6-8 Geometric monthly mean of dissolved zinc concentrations in the euphotic zone and in bottom waters (15–18 m depth) at C5 across 2015–2020. (A) Bottom waters versus euphotic waters at C5, (B) photic waters at C5 compared to C4, and (C) bottom waters at C5 compared to C4. SOURCE: Data courtesy of CDA Tribe and IDEQ and analyzed and graphed by the committee.

Thus, while zinc concentration in surface water at C4 peaks in January before starting to decline, at C5 the period of elevated discharge has low dissolved zinc concentrations, consistent with the proximity of C5 to unenriched inputs (Zn < 5 µg/L) from the St. Joe River. (The concentration of dissolved zinc in the St. Joe River averages about two orders of magnitude lower than in the CDA River.) Dissolved zinc in the C5 photic zone never exceeded 12 µg/L in April in any year during the period of record, in contrast, with average levels above 60 µg/L during April at C4 in many years (compare Tables 6-2 and 6-3). Low dissolved zinc concentrations (< 11 µg/L)

TABLE 6-3 Concentrations of Dissolved Zinc (in μg/L) by Month at C5

Photic zone												
	Jan	Feb	Mar	Apr	May	Jun	Jul	Aug	Sep	Oct	Nov	Dec
2007						32.8	29.3	38.9	23.0			35.6
2008			7.1	13.0	2.5	25.0	35.8		38.4			42.9
2009			12.0	2.5	25.5	34.6	36.9		28.2			31.3
2010			18.5	11.0	27.8	2.5	36.7	41.6	34.3		28.9	
2011	2.5			2.5	6.8	2.5	7.0	27.7	32.0		52.5	
2012			31.8		2.5	7.6	32.0	34.7	34.7	34.5	43.6	
2013				6.7	8.6	15.0	32.7	36.1	33.8	35.6	36.8	
2014			2.5	7.4	2.5	5.2	25.4	33.6	25.0	30.1	35.3	
2015		2.5	2.5		21.0	39.2	43.3	33.9	24.5	24.0	41.4	
2016				2.5	20.0	27.6	32.5	34.8	34.4	23.0	40.7	
2017				5.8	2.5	15.0	18.6	25.3	23.0	23.0	18.0	
2018			2.5	2.5	2.5	7.8	23.0	7.7	25.9		27.7	
2019	32.6			1.5	6.1	17.2	18.6	21.0	19.2	18.8	31.8	18.2
2020	23.6				10.9	18.3	22.7	19.1	16.1	16.7	19.2	12.5

Bottom												
	Jan	Feb	Mar	Apr	May	Jun	Jul	Aug	Sep	Oct	Nov	Dec
2007						72.2	72.8	71.5	80.1			31.6
2008				14.0	78.8	63.7	47.7	66.2		71.9		17.0
2009				68.2	57.6	50.2	56.9	75.5		66.0		27.3
2010			19.0	7.2	37.3	52.0	39.6	75.7	53.3		27.1	
2011		15.5		2.5	48.3	28.5	35.6	50.4	56.4		33.7	
2012			17.0		34.9	42.5	31.3	63.9	57.8	65.8	13.0	
2013				8.4	62.9	44.8	61.2	84.5	67.6	25.2	22.0	
2014			2.5	20.0	45.2	41.5	40.5	72.0	65.6	47.8	33.1	
2015		2.5	5.5		49.2	68.3	80.6	73.9	64.8	56.9	20.0	
2016				5.9	50.5	60.9	50.7	68.4	70.5	67.8	14.5	
2017				30.0	43.6	36.8	44.0	56.9	46.0	45.5	7.0	
2018			2.5	27.2	40.2	30.6	51.2	65.1	59.3		18.0	
2019	59.6			46.7	38.1	44.9	45.3	62.1	56.3	52.4	13.9	23.8
2020	33.0				51.7	30.2	36.6	50.8	60.8	49.9	10.9	14.3

NOTE: The colors follow a scale running from red for the highest values, to brown, then yellow, then green, and finally aqua for the lowest values.
SOURCE: Data courtesy of the CDA Tribe.

at C5 extend later into the highest flow years for the St. Joe River (2008, 2011, 2017, and 2018) than during drier years (e.g., 2010, 2015, and 2016). Through a combination of dilution and hydrodynamic advection from south to north, elevated discharges from the St. Joe River reduce zinc concentrations at C5.

An increase in dissolved zinc at C5 begins, on average, in May for surface waters and in April for bottom waters, as the influence of the St. Joe River wanes, although the actual month that concentrations begin to increase varies with the water year (Table 6-3). Concentrations increase progressively through August in both the surface and the bottom waters (Figure 6-8A). Dissolved zinc concentrations in bottom waters exceed those in surface waters by 30–40 μg/L by the end of the summer. The decline in zinc concentrations observed in the photic zone through the summer at C4 is not evident at C5, and the difference between surface and bottom waters is greater at C5 than at C1 and C4. The perception that C5 has lower zinc concentrations than C4 reflects differences between the two sites in the spring season; these differences recede later in the year as processes internal to the Lake become more influential than inputs.

There are several possible reasons for the summer increase in dissolved zinc concentrations in both bottom and photic waters at C5. First, declining river discharges could allow internal processes like lake hydrodynamics to move zinc from north to south (see Chapter 4). The occurrence of contaminated sediments in the southern Lake almost to the same degree as in the northern Lake (Horowitz et al., 1995) is evidence that north-to-south advection of particles can occur. Second, decomposition of phytoplankton in bottom waters during summer can increase zinc concentrations in three possible ways. The decomposition itself can lead to release of zinc from organic particles, it can cause production of enough carbonic acid to lower pH below 7.0, and it can lower dissolved oxygen levels. In Chapter 7, the committee presents evidence of the second mechanism at work; mean monthly pH at 15 m depth goes from greater than 7 in March–April to 6.3–6.9 between April/May and November. A pH reduction to less than 7.0 could facilitate zinc desorption from settling particles and from surficial bottom sediments.

Increases in zinc in the euphotic zone in summer suggest that incoming dissolved zinc is greater than what might be stripped from the water column by phytoplankton. The coincident timing of different processes and uncertainties about hydrodynamics make it difficult to differentiate their relative roles in the increase of dissolved zinc concentrations in both photic and bottom waters at C5 through the summer/fall.

Formal Trend Analysis of Dissolved Zinc

Statistical trends over the 14 years of record (considering all months in each year through the full water column) show a decline of dissolved zinc at C5 of −30 percent over the entire period (Figure 6-9), but the vast majority of that decline takes place in the first few years. Over the period 2010–2020, the downward trend is very small (less than 10 percent) and is particularly difficult to evaluate because a substantial number of surface water observations are below the limit of detection during times of high St. Joe River discharge. The deeper water samples are mostly

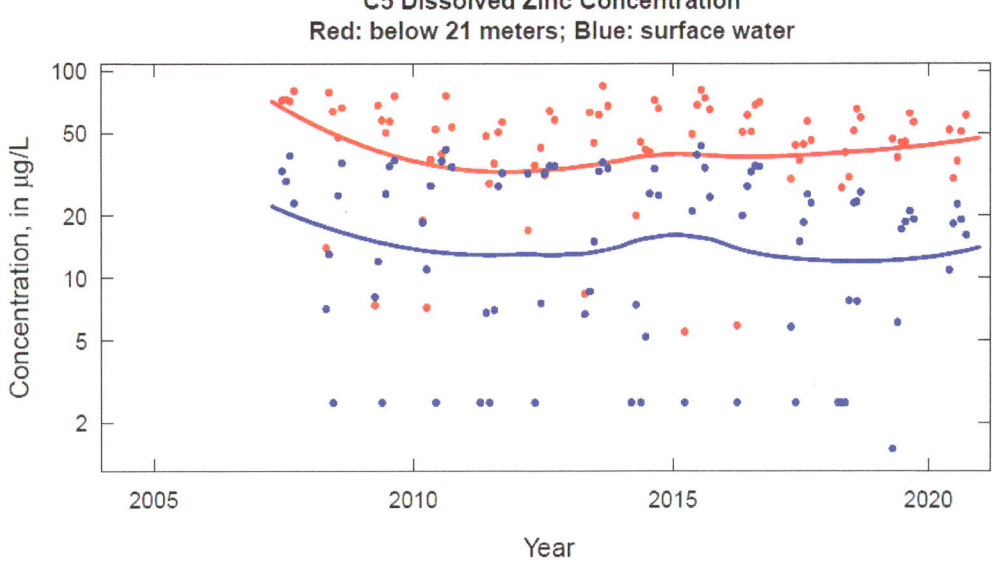

FIGURE 6-9 Dissolved zinc data at station C5 from 2007 to 2020. Surface water data is shown in blue, and bottom water data are shown in red. Smoothed estimates of the trends are shown in the red and blue lines. Note that the x-axis starts at 2004 to be consistent with Figures 6-4 and 6-7. SOURCE: Data courtesy of CDA Tribe and analyzed and graphed by the committee.

above the limit of detection and appear to have no significant trend over the decade 2010–2020. The differences between median surface water and bottom water concentrations are much more pronounced at C5 (20–30 µg/L) than they are at C1 or C4 (about 12 µg/L). This suggests increased importance of internal processes as a source of zinc at C5 compared to C1 and C4, including those processes already described (benthic flux from the sediments, mixing with bottom waters from the north, and biogenic recycling).

Table 6-3 shows dissolved zinc concentrations by year and month for 2007–2020 for both the euphotic zone and the bottom waters. It shows that the bottom waters in the summer and early fall have much higher concentrations than the euphotic zone water, but that all seasons and depths show declines in dissolved zinc over the 14-year period of record. Two of the years in this period of record—2008 and 2017—had particularly high discharge on the CDA and St. Joe Rivers (as measured by annual 30-day maximum discharge). In both of these years, Table 6-3 shows lower dissolved zinc concentrations for May through November as compared to most of the surrounding years. This is an indicator that high discharge events in the St. Joe River have a moderating influence on dissolved zinc at C5 in surface and deep waters.

The substantial differences between dissolved zinc concentrations in bottom waters and euphotic zone at C5 are most evident in the evaluation of exceedances of benchmarks in recent years (2015–2020). The 36 µg/L Lake Management Plan target was met in all but two months in euphotic waters at C5 between 2015 and 2020. However, the target was not met in bottom waters in 69 percent (35 of 51) of the months when data were available, particularly April through October. In particular, exceedances continued to occur in bottom waters from May through October from 2015 to 2020 (Table 6-3).

High discharge from the St. Joe River clearly drives water column zinc concentrations down in spring at C5, but this effect is mitigated as discharge recedes, such that zinc concentrations increase throughout the summer. Although one can postulate that the source of this high dissolved zinc at C5 is internal to the lake, it is not possible to quantitatively differentiate the contributions from advection of zinc-rich water from the north; influx of zinc into bottom waters from sediments following low pH or low dissolved oxygen conditions; or biogenic recycling and release of zinc by decomposition of organic material.

***

Table 6-4 summarizes the zinc trend results for the three sites. **In all cases, the trends in concentration were shown to be highly significantly ($p < 0.01$) downward for the period of record at all sites.** For C5, the slopes of these downward trends could not be reliably defined over the past ten years because of the large number of photic zone values that were less than the limit of detection of 5 µg/L. Over the most recent decade, dissolved zinc has declined by about 30 percent at C1 and C4—locations that represent areas directly influenced by zinc inflows from the CDA River. Trends in the past decade are more difficult to quantify at C5 because many of the surface water samples had values that were less than the limit of detection of 5 µg/L. C5 is not in the direct flow path of the CDA River and is subject to internal zinc inputs that may slow the response of the Lake to changing inputs.

TABLE 6-4 Trends for Dissolved Zinc at Sites C1, C4, and C5

Site	Total record evaluated	Total change	Change over past 10 years	2020 median deep water	2020 median surface water
C1	17 years	−38%	−30%	46 µg/L	33 µg/L
C4	17 years	−41%	−33%	48 µg/L	35 µg/L
C5	14 years	+30%	*	47 µg/L	14 µg/L

NOTES: See Appendix B for methods of calculation. Analysis uses both surface water and bottom water samples. The * for the 10-year period at C5 indicates substantial uncertainty about the amount of change, due to the relatively large number of euphotic zone values reported as less than the limit of detection.

ANALYSIS OF TRENDS IN DISSOLVED CADMIUM

The spatial and temporal patterns of dissolved cadmium in CDA Lake are very similar to those for zinc, and much of the process description presented on zinc will not be repeated here. Attempts were made to evaluate trends in dissolved cadmium at all five sites, but at sites C6 and SJ1 the data were all reported as less than the limit of detection (which was typically 0.1 µg/L), so no analysis was conducted for those two sites. At site C5, data below the detection limit were very common (25–50 percent of the data depending on depth and time period). Trend analyses are certainly possible for such datasets, but the complexity of determining the most appropriate statistical methods for doing so was judged to be beyond what was needed for this report. Nonetheless, in general, the bottom water samples at C5 show an indication of a downward trend over the 14 years evaluated. The records for 2004–2020 at sites C1 and C4 had no more than 10 percent of their samples below detection limits, so they were evaluated as described in Appendix B.

Dissolved Cadmium Concentrations at C4

The entire record for cadmium in the photic zone of C4 is shown in Table 6-5 as a heat map that shows the rapid declines through the late spring and summer as well as the rapid decline in the last few years of the record. Figure 6-10 shows the final six years of the photic zone data from C4. The seasonal pattern in the cadmium data is quite strong and similar to that for dissolved zinc, with the highest values occurring in the high-flow season of early spring. Indeed, the highest value in each of the years shown is always the first value in the year and the minimum is around September. The ratio of the maximum value for the year to the minimum is typically about 3:1. This seasonal pattern is related to the timing of maximum discharge for the year, with concentrations falling through the summer months as biological uptake and settling occurs.

Figure 6-11 shows the formal analysis of overall trends in dissolved cadmium at C4. The concentrations of cadmium in the photic zone average about 85 percent of the magnitude of those near the bottom. The trend in the

TABLE 6-5 Concentrations of Dissolved Cadmium in the C4 Photic Zone by Month, in ug/L

	Jan	Feb	Mar	Apr	May	Jun	Jul	Aug	Sep	Oct	Nov	Dec
2004		0.245		0.315	0.260	0.225	0.215	0.200		0.215		0.270
2005	0.265			0.310	0.235	0.205	0.225	0.210		0.165		0.200
2006		0.320	0.355		0.265	0.225	0.210	0.240				
2007							0.225	0.210	0.180	0.190		0.205
2008			0.310		0.390	0.245	0.190	0.190		0.245		
2009				0.320	0.265	0.260	0.220	0.225		0.190		
2010			0.210	0.245	0.220	0.235	0.220	0.235	0.175			0.230
2011		0.320		0.345	0.270	0.165	0.285	0.160		0.195		0.265
2012			0.305	0.375	0.270	0.250	0.210	0.210	0.235			
2013				0.295	0.240	0.245	0.265	0.260		0.215		0.240
2014			0.335	0.305	0.220	0.210	0.205	0.190	0.200			0.190
2015		0.360	0.350		0.220	0.200	0.215	0.210	0.165		0.165	
2016				0.275	0.205	0.195	0.200	0.190	0.160		0.190	
2017			0.290	0.270	0.200	0.165	0.170	0.160	0.140	0.155		0.165
2018			0.260	0.230	0.195	0.150	0.135	0.125	0.135		0.130	0.140
2019			0.160	0.180	0.150	0.145	0.140	0.134	0.123	0.130		0.120
2020				0.180	0.155	0.145	0.140	0.115	0.110			0.130

NOTE: The color scale progresses from bright red for highest values, through more muted reds, and then to light blue for the lowest values.
SOURCE: Data provided by IDEQ.

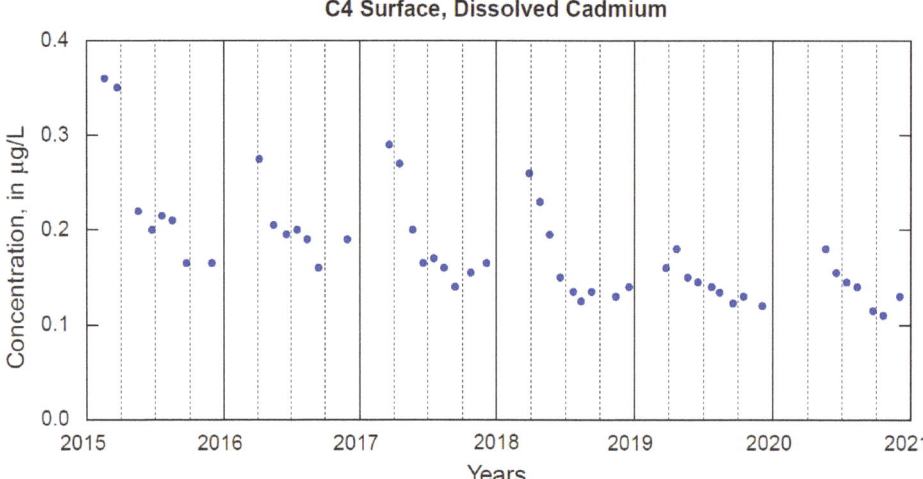

FIGURE 6-10 Time series of dissolved cadmium concentrations in the photic zone at C4. Winter gaps in the record occur in almost every year. The figure shows that the highest concentrations are in the earliest samples of the year (typically the first sample is in March). SOURCE: Data courtesy of IDEQ and plotted by the committee.

record shows a very sharp discontinuity, with virtually no trend from 2004 through about 2014, and then a rather steep downward trend after 2014. The trend over the full 17-year period is a change of about −49 percent. The change over the most recent ten years is also −49 percent, with most of that trend focused in the final five years of the record. The trends are all statistically highly significant ($p < 0.01$).

The temporal pattern of the trend in dissolved cadmium at C4 is similar to the temporal pattern of the trend in total cadmium flow-normalized flux for the CDA River near Harrison. That record shows almost no change from 2004 through 2013, followed by a steep decline to 2020 (see Figure 3-16B). At C4, the total change in dissolved

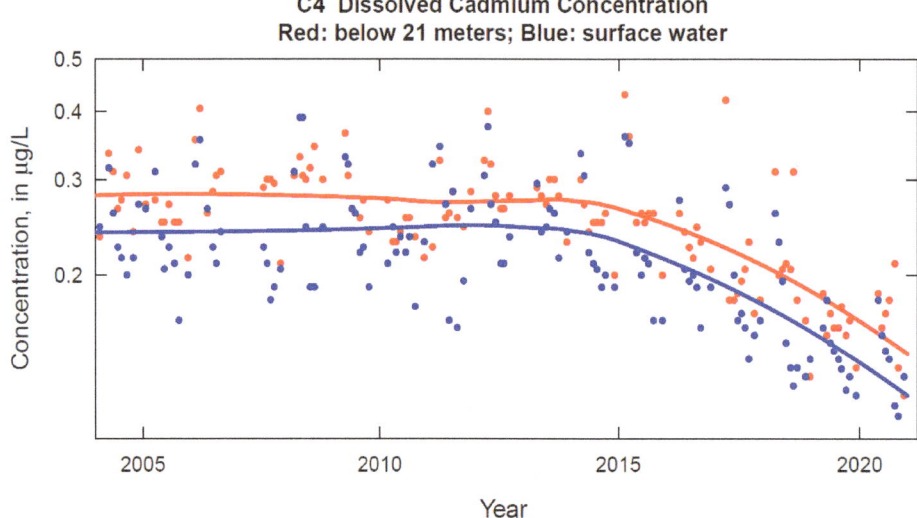

FIGURE 6-11 Dissolved cadmium concentrations at C4 from 2004 to 2020. Surface water data are shown in blue, and bottom water data are shown in red. Smoothed estimates of the trends are shown with red and blue lines. SOURCE: Data courtesy of IDEQ and analyzed and graphed by the committee.

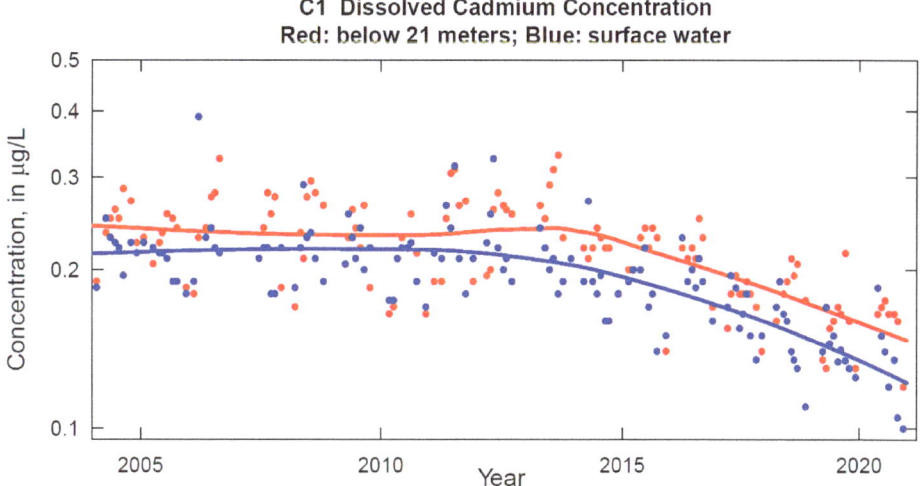

FIGURE 6-12 Dissolved cadmium concentrations at site C1 from 2004 to 2020. Surface water data are shown in blue, and bottom water data are shown in red. Smoothed estimates of the trends are shown in the red and blue lines. SOURCE: Data courtesy of IDEQ and analyzed and graphed by the committee.

cadmium concentration for the 17 years from 2004 to 2020 was a decline of 31 percent, while the total change from 2015 to 2020 was a decline of 20 percent. The flow-normalized flux of dissolved cadmium for the CDA River shows a similar pattern, but the declines are not as steep in percentage terms as those for the flow-normalized flux of total cadmium for the CDA River (see Chapter 3). Understanding the linkage between the river inputs and ambient concentrations in the Lake is an important topic that should be pursued in the future. Clearly, the Lake cadmium trends are highly responsive to changes in the inputs, and it is reasonable to project that the Lake cadmium concentrations will continue their steep decline if the river fluxes to the Lake continue to decline as they have over the decade 2010–2020.

Dissolved Cadmium Concentrations at C1

The trends and seasonal patterns in dissolved cadmium at site C1 (Figure 6-12) are similar to those at C4. The overall trend for 2004–2020 is a decline of 41 percent, with most of the trend happening in the final decade. For both trend periods (2004–2020 and 2011–2020) the declines are highly significant ($p < 0.01$). The ratio of photic zone concentrations to bottom water concentrations averaged about 0.85 but was slightly larger around 2010 (a value of about 0.93).

In summary, although only the two northern Lake sites could be evaluated for trends in dissolved cadmium concentrations, the trends are clear and summarized in Table 6-6. Evaluated over the full period, 2004–2020, dissolved cadmium is declining, but virtually all of the decline takes place after about 2014. The data also show that cadmium concentrations in deeper water (> 20 m) are greater than those in surface water, by about 15 percent.

TABLE 6-6 Trends in Dissolved Cadmium at Sites C1 and C4

Site	Total record evaluated	Change since start	Change over past 10 years	2020 median deep water	2020 median surface water
C1	17 years	−41%	−40%	0.15 µg/L	0.12 µg/L
C4	17 years	−49%	−49%	0.14 µg/L	0.12 µg/L

NOTES: See Appendix B for methods of calculation. The dark blue shading indicates that for all trend periods the observed trends are highly significant ($p < 0.01$). The analysis combines results from both surface and deep water.

ANALYSIS OF TRENDS IN TOTAL LEAD

The patterns in total lead concentration in the water column of CDA Lake were evaluated at three sites: C1, C4, and C5. This discussion is primarily based on the data from C4 because it is the most responsive to the river inputs of lead to the Lake, but the results at the other two sites are summarized at the end of this section.

Total Lead Concentrations at C4

Figure 6-13 shows the total lead data in surface water from C4. These data show a range of values over this 17-year period (2004–2020) of nearly three orders of magnitude. The data also are suggestive of a non-monotonic trend, rising from 2004 to sometime around 2015 and then downward to 2020. The trend is investigated in detail below. To gain some sense of the seasonality of the data, it is helpful to focus on a smaller number of years. Figure 6-14 shows only the data for 2015–2020.

An orderly pattern in the data is observed throughout the record (illustrated for 2015–2020 in Figure 6-14), starting with the highest values typically being in the late winter or spring, declining quite steeply every month, reaching an annual minimum around August or September, and rising slightly through the fall. The sampling is somewhat irregular in the months of September through March, with a few years missing values in April, May, and June. Only July and August have complete coverage over these 17 years. The irregularity of the sample collection is an impediment to data interpretation, but a number of clear patterns can be discerned in the data.

Table 6-7 is a heat map of the dataset (in the few cases where there are multiple sampling dates in the month, the heat map uses the median for the month). One of the notable things in Table 6-7 is that the months following some of the highest observed concentration values also tend to be high relative to values in that particular month in other years (e.g., 2011–2012; 2017). The final column in Table 6-7 shows the estimated annual flux of total lead for the CDA River. Generally, the highest concentration values in the table correspond to the years of highest estimated annual flux. Figure 6-15 shows the estimated monthly flux of total lead into the Lake, and it shows an obvious similarity to the pattern of concentration values in the Lake that are shown in Figure 6-14.

Figure 6-16 shows the relationship between the monthly mean values of observed surface water concentration of total lead at C4 (shown as a time series in Figure 6-14) in relation to the estimated average daily flux into the Lake for the months for which there are surface water concentration data at C4. The correlation between the

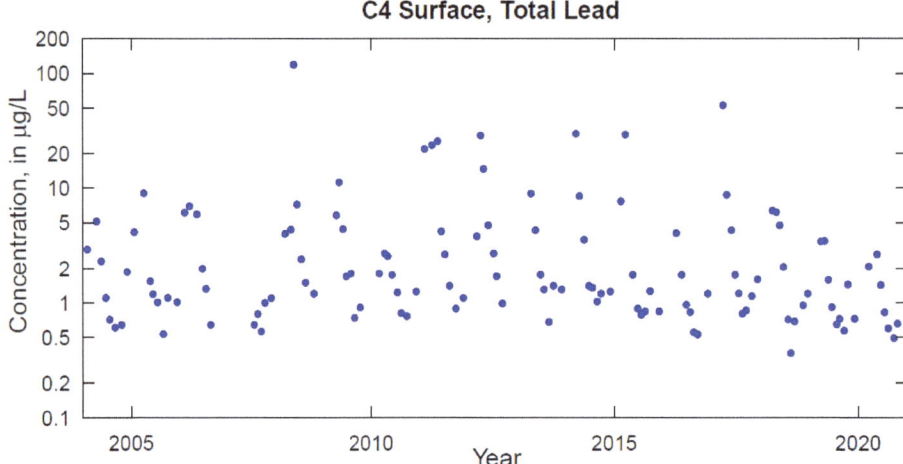

FIGURE 6-13 Time series of total lead concentrations in surface water (< 20 m) at C4. Winter gaps in record occur in almost every year. SOURCE: Data courtesy of IDEQ and plotted by the committee.

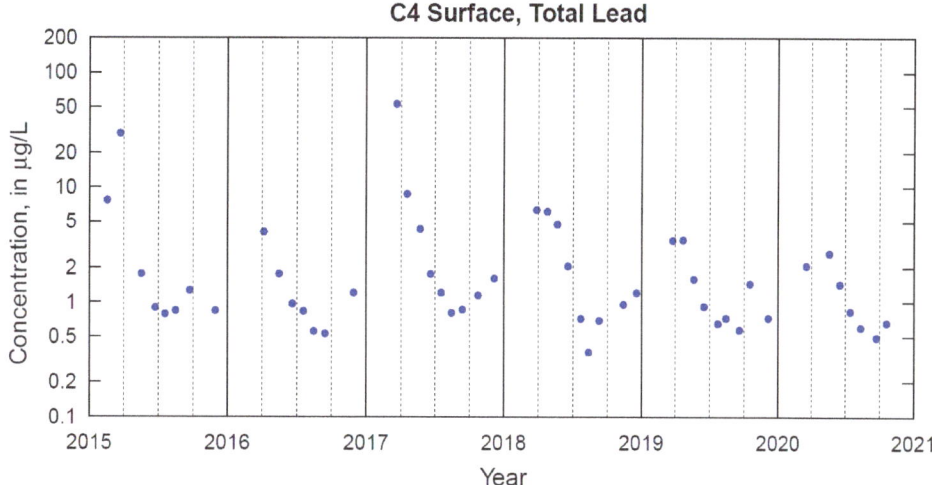

FIGURE 6-14 Time series of total lead concentrations in surface water (< 20 m) at C4 for 2015–2020. Solid vertical lines mark the start of the calendar year and the dashed lines separate the year into four quarters. SOURCE: Data courtesy of IDEQ and plotted by the committee.

TABLE 6-7 Concentrations of Total Lead in C4 Surface Waters by Month, in ug/L

	Jan	Feb	Mar	Apr	May	Jun	Jul	Aug	Sep	Oct	Nov	Dec	TPb flux
2004		2.9		5.1	2.3	1.1	0.7	0.6		0.6		1.9	69.0
2005	4.1			9.0	1.5	1.2	1.0	0.5		1.1		1.0	73.0
2006		6.1	7.0		5.9	2.0	1.3	0.6					256
2007							0.6	0.8	0.6	1.0		1.1	382.0
2008			4.0		61.7	7.2	2.4	1.5		1.2			378.0
2009			8.5		4.4	1.7	1.8	0.7		0.9			220.0
2010			1.8	2.7	2.6	1.8	1.2	0.8	0.8			1.3	65.0
2011	21.9		23.6	25.5	4.2	2.7	1.4		0.9			1.1	979.0
2012		3.8	28.7	14.7	4.8	2.7	1.7	1.0					493.0
2013			8.9	4.3	1.8	1.3	0.7		1.4			1.3	199.0
2014		29.7	8.5	3.6	1.4	1.4	1.0	1.2				1.3	460.0
2015	7.7	29.2		1.8	0.9	0.8	0.8	1.3		0.8			299.0
2016			4.1	1.8	1.0	0.8	0.6	0.5		1.2			254.0
2017		52.8	8.7	4.3	1.8	1.2	0.8	0.9	1.1			1.6	1009.0
2018		6.4	6.2	4.7	2.1	0.7	0.4	0.7		1.0		1.2	191.0
2019			3.4	3.5	1.6	0.9	0.6	0.7	0.6	1.4		0.7	62.0
2020			2.1		2.7	1.4	0.8	0.6	0.5	0.7			68.0

NOTES: The colors range from red for the highest values, through brown, then green, and then blue for the lowest values. The column to the right of the concentration data is the estimated annual flux values of total lead for the CDA River near Harrison in metric tons (MT)/yr. These flux values were computed using the Weighted Regressions on Time, Discharge, and Season (WRTDS) model described in Chapter 3. SOURCE: Data provided by IDEQ.

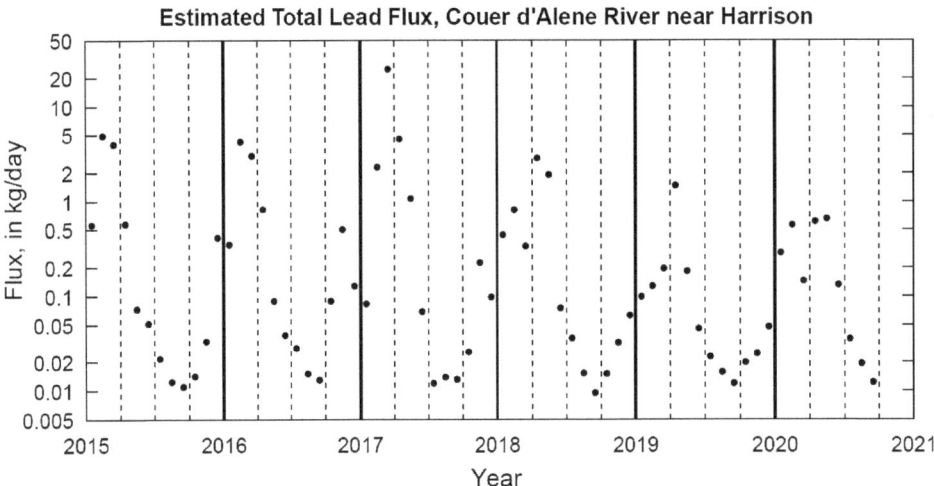

FIGURE 6-15 Time series of estimated flux of total lead into CDA Lake by month, 2015–2020. SOURCE: Based on WRTDS model of total lead for CDA River near Harrison (presented in Chapter 3). Data from U.S. Geological Survey (USGS) and model estimates by the committee.

natural logarithms of these two variables is 0.88, and estimated slope of the regression of ln(concentration) on ln(monthly flux) is 0.47. This means that concentration increases in proportion to the 0.47 power of flux into the Lake for that month. More complex models of this relationship are certainly possible (ones that might include lag effects) but this simple analysis demonstrates that **lead concentrations at C4 are very responsive to the current month's inputs of total lead.** The implication of this is that concentrations of total lead in the Lake should be responsive to future declines in total lead input to the Lake. The fact that estimated monthly flux of lead into the Lake explains a very large part of the variation in Lake lead concentrations is what one would expect given that 75–85 percent of the lead entering the Lake is particulate.

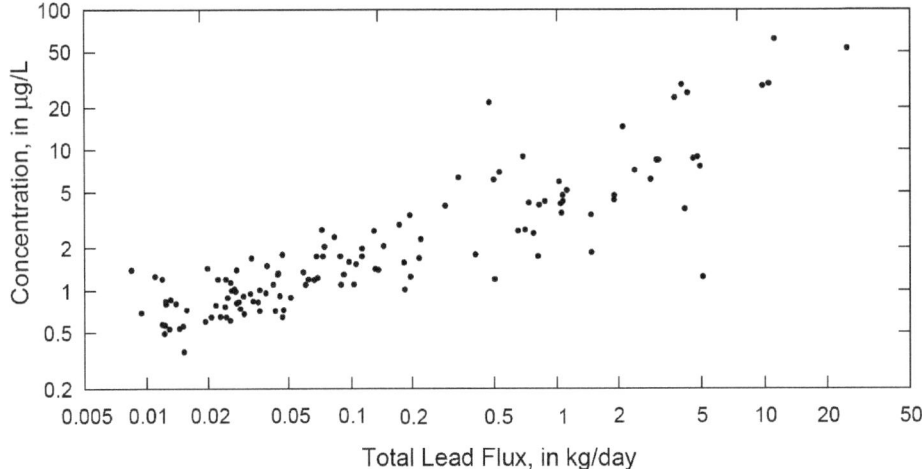

FIGURE 6-16 Scatterplot of monthly mean surface water total lead concentration at C4 versus the monthly estimated total lead flux for the CDA near Harrison. Data from 2004 to 2020. SOURCE: Data courtesy of IDEQ (concentration) and flux based on USGS National Water Information System (NWIS) data and WRTDS model estimated by the committee.

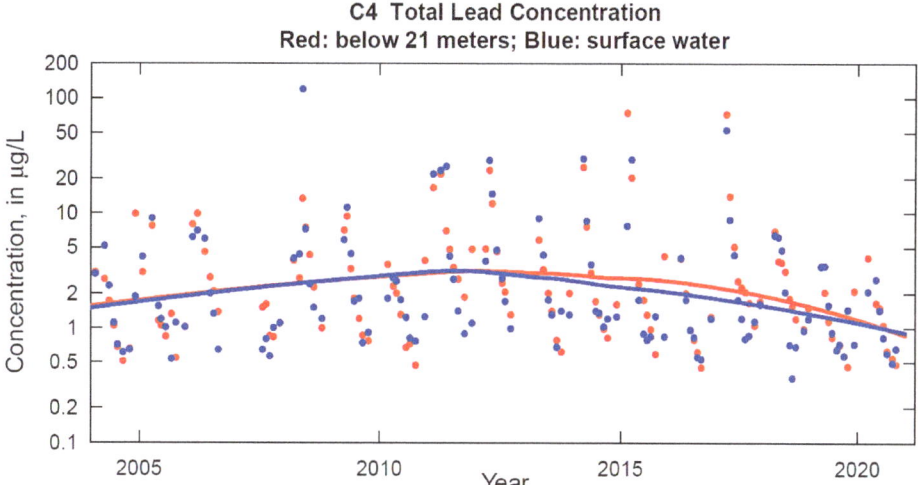

FIGURE 6-17 Total lead concentrations at station C4 from 2004 to 2020. Surface water data are shown in blue, and bottom water data are shown in red. Smoothed estimates of the trends are shown by the red and blue lines. SOURCE: Data courtesy of IDEQ and analyzed and graphed by the committee.

The surface and deep water concentrations were each modeled using a loess fit (as described in Appendix B), illustrated in Figure 6-17. The first observation is that, unlike zinc and cadmium, the lead data show much smaller differences between surface and deep water. At most, the deep water has concentrations averaging about 0.2 µg/L higher than the surface water, and for much of the record the differences are even smaller. The processes that drive the more substantial differences between surface- and deep-water concentrations of zinc and cadmium (such as potential benthic flux and biogenic cycling) have limited effects on lead. This is consistent with the low bioavailability of lead to phytoplankton (Chen et al., 2000) and the strong binding of lead to particulate material.

Another observation from these data is that the overall trend is distinctly non-monotonic, not unlike the temporal pattern of flow-normalized flux of total phosphorus at Harrison (described in Chapter 3). Over the period 2004–2020, the smoothed lake concentrations rise to a maximum in 2012 then fall from 2012 to 2020. The flow-normalized river flux rises from 2004 to 2009 and then falls to 2020. The estimated input flux record at a daily or monthly time scale could be used in conjunction with the observed Lake concentration data to calibrate a model of how the Lake lead levels respond to changing inputs. Such a model, using simplified physical process simulation with statistical parameter estimation, could be used to project future Lake concentrations as a function of various degrees of control of inputs to the Lake. What seems clear from the simple empirical analyses described in this report is that lead concentrations in the Lake are responsive to changing lead inputs at time scales of months to years. The aggregate trend over the entire 2004–2020 period, which is non-monotonic, is a change in mean concentration of −41 percent, or −3 percent per year. Even though it is a period of increase followed by decrease, the overall trend is statistically significant ($p = 0.04$). When evaluating the most recent ten years, the change is −74 percent, which is −13 percent per year and is highly significant ($p < 0.001$).

Dissolved Lead Concentrations at C4

A similar analysis of the *dissolved* lead concentration at C4 was also conducted (results not shown). Dissolved lead concentrations follow a similar non-monotonic trend pattern, but the rates of increase in the first part of the record and the rates of decrease in the most recent years are smaller in magnitude than those described above for total lead. The same pattern of seasonality exists, and the relationship of high winter and spring concentrations being associated with the years of high lead input also exists. Unlike zinc or cadmium, the dissolved lead annual mean concentrations are substantially lower than the total lead concentrations. Typically, the ratio of dissolved to total lead concentrations is in the range of 0.16 (also see Table 6-1).

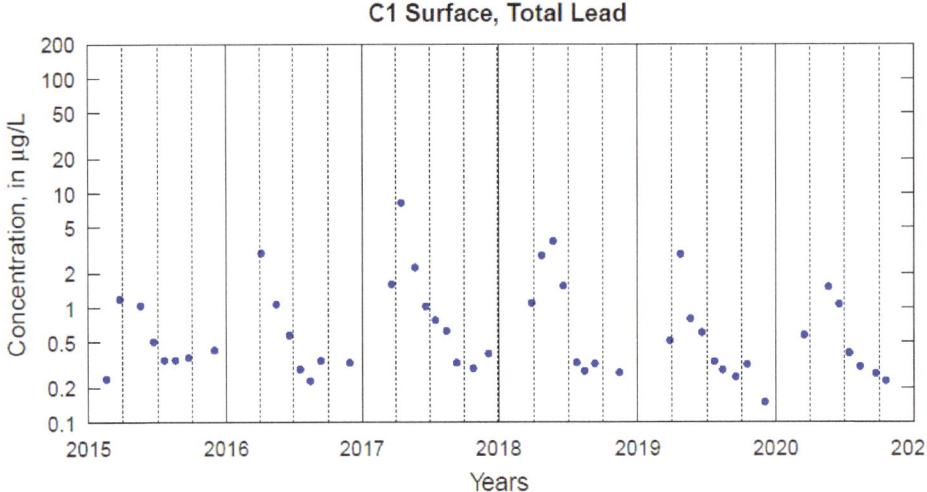

FIGURE 6-18 Time series of total lead concentrations in surface water (< 20 m) at C1 for 2015–2020. SOURCE: Data courtesy of IDEQ and plotted by the committee.

Total Lead Concentrations at C1

At C1 and farther from the mouth of the CDA River, the patterns of total lead seasonality and trends are similar to those observed at C4 in surface waters. Figure 6-18 shows the most recent six years of surface water data. Comparing it to Figure 6-14 (which uses the same vertical scale), one can see that the general seasonal pattern is similar to C4, with the highest concentrations typically in the earliest samples in the year, but the concentrations tend to be lower (mean of all of the annual mean values is 4.8 µg/L at C4 and 1.4 µg/L at C1) and the amplitude of the seasonal cycle is reduced (mean annual difference between maximum and minimum is 19.7 µg/L at C4, compared to 4.3 µg/L at C1). Figure 6-19 shows the patterns for both surface and deep water for the full period of 2004–2020. The overall trend for the full 17-year period is a change of −11 percent (−1 percent per year), but

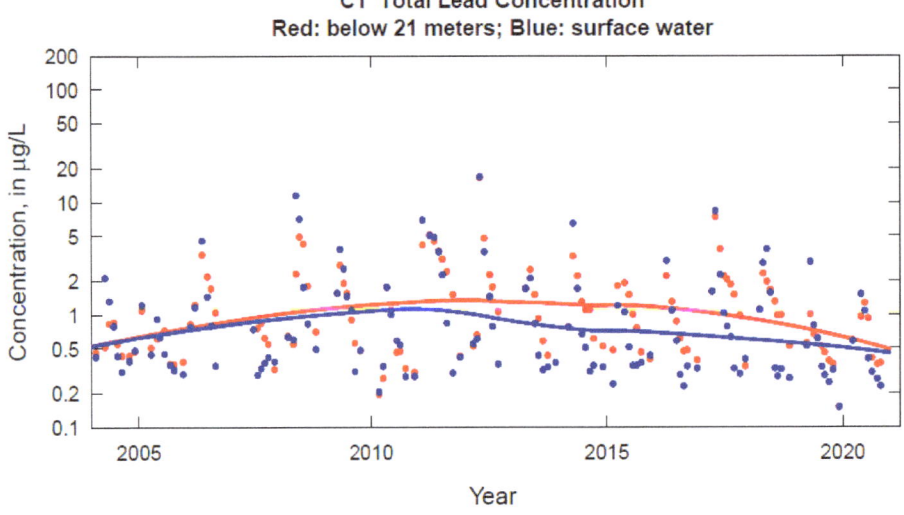

FIGURE 6-19 Total lead concentrations at site C1 from 2004 to 2020. Surface water data are in blue, and bottom water data are in red. Smoothed estimates of the trends are shown by the red and blue lines. SOURCE: Data courtesy of IDEQ and plotted by the committee.

it is not significant (p = 0.12). For the past ten years of the record, the change is −62 percent (or −9 percent per year), and it is highly significant (p < 0.001).

Total Lead Concentrations at C5

The only other site with a dataset of sufficient sample size and sufficient numbers of values above the detection limit is C5. The record for surface water concentrations for 2015–2020 is shown in Figure 6-20. The data are intentionally shown on the same scale as is used in Figures 6-14 and 6-18 to facilitate comparison with the two northern Lake locations.

The first thing to note is that annual maximum concentrations are consistently lower than those observed at C1 or C4. For example, over the years 2015–2020, the annual maximums at C4 are in the range of 2.5–50 µg/L, at C1 they are in the range of 1.2–8 µg/L, and for C5 they are in the range of 0.7–1.5 µg/L. In addition, the seasonal pattern at C5 is very different from that of the other two sites. The difference between the annual minimum and annual maximum is only 1.55 µg/L versus 4.3 (C1) and 19.7 (C4). What one generally sees at C5 is that in many years the trend through the year is rising values from the early spring through the summer months. The explanation for this is that the spring snowmelt season at C5 is dominated by water coming from the St. Joe River, which is relatively free of lead. Only later in the year, when the water movement is driven more by internal dynamics of the Lake, does one see increases in lead as water from the CDA River and northern portions of the Lake circulate south to C5.

Another important difference between C5 and either C1 or C4, apparent in Figure 6-21, is that the difference between total lead concentrations near the surface and concentrations at the bottom are more pronounced at C5. Statistical trends (Figure 6-21) indicate that total lead concentrations in deeper water are commonly more than double the concentrations observed in surface water. The mean total lead concentrations in 2021 near the bottom of C5 are similar to C4 (~ 1 µg/L) but higher than at C1 (~ 0.5 µg/L vs. 1 µg/L, respectively). Surface water mean concentrations at C5 in 2021 are similar to C1 but lower than C4.

The committee investigated the reason for the difference in lead concentration in surface versus bottom waters at C5. Given lead's chemistry and lack of biogenic cycling, the most likely process creating this difference is hydrodynamic advection of lead from the northern Lake, with desorption from suspended particles and/or upward transfer from the Lake bed to the water column being less likely processes. To confirm this, the committee graphed the ratio of *dissolved* lead concentrations (bottom water: surface water) using data from 2017 to 2020 grouped by month (see Figure 6-22). If there is desorption of lead from the bed sediment surface and sinking particulate

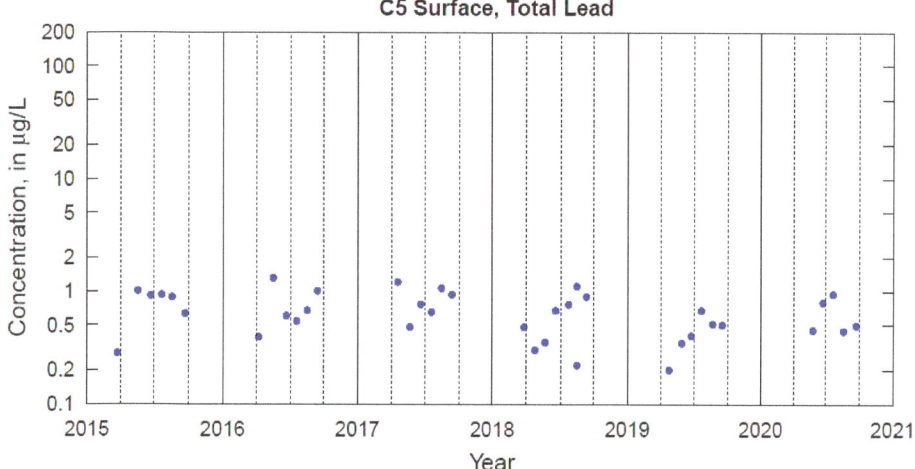

FIGURE 6-20 Time series of total lead concentrations in surface water (< 20 m) at C5 for 2015–2020. SOURCE: Data courtesy of CDA Tribe and plotted by the committee.

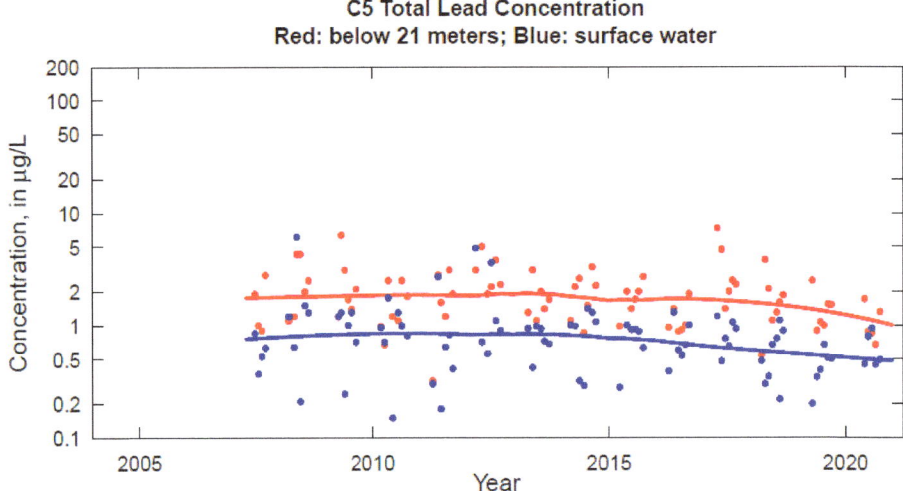

FIGURE 6-21 Total lead data at site C5 from 2007 to 2020. Surface water data are in blue, and bottom water data are in red. Smoothed estimates of the trends are shown by the red and blue lines. SOURCE: Data courtesy of CDA Tribe and plotted by the committee.

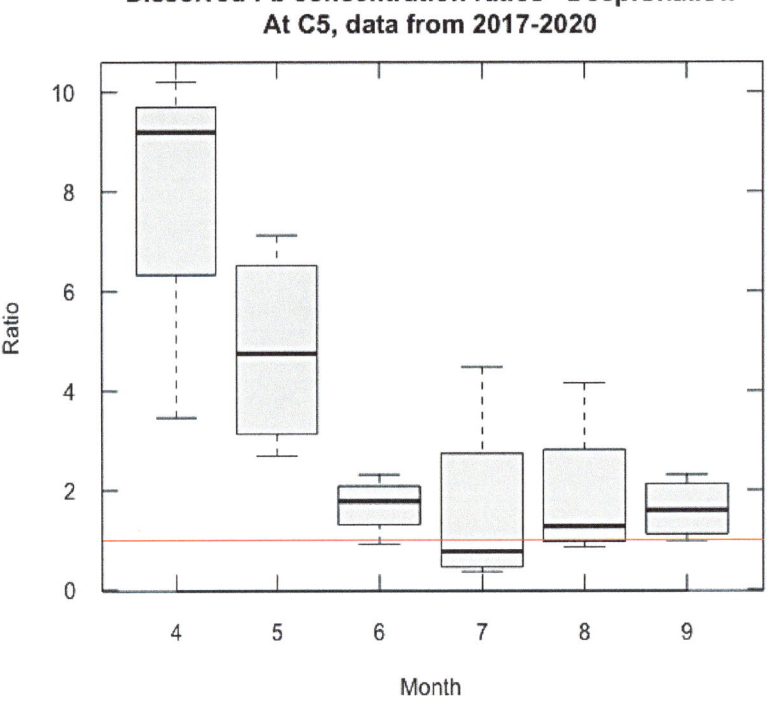

FIGURE 6-22 Dissolved lead concentration data at site C5 from 2017 to 2020. The bars are the ratio of bottom water concentrations to surface water concentrations by month. SOURCE: Data courtesy of CDA Tribe and analyzed and graphed by the committee.

material, concentrations of dissolved lead in bottom waters would be higher than in surface water during the summer period of stratification. The ratio near 1 for the summer months suggests that stratification (evidenced by dissolved oxygen and temperature) does not affect lead in June through September, such that there is little dissolved lead input to isolated bottom waters either from desorption from settling particles or from the bed sediments. This is very similar to what one observes at C1 and C4. The high ratios in April and May (see Figure 6-22) must, by the process of elimination, reflect advection of bottom waters to the south from the north as discharge rates from the St. Joe River begin to decline. Advection to the south of bottom waters would have to exceed that of surface waters to explain such results.

An important distinction that should be made across the sites is that the amplitude of seasonal and interannual variability at C5 is lower than at the other two sites. Peak concentrations originating from the CDA River are more damped before they make it to C5 as compared to C1. It is possible that a substantial amount of particulate lead has settled out before it reaches C5.

The trend over time at C5 does not show the same obvious non-monotonic pattern as is seen at the other two sites, but this may be an artifact of this being a shorter record than the other two. For the period for which there are data, one can generally say that there is very little change from 2007 to about 2016, and that the last few years of the record are suggestive of a downward trend. For the full 13-year record, the downward trend is a change of about –42 percent, and for the last ten years of the record it is similar (–45 percent). Most of this decline takes place in the last five years of the record. Both of these trends are significant, with p-values of 0.001 or lower. It is reasonable to assume that the decreased inputs from the CDA River are driving these downward trends at C5, but the influence of what appear to be sources that reflect a redistribution of lead within the Lake (internal sources) add to the complexity of the response and perhaps delays the decline.

***

A summary of the total lead trends at all three CDA Lake sites is shown in Table 6-8. For the longer two records (at C1 and C4), there is evidence of non-monotonic trends, i.e., rising values up to about 2012 and then declining to 2020. The downward trends over the period 2011–2020 are 9 percent per year at C1 and 12 percent per year at C4 (greater than observed for other metals) and statistically significant. At all three sites considered, the downward trend over the past ten years is categorized as statistically highly significant. These patterns are reminiscent of the non-monotonic pattern of trend in flow-normalized flux of total lead for the CDA River near Harrison (see Chapter 3).

The committee's hypothetical conceptual model for total lead dynamics is relatively simple in the northern part of CDA Lake (C4 and C1). Large loads of total lead enter the Lake with the rising limb of discharge in the spring, generating high concentrations in the water column. Currently, only 1 percent of the lead leaves the Lake via the Spokane River, such that most of the incoming lead is trapped in the Lake sediments. Lead concentrations in the water column progressively decline as discharge recedes and throughout the rest of the year, presumably as the 99 percent of the lead that remains in the Lake settles out of the water column to the sediments. Stratification makes little difference, as evidenced by similar total lead concentrations in surface water and bottom water

TABLE 6-8 Summary Table for Trends in Total Lead at Sites C1, C4, and C5

Site	Total record evaluated	Change since start	Change over past 10 years	2020 median deep water	2020 median surface water
C1	17 years	–11%	–61%	0.5 µg/L	0.4 µg/L
C4	17 years	–41%	–71%	0.9 µg/L	0.9 µg/L
C5	13 years	–42%	–45%	1.0 µg/L	0.5 µg/L

NOTES: See Appendix B for methods of calculation. The shading indicates the significance level of the result. Dark blue is highly significant ($p < 0.01$), light blue is significant ($0.01 < p < 0.1$), and no shading indicates a nonsignificant trend. Results are based on trends in both surface and bottom water data.

concentrations throughout the year. The spring peak in lead concentrations declines with distance away from the mouth of the river, evidenced by lower concentrations in spring at C1 than at C4. This decline is most likely driven by a combination of dilution and settling of particles with distance from the source. The dynamics of dissolved lead track total lead, but represent only 14–25 percent of the lead in the water column. Declining inputs of total lead from the watershed result in declining concentrations of total lead in the water column and declining flux to the sediments. Net lead flux from the sediments back to the water column at C4 and C1 is an undetectable fraction of the lead concentrations driven by inputs at this point in time, as evidenced by similar lead concentrations in surface and bottom waters.

The dynamics of total lead in the southern Lake are more complex. No spring peak in concentrations is observed. Discharge from the St. Joe River dilutes lead concentrations throughout the water column in spring, and then total and dissolved lead concentrations increase from May through October. Total lead concentrations in bottom waters exceed concentrations in surface waters after stratification begins, but dissolved lead concentrations are not different between bottom and surface waters for much of the summer/fall, consistent with the strong binding of lead to particulate material and a lack of benthic flux of lead from the sediments to the water column. Hence, where there are differences in dissolved lead between surface and bottom waters, it could be attributable to north-to-south advection of lead to C5 during the period of low discharge. If advection (internal redistribution) dominates the seasonal increase in lead at C5, then declines of lead inputs to the Lake will manifest at C5 in a similar way as they do at C1 and C4.

The most important conclusion to draw about the total lead trends in CDA Lake is that for the years from about 2003 to 2012, total lead concentrations were rising slowly but have declined over the past eight to ten years. The data are suggestive of an acceleration of the downward trends in the most recent five years (2016–2020), but the particular sequence of flow conditions (with 2017 being one of the highest discharge years and 2019 and 2020 among the lowest years in the record) makes the evidence rather weak. Frequent (e.g., every two years) reanalysis of the data should help to clarify if these apparent downward trends are meaningful.

The analysis also makes it clear that for C1 and C4, the highest observed total lead concentrations are directly linked to high flow conditions in the CDA River, which now deliver mostly legacy lead from past mining contamination that has been stored in the lower basin. The future trajectory of Lake concentrations will depend strongly on the extent to which this scouring of the lower basin declines. The analysis in Chapter 3 indicates that such declines are currently happening, but continued monitoring of the movement of sediment and lead and analysis of geomorphic change in the lower basin will be crucial to understanding the rate of these trends and their likely future development. Stronger inferences about the extent to which high winter and spring flows are delivering less lead than they had in past years can only come about with the accumulation of additional high-flow years and a high level of effort at river and Lake monitoring when such high-flow years take place, coupled with reanalysis of the CDA River and Lake water quality trends on a regular basis.

CONCLUSIONS AND RECOMMENDATIONS

This chapter analyzed water column data from CDA Lake over the past 30 years to reveal trends in zinc, cadmium, and lead concentrations. For many of the processes that control metals in the Lake, particularly hydrodynamic processes, benthic flux, and biogenic cycling, there were almost always insufficient data to distinguish between the processes. Based on the committee's analysis of available datasets to better understand changes in metal concentrations in CDA Lake over the past 30 years, the following detailed conclusions and recommendations are made. These conclusions apply to the monitored areas of the Lake and cannot be extended to embayments and niches along the shoreline or other Lake locations that are poorly monitored.

1. **Downward trends in dissolved zinc concentration in CDA Lake at sites C1 and C4 over the past 30 years are highly significant ($p < 0.01$).** Furthermore, zinc concentration at these sites is decreasing at a similar rate to declines in zinc inputs to CDA Lake from the CDA River. Internal cycling of zinc (see below) has not detectably slowed the response of the northern Lake to changes in zinc input from the CDA basin.

2. **Dissolved zinc concentrations have declined at C5 (in the southern Lake) over the past two decades, but the rate of decline is slower than at C1 and C4.** The dynamics of zinc at C5 are more complex than at C1 and C4 and reflect spring dilution from the St. Joe River, summer/fall hydrodynamic redistribution of metals from north to south, and other internal processes (see below).

3. **Dissolved cadmium concentrations at C1 and C4 declined from 2004 to 2020, with virtually all of the decline occurring after about 2014.** Like zinc, cadmium trends in the northern Lake are highly responsive to changes in cadmium inputs from the CDA River.

4. **Seasonal differences in dissolved zinc and cadmium concentrations between surface and bottom waters suggest there is an internal source of both metals to bottom waters** *during periods of stratification.* For zinc, after the onset of stratification, dissolved zinc concentrations decline in surface waters at C1 and C4 but increase in bottom waters. At C5 in summer, zinc concentrations in bottom waters increase faster than in surface waters, such that bottom-water zinc concentrations at C5 eventually equal those at C4. Possible bottom-water sources of zinc include desorption from sinking particles and the surface layer of bed sediments, decomposition of settling organic particles, or hydrodynamic redistribution (at C5). For cadmium, concentrations in deeper water (> 20 m) are greater than those in surface water, by about 15 percent, suggesting a role for internal processes (as for zinc). For both metals, there are insufficient data to quantify the relative importance of these processes.

5. **From about 2003 to 2012, total lead concentrations in CDA Lake rose slowly, but they have declined over the past eight to ten years.** Most lead enters the Lake during periods of high discharge, such that the future trajectory of Lake lead concentrations will depend strongly on the extent of scouring of legacy sediment from the bed and banks of the CDA River. Despite thermal stratification in summer, there were only small differences between total lead concentrations in surface and bottom waters at C1 and C4. This suggests minimal flux of lead from internal sources and is consistent with strong lead binding to particulates and the lower bioavailability to phytoplankton of lead compared to zinc and cadmium. At C5, hydrodynamic redistribution of lead from north to south is the most likely process to play a role in lead dynamics, with desorption of lead from suspended particulates settling to the bottom or from bed sediments of limited importance.

6. **Extremely large reservoirs of metal are present in particulate form (in suspension and in the Lake bed sediments) that could replenish dissolved concentrations in the water column whenever conditions are conducive to metal release from the particles.** Desorption, decomposition, and hydrodynamics (especially at C5) all can slow rates of decline in water column metal concentrations in lakes as inputs from the watershed decrease. To date, that does not appear to be the case north of the CDA River confluence (at C1 and C4), but those processes could be responsible for the slower rate of decline of cadmium, lead, and zinc at C5. Detailed understanding and monitoring of the processes that influence interactions between particulate metal and dissolved metal are critical to evaluating future changes in metals concentrations in the Lake as inputs from the watershed continue to decline.

REFERENCES

Balistrieri, L. 1998. Preliminary estimates of benthic fluxes of dissolved metals in Coeur d'Alene Lake, Idaho. USGS Open-File Report 98-793.

Balistrieri, L., S. Box, A. Bookstrom, R. Hooper, and J. Mahoney. 2002. Impacts of Historical Mining in the Coeur d'Alene River Basin. Chapter 6 *In:* Pathways of Metal Transfer from Mineralized Sources to Bioreceptors: A Synthesis of the Mineral Resources Program's Past Environmental Studies in the Western United States and Future Research Directions. USGS Bulletin 2191. L. S. Balistrieri and L. L. Stillings, eds.

Balistrieri, L. S., S. E. Box, and J. W. Tonkin. 2003. Modeling precipitation and sorption of elements during mixing of river water and porewater in the Coeur d'Alene River Basin. *Environ. Sci. Technol.* 37:4694–4701.

Chen, C. Y., R. S. Stemberger, B. Klaue, J. D. Blum, P. C. Pickhardt, and C. L. Folt. 2000. Accumulation of heavy metals in food web components across a gradient of lakes. *Limnol. Oceanogr.* 45(7):1525–1536.

Chess, D., G. Rothrock, G. Pettit, and B. Witherow. 2012. Coeur d'Alene Lake Monitoring Program 2009 Report. Coeur d'Alene Tribe and Idaho Department of Environmental Quality. 114 pp.

CH2M Hill and URS Corp. 2001. Final Ecological Risk Assessment. Coeur d'Alene Basin Remedial Investigation / Feasibility Study. Prepared by URS and CH2M HILL for EPA. Contract No. 86-W-98-228. Work Assignment No. 027-RI-CO-102Q. 1820 pp. May 18, 2001.

Davison, W. 1993. Iron and manganese in lakes. *Earth Science Reviews* 34(2):119–163.

Horowitz, A. J., K. A. Elrick, J. A. Robbins, and R. B. Cook. 1995. Effect of mining and related activities on the sediment trace-element geochemistry of Lake Coeur d'Alene, Idaho, USA Part II: Subsurface Sediments. *Hydrol. Process.* 9:35–54. https://doi.org/10.1002/hyp.3360090105.

Kuwabara, J. S., W. M. Berelson, L. S. Balistrieri, P. F. Woods, B. R. Topping, D. J. Steding, and D. P. Krabbenhoft. 2000. Benthic flux of metals and nutrients into the water column of Lake Coeur d'Alene, Idaho: Report of an August 1999 pilot study. U.S. Geological Survey Water-Resources Investigations Report 2000-4132. https://doi.org/10.3133/wri004132.

Kuwabara, J. S., B. R. Topping, P. F. Woods, J. L. Carter, and S. W. Hager. 2006. Interactive effects of dissolved zinc and orthophosphate on phytoplankton from Coeur d'Alene Lake, Idaho. USGS Scientific Investigations Report 2006-5091. http://pubs.usgs.gov/sir/2006/5091.

Langman, J. B., D. Behrens, and J. G. Moberly. 2020. Seasonal formation and stability of dissolved metal particles in mining-impacted, lacustrine sediments. *Journal of Contaminant Hydrology* 232:103655. https://doi.org/10.1016/j.jconhyd.2020.103655.

Lee, B. G., and N. S. Fisher. 1992. Degradation and elemental release rates from phytoplankton debris and their geochemical implications. *Limnol. Oceanogr.* 37(7):1345–1360.

Luoma, S. N., and P. S. Rainbow. 2008. Metal Contamination in Aquatic Environments: Science and Lateral Management. Cambridge, UK: Cambridge University Press.

Nriagu, J. O., G. Lawson, H. K. T. Wong, and V. Cheam. 1995. Dissolved trace metals in Lakes Superior, Erie, and Ontario. *Environ. Sci. Technol.* 30(1):178–187.

Sigg, L. 1985. Metal transfer mechanisms in lakes: the role of settling particles. Pp. 283–310 *In:* Chemical Processes in Lakes. W. Stumm (ed.). New York: John Wiley & Sons.

Smith, K. S., L. S. Balistrieri, and A. S. Todd. 2015. Using biotic ligand models to predict metal toxicity in mineralized systems. *Applied Geochemistry* 57:55–72.

7

Lake Bed Processes

This chapter examines the (bio)geochemical reactions that control the partitioning of lead (Pb), cadmium (Cd), zinc (Zn), and arsenic (As) between lake sediments and porewater, ultimately leading to the specific conditions of the Coeur d'Alene (CDA) Lake. After an overview of metal(loid) retention processes and the diagenetic (redox processes) reactions, metal deposition and migration within the Lake are examined. Using modeled and measured conditions, the operative processes controlling dissolved porewater metal(loid) contaminant concentrations and possible entry into Lake waters are further assessed. The chapter ends with an assessment of the potential for eutrophication and pH changes to release metals from the Lake sediments into the water column.

INTRODUCTION TO BIOGEOCHEMICAL PROCESSES IN LAKE SEDIMENTS

The heavy metals cadmium, lead, and zinc and the metalloid arsenic all have the commonality of being present within the CDA River and CDA Lake due to their presence in sulfidic minerals of the mined orebodies. They are also all subject to changes in partitioning between the solid and aqueous phases, with biogeochemical transformations occurring within the water column and the sediments. Under sulfidic conditions, they all form mineral sulfides of limited solubility, while under oxic conditions their dissolved concentrations are largely controlled by binding to metal (hydr)oxides, principally those of ferric iron [Fe(III)]. Owing to the dynamic cycling of Fe(III) phases and sulfides with changes in oxygenation and nitrate concentrations, the dissolved concentrations of all four of the metal(loid)s will be subject to changes in the oxygen status of the water column and underlying sediments. Further complicating the behavior of arsenic is its changes in oxidation state with the redox conditions of the water and sediments.

This section provides the principal biogeochemical factors that drive the cycling and dissolved concentrations of arsenic, cadmium, lead, and zinc, and the critical control that Fe(III) phases exert as a function of oxygen level and pH on the dissolved concentrations of all four contaminants in CDA Lake waters and sediments.

Cadmium, Lead, and Zinc

Cadmium, lead, and zinc are all hydrolyzable divalent cations within conditions of the surface environment. They all have a strong affinity for sulfide (they are chalcophiles), forming metal sulfides of limited solubility and making them highly sensitive to dissolved sulfide concentrations. The primary control on dissolved sulfide formation is microbial dissimilatory sulfate reduction (where sulfate is used in place of oxygen for respiration), driven largely by obligate anaerobic bacteria (Bradley et al., 2011). Dissimilatory sulfate-reducing bacteria (SRB) are nearly ubiquitous within lake sediments, and their activity and associated sulfide production is primarily dependent on oxygen levels, with upregulation of sulfate reductase at trace oxygen levels. Secondarily, their activity will then depend largely on temperature and microbially available organic carbon used in respiration (Rabus et al., 2013). SRB production of sulfide may then be scavenged by the differing metals within the sediments, making the total supply, as noted by dissolved concentrations, the important determinant of mineral sulfide formation. If, for example, the porewater sulfide concentration is 10 μM and the pH is 7, total dissolved Pb (largely as the neutral $Pb(HS)_2^0$ aqueous complex) and Zn (largely as Zn^{2+} and ZnS^0) will be approximately 1 pM (0.21 ng Pb/L, 0.065 ng Zn/L), while total dissolved Cd will be slightly higher at 10 pM [> 0.11 ng/L, largely as $Cd(HS)_2^0$] and be controlled through the formation of PbS (galena), ZnS (sphalerite), and CdS (greenockite) solids, respectively.

Under oxygenated conditions, sulfide production is limited and mineral sulfides undergo oxidative dissolution. Ferric oxides, oxyhydroxides, and oxides (hereafter collectively referred to as iron oxides) are instead formed and often become the dominant regulator of dissolved metal concentrations (Stumm et al., 1992). Adsorption reactions on iron oxides, however, are pH-dependent and vary for the three metal contaminants here. Typical of cationic metals, the extent of adsorption increases with increasing pH and undergoes a rapid shift in partitioning across a small pH change (Figure 7-1). Within 1 to 2 units of pH change, the metal contaminants will go from nearly entirely aqueous (low pH) to entirely adsorbed (high pH); the pH at which the transition occurs is termed the pH edge. The pH edge for lead adsorbing on iron oxides occurs between pH values of 4.5 and 5.5. Thus, at most pH values encountered in the environment, and clearly those within the Coeur d'Alene system, lead would be partitioned on iron oxide surfaces provided that (1) the sediments are oxidized, (2) sufficient iron is present to generate iron oxides with the capacity to adsorb the quantities of lead, and (3) competing ions for the iron oxide surface do not limit lead adsorption. Given the high affinity of lead for iron oxides (Dewey et al., 2021), competing ions typically do not limit uptake.

The adsorption edge for zinc on iron oxides is substantially higher than for lead, occurring in the range of 6.0–7.0. Additionally, the affinity of zinc for the iron oxides surfaces is less than for lead, making it subject to competitive displacement. Nevertheless, iron oxides often are regulators of dissolved zinc concentrations.

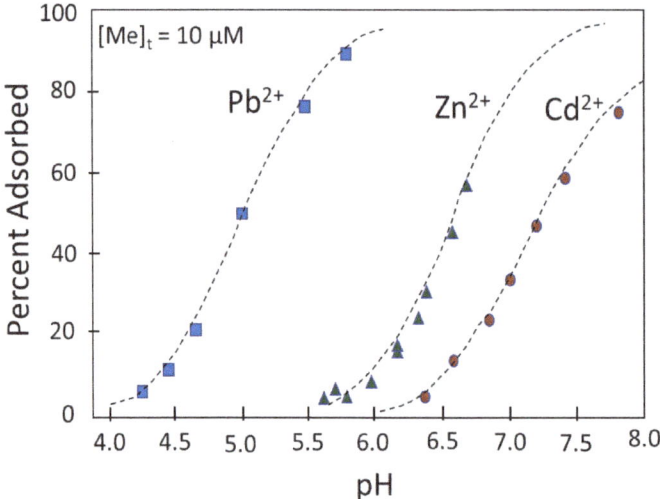

FIGURE 7-1 Example adsorption edges for lead, zinc, and cadmium on hematite ($Fe_2O_3 \cdot H_2O$) showing the fraction of metal retained as a function of pH. SOURCE: Benjamin and Leckie (1981). Reprinted from *J. Colloid Interface Sci.* 79. Benjamin, M. M., and J. O. Leckie. Multiple-site adsorption of Cd, Cu, Zn, and Pb on amorphous iron oxyhydroxide. Pp. 209–221. 1981, with permission from Elsevier.

Cadmium has the highest adsorption edge of the three heavy metal contaminants, occurring in the pH range of 6.5–7.5; it also has the lowest affinity for iron oxides and consequentially is not controlled by surface adsorption at the low concentrations noted for zinc and, most appreciably, lead. Critically, for systems such as CDA Lake that have pH values near neutral, slight changes in pH may have large impacts on the dissolved concentrations of zinc and cadmium under conditions where adsorption to iron oxide minerals is important.

Cadmium, lead, and zinc may also form metal carbonates, and although more soluble than their sulfide mineral cousins, they may regulate the dissolved concentrations of these metals. Although metal coprecipitates or hydroxo carbonates, such as hydrocerussite, may lead to lower concentrations, considering the pure metal carbonate phases provides a useful reference point for the comparison in solubility of differing phases. At pH 7 and total dissolved carbonate concentrations of 2 mM, total dissolved lead and zinc would be near 1 μM (202 ppb and 65 ppb, respectively) while cadmium would be at 0.1 μM (66 ppb) if in equilibrium with $PbCO_3$ (cerussite), $ZnCO_3$ (smithsonite), and $CdCO_3$ (otavite), respectively. The metal carbonates should serve to regulate the upper limits of dissolved lead, zinc, and cadmium concentrations. Metal carbonates have, in fact, been noted to control the dissolved concentrations of contaminants in mining-influenced sites. For example, cerussite ($PbCO_{3, crystalline}$) was noted as a major secondary weathering phase in sediments influenced by a historic mining district of Wanlockhead, Scotland (Hillier et al., 2001).

Arsenic

Two oxidation states of arsenic, arsenate [As(V)] and arsenite [As(III)], predominate in surface and near-surface environments. At circumneutral pH, the partially protonated forms of arsenate, $H_2AsO_4^-$ and $HAsO_4^{2-}$, and fully protonated form of arsenite, $H_3AsO_3^0$, dominate. Microbial activity may methylate As(V) or As(III), forming, for example, dimethylarsenic acid and monomethylarsonous acid (Cullen and Reimer, 1989). Thio- and carbonato-complexes of arsenic also exist within anaerobic systems; thiolated forms of arsenic may, in fact, represent an important reactive component within sulfidic environments (Wilkin et al., 2003; Wang et al., 2020).

Partitioning of arsenic onto soil solids is foremost dependent on its oxidation state. In general, As(V) binds extensively and strongly to most mineral constituents of soils and sediments, often partitioning most appreciably on iron oxides, while As(III) retention is more convoluted and dependent on specific soil chemical conditions. As a consequence, under oxidizing conditions where arsenate predominates, dissolved arsenic concentrations are limited except at very high pH (> 8.5). Phosphate, having analogous binding properties to arsenate, also is strongly retained by mineral surfaces except at very high pH. The surface complexes of arsenite, although extensive, are far more labile than for its oxidized counterpart, leading to higher dissolved arsenite concentrations and a greater propensity for migration (Kocar et al., 2006; Tufano et al., 2008).

In surface and subsurface environments, changes in water chemistry often result in release of arsenic from solid phases through various desorption pathways. Processes leading to arsenic desorption can broadly be grouped into four categories: (1) ion displacement, (2) alkalinity (pH values > 8.5), (3) reductive dissolution of As(V)-bearing Fe(III) oxides, and (4) oxidative dissolution of arsenic sulfide phase. Although the latter process will increase dissolved concentrations of arsenic, the typical concomitant generation of Fe(III) oxides re-partitions arsenic to the solid phase as As(V) adsorbed on Fe oxides. Furthermore, the pH values of the Coeur d'Alene system negate the likelihood of alkalinity-promoted desorption. Competitive ion displacement can represent an important means by which arsenic is released to the aqueous phase, but it is most appreciable in regions where extensive fertilizer or pesticide runoff occur (Jain and Loeppert, 2000; Peryea and Kammerack, 1997), and hence it is likely to be limited in comparison to that imposed by the onset of anaerobic conditions. Thus, the process of greatest concern is the reductive dissolution of As(V)-bearing Fe(III) oxides.

Within most soils and sediments, total arsenic levels correlate with iron content rather than aluminum or clay content (Smedley and Kinniburgh, 2002), and thus reductive dissolution/ transformation of Fe(III) phases should have a major impact on arsenic. As a consequence, the greatest likelihood for arsenic release in soils and sediments typically occurs upon a transition from oxidizing (oxygenated/aerobic) to reducing (deoxygenated/anaerobic) conditions. Arsenic may be displaced either through reduction of arsenate to arsenite or through reductive dissolution of Fe(III) oxides. Microbially driven oxidation of organic carbon coupled to the dissimilatory reductive dissolution of As-bearing Fe oxides causes the transfer of arsenic from sediment solids to groundwater (Islam et al., 2004; Fendorf et al., 2010), as shown below:

$$CH_2O + 4FeOOH\text{-}(H_2AsO_4)_x + (7+3x)H+ \Leftrightarrow 4Fe^{2+} + HCO_3^- + (6+x)H_2O + xH_3AsO_3$$

where CH_2O generically represents organic carbon and may include other fermentation products such as $H_{2(aq)}$, As (as arsenate) is bound to sedimentary iron oxide (goethite as written in reaction 1), and x is the stoichiometric coefficient of arsenic content associated with the iron oxides. Dissimilatory As(V)/Fe(III) reduction requires (1) anaerobic conditions with low sulfate supply, (2) reactive As-Fe complexes, and (3) microbially available organic carbon.

Microbial Transformation of Iron and Arsenic

Under oxidizing conditions created largely by oxygen or nitrate, iron generally exists as Fe(III), forming Fe(III) oxides sparingly (Kappler et al., 2021). Cycling of iron oxides can exert a dominant control on nutrient and contaminant mobility and bioavailability due to their large surface areas and high reactivity (Cornell and Schwertmann, 2003). For example, within the large watersheds of the Ganges-Brahmaputra, Mekong, and Red Rivers in southeast Asia, iron oxides are the dominant hosts of arsenic in oxygenated surface environments (Fendorf et al., 2010). Accordingly, dissolution and transformation of iron oxides can have major implications for the fate and transport of metal(loid) contaminants (see, e.g., DeLemos et al., 2006); concomitant with the dissolution (inclusive of transformation) of iron oxides, adsorbed or coprecipitated contaminants may be mobilized.

Under oxygen-limiting conditions, microorganisms have the capacity to use alternate electron acceptors (to oxygen) in respiration, such as Fe(III). The general sequence of electron acceptors is O_2 followed by nitrate, manganese (Mn) dioxides, iron oxides, sulfate, and methanogenesis (Figure 7-2)—although the potential for iron oxides and sulfate often overlap, with As(V) reduction occurring concomitantly with Fe(III) reduction (Postma and Jakobsen, 1996; Kocar et al., 2008).

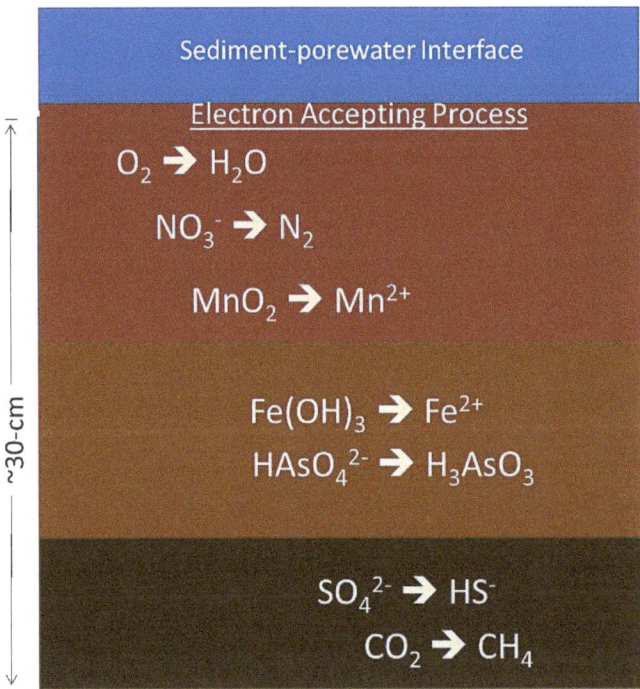

FIGURE 7-2 Idealized succession of terminal electron acceptors used in microbial respiration of organic carbon.

A host of bacteria and archaea can respire by coupling organic carbon or a fermentation product with Fe(III) oxides (processes known as dissimilatory iron reduction), reducing Fe(III) to Fe(II) (Gorby and Lovley, 1991; Nealson and Saffarini, 1994; Kappler et al., 2021) and releasing Fe^{2+} and adsorbed phases to porewater. Respiratory reduction of iron in sediments generally occurs in zones where O_2, NO_3^-, and Mn(IV) are diminished (Lovley, 1997), as depicted in Figure 7-2.

Dissimilatory iron-reducing bacteria upregulate iron reductases at low concentrations of O_2 and depletion of nitrate (Kappler et al., 2021). Given atmospheric supply of O_2, and nitrate from the water column, sediments or bottom waters will have a depth-stratified succession in iron oxidation and reduction. Surface layers supplied with O_2 and nitrate will lead to the persistence or generation of Fe(III) oxides. Microbial utilization of O_2, followed by nitrate and then Mn(III/IV) oxides, will lead to anaerobic conditions and concomitant Fe(III) and As(V) reduction (Figure 7-2). The transition depth at which iron goes from being oxidized (by O_2 or nitrate) to being reduced (by iron-reducing bacteria) is largely dependent on the supply and demand of O_2 (Figure 7-2). The supply side is controlled by the diffusion through the water column while demand is dependent on microbially available organic carbon, temperature, and limiting nutrients (if any are limiting). Increased biological demand in the water column or sediments will lead to the drawdown of oxygen higher in the profile, moving the iron oxide reduction boundary higher in the sediment profile. If oxygen demand is limited by nutrient or organic carbon supply, then the reduction boundary will move lower in the profile and a greater zone of iron oxides will be generated. Provided sufficient iron oxide content and pH values that promote metal(loid) retention, contaminants such as arsenic, lead, and zinc, and potentially cadmium, will be retained by the iron oxide via adsorption. By contrast, if bottom waters become anoxic, the iron-reduction zone can move close to the water–sediment interface. In such cases, Fe^{2+} along with contaminants on the surface of the Fe oxides may then freely diffuse into the water column.

Possibly limiting metal(loid) release under anaerobic conditions is the generation of sufficient sulfide (from dissimilatory sulfate reduction) to result in metal(loid) sulfide formation, as noted above. However, many freshwater systems are sulfur-limited (compared to metal demand), which can limit secondary sequestration of contaminants in sulfidic phases (e.g., O'Day et al., 2004); this appears to be the case for CDA Lake (Toevs et al., 2006).

METAL(LOID) DYNAMICS WITHIN CDA LAKE SEDIMENTS

This section examines metal deposition and migration within the CDA Lake sediments using modeled and measured conditions, in order to further assess the operative processes controlling dissolved porewater metal(loid) contaminant concentrations and possible entry into Lake waters.

Information from Sediment Core Studies

Metal Deposition

Effective impoundment of milling wastes was largely completed by 1968 (Morra et al., 2015), but the historic deposition along the river corridor remains a source of metal deposition to CDA Lake. The stockpile of contaminated sediments upstream from the Lake remains 100 times more enriched than background conditions 52 years after the first effective constraints on primary waste inputs were imposed (Sprenke et al., 2000; Langman et al., 2020). Bookstrom et al. (2013) estimated that in 1998 about one-third of the lead load entering CDA Lake in any year was derived from the upper basin, with a small fraction derived from erosion of mainstem river banks (lower basin floodplains); the large majority entering the Lake was probably from the mainstem river bed. Bookstrom et al. (2013) note that as the metal-enriched sediments in the riverbed progressively undergo remobilization and dilution during transport, they will re-deposit to form and re-form persistent "hot spots." The hot spots will continue to function as secondary sources of lead and other metals for the Lake.

In depositional environments of the Lake, contaminated sediments are slowly buried by successively younger layers of progressively diluted and less metal-enriched sediments. Indeed, sediments deposited on lower basin floodplains averaged 5,400 ± ~ 2,000 µg/g lead between 1903 and 1980, 4,000 ± 1,000 µg/g lead between 1968 and 1980, and 3,100 ± 1,950 µg/g lead between 1980 and 1998 (Figures 21, 23 in Bookstrom et al., 2013).

Furthermore, on the basis of two cores collected in 2010, Morra et al. (2015) illustrate that there was a distinct drop in sedimentation rates in the Lake after impoundment was completed in 1968. Those rates remain low, at 0.1 g cm^{-2} yr^{-1} compared to the maximum of 0.8 g cm^{-2} yr^{-1}. Using distinct markers within the sediment profile, such as the eruption of Mount St. Helens in 1980, provides easy demarcation of sedimentation rates and metal loading over different time intervals (see Bookstrom et al., 2013; Morra et al., 2015). Zinc and cadmium have declined in concentration in lake sediment cores over the past few decades. Cadmium concentrations were 80 µg/g prior to 1968 and dropped to an average of 25.6 µg/g in the 1980–2010 interval; zinc concentrations similarly declined, to an average of 3,350 µg/g in the 1980–2010 interval. Concentrations of arsenic were at a maximum in the surface sediments, probably as a result of diagenesis. Lead concentrations, however, were not different in the 1980–2010 segment compared to the 1980–1998 segment reported by Bookstrom et al. (2013; ~ 3,000 µg/g dw).

Estimates using different methods show that CDA Lake continues to be a net sink for metals coming in from the watershed. Using cores from 2010, Morra et al. (2015) estimated that annually 91 percent of incoming lead was retained, and 56–58 percent of the zinc and 30 percent of dissolved cadmium that enter the Lake are retained. Based upon fluxes as calculated in Chapter 3, trapping of zinc ranged from ~ 45 to 65 percent between 2010 and 2015; furthermore, although lead retention in the Lake is highly variable from year to year, the ratio of lead export to import is decreasing. Trapping has increased since the 1990s from as low as 80 percent in 1991 to 93–98 percent between 2015 and 2020 (consistent with Morra et al., 2015). This is a positive development for lead not being exported downstream, but its meaning for future trends in dissolved lead concentrations in the Lake is unclear, depending upon the degree of internal flux from sediments to the water column. Box 7-1 further explores the trend of increasing CDA Lake lead trapping efficiency.

BOX 7-1
Trends in Lead Trapping Efficiency of CDA Lake

The Weighted Regressions on Time, Discharge, and Season (WRTDS) model described in Chapter 3 for total lead for the CDA River at Harrison and for the Spokane River below the Lake outlet was used to estimate how much incoming lead is trapped by the Lake. Here, instead of using the flow-normalized values (as was done in Chapter 3), the committee used its best estimate of the true flux values at a daily time step throughout the entire 30-year record from 1991 to the present. The results expressed as cumulative loads are shown in Figure 7-1-1.

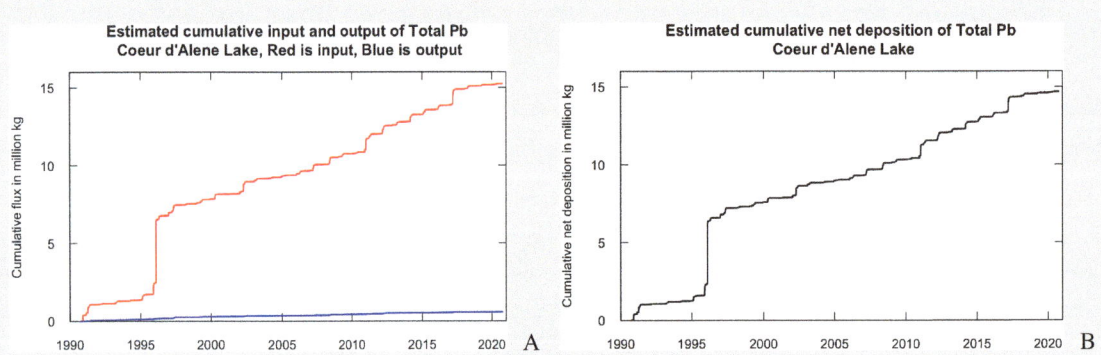

FIGURE 7-1-1 Estimated cumulative deposition of total lead in CDA Lake. (A) shows input (lead flux at Harrison) and output (lead flux at the Spokane River outlet), while (B) shows the net deposition.

One thing that stands out in Figure 7-1-1 is just how large an event the 1996 flood was in terms of transporting lead. It is also interesting to note that although 1996 was the year of the largest annual water discharge (119 m^3/s average flow), it was really just one of four very high discharge years in the past 30 years (the others were 2011 at 117 m^3/s, 1997 at 107 m^3/s, and 2017 at 104 m^3/s). As shown in Table 7-1-1, the input of lead for each of those other years was vastly smaller than the input in 1996. For 1996 the estimated input was 5.04 × 10^6 kg, and the other three (in order of their discharge magnitude) were 1.17 × 10^6 kg, 0.68 × 10^6 kg, and 1.03 × 10^6 kg. So, in the years after 1996, when these very high flow years took place, the lead flux into the Lake was much lower. One would need to look more closely at the hydrographs of each of these three years to understand why they differ so much in terms of lead flux.

TABLE 7-1-1 Ratios of Annual Fluxes in and out of CDA Lake

Water Year Mean	Q	Input	Output	Ratio
1991	96	1.09	0.070	0.06
1992	42	0.06	0.012	0.22
1993	64	0.16	0.032	0.20
1994	38	0.06	0.010	0.18
1995	74	0.37	0.024	0.06
1996	119	5.04	0.049	0.01
1997	107	0.68	0.063	0.09
1998	57	0.09	0.013	0.14
1999	86	0.29	0.020	0.07
2000	77	0.33	0.018	0.05
2001	30	0.04	0.004	0.11
2002	94	0.78	0.027	0.04
2003	52	0.20	0.007	0.04
2004	55	0.08	0.006	0.08
2005	60	0.14	0.005	0.04
2006	81	0.27	0.012	0.04
2007	73	0.40	0.011	0.03
2008	88	0.46	0.041	0.09
2009	71	0.23	0.014	0.06
2010	55	0.10	0.007	0.08
2011	117	1.17	0.032	0.03
2012	93	0.53	0.034	0.06
2013	75	0.24	0.007	0.03
2014	84	0.47	0.012	0.03
2015	60	0.31	0.005	0.02
2016	68	0.27	0.006	0.02
2017	104	1.03	0.015	0.01
2018	86	0.21	0.010	0.05
2019	49	0.07	0.004	0.05
2020	63	0.08	0.006	0.07

Figure 7-1-2 is based on the values in Table 7-1-1. It splits the values into three decades to see if the relationship between input and output is changing over time. Also shown are loess smooth curves for each of the decades, using the same color scheme. The extreme high value of input (around 5 billion kg in 1996) is not included in the loess smooth. That year was so unusual that including it in the smoothed curve would be distracting from the main message, which is that for any given amount of input flux, the output flux diminishes decade by decade. That means that the Lake is more efficient at trapping lead from one decade to the next. One possible explanation could be that the particles carrying lead are getting coarser over time. Another possibility is that more lead entered the lake in 1996 because the CDA River cut into lead-rich sediments that are no longer there.

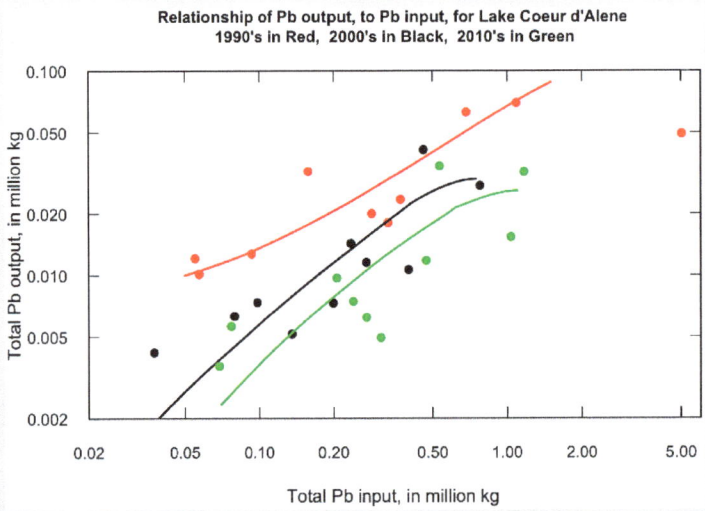

FIGURE 7-1-2 Total lead output versus input for CDA Lake. Red is the decade 1991–2000, black is the decade 2001–2010, and green is the decade 2011–2020.

The Bookstrom et al. (2013) and Morra et al. (2015) studies suggest some progress in declining metal inputs to the sediments of CDA Lake. However, as noted in Morra et al. (2015), the surface sediment concentrations still exceed EPA's 50th percentile probability of adverse biological effect by approximately tenfold or more and exceed background concentrations by 10–100 times (Table 7-1). Bookstrom et al. (2013) noted that natural processes that would be expected to reduce inputs to CDA Lake will take centuries, "depending on the size and metal contents of secondary sources, and the size and discharge of the drainage basin they occupy." Coulthard and Macklin (2003) modeled long-term contamination in river systems from historical metal mining showing that more than 70 percent of the deposited contaminants remain within river systems for more than 200 years after mine closure. Natural recovery of the Rio Tinto in Spain and its floodplain, a river originally mined by the Romans (during the Copper Age), took 450 years.

Another example is provided by Diamond (1995), who used a mass-loading model to illustrate one scenario for the fate of water quality in Lake Moira, Ontario, a lake subject to high mass inputs of metals from mining wastes (Box 7-2). In this case, metal trapping, mixing and redistribution within the lake greatly delayed responses of the lake to changes in arsenic inputs once the legacy of contamination was established in lake sediments. CDA Lake is a deeper system, with longer residence times, than Lake Moira. The fate of arsenic is to some degree specific to that metalloid as well. Nevertheless, the example presented by Diamond (1995)

TABLE 7-1 Concentrations of Selected Metal(loid)s in Cores from Peaceful Point Showing Concentrations within Surface Sediments

Metal(loid)	Natural Background[a] µg/g	Mean Surface Concentration (2010)[b] µg/g	Potential 50% Toxicity[b] µg/g
As	5	465	32.6
Cd	3	25.3	2.5
Pb	24	3,000–4,000	161
Zn	110	3,326	384

[a] SOURCE: Horowitz et al. (1993).
[b] SOURCE: Morra et al. (2015).

clearly demonstrates that a lake's response to changes in inputs can lag by a considerable period (decades). Moreover, CDA Lake remains in a phase of mass metal influx exceeding mass metal export (see Box 7-1). Thus, the Lake remains a net metal sink and is accumulating lead, zinc, cadmium, and arsenic. Morra et al. (2015) concluded that "without a concerted effort on remediation to prevent silt-sized (Toevs et al., 2008), lead-laden particles from entering the lake, lead concentrations in the sediment will remain at toxic levels with substantial potential to cause negative environmental impacts."

BOX 7-2
Water Quality in Lake Moira, Ontario

Diamond (1995) used a mass loadings model to quantify mass fluxes during and after Lake Moira (Ontario, Canada) was subjected to arsenic releases from a mine established in the 1860s. Diamond used sediment cores to model conditions during mining and then compared conditions after mining ceased. During the mining era, 37 tons per year of arsenic entered the lake from the river draining from the mine-affected watershed. On average, 32 tons per year (86 percent) of arsenic inputs were exported from the lake; only an average of 14 percent of the arsenic was retained in the lake per year. Although most of the arsenic in the lake occurred in dissolved form, the proportion of arsenic that was not exported was associated with particulate material that settled to the lake sediments. Even though only a small percentage of the incoming load was trapped by the lake, the mass of arsenic in the sediments progressively increased during the mining era. Sediment cores showed that arsenic concentrations in the sediments increased from an estimated 0.2 µg/g arsenic in pre-mining sediments (pre-1860) to 360 µg/g arsenic in the 1970s.

After mining ceased, arsenic loadings from the river were reduced by remediation activities. However, high concentrations of arsenic in the lake water were sustained by inputs from contaminated sediments within the lake. The lake was relatively shallow, so more arsenic was released from re-suspended sediments than from bed sediments. Over the decades a slow burial of sediments occurred in the lake as cleaner incoming sediments from the remediated watershed deposited on top of the historically contaminated sediments. However, diagenetic flux of arsenic toward the surface of the sediment column slowed the rate of decline in arsenic concentrations at the surface. By the 1990s, export of arsenic from the lake was up to 1.63 times higher than imports. Diamond concluded that "sediment burial and reaction will permanently remove arsenic from the lake, albeit at a slow rate, whereas sediment–water exchange will maintain the presence of in-place pollution in the lake." But the response of the lake to the reduction in inputs was greatly slowed by the large mass of highly concentrated arsenic retained in the sediments. Diamond (1995) concluded that elevated concentrations of arsenic in the water column, remobilization and downstream export of arsenic would occur in this lake as long as the in-place arsenic-enriched sediments were exposed to the water column, which is through the foreseeable future.

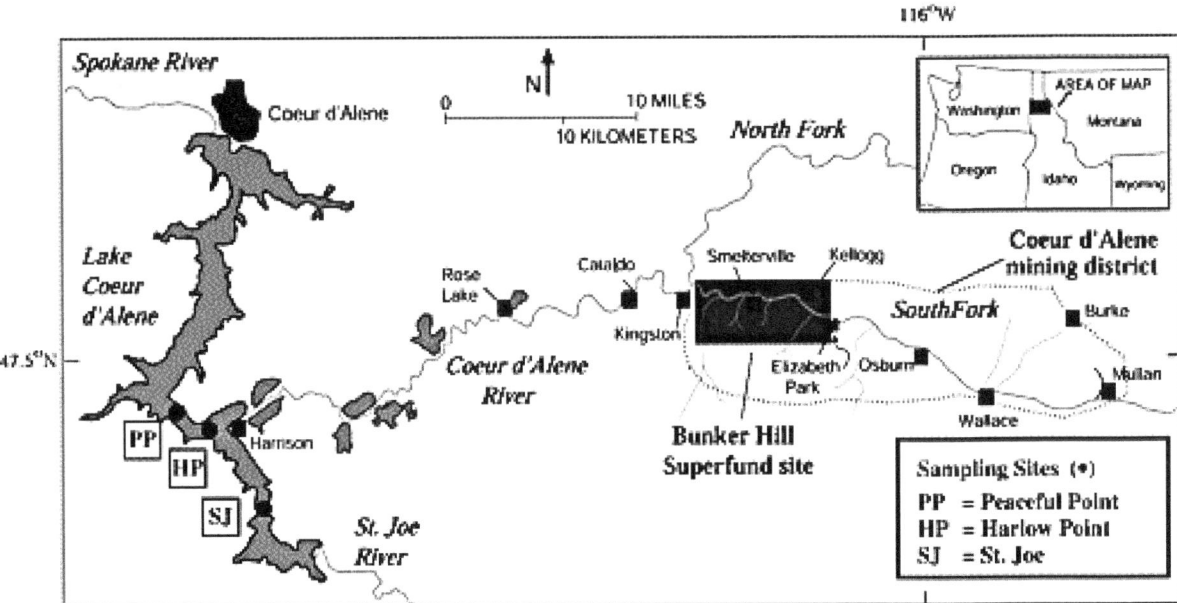

FIGURE 7-3 Regional and local map of CDA Lake including sampling locations of Morra et al. (2015) and Toevs et al. (2006). SOURCE: Toevs et al. (2006). Reprinted (adapted) with permission from Toevs, G. R., M. J. Morra, M. L. Polizzotto, D. G. Strawn, B. C. Bostick, and S. Fendorf. 2006. Metal(loid) diagenesis in mine-impacted sediments of Lake Coeur d'Alene, Idaho. *Environ. Sci. Technol.* 40(8):2537–2543. https://doi.org/10.1021/es051781c. Copyright 2006 American Chemical Society.

It is worth noting that the core dating from Morra et al. (2015) and Horowitz et al. (1993) was performed from localized, highly contaminated areas near the delta region of CDA River (see Figure 7-3 for a map). Areas that see more frequent anoxia relative to Peaceful Point could have a more depleted Fe(III) oxide layer or greater concentration and migration of redox active elements within the upper 10 cm of sediments. Likewise, areas with differing sedimentation influences (stream inputs or bays) may show decreasing, rather than constant, deposition of lead since 1968.

Metal Associations within the Lake Sediments

The metal(loid)s deposited in the lake sediments may associate with different solids or partition into the porewater, with the possibility of dynamic cycling between different solids and the aqueous phase. A number of studies have examined the partitioning of the metals and their solid phase associations in the sediments of CDA Lake (see Balistrieri and Blank, 2008; Bostick et al., 2001; Harrington et al., 1998a,b; Haus et al., 2008; Horowitz et al., 1993; Kuwabara et al., 2003; Moberly et al., 2009, 2016). Horowitz et al. (1995) found lead, cadmium, zinc, and arsenic in extractions that target iron oxides, with only minor amounts of these metals in extractions that seek to target sulfide minerals (Horowitz et al., 1993, 1995). Additionally, studies of mixing experiments of acidic and metal-contaminated water with uncontaminated surface waters illustrate metal binding to iron oxides (Balistrieri et al., 1999, 2003; Paulson and Balistrieri, 1999; Tonkin et al., 2002). By contrast, Harrington et al. (1998a) found that heavy metals in the delta region of CDA Lake were extracted dominantly using a procedure targeting sulfide minerals (reaction with $KClO_3$ as an oxidant followed by reaction with concentrated HCl and then 4 M HNO_3 while boiling). Examining the molar masses of the potential sulfur-binding metals (Pb, Cd, Zn, and Fe) and arsenic relative to sulfur helps illustrate the likely dominant phases controlling the contaminants. Table 7-2 provides the elemental abundance of the heavy metals along with iron, arsenic, and sulfur in sediment taken from two locations within the Lake in 2002. These are Peaceful Point and Harlow Point, less than 1 km from the mouth of the CDA River, as shown in Figure 7-3.

TABLE 7-2 Elemental Concentrations in Near-Surface Sediments of CDA Lake

Depth (cm)	Element Concentration (mmol/Kg)					
	As	Cd	Pb	Zn	Fe	S
Peaceful Point						
0–3	4.01	0.227	16.53	48.05	1259	46.51
12–18	2.81	0.219	16.12	44.08	1717	110.5
24–30	1.83	0.2	28.34	56.74	1618	121.8
Harlow Point						
0–3	1.9	0.206	20.96	51.71	1377	95.12
12–18	1.89	0.247	22.35	56.05	1368	135.3
24–30	3.04	0.293	18.76	49.88	1529	140.9

SOURCE: Toevs et al. (2006). Reprinted (adapted) with permission from Toevs, G. R., M. J. Morra, M. L. Polizzotto, D. G. Strawn, B. C. Bostick, and S. Fendorf. 2006. Metal(loid) diagenesis in mine-impacted sediments of Lake Coeur d'Alene, Idaho. *Environ. Sci. Technol.* 40(8):2537–2543. https://doi.org/10.1021/es051781c. Copyright 2006 American Chemical Society.

Summing the molar abundances of arsenic, cadmium, lead, and zinc (and neglecting iron) yields values ranging from 60 to 80 mmol/kg across depths for the two sites, which are slightly less than the range of total sulfur that varies from ~ 100 to 120 mmol/kg. Thus, there *is* sufficient total sulfur such that if it is all in sulfidic form, monosulfide phases such as CdS, PbS, and ZnS could form and control dissolved metal concentrations. However, when considering the molar abundance of iron in the sediments, the projections change substantially. If one assumes all of the sulfur were present in sulfidic form and FeS_2 is considered as a dominant sulfur product, there is only sufficient sulfur to react with ~ 2–4.6 percent of the total iron. Furthermore, Toevs et al. (2006) found that the maximum amount of total sulfur in sulfidic form (within the upper 36 cm) was just less than 50 percent, indicating that there is sufficient sulfide to react with only 1–2.3 percent of the total iron. Consistent with this estimate based on mass ratios, Toevs et al. (2006) found that FeS_2 was the main sulfide mineral in contaminated areas, increasing with sediment depth to a maximum of ~ 50 percent of total sulfur at ~ 36 cm, and that only 2–2.5 percent of the total iron is associated with pyritic minerals. Winowiecki (2002) reports similar trends in pyrite with depth, and in iron association with pyrite, using acid volatile sulfide and sulfur X-ray absorption near-edge structure spectroscopy for lake sediments from contaminated and uncontaminated sites sampled between 2000 and 2002. Thus, it appears that sulfidic phases are unlikely to be major hosts of heavy metal contaminants within CDA Lake, although they may vary and have locally important impacts.

Rather than a dominance of metal sulfides, Toevs et al. (2006) show that Fe(III) oxides and Fe(II) as siderite ($FeCO_3$) are present down to at least 36 cm, with the proportion of Fe(III) (hydr)oxides, relative to siderite, decreasing with depth. Although the redox gradient produces dominant Fe(II) forms, Fe(III) phases persist beyond the oxic upper layer (Toevs et al., 2006). The reactive transport modeling of Şengör et al. (2007) suggests the dominant role of Fe(III) oxides in metal and arsenic partitioning with the solids of the upper layer (top 30 cm) of the sediments. Additionally, in experimental work conducted by Balistrieri et al. (2003), correlation between metal concentrations in sediments and those predicted by the formation of ferrihydrite (nominally ferric hydroxide) suggest that arsenic is predominantly associated with iron oxides, with lead and cadmium also associating largely with iron oxides at low to moderate loadings, while other phases (such as metal sulfides or carbonates) contribute at the highest metal loadings.

Modeling and Measurement of the Processes Controlling Metal Binding

The data from sediment cores suggest that Fe(III) oxides have a dominant role in controlling the dissolved concentrations of the heavy metal contaminants. The iron mineral ferrihydrite is likely the predominant phase within the surface and thus, based on ion affinities, is a prime determinant of dissolved metal concentrations.

TABLE 7-3 Surface Complexation Reactions and Optimized Binding Constants

	Reaction			Log K_{int}
$\equiv$FeOH + H$^+$	$\Leftrightarrow$	$\equiv$FeOH$_2^+$		7.0 (K$^+$)
$\equiv$FeOH	$\Leftrightarrow$	$\equiv$FeO$^-$ + H$^+$		−9.2 (K$^-$)
2$\equiv$Fe$_s$OH + Cd^{2+}	$\Leftrightarrow$	$\equiv$Fe$_s$OH$\equiv$Fe$_s$OCd$^+$ + H$^+$		0.14 (.040)
2$\equiv$Fe$_w$OH + Cd^{2+}	$\Leftrightarrow$	$\equiv$Fe$_w$OH$\equiv$Fe$_w$OCd$^+$ + H$^+$		−3.55 (0.026)
$\equiv$Fe$_w$OH + Cd^{2+}	$\Leftrightarrow$	$\equiv$Fe$_w$OCd$^+$ + H$^+$		−5.47 (0.15)
$\equiv$Fe$_w$OH + Cd^{2+} + H$_2$O	$\Leftrightarrow$	$\equiv$Fe(OH)$_3$(s) + $=$Cd$_w$OH$_2^+$ + H$^+$		−2.97
$\equiv$Fe$_s$OH + Pb^{2+}	$\Leftrightarrow$	$\equiv$Fe$_s$OPb$^+$ + H$^+$		4.13
2$\equiv$Fe$_w$OH + Pb^{2+}	$\Leftrightarrow$	$\equiv$Fe$_w$OH$\equiv$Fe$_w$OPb$^+$ + H$^+$		0.59 (.0375)
2$\equiv$Fe$_s$OH + Zn^{2+}	$\Leftrightarrow$	$\equiv$Fe$_s$OH$\equiv$Fe$_s$OZn$^+$ + H$^+$		1.04 (.045)
2$\equiv$Fe$_w$OH + Zn^{2+}	$\Leftrightarrow$	$\equiv$Fe$_w$OH$\equiv$Fe$_w$OZn$^+$ + H$^+$		−2.33 (0.051)
2$\equiv$Fe$_w$OH + 2Zn^{2+} + H$_2$O	$\Leftrightarrow$	$\equiv$Fe$_w$OH$\equiv$Fe$_w$OZnOZn$^+$ + 3H$^+$		−14.3 (0.147)
Specific surface area (SSA) = 650 m^2/g				
Site Density				
High-affinity Sites (mol sites/mol Fe)				0.01
Low-affinity Sites (mol sites/mol Fe)				0.68

SOURCE: Nomaan et al. (2021). Reprinted from *Chemical Geology* 573, Nomaan, S. M., S. N. Stokes, J. Han, and L. E. Katz. Application of spectroscopic evidence to diffuse layer model (DLM) parameter estimation for cation adsorption onto ferrihydrite in single-and bi-solute systems. https://doi.org/10.1016/j.chemgeo.2021.120199. Copyright 2021, with permission from Elsevier.

The binding and related dissolved concentrations of metal(loid) contaminants resulting from reaction with iron oxides was examined by the committee using the surface complexation model (SCM) described in Nomaan et al. (2021) with the physical-chemical characteristics of CDA Lake sediments described in Şengör et al. (2007). The surface reactions and binding constants for cadmium, lead, and zinc are provided in Table 7-3 and are taken from Nomaan et al. (2021); the reaction and binding constants for arsenic are from Dzombak and Morel (1990). The model includes the simultaneous reaction of all three metals and arsenic with the iron oxide surface. Although ignoring the full suite of reactions that control solute concentrations, the SCM illustrates the extent to which iron oxides may control heavy metal and arsenic concentrations in porewater (Figure 7-4). (Note that this exercise could also have been done with other models that capture transport processes, such as the USGS software PHREEQC.)

There are two striking outcomes from the SCM (Figure 7-4). First is the role pH plays in the binding of cadmium, lead, and zinc to ferrihydrite. The near-neutral pH of the Lake sediments is a critical threshold, such that even slight changes in pH will have pronounced impact on dissolved concentrations for cadmium and zinc. Second, arsenic, when residing as As(V), is completely bound to the ferrihydrite, with nondetectable dissolved concentrations. While highly sensitive to pH change, at pH 7.2, dissolved lead (< 1 µg/L) and cadmium (~ 1 µg/L) concentrations are also predicted to be near the limits of detection. By contrast, even at pH 7.2, a zinc concentration of 1,100 µg/L (16.8 µM) is projected, consistent with measured porewater concentrations (Harrington et al., 1998a; Balistrieri, 1998).

The projected regulation of dissolved metal concentrations by Fe(III) [and Mn(III/IV)] oxides is consistent with measured porewater concentration (Balistrieri, 1998; Winowiecki, 2002; Toevs et al., 2006). Moreover, Balistrieri (1998) allows estimates of benthic fluxes from the sediment into the Lake bottom water. Albeit with sparse temporal and spatial data, Balistrieri (1998) projected that sediments would provide a positive flux only for zinc. However, due to the dynamic biogeochemical cycling of the Lake sediments, seasonal alteration of environmental conditions and substantial presence of various metals and inorganic and organic sulfur and carbon create a dynamic source of potential metals for the overlying Lake water, in both dissolved and colloidal forms.

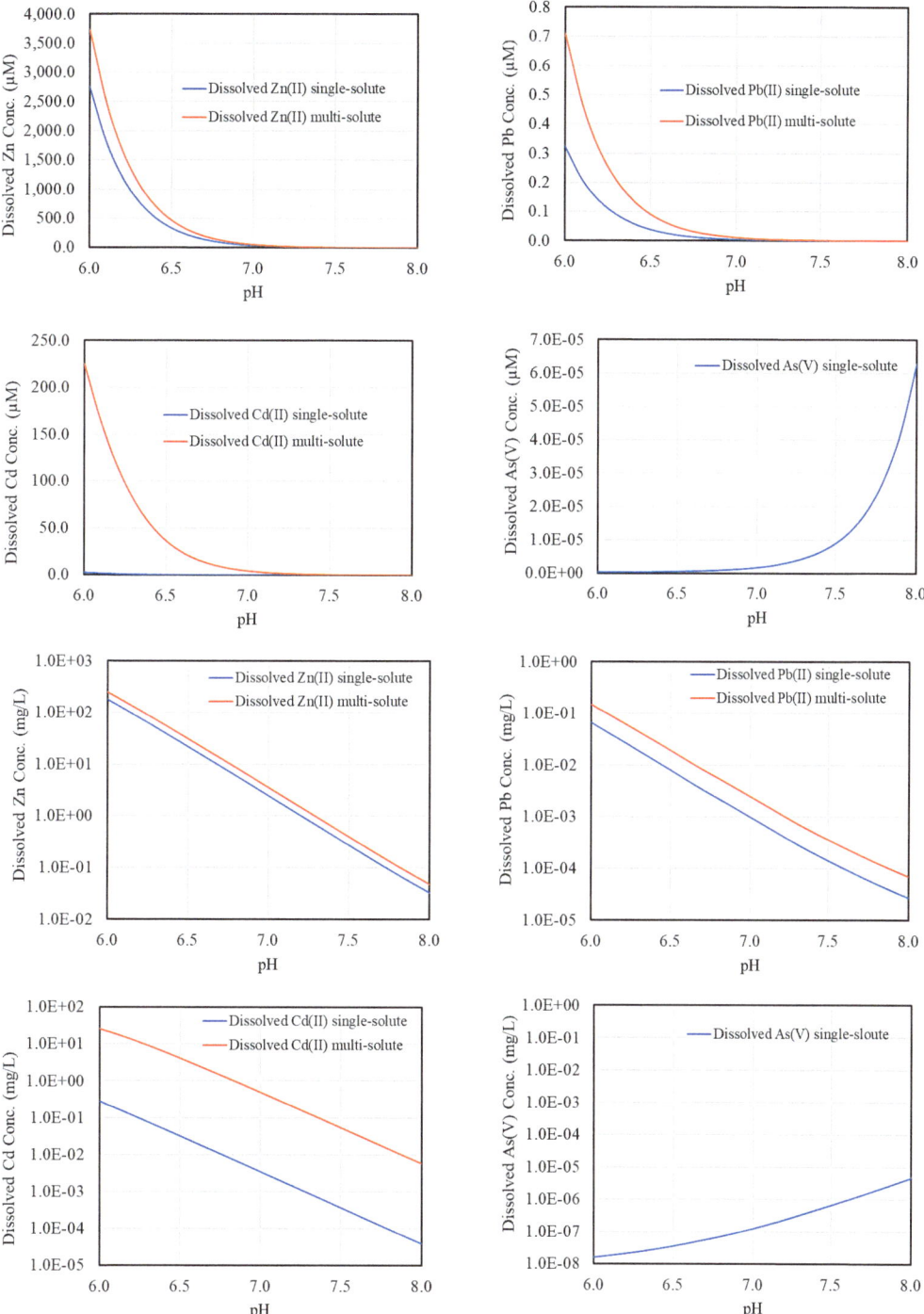

FIGURE 7-4 Predicted dissolved concentrations resulting from the surface complexation model of Nomaan et al. (2021) for individual (blue line) and concurrent (orange line) arsenic, cadmium, lead, and zinc reacting with ferrihydrite under the physical-chemical conditions of CDAA Lake taken from Şengör et al. (2007). The top four panels have an arithmetic y-axis in molar concentrations, while the bottom four panels have a logarithmic y-axis in mg/L, to show changes not apparent in the top four panels. SOURCE: VMINTEQ Software was used by the committee to generate these plots.

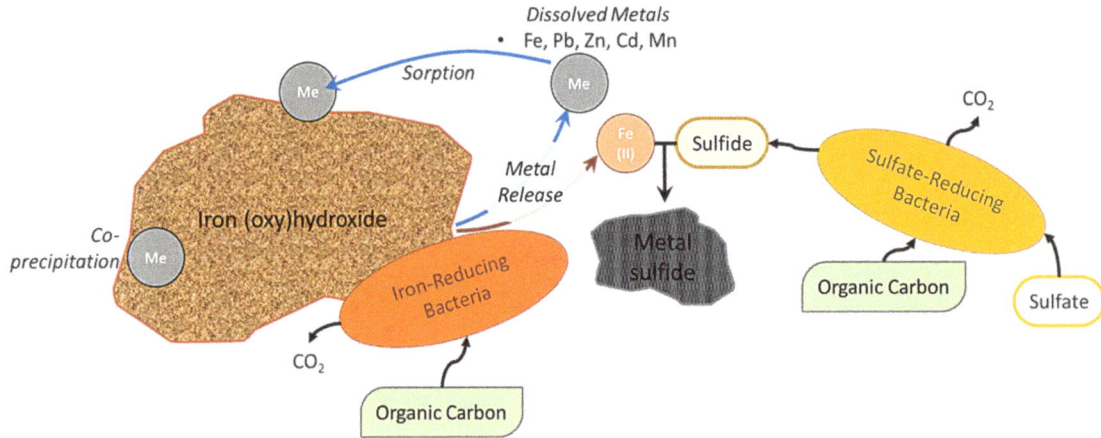

FIGURE 7-5 Simplified diagenetic model for heavy metal retention and release within CDA Lake sediments.

Despite being highly simplistic, the sediment diagenesis reactive transport model developed by Şengör et al. (2007) provides a useful means of further examining the processes likely controlling metal contaminant fate within CDA Lake. The predominant biogeochemical processes for anaerobic diagenesis encapsulated within the model are shown in Figure 7-5 and the biogeochemical outcomes are presented in Figure 7-6. The model is predicated largely on the field analysis conducted by Cummings et al. (2000) on microbial Fe(III) reduction in CDA Lake sediments, the sediment characterization by Toevs et al. (2006), and porewater data from Winowiecki (2002) and Balistrieri (1998).

The biogeochemical (diagenetic) processes (as modeled in Şengör et al., 2007; Figure 7-6), supported by depth profiles (particularly the detailed core analysis of Morra et al., 2015, shown in Figure 7-7) paint a picture of metal(loid) cycling within the near-surface sediments. As shown schematically in Figure 7-8, over the approximate upper 30 cm, oxygen introduced from the Lake water (which in turn is introduced from the atmosphere) is consumed, increasingly leading to anaerobic processes (with depth) in the sediments. Iron(III) oxides occur throughout the upper sediments and largely dominate the retention of lead, zinc, and cadmium. Despite dissimilatory Fe(III) reduction proceeding with depth, and the generation of diagenetic pyrite, lead, zinc, and cadmium depth abundances are unchanged (Figure 7-7), indicating that metal repartitioning within the profile is not a major factor. Porewater concentrations of these three metals are unlikely to vary greatly with depth (as noted by the SCM results; Figure 7-4), where only zinc is appreciable in the aqueous phase.

The profiles for the redox active elements stand in stark contrast to lead, zinc and cadmium, which do not undergo redox transformation. The profiles of iron and manganese (Figure 7-7) indicate upward migration within the near-surface, oxygenated sediment, which is particularly apparent relative to depositional fluxes (Morra et al., 2015). The depth profile of arsenic is particularly striking in its surface accumulation, even in comparison to iron and manganese, which results from a combination of ferrihydrite accumulation and the formation of As(V), which has a greater affinity for Fe(III) minerals (Tufano et al., 2008). The combined reduction of As(V) [to As(III)] and Fe(III) results in a weaker partitioning of As(III) and increased dissolved concentrations. Upward diffusion of arsenic leads to As(III) oxidation to As(V) [coupled with Fe(II) oxidation] and the concomitant strong partitioning back onto the solids (see Figure 7-8). As an outcome, arsenic has a distinct peak abundance in the upper profile (Figure 7-7) that results from biogeochemical processes rather than deposition (Morra et al., 2015). The buildup of arsenic in the near-surface sediments also illuminates the concern of bottom-water anoxia. If the sediments become devoid of oxygen, arsenic will no longer repartition in the near-surface and would rather diffuse into the overlying lake water.

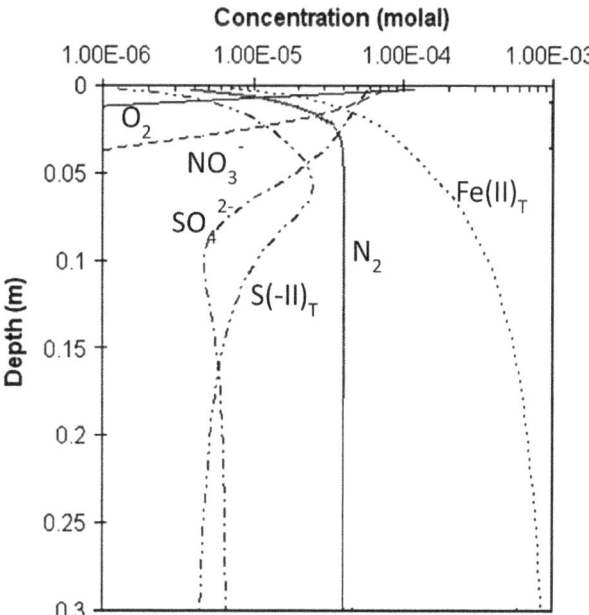

FIGURE 7-6 Simulated diagenetic trends of porewater concentrations for major redox active elements of CDA Lake sediments. SOURCE: Adapted from Şengör et al. (2007). Reprinted from *Applied Geochemistry* 22(12), Şengör, S., N. F. Spycher, T. R. Ginn. R. K. Sani, and B. Peyton. Biogeochemical reactive–diffusive transport of heavy metals in Lake Coeur d'Alene sediments. 2569–2594. Copyright (2007), with permission from Elsevier.

Manganese is more sensitive to low oxygen levels than arsenic or iron (Stumm and Morgan, 1981; Davison, 1993) owing to a higher redox potential (Figure 7-2). Reduction of manganese oxides, prevalent under oxygenated conditions, leads to their dissolution and resulting Mn(II). Accordingly, dissolved concentrations of Mn(II) within bottom waters may serve as a sentry for the potential release of arsenic, indicating when the oxygen-created Fe/Mn oxide barrier at the surface sediments is breached. Indeed, large spikes in filterable[1] manganese in bottom waters occur in association with anoxia at C6 in CDA Lake (Figure 7-9A). Low-concentration spikes of filterable manganese also occur coincident with hypoxia[2] in the Lake bottom at C5 (e.g., Figure 7-9B).

Arsenic is also seen in bottom waters coincident with seasonal anoxia. Chess (2021) showed that release of arsenic occurred from the lateral lakes during periods of anoxia. Figure 7-10 shows a peak in arsenic concentrations in bottom waters of C6 coincident with anoxia in two examples (2012 and 2019). Concentrations of arsenic in bottom sediments of the St. Joe River watershed are not enriched in arsenic compared to locations in CDA Lake affected by mine wastes (Toevs et al., 2006). Nevertheless, dissolved arsenic in C6 bottom waters reached 5–10 μg/L during anoxia (see Figure 7-10).

C5 is the mine-effected monitoring location in CDA Lake where the seasonal sag in oxygen concentration is the largest. The cycle of arsenic in the waters at C5 also reflects hypoxia to some degree (Figure 7-11). Elevated concentrations of dissolved arsenic in bottom waters at C5 also coincide with the fall oxygen sag, but peak concentrations are ~ 10 percent of those at C6, coincident with less hypoxia (Figure 7-11). Detectable arsenic concentrations were consistently available only from 2018 to 2020. Concentrations of arsenic in mine waste-impacted surface sediments in CDA Lake are typically 10–20 times higher at C5 than at the mouth of the St. Joe River. It would be important to understand how these higher arsenic concentrations in sediment would affect peak arsenic concentrations in bottom waters under anoxic conditions.

---

[1] Filterable can include dissolved Mn(II) or nano-, colloidal-size Mn particles.

[2] The term "hypoxia" is used to describe situations in which oxygen is depleted below saturation levels (saturation refers to the maximum amount of oxygen water can hold as a function of temperature and pressure), and that depletion is detrimental to organisms of interest.

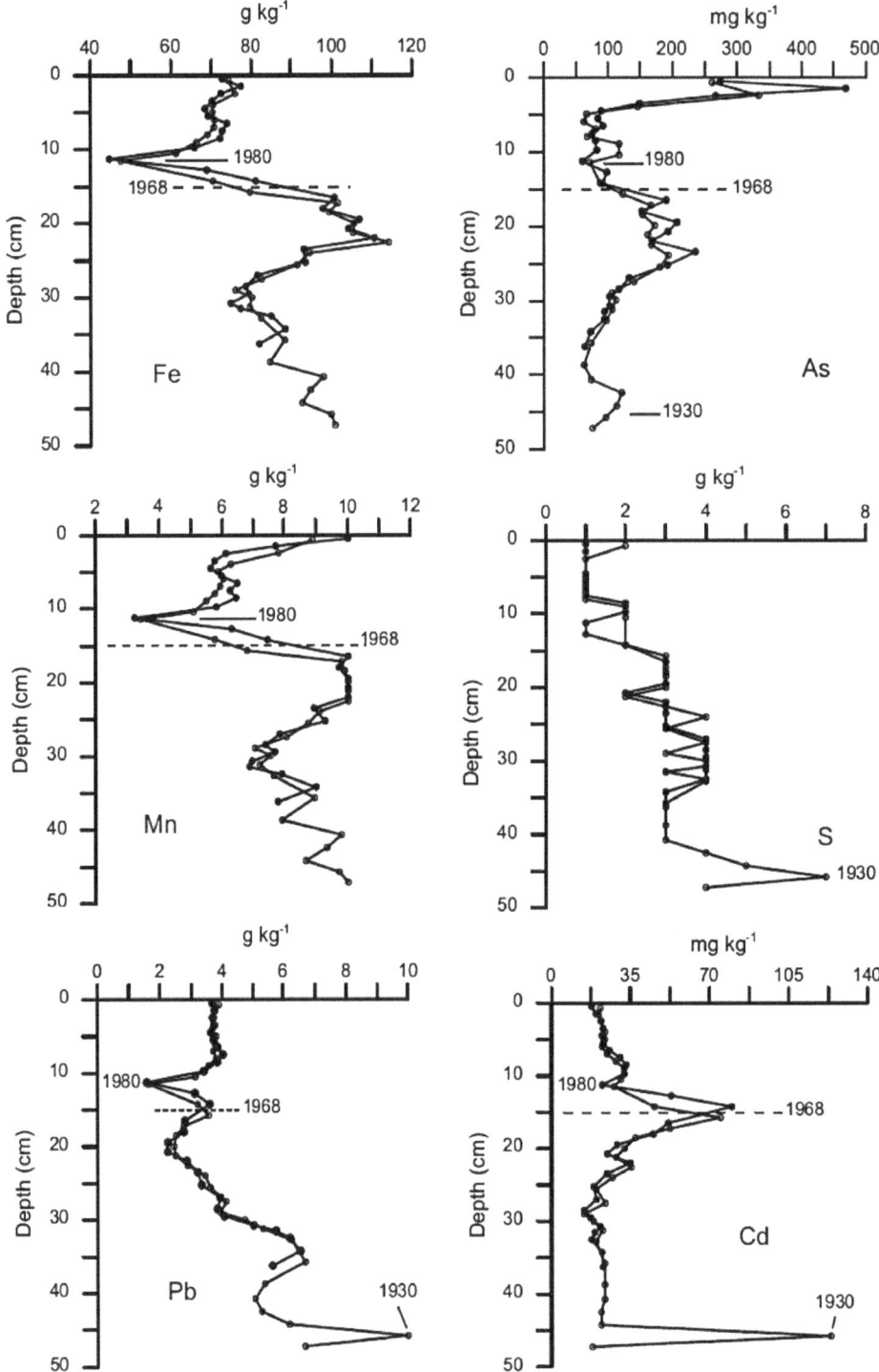

FIGURE 7-7 Sediment concentration depth profiles for redox active elements (Fe, As, Mn, and S) and inactive elements (Pb and Cd) within sediments from Peaceful Point of CDA Lake (see Figure 7-3 for sampling location). SOURCE: Morra et al. (2015). Reprinted from *Chemosphere* 134, Morra, M. J., M. M. Carter, W. C. Rember, and J. M. Kaste. Reconstructing the history of mining and remediation in the Coeur d'Alene, Idaho Mining District using lake sediments. Pp. 319–327. Copyright (2015) with permission from Elsevier.

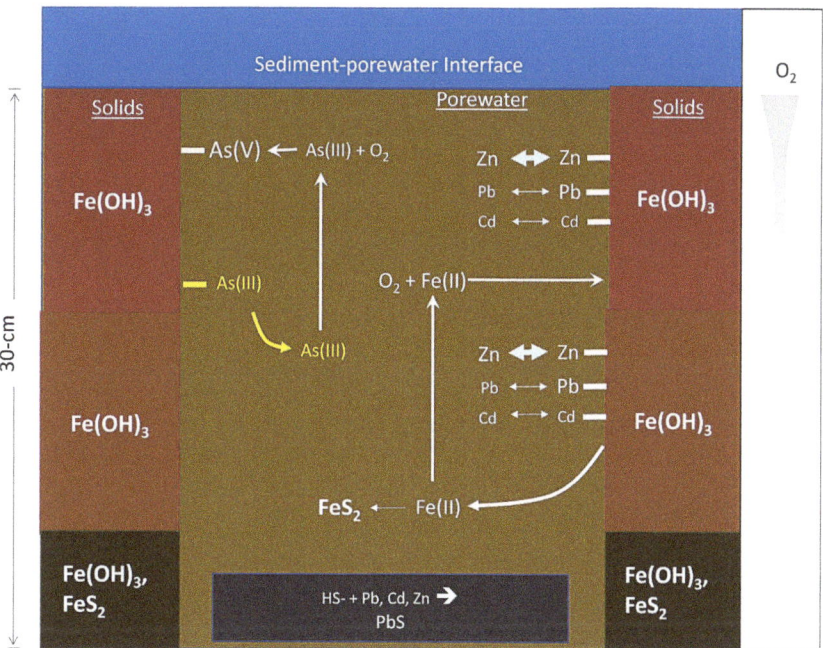

FIGURE 7-8 Arsenic and metals migration and partitioning within CDA Lake sediments.

The biogeochemical cycling drawn from modeling and measurement lead to four principal conclusions. First, Fe(III) oxides are dominant phases restricting the dissolved concentration of heavy metals in the upper 35 cm of the lake column. Second, reductive dissolution of the Fe(III) phases transpires in the sediments, becoming increasingly prominent with depth. Third, generation of sulfide (through sulfate reduction) would likely fail to sufficiently sequester heavily metals owing to the large Fe:S ratio within the CDA Lake system, consistent with the conclusions of Toevs et al. (2006). Fourth, alterations to lake chemistry that (1) further promote Fe(III) reduction (such as periods of anoxia) and/or (2) stimulate processes that acidify (even slightly) the sediment porewater, may lead to release of metal(oid)s from the sediment and thus enhance their concentrations in the water column.

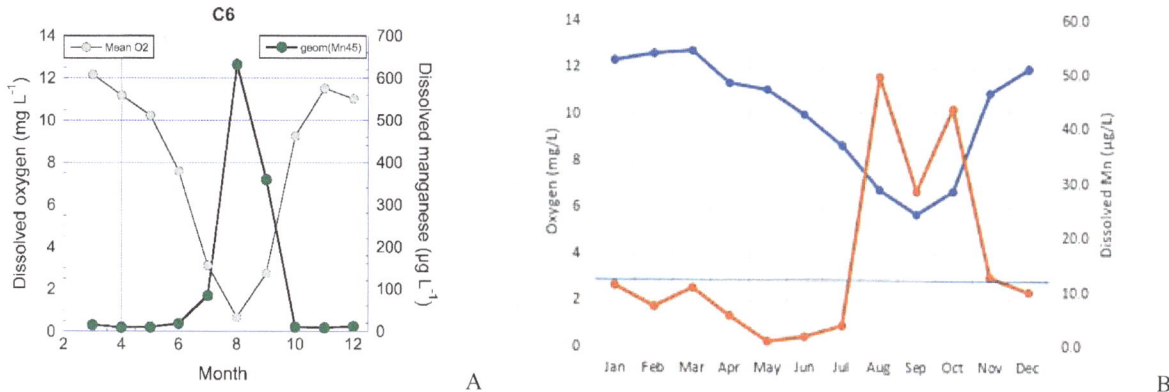

FIGURE 7-9 Monthly mean concentrations of dissolved oxygen and filterable manganese 2013–2020 at (A) C6 and (B) C5. In Panel B, oxygen is the blue line and Mn is the orange line. NOTE: Panel B of Figure 7-9 was replaced after report release to reflect correction of Table 5-3, the previous version of which had data inadvertently shifted by 2 months. Note the differences in the manganese scale in the two panels. SOURCE: Data courtesy of the CDA Tribe.

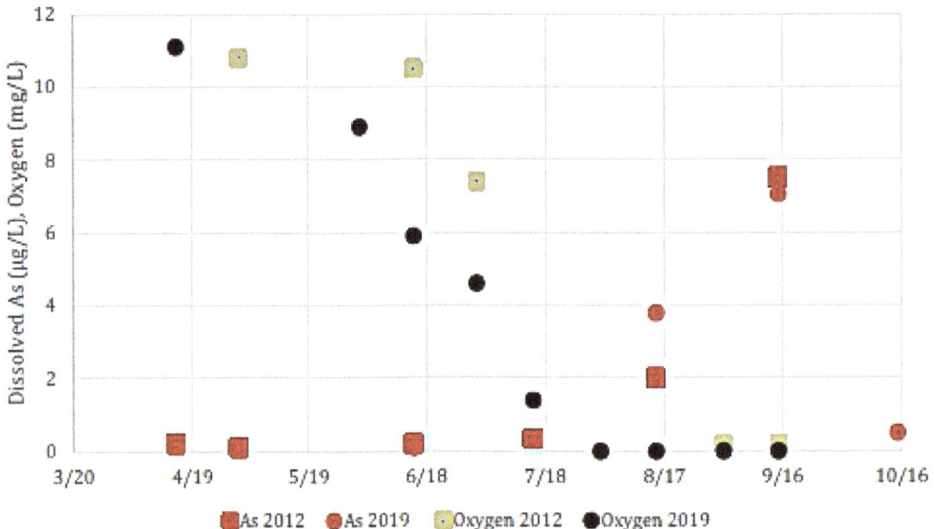

FIGURE 7-10 Dissolved arsenic and oxygen concentrations in near bottom waters (11–12 m depth) at C6 in spring–summer 2012 and 2019. SOURCE: Data courtesy of CDA Tribe.

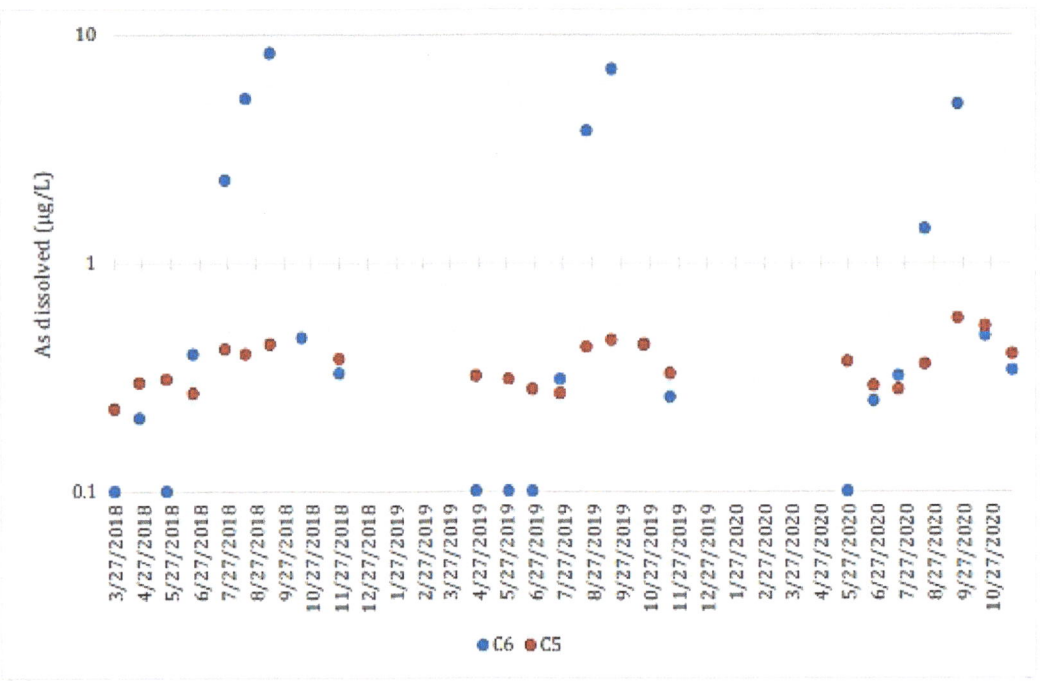

FIGURE 7-11 Arsenic concentrations in bottom waters (µg/L) at C5 and C6 in 2018–2020. Peaks at both locations are coincident with hypoxia, but higher oxygen concentration at C5 than at C6 results in less arsenic mobilization, despite the higher arsenic concentrations in sediments at C5.

KEY ISSUES ABOUT SEDIMENT PROCESSES UNDER CHANGING CONDITIONS

The previous sections suggest that CDA Lake sediments could affect Lake water quality in the future if either enhanced anoxia or lowering of pH of the Lake sediments were to occur. Both enhanced anoxia and lowering of pH have the potential to release phosphorus and/or metals from sediments, as discussed below.

Changes Promoting Anoxia

Loss of oxygen from the bottom waters and sediment porewaters of CDA Lake could lead to geochemical changes that release metals from the solid phases in which they reside in the sediment. The primary concern is the upward migration of arsenic into the water column if oxygen levels were to decline at the sediment–water interface. The extent of release is proportional to the concentration of the arsenic in the sediment and length of time the bottom waters and sediment–water interface are under anoxic conditions. Because of the abundance of iron oxide in the sediments, oxygen levels need to be depleted (creating anoxia) at the sediment–water interface before arsenic is released to the water column.

To put this in perspective, the relatively uncontaminated site C6 has released arsenic into the water column, reaching concentrations of 5–10 µg/L, under transient anoxia with an estimated sediment arsenic concentration of 12–15 mg/kg (Toevs et al., 2006). What would a similar period of anoxia produce at other CDA Lake sites with much higher sediment concentrations of arsenic? For example, arsenic concentrations of 140–330 mg/kg were observed in surface sediments at two sites in 2002 (Toevs et al., 2006). Mercury methylation can also be concerning under anoxia, but is not considered further here given the low abundance of mercury within the sediments.

There are two primary ways to enhance anoxia of bottom waters and underlying sediments: lake eutrophication and lake stratification. Eutrophication (or increased biological productivity) within the lake water results in an enlarged supply of decaying algal material to the lake bottom that increases the microbial oxygen demand and can lead to anoxia. The committee's analysis of anoxia trends in Chapter 5 found that anoxia is not happening more frequently or for longer periods than in the past, but if lake water quality conditions changed, this trend could be reversed. These changes include (1) alterations in nutrient supply to the water column (Kuwabara et al., 2007; Moberly et al., 2010), (2) increases in water temperature from climate change (Rigosi et al., 2014, Carey et al., 2012; O'Neil et al., 2012; Paerl and Scott, 2010; Dziallas and Grossart, 2011; Moss et al., 2011), and (3) future nutrient release from the lake sediments into the water column. The phosphorus input analysis in Chapter 3 indicates that increases in external nutrient supply to the water column are unlikely because phosphorus loading from the watershed is trending down. Furthermore, phosphorus is likely transporting as an adsorbed phase on Fe(III) oxides and thus has limited bioavailability. An increase in lake water temperature that might lead to oxygen depletion is considered in Chapter 10. Here, the focus is on phosphorus release from the lake sediments (#3), which is of particular concern because primary productivity within CDA Lake is mainly phosphorus-limited (Woods and Beckwith, 1997; but see Chapter 5).

It is possible that phosphorus release from sediments during low oxygen conditions could create a positive feedback loop that leads to greater productivity and subsequently lowers dissolved oxygen even more. However, there is currently no evidence of this occurring in CDA Lake. It is possible that the Lake is nitrogen-limited during the summer, which would prevent the creation of a positive feedback loop between eutrophication and phosphorus release from the sediments (see Chapter 5). Also, iron (and manganese) oxides within the surface sediment strongly bind and limit phosphorus exports into the water column. Given the conditions of the CDA sediments, the committee's model projections [based on surface complexation modeling using the diffuse layer model developed by Dzombak and Morel (1990) and using phosphorus and ferrihydrite concentrations for the sediment from Toevs et al. (2006)] are that porewater phosphate concentrations would vary only from 0.8 to 0.19 µM, which is lower than the reported mean concentration (13.1 µM). As reported by Morra et al. (2015), total phosphorus in the top 5 cm of sediment ranges from 0.94 to 1.44 g kg^{-1}, thus residing within the range of 0.5–1.6 g kg^{-1} measured in 1992 (Woods and Beckwith, 1997), which is low compared to those reported for sediments of other oligotrophic lakes (Carey and Rydin, 2011). Nonetheless, the potential for phosphorus release from sediments during anoxia cannot be ruled out. The CDA Tribe has collected data showing elevated phosphorus and nitrogen concentrations in the hypolimnion at site SJ1 during anoxic conditions (Chess, 2021).

Lowering pH in Lake Sediments

Another master variable controlling metal partitioning to the sediment solids is pH. However, as illustrated by the surface complexation modeling presented in Figure 7-4, for the pH range of the CDA Lake sediments (pH 6.1–8), while all the metals increase in concentration, the magnitude of change for zinc dwarfs that of the other metals (by more than ten times).

Although lake sediments are typically well buffered due to the presence of oxide and carbonate minerals (Stumm and Morgan, 1981), and their pH is somewhat decoupled from fluctuations within the water column, the sediment–water interface would be influenced by changes in lake water pH. The bottom-water pH data at station C1 from 2005 through 2020 illustrate the wide range of values, spanning from 5.8 to 8.3 (Figure 7-12). Examining the dissolved zinc concentration in relation to pH (at C1 from 2005 to 2020) shows a rather poor correlation (Figure 7-12A). However, if the graph is restricted to the periods when stratification was likely (May through September) and it excludes early summer data from 2011 and 2017 when inflows included low pH waters, then a stronger correlation emerges that is statistically significant ($p < 0.01$) (Figure 7-12B). When flows subside in the

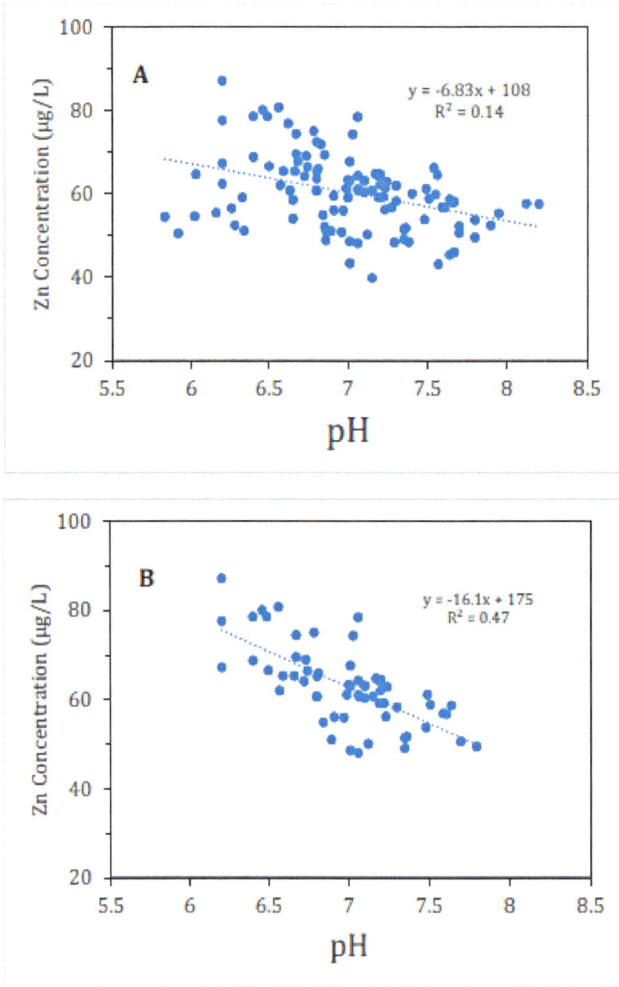

FIGURE 7-12 pH and dissolved zinc concentrations in bottom waters at C1 from 2005 through 2020. (A) All data for the time period, and (B) data for only stratified season (May through September) and excluding the early summer data from 2011 and 2017 when inflows included low pH waters. For (B), the linear relations between pH and dissolved zinc concentration are statistically significant ($p < 0.01$). SOURCE: Data courtesy of the IDEQ and graphed by the committee.

spring and early summer, the impact of dissolved zinc inputs from the watershed on dissolved zinc concentrations in the Lake wanes; during these more quiescent, stratified periods, decreasing pH may lead to zinc desorption from sediments at the sediment–water interface, leading to increases in dissolved zinc concentrations in bottom waters.

At depths below the sediment–water interface (likely a few centimeters), the largest influence on pH results from changes in oxygenation of sediments. Under oxygen-limited conditions, anaerobic processes generally move the pH toward neutrality, with reductive dissolution of iron and manganese oxides pushing pH upward, while being offset by metal carbonate precipitation, increases in carbonic and organic acids from microbial decomposition of organic matter, and sulfate reduction (McBride, 1994). An increase in oxygenation, by contrast, can lead to decreases in pH; the principal reactions that acidify lake sediments are oxidation of Fe(II) and reduced sulfur. Seasonal changes in pH resulting from shifts in oxygenation would thus be expected to impact porewater zinc concentrations and efflux to overlying waters, but due to complex and competing processes, it is difficult to predict the net effect. Lake conditions that lead to greater fluctuation in oxygen levels will lead to larger changes in pH, including periods of lower pH and release of zinc.

Despite the complexity of oxygen dynamics and pH within the Lake sediments, major patterns in these two drivers and their effects on dissolved metal and arsenic concentrations are observed. As illustrated (and described) in Figure 7-13, Lake stratification in late spring and summer limits oxygen supply from the atmosphere (and upper waters) into the bottom waters. Microbial respiration of dissolved and suspended organic matter leads to a progressive decrease in pH. Although the sediments are well buffered, acidification of the sediment–water interface leads to release of zinc (and in some cases cadmium) into the water column (Figure 7-13A), as shown in Chapter 6. If microbial respiration is sufficient to completely consume oxygen in the water column (Figure 7-13B), anoxia at the sediment–water interface results in microbial Fe(III) and As(III) reduction, which has two consequences: (1) As(III) is released into the water column and (2) the pH increases abruptly, limiting (acid-promoted) desorption of zinc. Figure 7-14 is a finer scale schematic of the sediment–water interface showing both drivers of metals release.

CONCLUSIONS AND RECOMMENDATIONS

Unlike the previous three chapters, which analyzed trends in long-term monitoring data in the watershed and the Lake, this chapter relied on data collected during a handful of special studies of Lake sediments coupled with reactive transport modeling (conducted by the committee) and diagenetic modeling (conducted by Şengör et al., 2007), along with known biogeochemistry of elements, to evaluate the current status of zinc, lead, cadmium, and arsenic cycling. Bottom-water data revealed that certain metals are periodically released from Lake sediments during periods of thermal stratification at some Lake sites. Until cleaner materials are deposited in the Lake system, the potential for release of metals from sediments into the water column will persist. The exceptions are redox active elements or elements affected by redox processes, which will continue to migrate toward the sediment–water interface under anoxic sediment conditions, regardless of decreases in the metals concentration of overlying material.

1. **Iron(III) (hydr)oxides are an abundant and dominant control on metal and arsenic concentrations in the Lake sediments** (at the sediment–water interface). Data from multiple studies illustrate the preponderance of iron within the sediments, with Toevs et al. (2006) showing that it exists in the near-surface sediments as the poorly crystalline Fe(III) hydroxide ferrihydrite. With the pH conditions of the sediments, the Fe(III) oxides serve as principal adsorbents of arsenic, cadmium, lead, and zinc that regulate dissolved concentrations as shown through measurement and modeling.

2. **The mass of sulfur relative to the composite of zinc, lead, cadmium, and iron would not allow metal sulfides to be dominant phases within the CDA Lake sediments.** Due to the abnormally high metal concentrations (particularly iron, which makes up more than 10 percent of the solids) within the sediments, sulfur occurs in quantities insufficient to control the full suite of heavy metal concentrations and may exert local and selective controls, but not universal control, on metal retention in the sediments. Iron sulfide in the form of pyrite (FeS_2) accounts for the majority of the sulfur, leaving the possibility of small quantities of ZnS, PbS, and CdS that may have local impacts on metal contaminant concentrations but not sweeping controls.

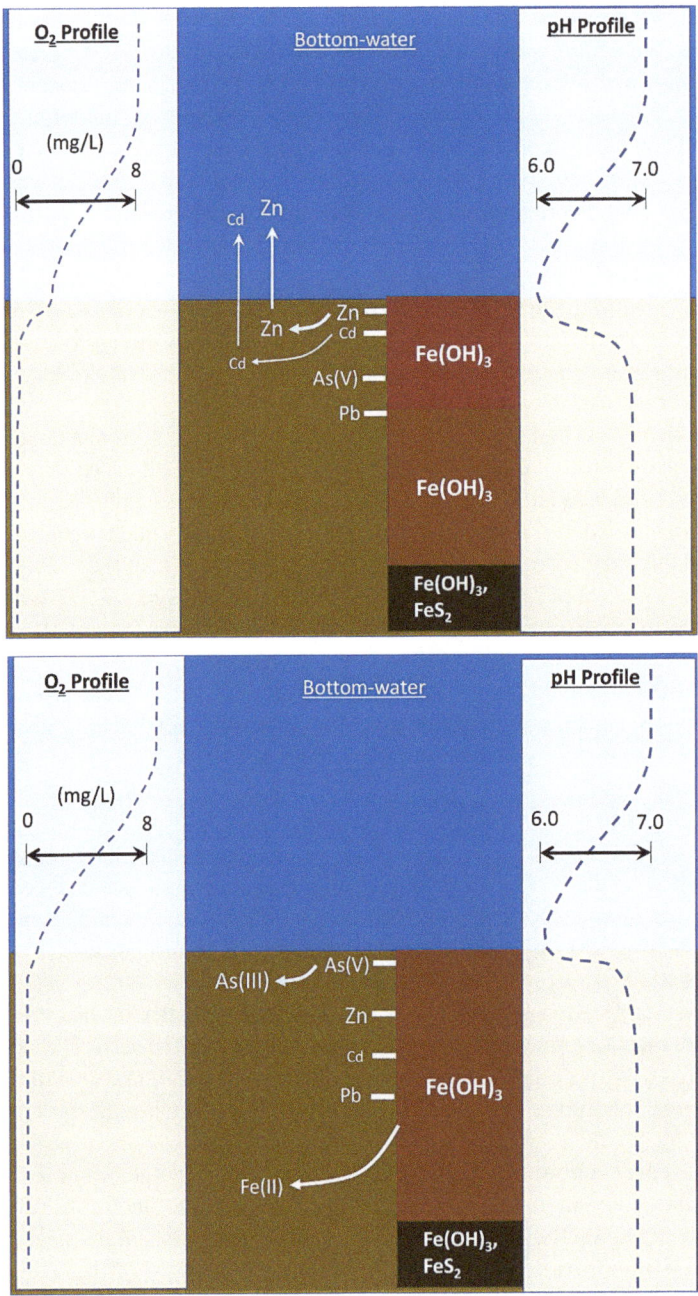

FIGURE 7-13 Oxygen and pH depth profiles, and concomitant impacts on contaminant release, resulting from microbial respiration. Limited oxygen supply during lake stratification coupled with microbial respiration leads to a progressive depletion of oxygen with depth. Paralleling oxygen depletion within the water column is a decrease in pH resulting from microbial production of carbonic acid from organic matter decomposition. (A) For conditions where microbial activity does not lead to complete depletion of oxygen within the water column (usually due to limitations in dissolved or suspended organic carbon), the acidified waters intersect with bottom sediments, leading to the acid-promoted desorption of Zn and Cd into porewater and subsequent upward diffusion into the water column. (B) When microbial activity is sufficient to completely deplete bottom-water oxygen, anoxia at the sediment–water interface leads to dissimilatory Fe(III) reductive dissolution, rapidly increasing the sediment pH. In addition to Fe(III) reduction, anoxia also leads to As(III) reduction, the combination of which results in As(III) release into the porewater and upward diffusion into the water column.

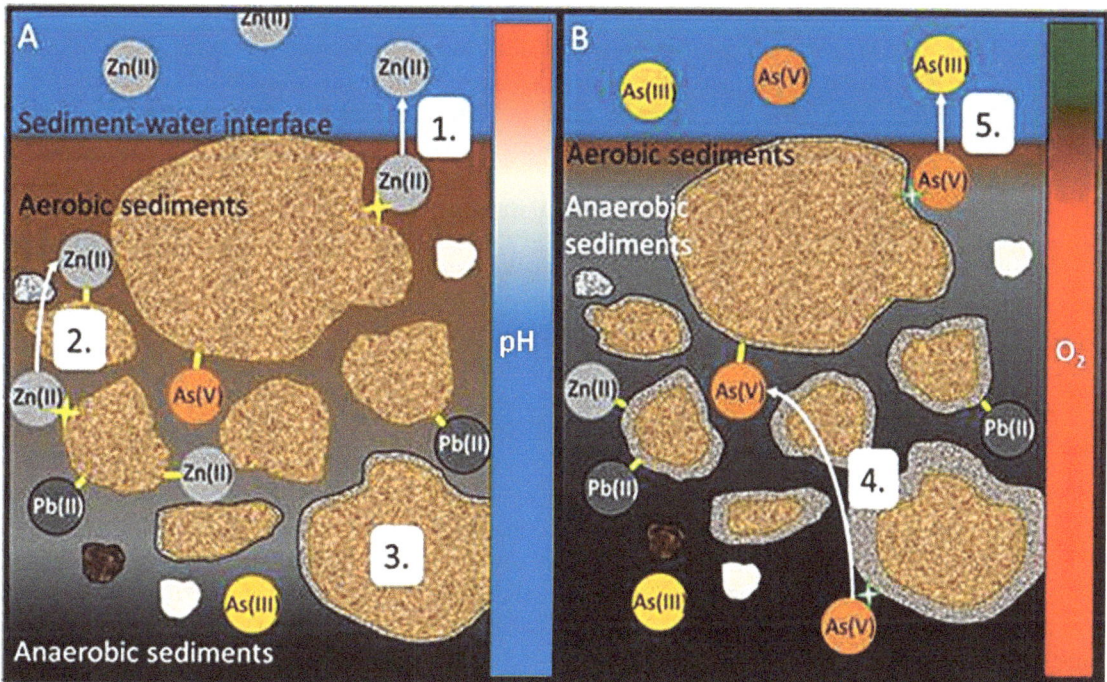

FIGURE 7-14 Metals release from sediments under (A) decreasing pH and (B) decreasing oxygen concentration. Often processes which result in a decrease in oxygen also result in a decrease in pH. NOTES: 1. Desorption of zinc from iron (oxy) hydroxides. 2. Readsorbtion/repartitioning of zinc to available iron (oxy)hydroxides. 3. Biological processes catalyze reduction of iron (oxy)hydroxides (shown as light colored ring). 4. Desorption of arsenic (V) (arsenate) followed by readsorbtion to iron (oxy)hydroxides. 5. Reduction of arsenic(V) (arsenate) to arsenic(III) (arsenite) followed by release into the water column.

3. **The greatest threat of enhanced anoxia, should it occur, is release of arsenic into the Lake water column. However, there is no evidence that anoxia is getting worse** (see Chapter 5). Because As(V) adsorption onto Fe(III) oxides is the dominant control on arsenic concentrations within the porewaters of lake sediments, decreased oxygen concentrations leading to anoxia in the bottom waters will promote the reductive dissolution of arsenic within the upper sediments and arsenic release into overlying lake waters. Presently, albeit with a few exceptions temporally and spatially, oxygenation of the upper sediments appears to provide a protective cap, where As(V) retention on Fe(III) oxides limits the dissolved concentrations of arsenic within the porewater and efflux from the sediments into the overlying lake waters.

4. **Because phosphorus chemistry is similar to As(V), a second threat of bottom water anoxia is release of phosphorus.** Although not redox-active itself, phosphate is largely bound to Fe(III) oxides; anoxia leads to the reductive dissolution of the Fe(III) phases and thus the concomitant release of phosphate. If the Lake is phosphorus-limited, it is possible that such release of phosphorus from the Fe(III) oxides could create a positive feedback loop that further promotes biological productivity, anoxia, and arsenic release from the sediment.

5. **Because the adsorption edge for zinc on iron oxides occurs in the pH range of 6.0–7.0 (typical of CDA Lake), local lowering of pH can cause release of dissolved zinc from the sediment.** A decline in pH to less than 7 occurs in bottom waters of CDA Lake in some locations and some years, with the onset of stratification. Of the metal(loid) contaminants, zinc has the highest dissolved concentration within the sediment porewater, while cadmium is present at much lower concentrations than zinc. Upward diffusion of zinc from porewaters and desorption from surficial layers of the sediment is possible at pH less than 7. Less of a concern for zinc release would be an alteration in oxygen concentrations within the bottom waters and top few centimeters of

the sediment. Although this could lead to a transient increase in dissolved zinc concentrations, zinc largely shifts fluidly between solids, partitioning to Fe(III) oxides under periods of oxygenation or to sulfide phases (precipitates such as ZnS or adsorbed on FeSx) and organic matter under periods of anoxia. The lack of upward zinc migration noted within the sediment profiles illustrates zinc partitioning to the solids under both oxic and anoxic conditions.

6. **Lead partitions strongly to the Lake sediment solids both under oxygenated and anoxic conditions.** Similar to zinc, it partitions strongly to iron oxides under oxygenation and to sulfide phases (PbS precipitates and is adsorbed on FeSx solids), along with organic matter, under anoxia. Lead adsorption on iron oxides is stronger than zinc adsorption, maintaining a lower dissolved lead concentration in porewaters, and lead desorption is unlikely to occur within the pH range of the CDA Lake system.

7. **There are limited datasets of sediment cores from within CDA Lake, making predictions about the future of the Lake water–sediment interactions difficult.** The addition of periodic core sampling coupled to existing sampling efforts would provide needed information on in-lake processes, such as redox changes in the sediment leading to metals migration and trends in metal deposition over time. Core sampling should include metal(loid) and nutrient (phosphorus) profiling, metal(loid) and nutrient speciation or phase association, depositional dating, and grain-size fractionation with metals content. Carefully designed collection of porewater data could also improve understanding of sediment–water exchanges and other in-lake processes. An initial core sampling campaign could also provide data needed to develop and calibrate a coupled hydrodynamic–biogeochemical model with specific information tailored to spatial regions within CDA Lake.

REFERENCES

Balistrieri, L. S. 1998. Preliminary estimates of benthic fluxes of dissolved metals in Coeur d'Alene Lake, Idaho. USGS Open-File Report 98-793.

Balistrieri, L., S. E. Box, A. A. Bookstrom, and M. Ikramuddin. 1999. Assessing the influence of reacting pyrite and carbonate minerals on the geochemistry of drainage in the Coeur d'Alene mining district. *Environ. Sci. Technol.* 33:3347–3353.

Balistrieri, L. S., S. E. Box, and J. W. Tonkin. 2003. Modeling precipitation and sorption of elements during mixing of river water and porewater in the Coeur d'Alene River Basin. *Environ. Sci. Technol.* 37:4694–4701.

Balistrieri, L. S., and R. G. Blank. 2008. Dissolved and labile concentrations of Cd, Cu, Pb, and Zn in the South Fork Coeur d'Alene River, Idaho: Comparisons among chemical equilibrium models and implications for biotic ligand models. *Appl. Geochem.* 23:3355–3371.

Benjamin, M. M., and J. O. Leckie. 1981. Multiple-site adsorption of Cd, Cu, Zn, and Pb on amorphous iron oxyhydroxide. *J. Colloid Interface Sci.* 79:209–221.

Bookstrom, A. A., S. E. Box, R. S. Fousek, J. C. Wallis, H. Z. Kayser, and B. L. Jackson. 2013. Baseline, Historic and Background Rates of Deposition of Lead-Rich Sediments on the Floodplain of the Coeur d'Alene River, Idaho. U.S. Geological Survey Open-File Report 2004-1211. http://pubs.usgs.gov/of/2004/1211/.

Bostick, B. C., C. M. Hansel, M. J. La Force, and S. Fendorf. 2001. Seasonal fluctuations in zinc speciation within a contaminated wetland. *Environ. Sci. Technol.* 35(19):3823–3829. doi:10.1021/es010549d.

Bradley, A. S., W. D. Leavitt, and D. T. Johnston. 2011. Revisiting the dissimilatory sulfate reduction pathway. *Geobiology* 9:446–457. https://doi.org/10.1111/j.1472-4669.2011.00292.x.

Carey, C. C., and E. Rydin. 2011. Lake trophic status can be determined by the depth distribution of sediment phosphorus. *Limnol. Oceanogr.* 56(6):2051–2063.

Carey, C. C., B. W. Ibelings, E. P. Hoffmann, D. P. Hamilton, and J. D. Brookes. 2012. Eco-physiological adaptations that favour freshwater cyanobacteria in a changing climate. *Water Research* 46(5):1394–1407.

Chess, D. 2021. Water Quality Data Summary. Presentation to the NASEM Committee. February 26, 2021.

Cornell, R. M., and U. Schwertmann. 2003. The Iron Oxides: Structure, Properties, Reactions, Occurrence and Uses. New York: Wiley-VCH.

Coulthard, T. J., and M. G. Macklin. 2003. Modeling long-term contamination in river systems from historical metal mining. *Geology* 31(5):451–454. https://doi.org/10.1130/0091-7613(2003)031<0451:MLCIRS>2.0.CO;2.

Cullen, W. R., and K. J. Reimer. 1989. Arsenic speciation in the environment. *Chemical Reviews* 89:713–764.

Cummings, D. E., A. W. March, B. Bostick, S. Spring, F. Caccavo, S. Fendorf, and R. F. Rosenzweig. 2000. Evidence for microbial Fe(III) reduction in anoxic, mining-impacted lake sediments (Lake Coeur d'Alene, Idaho). *Appl. Environ. Microbiol.* 66(1):154–162. https://doi.org/10.1128/AEM.66.1.154-162.2000.

Davison, W. 1993. Iron and manganese in lakes. *Earth Science Reviews* 34(2):119–163.

DeLemos, J. L., B. C. Bostick, C. E. Renshaw, S. Sturup, and X. H. Feng. 2006. Landfill-stimulated iron reduction and arsenic release at the Coakley Superfund Site (NH). *Environ. Sci. Technol.* 40(1):67–73.

Dewey, C., J. R. Bargar, and S. Fendorf. 2021. Porewater lead concentrations limited by particulate organic matter coupled with ephemeral iron(III) and sulfide phases during redox cycles within contaminated floodplain soils. *Environ. Sci. Technol.* 55(9):5878–5886.

Diamond, M. L. 1995. Application of a mass balance model to assess in-place arsenic pollution. *Environ. Sci. Technol.* 29:29–42.

Dziallas, C., and H. P. Grossart. 2011. Increasing oxygen radicals and water temperature select for toxic *Microcystis* sp. *Plos One* 6(9):8. https://doi.org/10.1371/journal.pone.0025569.

Dzombak, D., and F. Morel. 1990. Surface Complexation Modeling: Hydrous Ferric Oxide. New York: Wiley Interscience.

Fendorf, S., H. A. Michael, and A. van Geen. 2010. Spatial and temporal variations of groundwater arsenic in South and Southeast Asia. *Science* 328:1123–1127. https://doi.org/10.1126/science.1172974.

Gorby, Y. A., and D. R. Lovley. 1991. Electron transport in the dissimilatory iron reducer, GS-15. *Appl. Environ. Microb.* 57(3):867–870.

Harrington, J. M., M. J. LaForce, W. C. Rember, S. E. Fendorf, and R. F. Rosenzweig. 1998a. Phase associations and mobilization of iron and trace elements in Coeur d'Alene Lake, Idaho. *Environ. Sci. Technol.* 32(5):650–656.

Harrington, J. M., S. E. Fendorf and R. F. Rosenzweig. 1998b. Biotic generation of arsenic(III) in metal(loid)-contaminated freshwater lake sediments. *Environ. Sci. Technol.* 32(16):2425–2430.

Haus, K. L., R. L. Hooper, L. A. Strumness, and J. B. Mahoney. 2008. Analysis of arsenic speciation in mine contaminated lacustrine sediment using selective sequential extraction, HR-ICPMS and TEM. *Applied Geochemistry* 23(4):692–704.

Hillier, S., K. Suzuki, and J. Cotter-Howells. 2001. Quantitative determination of cerussite (lead carbonate) by X-ray powder diffraction and inferences for lead speciation and transport in stream sediments from a former lead mining area in Scotland. *Applied Geochemistry* 16:597–608. doi.org/10.1016/S0883-2927(00)00059-7.

Horowitz, A. J., K. A. Elrick, and R. B. Cook. 1993. Effect of mining and related activities on the sediment trace element geochemistry of Lake Coeur d'Alene, Idaho, USA. Part I: surface sediments. *Hydrol. Process.* 7:403–423.

Horowitz, A. J., K. A. Elrick, J. A. Robbins, and R. B. Cook. 1995. Effect of mining related activities on the sediment trace element geochemistry of Lake Coeur d'Alene, Idaho. Part II: subsurface sediments. *Hydrol. Process.* 9:35–54. https://doi.org/10.1002/hyp.3360090105.

Islam, F. S., A. G. Gault, C. Boothman, D. A. Polya, J. M. Charnock, D. Chatterjee, and J. R. Lloyd. 2004. Role of metal-reducing bacteria in arsenic release from Bengal delta sediments. *Nature* 430:68–71.

Jain, A., and R. H. Loeppert. 2000. Effect of competing anions on the adsorption of arsenate and arsenite by ferrihydrite. *J. Environ. Qual.* 29:1422–1430.

Kappler, A., C. Bryce, M. Mansor, U. Lueder, J. M. Byrne and E. D. Swanner. 2021. An evolving view on biogeochemical cycling of iron. *Nat. Rev. Microbiol.* 19:360–374. https://doi.org/10.1038/s41579-020-00502-7.

Kocar, B. D., M. J. Herbel, K. J. Tufano, and S. Fendorf. 2006. Contrasting effects of dissimilatory iron(III) and arsenic(V) reduction on arsenic retention and transport. *Environ. Sci. Technol.* 40:6715–6721.

Kocar, B. D., M. L. Polizzotto, S. G. Benner, S. Ying, M. Ung, K. Ouch, S. Samreth, B. Suy, K. Phan, M. Sampson, and S. Fendorf. 2008. Integrated biogeochemical and hydrologic processes driving arsenic release from shallow sediments to groundwaters of the Mekong Delta. *Appl. Geochem.* 23:3059–3071.

Kuwabara, J. S., P. F. Woods, W. M. Berelson, L. S. Balistrieri, J. L. Carter, B. R. Topping, and S. V. Fend. 2003. Importance of sediment-water interactions in Coeur d'Alene Lake, Idaho, USA: management implications. *Environmental Management* 32:348–359. https://doi.org/10.1007/s00267-003-0020-7.

Kuwabara, J. S., B. R. Topping, P. F. Woods, and J. L. Carter. 2007. Free zinc ion and dissolved orthophosphate effects on phytoplankton from Coeur d'Alene Lake, Idaho. *Environ. Sci. Technol.* 41(8):2811–2817. https://doi.org/10.1021/es0629231.

Langman, J. B., D. Behrens, and J. G. Moberly. 2020. Seasonal formation and stability of dissolved metal particles in mining-impacted, lacustrine sediments. *Journal of Contaminant Hydrology* 232:103655. https://doi.org/10.1016/j.jconhyd.2020.103655.

Lovley, D. R. 1997. Microbial Fe(III) reduction in subsurface environments. *FEMS Microbiol. Rev.* 20(3–4):305–313.

McBride, M. B. 1994. Environmental Chemistry of Soils. New York: Oxford Press.

Moberly, J., T. Borch, R. Sani, N. Spycher, S. Şengör, T. Ginn, and B. Peyton. 2009. Heavy metal-mineral associations in Coeur d'Alene River sediments: a synchrotron-based analysis. *Water Air Soil Pollut.* 201(1):195–208.

Moberly, J. G., A. Staven, R. K. Sani, and B. M. Peyton. 2010. Influence of pH and inorganic phosphate on toxicity of zinc to *Arthrobacter* sp. isolated from heavy-metal-contaminated sediments. *Environ. Sci. Technol.* 44(19):7302–7308. https://doi.org/10.1021/es100117f.

Moberly, J., S. D'Imperio, A. Parker, and B. Peyton. 2016. Microbial community signature in Lake Coeur d'Alene: association of environmental variables and toxic heavy metal phases. *Appl. Geochemistry* 66:174–183. https://doi.org/10.1016/j.apgeochem.2015.12.013.

Morra, M. J., M. M. Carter, W. C. Rember, and J. M. Kaste. 2015. Reconstructing the history of mining and remediation in the Coeur d'Alene, Idaho Mining District using lake sediments. *Chemosphere* 134:319–327. https://doi.org/10.1016/j.chemosphere.2015.04.055.

Moss, B., S. Kosten, M. Meerhoff, R. W. Battarbee, E. Jeppesen, N. Mazzeo, K. Havens, G. Lacerot, Z. Liu, L. De Meester, H. Paerl, and M. Scheffer. 2011. Allied attack: climate change and eutrophication. *Inland Waters* 1(2):101–105.

Nealson, K. H., and D. Saffarini. 1994. Iron and manganese in anaerobic respiration: Environmental significance, physiology, and regulation. *Annual Review of Microbiology* 48:311–343.

Nomaan, S. M., S. N. Stokes, J. Han, and L. E. Katz. 2021. Application of spectroscopic evidence to diffuse layer model (DLM) parameter estimation for cation adsorption onto ferrihydrite in single-and bi-solute systems. *Chemical Geology* 573:120199. https://doi.org/10.1016/j.chemgeo.2021.120199.

O'Day, P. A., D. Vlassopoulos, R. Root, and N. Rivera. 2004. The influence of sulfur and iron on dissolved arsenic concentrations in the shallow subsurface under changing redox conditions. *Proc. Nat. Acad. Sci.* 101:13703–13708. https://doi.org/10.1073/pnas.0402775101.

O'Neil, J. M., T. W. Davis, M. A. Burford, and C. J. Gobler. 2012. The rise of harmful cyanobacteria blooms: The potential roles of eutrophication and climate change. *Harmful Algae* 14:313–334.

Paerl, H. W., and J. T. Scott. 2010. Throwing fuel on the fire: synergistic effects of excessive nitrogen inputs and global warming on harmful algal blooms. *Environ. Sci. Technol.* 44(20):7756–7758.

Paulson, A. J., and L. Balistrieri. 1999. Modeling removal of Cd, Cu, Pb, and Zn in acidic groundwater during neutralization by ambient surface waters and groundwaters. *Environ. Sci. Technol.* 33(21):3850–3856.

Peryea, F. J., and R. Kammerack. 1997. Phosphate-enhanced movement of arsenic out of a lead arsenate-contaminated topsoil and through uncontaminated subsoil. *Water, Air, Soil Pollut.* 93:243–254.

Postma, D., and R. Jakobsen. 1996. Redox zonation: equilibrium constraints on the Fe(III)/SO_4-reduction interface. *Geochim. Cosmochim. Acta* 60:3169–3175. doi.org/10.1016/0016-7037(96)00156-1.

Rabus, R., T. A. Hansen, and F. Widdel. 2013. Dissimilatory sulfate- and sulfur-reducing prokaryotes. *In:* The Prokaryotes. E. Rosenberg, E. F. DeLong, S. Lory, E. Stackebrandt, and F. Thompson (eds.). Berlin, Heidelberg: Springer. https://doi.org/10.1007/978-3-642-30141-4_70.

Rigosi, A., C. C. Carey, B. W. Ibelings, and J. D. Brookes. 2014. The interaction between climate warming and eutrophication to promote cyanobacteria is dependent on trophic state and varies among taxa. *Limnol. Oceanogr.* 59(1):99–114.

Şengör, S., N. F. Spycher, T. R. Ginn. R. K. Sani, and B. Peyton. 2007. Biogeochemical reactive–diffusive transport of heavy metals in Lake Coeur d'Alene sediments. *Applied Geochemistry* 22(12):2569–2594.

Smedley, P. L, and D. G. Kinniburgh. 2002. A review of the source, behaviour and distribution of arsenic in natural waters. *Appl. Geochem.* 17:517–568.

Sprenke, K., W. Rember, S. F. Bender, M. L. Hoffmann, F. Rabbi, and V. E. Chamberlain. 2000. Toxic metal contamination in the lateral lakes of the Coeur d'Alene River valley, Idaho. *Environmental Geology* 39:575–586. https://doi.org/10.1007/s002540050469.

Stumm, W., and J. J. Morgan. 1981. Aquatic Chemistry. New York: Wiley.

Stumm, W., B. Sulzberger, and L. Sigg. 1992. Chemistry of the solid-water interface: Processes at the mineral-water and particle-water interface in natural systems. United Kingdom: Wiley.

Toevs, G. R., M. J. Morra, M. L. Polizzotto, D. G. Strawn, B. C. Bostick, and S. Fendorf. 2006. Metal(loid) diagenesis in mine-impacted sediments of Lake Coeur d'Alene, Idaho. *Environ. Sci. Technol.* 40(8):2537–2543. https://doi.org/10.1021/es051781c.

Tonkin, J., L. Balistrieri, and J. Murray. 2002. Modeling metal removal onto natural particles formed during mixing of acid rock drainage with ambient surface water. *Environ. Sci. Technol.* 36(3):484–492.

Tufano, K. T., C. W. Reyes, C. Saltikov, and S. Fendorf. 2008. Reductive processes controlling arsenic retention: Revealing the relative importance of iron and arsenic reduction. *Environ. Sci. Technol.* 42:8283–8289. https://doi.org/10.1021/es801059s.

Wang, J., C. F. Kerl, P. Hu, M. Martin, T. Mu, L. Brüggenwirth, G. Wu, D. Said-Pullicino, M. Romani, L. Wu, and B. Planer-Friedrich. 2020. Thiolated arsenic species observed in rice paddy pore waters. *Nat. Geosci.* 13:282–287. https://doi.org/10.1038/s41561-020-0533-1.

Wilkin, R. T., D. Wallschlager, and R. G. Ford. 2003. Speciation of arsenic in sulfidic waters. *Geochemical Transactions* 4:1–7.

Winowiecki, L. 2002. Geochemical Cycling of Heavy Metals in the Sediment of Lake Coeur d'Alene, Idaho. Masters Thesis, University of Idaho, Moscow, Idaho.

Woods, P. F., and M. A. Beckwith. 1997. Nutrient and trace-element enrichment of Coeur d'Alene Lake, Idaho. USGS Water Supply Paper 2485. https://doi.org/10.3133/wsp2485.

8

Gaps in Lake and Watershed Monitoring

Chapter 2 summarized the various long-term water quality data collected throughout the Coeur d'Alene (CDA) watershed, including the Lake and its tributaries, as well as the CDA and St. Joe Rivers, by the U.S. Geological Survey (USGS), the Idaho Department of Environmental Quality (IDEQ), and the CDA Tribe. Those data were analyzed by the committee in Chapters 3 through 6 to reveal trends in water quality parameters and highlight particular issues of concern. Subsequent to those analyses and building off of their limitations, this chapter describes gaps in the long-term monitoring networks and in the evaluation of monitoring data that, if filled, would allow for more comprehensive understanding of water quality in CDA Lake into the future.

There are five major aspects of the Lake and river monitoring programs considered here: (1) where monitoring should take place; (2) which analytes should be monitored (including questions of detection and precision); (3) how samples should be collected; (4) how many samples should be collected each year; and (5) how to select the times when samples should be collected. Gaps in the current long-term monitoring programs that relate to these five areas are discussed below, first for the watershed and then for the Lake.

IMPROVEMENTS TO RIVER MONITORING

It is worth reflecting on the reasons why the river monitoring program is important to the future of CDA Lake. Monitoring the movement of the relevant metals and nutrients through the tributaries of the Lake is a crucial part of an adaptive management approach to enhancing the water quality of the Lake. Very substantial cleanup efforts have already taken place in the watershed (expenditures of over $600 million to date), and very substantial investments will be made in the future. Monitoring the movement of the major contaminants through the system along with ongoing data analysis are vital to understand the effectiveness of the investments made up to the present and to help prioritize and plan for the remediation efforts to take place in the future.

Where Monitoring Should Take Place

The current system of regular fixed-station monitoring of the rivers that is carried out by the USGS (shown in Figure 2-3) seems highly appropriate for covering the monitoring of the metals of concern and obtaining mass balance information for the whole Lake watershed (by including two major tributaries that are not major contributors of metals—the St. Joe River and the North Fork of the CDA River). The primary feature that is lacking is coverage of the nutrient fluxes from the portions of the CDA Lake watershed that are not within the CDA River or

St. Joe River watersheds. Of the total watershed of the Lake, 16 percent of the land area has not been monitored on a systematic basis. Although it is safe to assume that the metal input from these areas is a very small portion of the total metal input to the Lake, it is not safe to assume that these areas are unimportant sources of nutrients.

A substantial amount of the population growth in the watershed in recent decades is in this unmonitored portion of the watershed. Given the level of concern expressed by stakeholders regarding the potential importance of nutrients derived from these areas, it is important that a program of monitoring and modeling (some combination of statistical and deterministic modeling) be developed to assess these nutrient inputs and their changes over time. This is not a simple task because the sources are so dispersed and there is no single location that can be monitored to evaluate these nutrient inputs. There needs to be some combination of fixed station sampling coupled with synoptic sampling of locations in the 16 percent of the watershed that is currently unmonitored. (The committee is aware that the IDEQ has started a new monitoring effort for nutrients in these areas and has reported preliminary results in early 2022. Because this effort is very new and the committee had no documents about its design, it is not considered in this report. However, the committee believes it to be an important addition to the overall watershed monitoring efforts.)

In the simplest terms, what is needed is perhaps two locations that can be monitored on a frequent basis (e.g., 18 times per year) for several years, with the samples allocated to cover much of the range of variation in streamflow as well as season. To be useful, these locations need to be monitored for streamflow as well, just as is the case with the existing fixed-station monitoring locations. Ideally these two sites would have rather different land use characteristics, focusing on agriculture and residential land use. Then, in conjunction with these fixed sites, a set of synoptic sampling sites should be developed where a few samples are collected over a period of perhaps three years, including samples under a range of different hydrologic conditions (e.g., snowmelt, rainstorms, and base flow). These sites would not require stream gages but should be coupled with onsite discharge measurements at the time of sampling. The strategy should be developed such that the more intensive sampling of the fixed stations can put the infrequently monitored synoptic sites into context so that watershed-wide estimates of nutrient inputs to the Lake can be made and trends in these inputs can be assessed. This kind of sampling strategy is not a simple matter, but given the level of concern over potential increases in nitrogen and phosphorus inputs to the Lake, some focused monitoring and modeling efforts must be a part of an overall adaptive management strategy for the Lake.

Sampling Frequency

At many of the river monitoring locations, the frequency of sampling in the past decade has been on the order of five samples per year. Given the significant investment made in the Superfund cleanup and the challenges that remain in the future for the protection of CDA Lake, this is considered by the committee to be an inadequate frequency of monitoring. The purpose of sampling the rivers is to characterize how water quality varies with respect to season and hydrologic conditions and how these relationships are changing over time. Describing these trends (or lack of trends) is crucial to the ongoing evaluation of the effectiveness of current and past remediation efforts, including targeting the locations that need more aggressive remediation and the types of conditions (e.g., seasons and flow conditions) that are in the greatest need of improved control. In those situations where the transport of the contaminants of interest is focused on times of high discharge, the sampling required to characterize the behavior of the system demands that these high-discharge conditions be sampled frequently. For lead and phosphorus, the concentrations tend to increase with increasing discharge at all of the monitored sites. For cadmium and zinc, concentrations tend to decline with increasing discharge for sites in the South Fork of the CDA watershed, but for the main stem of the CDA River near Harrison, concentrations rise with increasing discharge. For those contaminants for which concentration declines with increasing discharge, one could argue that long-term trends and average transport can be characterized well with small numbers of samples per year (less than 12). However, lead and phosphorus are sampled at the same sites and they need substantially more intensive sampling to characterize their trends; given that the largest part of the monitoring program cost is related to traveling to and from the river site and collecting the samples (versus the cost of laboratory analysis), consideration needs to be given to more intensive sampling of all constituents at the river sampling sites.

The biggest challenges for further reduction of metals input to the Lake will be to reduce the future transport of the metals that are stored in the alluvial material of the lower CDA River. Finding effective control strategies for these kinds of materials is not a routine environmental cleanup task. It is much more scientifically challenging than dealing with the tailings, tunnels, and settling ponds, which have been central features of the successful ongoing cleanup of the mined areas of the South Fork of the CDA River. The strategies for remediating the lower basin will require much more of an adaptive management approach, implementing novel remediation strategies, attempting to build predictive models of their likely effectiveness, monitoring their performance over a number of years, and using the monitoring data to provide feedback to the modeling and to enhance these remedial measures. The U.S. Environmental Protection Agency (EPA) has published an Adaptive Management Plan (EPA, 2020) that has a section on "monitoring and assessment," but with no specifics regarding the approaches to monitoring and assessment that will be used. Monitoring strategies will require mass balances by event and by year for the entire segment (the upstream end being the inputs from the South Fork near Pinehurst and the North Fork at Enaville, the downstream end being the CDA River near Harrison), which will require several high-flow samples at each site as well as several samples at moderate to low flows over a period of several years. Indeed, more frequent sampling of high-flow events is needed to capture the "first flush" of contaminants from the landscape, which happens upon the first snowmelt and other early storm events. Understanding first flush is particularly relevant to metals contamination, because it is during first flush that a large mass of particulate metal is transported and acute water quality criteria are most likely to be violated.

Looking to another watershed-scale restoration program as an example for CDA Lake, the Chesapeake Bay watershed has a monitoring network of more than 100 monitoring sites that are sampled about 20 times per year (Chanat et al., 2016; see Box 2-1). The sampling design is discussed in Gilroy et al. (1990), Sprague (2001), and Oelsner et al. (2017); the latter contains extensive analysis of the impact of sampling frequency on the analysis of decadal nutrient flux trends. In general, the findings were that in some cases sampling frequencies as low as four per year could provide fairly reliable estimates of trend (errors mostly less than 10 percent), but in other cases, sampling frequencies even as high as nine samples per year could result in flux trend errors of as much as 60 percent. Only when frequencies reached 12 per year were the errors in trend results consistently small (i.e., less than 10 percent error). Applying these results to the CDA River, one can conclude that, particularly for evaluating trends in total lead or total phosphorus, sampling frequencies of less than about 12 per year could be problematic. The consequence of continuing a low sampling frequency, particularly at the upstream and downstream ends of the lower basin, is that the time it will take to demonstrate if the strategies are working will be stretched out over many years. The evaluation of very costly future cleanup strategies, particularly in the lower CDA River, will need a higher degree of accuracy than what the past records have provided. For example, the 90 percent confidence interval for total lead flux for the CDA River near Harrison in the year 2020 covers the range of about 400–1,400 metric tons (MT)/yr (data not shown).

As mentioned in Chapter 3 in the discussion of the suspended sediment budget of the lower CDA River, the accuracy of load estimates for sediment and for total lead and total phosphorus would be greatly enhanced if continuous turbidity monitors were placed at the inflow points (from the North Fork and South Fork) and at the outflow at Harrison. These turbidity sensor data could then be related to sediment, lead, and phosphorus concentrations measured in the discrete samples. Using turbidity as a surrogate measurement,[1] a continuous record of concentration could be estimated. This would greatly improve the ability to describe the dynamic nature of the deposition, resuspension, and transport of these three constituents. This type of information will be crucial to assessing the success of any strategies to reduce lead and phosphorus inputs to the Lake. Use of such surrogate relationships will still require the collection of the discrete samples, and those relationships will require recalibration over a period of years at each location, as the size and type of particulate material changes over time. Improved understanding and predictive capabilities would also be enhanced if the samples that are collected were analyzed in more detail, providing information on particle size distribution and the major forms of the metals and nutrients in the suspended and dissolved loads.

---

[1] https://www.usgs.gov/mission-areas/water-resources/science/sediment-surrogate-techniques#overview

Sample Timing

Evaluating the effectiveness of remediation strategies in situations where concentrations increase with discharge depends on placing extra effort on sampling multiple high-flow events before the remediation efforts get underway and multiple events after they have been put in place. In the case of the Chesapeake Bay river monitoring system (see Box 2-1), for the past 30 years the sampling strategy used a stratified approach that calls for monthly sampling, typically on a calendar basis (e.g., the first Monday in the month) supplemented by about eight high-flow samples per year. Managing that sampling process can be challenging because one cannot know the timing and extent of the high-flow conditions for the year in advance. For example, a large storm event may trigger a major high-flow sampling effort perhaps including three to four samples taken, some on the rising limb of the hydrograph and some on the falling limb. Then later in the year an extraordinarily large event may happen, but the sampling budget may be nearly exhausted, and this crucial event may be poorly sampled. Conversely, the first high-flow event of the year may be the only one that happens, and the remaining samples for the year may be used on less-important lower-flow conditions. The bottom line is that this type of sampling can be a management challenge and the results will be imperfect, but there must be enough sampling effort allocated so that seasonal variation and the role of high-flow events can be characterized.

The simulations documented by Oelsner et al. (2017) also explore the question of the importance of high-flow sampling to the accuracy of flow-normalized fluxes. For these simulations, the high-flow threshold is defined as the 85th percentile on the daily flow duration curve (i.e., the highest 15 percent of daily discharges). The study found that having at least 14 percent of the samples being above the high-flow threshold was sufficient to obtain a fairly high accuracy of flow-normalized flux. The USGS monitoring program in the CDA watershed has used this kind of sampling design that focuses on high-flow conditions, even though the overall sampling frequency is much lower than 20 per year. Using the dataset for total lead samples collected at the CDA River near Harrison as an example, about 27 percent of the samples over the period of record were collected at discharges that exceeded the 85 percent high-flow threshold. That is a good result, reflecting a high degree of attention to sampling at the higher discharges. Lee et al. (2019) also explored the question of the impact of sampling frequency on the accuracy of flux estimates for a variety of constituents at a number of sites. Those results showed that below frequencies of about six per year, the accuracy of estimates diminished rather sharply, but there is little difference in accuracy between sampling strategies that used 12 scheduled samples plus about six high-flow samples versus strategies that used as many as 26 samples per year.

To illustrate how important the days of high discharge can be in the lower basin, particularly for total lead transport, the committee conducted a simple numerical experiment. The 3,653 days in the water years 2011–2020 were ranked from the lowest discharge to the highest. For each of these days, there is also an estimate of the flux of total lead transported on that day (these estimates come from the Weighted Regressions on Time, Discharge, and Season model described in Chapter 3). For purposes of this illustration, these flux estimates are treated as if they were the true value for that day. (Of course, they are not true values, but they are a reasonable representation of how total lead flux varies with season and discharge.) The following question is then posed: What fraction of all the total lead transported during this decade was transported on the top half of the days? The answer is 98.5 percent. Thus, when it comes to estimating total lead, the lower half of the days (from a discharge perspective) are responsible for a tiny fraction of all the transport. This suggests that collecting samples on those lower discharge half of the days is almost irrelevant to the ability to estimate the total transport of lead. Looking at the days in the top 10 percent of the discharge distribution, the total transport on those days is 76 percent of the total for the decade. Finally, the top 1 percent of the discharge days (36 days in the ten-year period) constitutes 34 percent of the total transport. The results of such an analysis for zinc or cadmium are not as extreme as the results for lead, but management of the lead coming out of the lower CDA River is such a crucial aspect of managing the future quality of CDA Lake. Thus, sampling for this system needs to be designed with this discharge versus lead flux relationship in mind.

How well does the current sampling design reflect this issue of the importance of sampling the high-flow days? Using the same 2011–2020 decade as an example, there were only 51 days sampled, but they were sampled with an appropriate bias toward the higher discharge days (see Figure 8-1). The boxplot on the left shows (on a log scale) the discharge on the 51 sampled days, while the one on the right shows the discharges on all 3,653 days.

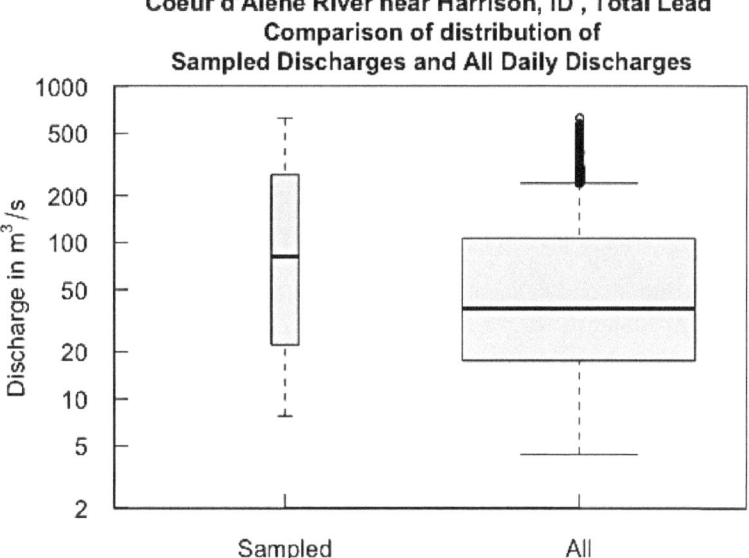

FIGURE 8-1 Boxplots showing the distribution of discharge values. Box on the left is for the sampled days in the 2011–2020 period. Box on the right is for all days in the 2011–2020 period. Box widths are proportional to the square root of the sample size. SOURCE: Data courtesy of USGS and analyzed and plotted by the committee.

(Note that the width of the boxes is proportional to the square root of the sample size). Figure 8-1 shows that the median of the sampled dates is only slightly less than the upper quartile of the full distribution, and the highest sampled date is very nearly equal to the highest discharge of the entire ten-year record. This shows an effective distribution of sampling effort, even though the total amount of sampling is quite low.

In summary, the overall sampling frequencies employed, particularly at the upstream and downstream ends of the lower CDA River segment, are simply too low given the role that this segment plays in the future delivery of contaminants to the Lake. However, the allocation of sampling efforts toward high-flow conditions is appropriate, and even if overall sampling frequencies can be increased in the future, this high-flow biased sampling strategy should continue to be used.

IMPROVEMENTS TO LAKE MONITORING

Spatial Resolution

As in many limnological monitoring programs, the focus of long-term monitoring within CDA Lake has been on pelagic sampling, with the main routine sampling sites (C1–C6) aligned longitudinally from the dam to the stream inflows. Although this has meant good spatial representation of the main Lake, it means that changes in the littoral zones of the Lake are unmonitored. Littoral regions are where the greatest interactions with the public occur and the greatest watershed impacts manifest. They are difficult to monitor because the shoreline is highly heterogeneous, and no one location can be considered representative.

Approaches that have proven useful at other lakes include variations and combinations of the following four approaches. First, it is valuable to establish a finite number of fixed sites that encompass the range of variability. This could include littoral areas adjacent to highly populated regions, undisturbed regions, creek mouths, and wetland areas to name a few. These fixed sites could be monitored on a regular (monthly) basis for the usual limnological variables, but with added emphasis on littoral algae, coliforms, etc. In addition, it is possible to install automated sampling stations that can measure key physical and chemical factors at high resolution (every 1–10 minutes). Such variables include water temperature, wave height, chlorophyll fluorescence, phycocyanin fluorescence (an indicator of cyanobacteria), water clarity, chromophoric dissolved organic matter (CDOM), dissolved oxygen, pH and turbidity.

The second concept is synoptic sampling of a broader range of sites at one time of year, particularly when littoral algae are near their peak. Synoptic sampling is the collection of samples from many locations during a short period of time, typically a few hours (Kimball, 1997). Such a "snapshot" can help identify spatial trends more effectively than the smaller number of routine stations. For CDA Lake, synoptic sampling could include 30–40 sites, whereas the routine littoral sites would typically include fewer than ten.

Third, the use of remote sensing should be considered to better understand dynamics in the littoral zones. Although spaceborne remote sensing does not currently have the resolution needed for littoral monitoring,[2] the use of drones, fixed-wing aircraft and helicopters can overcome the issue of resolution. These platforms can be equipped with something as simple as GoPro cameras to something more complex, such as multispectral or even hyperspectral cameras. The images obtained can provide quantifiable data on variables such as the extent of littoral algae, inflow plumes, and sediment resuspension. Such data could be valuable prior to the establishment of the fixed sites.

Fourth, the engagement of private citizens in data collection is a valuable and often untapped resource. Simple reporting of both ordinary and extraordinary events can provide researchers and lake managers with data over a spatial range that is impossible to achieve in other ways. Such observations can also serve as an important early warning system. The near universality of high-quality cameras on personal cell phones and the simultaneous and automatic recording of GPS location can provide valuable data. What is required to make such data useful is the creation of an app or other data portal that makes uploading simple for the private citizen, and structures the data in a way that minimizes the time and effort required by the user. There are now examples of such app at many lakes.

A cross-lake transect, which is distinct from the littoral monitoring program, could be added to the routine pelagic program. It would not be prohibitively expensive and would provide information on the connections between the littoral and pelagic zones.

The Lake's vertical profile monitoring program is generally well designed. The depth resolution for the vertical profiles is very good, with contemporary intervals of < 1 m generally. The only major suggestion regarding the location of the vertical profile sampling program is that it would be very advantageous to have an observation collected during each sampling event from a depth relative to the sediment–water interface (e.g., 0.5 or 1.0 m above the sediments). As it currently stands, it is not clear how far the deepest profile observation is from the sediment interface for any of the historic data. This is very important because the most important redox reactions (i.e., nutrient and metals release from the sediments) are directly related to oxygen concentrations at the immediate sediment–water interface.

Temporal Resolution

The temporal sampling resolution of the program is adequate, with monthly sampling most of the year and most recently twice monthly sampling during the summer months. Nonetheless, the profile sampling is incomplete during much of the November to March period. It would be good if the Lake was monitored monthly during this period, or, if that is not possible, if it was consistently sampled every other month (e.g., December and February) to facilitate long-term trend analyses for those winter months that have the most data. Most important is being more consistent in which months are sampled during the winter—for example, sampling October, December, and February every year and November and January when feasible.

It should be noted that monthly sampling can miss limnological events such as the onset of algal blooms that signal water quality degradation. Although intensifying sampling frequency can add costs, especially for samples that are intensive to process (such as phytoplankton community samples), higher temporal resolution will increase the chances of capturing such short-lived events and allow them to be placed into a broader context of long-term trends.

---

[2] Regarding the potential for satellite remote sensing on CDA Lake, this could be limited by the narrowness of the Lake, and satellite algorithms for chlorophyll and suspended sediment would need to be groundtruthed with traditional measurements for some time.

Sampling Methods

When using the profile data collected from CDA Lake it is very important to screen for any data collected during sonde upcasts. An upcast is when the sonde is retrieved from the bottom of the profile to the top while recording data. This is evident by comparing the time stamp for sonde readings to the corresponding depths. Upcast data are invalid because when the probe is being pulled upward through the water column it creates a void behind it that gets filled with water that has been turbulently mixed and displaced upward. Conversely, downcast data are collected as the probe is being lowered through the water column; therefore, the sensors, which are located on the bottom of the probe, encounter laminar conditions. Upcast data are very noisy and offset compared to downcast data. Upcast data commonly occur when, for example, a cable gets tangled and needs to be lowered 1–2 m to resolve the tangle. Upcasts also occur if a regular downcast is collected and the sonde is left on when retrieving the sonde to the surface. In both of these cases, the upcast causes no serious problem, and the upcast data can just be discarded. In most cases, only a small portion of the profile data collected from CDA Lake was discarded. However, many profiles collected during 2016 were solely taken as upcasts, such that downcast data were not available for these profiles. Similarly, most of the 2015–2021 data collected at the Cougar 2 station were collected as upcasts. Upcast data should not be used, and those collecting profile data from CDA Lake in the future should be cautioned to collect sonde data only as downcasts.

Better Monitoring of Nutrients

As mentioned in Chapter 2, the large majority of the phosphorus data in the IDEQ and CDA Tribe datasets are currently presented only for the total phosphorus fraction. Many fewer data points are available on dissolved phosphorus or soluble reactive phosphorus, and no particulate phosphorus data have been reported. The *Total Phosphorus Nutrient Inventory, 2004–2013* (IDEQ and CDA Tribe, 2020) presented a comparison of total phosphorus, total dissolved phosphorus, and soluble reactive phosphorus as a function of river flow for the CDA River at Harrison. The report showed that dissolved phosphorus concentrations are nearly constant with regard to flow, whereas total phosphorus concentrations increase sharply at river flows > 5,000 cfs. Since much of the phosphorus loading to CDA Lake is associated with high river flow, the report indicated that much of the phosphorus transported from the CDA River to the Lake is in the particulate phase.

The bioavailability of the particulate phosphorus fraction can vary greatly depending on whether this phosphorus is associated with organic matter (e.g., leaf litter or sloughed-off algae) or is part of mineral complexes with iron, aluminum, calcium, or clays. Particulate phosphorus associated with high-sediment transport events is less likely to be bioavailable, and more likely to rapidly settle to the lake sediments, than particulate phosphorus that is not associated with high-sediment transport events. Future research on the CDA basin would benefit from consistently quantifying both the total and dissolved phosphorus fractions, including the mineral composition and bioavailability of the particulate fraction. Indeed, on a few occasions when particulate phosphorus concentrations are particularly high, the mineral composition of the particulate phosphorus should be characterized to determine whether this phosphorus is associated with organic matter or is in mineral associations with calcium, aluminum, iron, etc.

Most long-term limnological monitoring programs on oligotrophic lakes have total phosphorus reporting limits of approximately 1 µg/L. These low reporting limits are achieved by using 4- or 10-cm length spectrophotometer cells (as opposed to the standard 1-cm cell) and detailed protocols for washing the glassware. The Lake Tahoe Environmental Research Center, which consistently deals with lake phosphorus concentrations close to the detection limit of 1 µg/L, washes its phosphorus glassware between each phosphorus analysis by autoclaving in a dilute acid solution.

In addition to the above considerations for phosphorus, the long-term monitoring program needs to consider regular monitoring of nitrogen, including the entire suite of species (total N, NH_3/NH_4^+, NO_3^-) in the epilimnion or euphotic zone. Given the fact that CDA Lake may be nitrogen-limited for part of the year, both nutrients are important. By analyzing the total nutrient data set in the Lake Tahoe watershed, for example, it could be concluded that decreasing trends in nitrogen and phosphorus were related to fire suppression and forest growth (Coats et al., 2016).

Monitoring of Physical Parameters

The physical characteristics of a lake provide information on the transport, mixing, and dilution of dissolved and particulate material (both horizontally and vertically) and can provide insights into the spatial distributions of variables such as dissolved oxygen, metals, nutrients, and the biota. With an understanding of these characteristics, it is possible to infer the future trajectory of lake conditions in the face of factors such as changing watershed characteristics and climate change.

As physical characteristics vary over a very broad range of timescales (from seconds to years), intermittent sampling at monthly or similar intervals does little to improve understanding. Instead, time series data provided by in situ sensors are more useful. Thermistor chains, where a set of high-resolution temperature sensors are distributed along a taut vertical mooring line, are the most ubiquitous, economical and useful for providing such data. Including sensors for electrical conductivity, turbidity, and dissolved oxygen make such an installation highly informative of the conditions that drive water quality change. Typically sampling at intervals between 30 seconds and 10 minutes, with the instruments staying in place throughout the year, will yield a data record with information on a broad range of important processes, including duration and extent of thermal stratification, occurrence of episodic mixing events triggered by storms, insertion depth of river plumes, causes of vertical mixing that can transfer oxygen to the lake sediments, and year-to-year differences in these processes. When combined with water sampling, these measurements can provide a phenomenological explanation and understanding of changes in lake chemistry and biology (see, e.g., MacIntrye et al., 2009; Marcé et al., 2016; Meinson et al., 2016; Roberts et al., 2020; Ward et al., 2022). Such data are critical for calibrating and validating the performance of lake models, which are valuable tools for exploring the future state of the Lake.

Typically, the data from these instruments would be downloaded two to three times a year. The moorings can be enhanced to provide real-time data, which would require a surface buoy to transmit the data. Real-time data provide the unique advantage of observing critical phenomena as they occur, thus allowing for additional measurements to be taken as needed by dispatching a sampling vessel. A minimum of three stations (C1, C4, and C5) should be considered; if resources are available for a fourth station, C3 could be added.

Although velocity can in part be inferred from thermistor chain records, it is better to directly measure lake velocities. Water velocities are universally measured at present with acoustic Doppler current meters—instruments that can be deployed at the bottom of the Lake and can measure the velocity from the bottom to the surface of the Lake. The information they yield would help managers better understand periods when sediments are resuspended (during high inflow events and possibly during internal seiches events) and could also be useful for calibrating and validating Lake models.

The final element of determining the Lake's physical characteristics would be the use of high-resolution, multiparameter profilers. These would be used to take a profile at each station visited during a water-sampling day, falling through the water column at 0.5 m/s and sampling at 4-8 hz. The profilers typically measure temperature, electrical conductivity, dissolved oxygen, turbidity, chlorophyll fluorescence, pH, photosynthetically active radiation, and beam attenuation. They provide high spatial resolution (5–10 cm) data albeit at monthly intervals. As such, they complement the vertical instrument chain data that have lower spatial resolution (possibly located at 10 depths in the water column) but very high temporal resolution.

Improvements to Metals Sampling

Measurements of particulate and dissolved lead, zinc, and cadmium including metal(loid)s in the water column are reported on a near-monthly basis, along with special water column studies (e.g., profiles, intense studies at selected sites). The existing continuous dataset is less than 15 years old—a short window of information on the scale at which the Lake has changed and continues to change. It is critical to continue the heavy metal monitoring as well as to collect data that aid interpretation of the metal data (e.g., manganese, iron, dissolved oxygen, hardness). Although resource constraints always require choices in monitoring, the suggestions below should not come at the expense of maintaining long-term water column monitoring.

Although monthly sampling is adequate for some interpretations, the gap in monitoring in the winter prevents a full understanding of metal dynamics. More frequent sampling or well-designed special studies targeted at specific questions about winter influences or intramonthly temporal variability at critical times could help fill in gaps.

Methods for Determining Particulate versus Dissolved Metals

Metal concentrations measured by USGS, IDEQ, and the CDA Tribe (i.e., both inputs and in CDA Lake) are reported as "total," "filterable," and "total recoverable" concentrations. There are two major issues with these methods that must be considered when interpreting the metals data. First, the water column metal concentrations reported as "filterable" by the CDA Tribe and IDEQ refer to anything that passes through a 0.45-μm filter, and this concentration is interpreted to be the dissolved metal concentration. However, metals bound to colloid-sized particles have been observed on occasion to pass through the 0.45-μm filters; at times, the quantities were sufficient to make the filtrate cloudy (Chess, 2021). These cloudy filtrates can be decomposed with a strong acid treatment (pH < 2) to free the colloid-bound metal (resulting in "total recoverable metal" according to EPA method 200.7). **Hence, the "filtered" sample includes dissolved + colloidal metal in the filtrate, and thus may overstate at least dissolved lead concentrations.** This is a more important issue for lead than for zinc and cadmium (because very little zinc and cadmium associate with colloids in CDA waters—Balistrieri and Blank, 2008) and is most important during periods of high runoff. It is not clear in the metals datasets when the total recoverable treatment has been done, but it is not routine. Clear designation of samples where total recoverable digestions were conducted would be a minimal correction for these issues. Including both "dissolved" determinations and "dissolved after total recoverable digestion" would help quantify where such issues exist.

The second problem has to do with the measurement of "total" metal concentrations, which are unfiltered samples that have been preserved at pH 2—a pH low enough to desorb much (but not all) metal from particulate material (Balistrieri et al., 1994). Hence, "total metal concentration" in the water column does not actually represent the total metal concentration on the particles, because only particulate metals desorbed at pH 2 are included. **Thus, the actual mass of particulate metal in a sample is larger than represented by "total" minus "dissolved" metal determinations.** The actual concentration of metal on particulate materials can only be determined from a completely decomposed particulate sample (as was done in Balistrieri et al., 2002, and Kuwabara et al., 2006). Metal concentration on particulates is then determined by dividing this value by the weight of the particulate in the sample (reported as $\mu g_{metal}/g_{particulate}$). This is termed total particulate metal, and it is the measure that influences reactions with dissolved metal and effects on biota.

Conventional practice is to use total metal and not total particulate metal, when doing flux calculations, which is the approach taken by this committee and throughout the literature. This will have little effect on trends, which are of interest in this report, because total metal is typically a function of total particulate metal for each element. But this issue will lead to understating the absolute masses of metal deposition or transport based upon fluxes versus those determined from dated cores that used total particulate metals. This issue becomes important if a budget of deposited lead across the entire system is needed (e.g., historically, at present, or into the future).

Need for Additional Sediment Cores

One of the most important gaps in the collection of data critical to understanding the influence of mining legacy inputs on water quality in CDA Lake is data on lateral and vertical distribution of metals in the Lake sediments. A number of studies have measured lead, zinc, cadmium, arsenic and other metals in sediments since the comprehensive studies of Horowitz et al. (1993), including Horowitz et al. (1995), Harrington et al. (1998), Toevs et al. (2006), Morra et al. (2015), and Kuwabara et al. (2003, 2007). Nevertheless, there is not a systematic understanding of the spatial distribution of metals in sediments. Areas of extreme metal enrichment (10,000 μg Pb/g dw) are still present, and in some locations, in-shore areas with much less contamination (< 350 μg/g dw) are also present (EcoAnalysts, Inc., 2017; Scofield et al., 2021). Few studies have been sufficiently intensive to allow generalizations or identification of areas of greatest and least concern. Better understanding the patterns of distribution and locations of "hot spots" is critical to better understanding the Lake-wide flux of metal from the sediments to the water column, ecological risks to the benthic food web, and human exposures to lead during activities in the Lake.

A spatial assessment of lakeside soils is also essential to evaluate if there are human health risks in the shoreline area, especially associated with areas where flooding might have occurred in the past. Given the extreme concentrations of lead in the particulate material in flood waters from the CDA River, soil monitoring of lakeside areas that can be flooded would also fill an important gap.

Dated vertical cores of sediments in many different locations (including but not limited to the CDA River mouth and Harrison Slough) will be critical to understanding the degree to which metal enrichment in Lake sediments is receding as remediation continues to reduce inputs of metals from the watershed. Only one dated core has been studied since 2010 (Morra et al., 2015). Thus, very little is known about recent changes in sediments. Such understanding is crucial to interpreting future changes in the Lake. Well-studied cores in critical locations would be an effective means to monitor changes in sediment enrichment in different pools of the Lake and might aid understanding of sediment transport—another important gap. Once a sense of the time series in sediments in different locations in the Lake is known, monitoring of surface sediments (e.g., every three years) might be sufficient to evaluate sediment changes over time.

Lack of Ecological Data

Although CDA Lake is heavily enriched with potentially toxic metals, the ecological implications of that contamination remain ambiguous. In many locations, lead, zinc, cadmium, and arsenic concentrations in water and particulate material (the food of many invertebrates) are higher than concentrations that cause toxicity in controlled studies. Yet, lack of ecological monitoring makes it difficult to extrapolate from such high concentrations to changes in the pelagic or benthic food webs of the Lake. The ambiguities point to the challenges in establishing how ecosystem services in the Lake have been affected by the legacy wastes of mining. As discussed in Chapter 9, building a body of understanding at different levels of ecological organization, from exposure through effects, is crucial to determining what benefits might arise from remediation of the metal contamination in the system. Specifically, systematic pilot studies and monitoring of metal enrichment in muscle and livers of sport fish and fish from the pelagic and benthic food webs are major gaps. Stable isotopes studies (^{15}N and ^{12}C) along with studies of metal bioaccumulation could allow evaluation of exposures of pelagic and benthic food webs. Possible effects on the benthic food web are of particular concern given the extreme concentrations of metals in sediments. Only two studies have evaluated the benthic community of the Lake in association with metal exposures in the past two decades (Kuwabara et al., 2006; EcoAnalysts, Inc., 2017), and those studies illustrate the complexity of attempting to assess lake-wide effects of contaminated sediments on the benthos (see Chapter 9). Given the importance of benthic food webs to lake ecosystems in general, this is a major gap (Vander Zanden and Vadeboncouer, 2002). Mesocosm studies, as described in Chapter 9, could also greatly aid understanding of pelagic influences of metal contamination in particular; but would also be useful with benthos.

Although the scientific program at CDA Lake is impressive in some ways, one of its striking shortfalls is the relative paucity of monitoring data for higher trophic levels (zooplankton, fish) as well as focused studies of ecological interactions such as food web structure or benthic–pelagic coupling. In a region where tourism, boating, and fishing are important, a comprehensive understanding of lake food webs could enhance these ecosystem services, and facilitate future assessment of shifts in these services as watershed remediation efforts, climate change, and shoreline development proceed.

***

The appropriate monitoring program for a lake is highly site-specific and dependent on many local considerations; however, much can be learned from existing long-term lake-monitoring programs. Boxes 8-1 and 8-2 provide summaries of the monitoring programs at Flathead Lake and Lake Tahoe. Compared to CDA Lake, the Flathead Lake Monitoring Program (FMP, Box 8-1) has more extensive characterization of nutrient chemistry, direct assessment of phytoplankton productivity via the carbon-14 method, and a thorough quantification of zooplankton biomass and community composition. Another advantage of the FMP is a more frequent and regularized

> **BOX 8-1**
> **Flathead Lake Monitoring Program**
>
> Flathead Lake in western Montana is the largest (by surface area) natural freshwater lake in the western continental United States and is similar in many ways to CDA Lake. Flathead is elevated (882 m asl vs. 649 m asl for CDA Lake), large (510 km^2 vs. 129 km^2 for CDA Lake), is relatively deep (116 m maximum vs. 67 m for CDA Lake), and has a short residence time (2–3 years vs. 0.5 years for CDA Lake).
>
> Of interest to the situation of CDA Lake, the Flathead Lake Monitoring Program (FMP) was established in 1977 with funding from the state of Montana to establish baseline data for the lake and watershed prior to the potential opening of coal mines in the headwaters of the North Fork of the Flathead River in Canada. (The mines were not established.) These concerns were at least in part related to observed impacts of mining on other regional lakes, including CDA Lake. The core FMP program was designed in consultation with researchers from Lake Tahoe, where long-term monitoring had been established in the early 1960s.
>
> FMP plays a critical role in an ongoing total maximum daily load (TMDL) process and voluntary nutrient reduction program, as the lake was classified as an impaired water in the late 1990s due to observed increases in lake productivity since the 1970s. Parameters relevant to Flathead's TMDL process include phosphorus and nitrogen concentrations, deep-water oxygen, chlorophyll, proliferation of phytoplankton indicative of low water quality (e.g., cyanobacteria), and levels of periphyton biomass. Supplemental funding from philanthropic sources has allowed the program to include upper food web components (zooplankton, opossum shrimp). However, like CDA Lake, fish sampling is not included in the FMP (such studies are performed by state of Montana biologists and by tribal fisheries biologists), nor is monitoring of microbial (prokaryotic) communities, and littoral habitats are not included.
>
> FMP has two main components: lake monitoring and river monitoring. The primary goal of the river monitoring is to quantify nutrient (N, P) loading to the lake, including using high-sensitivity methods for phosphorus. This is accomplished by sampling of the main inflows to the lake (the mainstem of the Flathead River and the Swan River, ~ 95 percent of inflow volume) as well as the outlet. River sampling for chemical properties occurs monthly, with intensified sampling in response to flood events and peak flow periods. Measured parameters are total phosphorus, total nitrogen, nitrate (NO_3), nitrite (NO_2), ammonium (NH_4), soluble reactive phosphorus, major ions, and SiO_2 as well as total and dissolved organic carbon, benchtop turbidity, and total suspended solids. Particulate carbon, nitrogen, and phosphorus were added to the program in recent years. The primary goal of the lake monitoring is to establish a basis to assess changes in water quality in the lake with a focus on lake physical structure (temperature stratification), water transparency, oxygen levels, phytoplankton abundance and productivity, and nutrient forms and concentrations.
>
> Since 1977, the lake has been sampled at a fixed point (mid-lake deep [MLD]) in the deepest part of the lake (110 m) adjacent to the Flathead Lake Biological Station. Under FMP, the lake is generally sampled 15 times per year (e.g., monthly with additional sampling in June, July, and August). Besides measures of water transparency, light penetration, and profiles of physical/chemical properties (including oxygen, via Hydrolab), FMP also takes discrete water samples (5 m, 90 m, and 0–30 m composite) that are analyzed for concentrations and forms of phosphorus (total phosphorus [TP], soluble reactive phosphorus [SRP]), nitrogen (total nitrogen [TN], $NO_{3/2}$, $NH_{4/3}$), and alkalinity. Biological assessments include chlorophyll concentrations and primary productivity (via ^{14}C) and microscopic assessment of the phytoplankton community. In response to queries from the local community, in 2018 a second monitoring station in the shallow (9 m) south end of the lake in Polson Bay was established and monitored on the same schedule as for MLD. Finally, assessments of periphyton biomass involve chlorophyll analysis for ten samples at 5-m depth at each of two fixed stations located on islands.
>
> FMP data have been essential in assessing the ongoing effects of Mysis invasion on the lake ecosystem (Spencer et al., 1991; Ellis et al., 2011) as well as in documenting the lake's persistent N:P stoichiometric imbalance (J. J. Elser, unpublished).

sampling program that provides data on at least 15 dates per year. However, since Flathead Lake is less like a "run-of-the-river" system than CDA Lake, FMP focuses on a single deep-water sampling station with much less assessment of spatial variation than occurs at CDA Lake. Similarly, the water quality monitoring program at Lake Tahoe (Box 8-2) provides an excellent example of more effective time series measurements of temperature and dissolved oxygen, it conducts separate monitoring of littoral zones, it monitors both nitrogen and phosphorus and multiple forms of these nutrients, and it uses calibrated models in conjunction with the monitoring data to inform management decisions.

BOX 8-2
Monitoring in Lake Tahoe

Lake Tahoe, the second deepest lake in the country, straddles the border between California and Nevada at a latitude of 39°N. It is a sub-Alpine lake, situated at an elevation of 1,897 m, with a maximum depth of 501 m (average depth 305 m). With a lake surface area of 495 km^2 contained within a relatively small watershed area of 800 km^2, it has a very long mean retention time of about 600 years.

Stream monitoring at Lake Tahoe is conducted on a representative subset of the 63 streams that enter the lake. Currently seven streams are monitored by the USGS and UC Davis, with those streams cumulatively representing about 40 percent of the total stream inflow volume. All these streams have gaging stations for flow and stage. Over the past seven years they have been upgraded with continuous temperature and turbidity sensors that permit improved estimation of nutrient and sediment loads (Smith and Naranjo, 2021). Approximately 25 times each year, water samples are taken for laboratory analyses of a suite of nutrients, particle size analysis, and total suspended solids. Funding for the stream program (Lake Tahoe Interagency Monitoring Program) is provided by the bi-State Tahoe Regional Planning Agency (TRPA), USGS, and UC Davis.

The continuous, routine monitoring of Lake Tahoe's water has primarily been conducted by UC Davis since 1968. At two stations within the lake (the Mid-lake station with a water depth of 500 m, and the Index station with a water depth of 125 m), water samples are taken from 13 depths each month (the samplings are staggered in time) and analyzed for soluble reactive phosphorus, total phosphorus, nitrate, nitrite, ammonium, total Kjeldahl nitrogen, dissolved inorganic carbon, and particle size analysis. At a subset of depths at the Index station, samples are also taken for phytoplankton speciation and enumeration. Primary productivity, measured using the ^{14}C technique, has been monitored at least monthly. At both stations, vertical profiles of temperature, electrical conductivity, dissolved oxygen, pH, chlorophyll fluorescence, beam transmissivity, turbidity, and photosynthetically active radiation (PAR) are taken at 8 Hz. These measurements are made every 10 days at the Index station, along with a Secchi depth measurement. Monthly *in situ* vertical profiles of particle-size distribution are made at the Index station. Measurements of nitrogen and phosphorus deposition on the lake are made at one station every two weeks.

Continuous measurements at high frequency come from a number of UC Davis–maintained stations. These include two thermistor chains (one in the east with 450 m water depth, one in the west with 120 m water depth) with a dissolved oxygen sensor at the bottom, and a network of 10 real-time nearshore stations measuring temperature, conductivity, dissolved oxygen, and turbidity around the lake. The funding for this program, which has been scaled back to its present status over the years, comes from the TRPA, the Lahontan Regional Water Quality Control Board (LRWQCB, part of the state EPA), and UC Davis. For details of the measurements refer to UC Davis (2021).

Intermittently since 1982, but continuously since 2000, the attached algae (periphyton) that grow on rocks at the edges of the lake have been monitored. This program, funded by the LRWQCB, has monitored dozens of sites around the entire lake perimeter using snorkel surveys and collecting samples of algae for biomass and chlorophyll assessment.

Important additional parts of the monitoring program have been maintained for the past 20 years using non-traditional funding support. These include on-lake and shoreline meteorological stations that have been supported by the National Aeronautics and Space Administration's Jet Propulsion Laboratory and philanthropic support; time-continuous measurement of temperature and dissolved oxygen from instrument chains at multiple locations supported primarily by philanthropic support; real-time monitoring of nearshore temperature, dissolved oxygen, turbidity, and chlorophyll fluorescence; and quarterly monitoring of zooplankton.

This extensive data repository does not answer all questions about the condition of Lake Tahoe, particularly as new issues emerge over time. It does, however, serve several important functions. First, it allows for the evaluation of the trajectory of long-term change. Many such changes are clearly linked to climate change and cannot be identified with datasets that cover only short time ranges. Second, they provide a contextual background for more intensive, short-term studies that are conducted by researchers from various institutions. Third, with the trajectory of change seeming to accelerate, and with the occurrence of novel events (e.g., harmful algal blooms), there is a growing recognition of the need for predictive modeling to help inform lake management decisions. Such datasets are a necessary component for developing and applying such models. Although funding this long-term database over time has been challenging, its existence now, at a time when rapid changes are being experienced, makes it invaluable for both advancing scientific understanding and providing guidance and support for planning and management decisions.

INSTITUTIONAL CONSIDERATIONS

Chapter 2 discussed the reports that IDEQ, the CDA Tribe, the USGS, and EPA produce on a regular basis to summarize and analyze their long-term monitoring data to determine whether the Superfund remedy and the Lake Management Plan (IDEQ and CDA Tribe, 2009) are effective. Trend and special study results need to be published with minimal latency and coordinated among agencies and localities in the watershed and the Lake. That is, they should be published within one year after the end of the time period being evaluated (e.g., trends through water year 2020 should be published sometime in 2021). The publication of trend results needs to be very explicit about the analysis methods that were used, including issues such as treatment of censored data ("less-than" values). This kind of publication is common in other highly visible and highly complex systems such as the Chesapeake Bay, Lake Champlain, Lake Erie, the Mississippi River, and Lake Tahoe.

Chapter 2 also mentioned the many benefits of making water quality data widely available to interested parties, and the agencies involved in data collection are encouraged to provide a mechanism to make the relevant data available to the wider community of stakeholders, agencies, and scientists. Placing the data in an easily accessible platform can increase public trust, and it has the advantage of giving external scientists the chance to do analysis that could provide confirmation of the agency's own conclusions or open up discussion because of differences in findings among scientists. The river data are already available through such a system (USGS, 2020), but the Lake data are not in a common system. A unified data-management and data-sharing system would need to be created.

Finally, a system such as the Coeur d'Alene should have a process for regularly producing data synthesis products that interpret the spatial and temporal characteristics of the data, with particular attention to trends. The required synthesis tasks are regularly evaluating mass balances, evaluating and explaining concentration trends, estimating exceedances of criteria, generating hypotheses about system drivers, and evaluating the ecological and human health implications of findings. This regular synthesis analysis depends on a robust and long-term monitoring program that collects a variety of physical, chemical and biological data as discussed in this report. The analysis of trends should be applied to the various datasets frequently (e.g., once every two years) using agreed-on methods that have been published in scientific journals. The data analysis methods should have at least these characteristics: (1) they need to be able to capture and evaluate non-monotonic trends, (2) they need to use methods appropriate to censored data (non-detects) where censoring exists, (3) they need to use methods appropriate to seasonal datasets, and (4) they should be designed to resist the potentially confounding effects of sequences of wet and dry years. The methods should be agreed to by a consensus of the organizations involved in long-term action and/or monitoring of the water quality of the Lake and watershed. This does not preclude subjecting the data to new and experimental data-analysis methods, nor should it preclude evolving toward better methods as the field of environmental data-analysis progresses, but there needs to be a basic set of assessment methods that are applied regularly over time.

Evaluation of the success of management actions should be based on monitoring and analysis designed to quantify the impact of those actions on ecosystem health. This is crucial to supporting future steps needed for the improvement of water quality in the Lake and understanding the Lake responses to remediation and shoreline development. Bringing the agencies and stakeholders together on a regular basis (every five years, at minimum) to publish a synthesis on the state of the CDA watershed and Lake and the regional/global changes in conditions would enhance public trust and help point toward agreements among the agencies and stakeholders about future priorities for protection of the Lake. The institutional structure could take many forms, and it should enhance collaboration among federal (USGS, EPA), state (IDEQ, Fish and Game) tribal (CDA Tribe) and academic (University of Idaho and Idaho State University) parties to ensure that resources can be maximized.

CONCLUSIONS AND RECOMMENDATIONS

The main features of the long-term monitoring program for the CDA watershed are carried out in a highly professional manner with good sampling protocols and quality-assurance/quality-control procedures. Although the river sampling has an appropriate level of focus on high-flow sampling and the Lake sampling provides good coverage of the main body of the Lake, the following improvements to the long-term monitoring program are needed to better understand the water quality changes likely to accompany growth in the region and climate change. This is especially critical given the Lake's historical importance to the CDA Tribe, its proximity to a major Superfund site, and as the focus of a rapidly expanding regional center (around the city of Coeur d'Alene).

1. **Understanding of the water quality status of CDA Lake could be improved by increasing the spatial and temporal intensity of sampling in the Lake.** The Lake program could expand to encompass littoral areas by (1) establishing a finite number of fixed littoral sites, (2) conducting synoptic sampling of a broader range of littoral sites at one time of year, (3) considering use of remote sensing to better understand dynamics in the littoral zones, and (4) engaging private citizens in data collection. In terms of temporal improvements, monthly sampling can miss important limnological events such as the onset of algal blooms that signal water quality degradation. Although intensifying sampling frequency can add costs, especially for samples that are intensive to process (such as phytoplankton community samples), higher temporal resolution will increase the chances of capturing such short-lived events and allow them to be placed into a broader context of long-term trends. A strategy to increase temporal resolution in Lake sampling can also be achieved by implementation of carefully chosen sensors targeting physical (temperature), chemical (oxygen), and biological (chlorophyll, fluorescence) properties of interest.

2. **The monitoring strategy for rivers in the CDA watershed could be improved by increasing the sampling frequency, particularly at the upstream and downstream ends of the lower basin, with continued attention to sampling high-flow events (including the first flush).** Use of new continuous monitoring strategies, particularly for turbidity, can be of great value to the estimation of transport into and out of the lower CDA River. Knowledge of transport (particularly for lead and phosphorus) will be critical to the design, evaluation, and enhancement of whatever strategies are employed in the future to limit pollutant delivery to the Lake.

3. **An efficient sampling strategy designed to better understand inputs of nutrients from the lakeshore tributaries would benefit lake management as the population in nearshore areas grows.** Since the time this study began, the IDEQ has initiated a strategy to fill this information gap. It will be crucial after a few years of data have been collected to undertake and publish a synthesis of what can be learned from this monitoring of Lake tributaries, and to use that to improve the monitoring network and ongoing analyses. The goal would be to use these new data and geospatial landscape and development data to produce estimates of nutrient fluxes across the majority of the previously unmonitored 16 percent of the CDA Lake watershed.

4. **The dissolved phosphorus data available for the major tributaries to CDA Lake are of limited quantity and quality due to methodological constraints.** Better assessment of the chemical composition and bioavailability of phosphorus entering the Lake from tributaries is needed. That is, knowing what proportion of the total phosphorus load to CDA Lake comes as inorganic particulate phosphorus and the mineralogy of this phosphorus (i.e., if it is bound to aluminum, iron, calcium, lead, etc.) could be very informative for understanding whether these inputs are likely to stimulate primary production and generate organic matter in the Lake. Hence, both total and dissolved phosphorus should be quantified for all samples so that particulate phosphorus concentrations can be estimated. On a few occasions, the mineral composition of the particulate phosphorus fraction should be characterized for each of the main tributaries. The river and Lake sampling programs need to consistently meet a 1 µg/L detection limit for both total and dissolved phosphorus.

5. **More monitoring of the physical characteristics of CDA Lake** would provide information on the transport, mixing, and dilution of dissolved and particulate material (both horizontally and vertically) and insights into the spatial distributions of key variables such as dissolved oxygen, metals, nutrients, and the biota. Time series data provided by in situ sensors on vertical chains, particularly for temperature and dissolved oxygen, would be highly informative. Acoustic Doppler current profilers deployed at the bottom of the Lake could be used to measure water velocity from the bottom to the surface of the Lake. The use of high-resolution, multiparameter profilers that measure temperature, electrical conductivity, dissolved oxygen, turbidity, chlorophyll fluorescence, pH, PAR, and beam attenuation would complement the vertical instrument chain data.

6. **Monitoring of water-column metal enrichment is adequate and must be sustained.** Greater attention to detection limits, interlaboratory comparisons, periodic evaluation of total particulate metal, and "total recoverable" treatments would benefit the existing programs. Quantifying the differences between traditional dissolved concentrations and colloidal concentrations, at least during periods of elevated discharge, could reduce uncertainties about the forms of metal in the dissolved fraction. Periodic calculation of dissolved metal speciation might also reduce uncertainties about metal bioavailability during high discharge events. Support for the monitoring program and coordination among jurisdictions are necessary priorities into the future as changes in the Lake manifest.

7. **As mentioned in Chapter 7, new data on sediment cores could help to better understand the spatial distributions of metal contamination and trends through time.** Understanding the effects of metals on the benthic food web of the Lake requires more granular information on metals contamination of sediments as a function of water depth, hydrodynamics and other drivers, including the location of hot spots. Knowledge on the spatial patterns of lead contamination in littoral sediments and historically flooded soils could aid anticipating human exposure pathways.

8. **Important ecological components of the Lake are understudied.** Targeted expansion of ecological monitoring beyond the phytoplankton community could help identify how ecological processes in the Lake are responding to changing metal and nutrient concentrations. Obvious questions relate to the effects of the legacy of metal enrichment on the pelagic and benthic food webs as well as the status of fish populations in parts of the Lake with different exposures to metals and different water-column communities (e.g., algal communities). Data on zooplankton, phytoplankton, and bacterioplankton taxonomic composition and dynamics are notably limited. There is very little monitoring of higher trophic levels (zooplankton, fish), nor are there focused studies of ecological interactions such as food web structure or benthic–pelagic coupling.

9. **The agencies involved in data collection are encouraged to provide a mechanism to make the relevant data available to the wider community of stakeholders, agencies, and scientists.** The river data are already available through such a system, but the Lake data are not. A common data repository, mechanisms to facilitate data sharing, and regular discussions of data interpretations among interested parties would benefit the lake-wide syntheses necessary to anticipate changes into the future (see below).

10. To succeed at adaptively managing the Lake for decades into the future, **a scientific and institutional structure for carrying out data synthesis, coordinated among jurisdictions and interest groups, is needed.** The synthesis tasks undertaken should be to regularly evaluate mass balances, relate concentration trends to lake processes and inputs, generate hypotheses about system drivers, and periodically evaluate the ecological and human health implications of the findings. Lake-wide monitoring with the goal of evaluating change and management into the future can only benefit from improved coordination among all the parties.

REFERENCES

Balistrieri, L. S., J. W. Murray, and B. Paul. 1994. The geochemical cycling of trace elements in a biogenic meromictic lake. *Geochim. Cosmochim. ACTA* 58(19):3993–4008.

Balistrieri, L., S. Box, A. Bookstrom, R. Hooper, and J. Mahoney. 2002. Impacts of Historical Mining in the Coeur d'Alene River Basin. Chapter 6 *In:* Pathways of Metal Transfer from Mineralized Sources to Bioreceptors: A Synthesis of the Mineral Resources Program's Past Environmental Studies in the Western United States and Future Research Directions. USGS Bulletin 2191. L. S. Balistrieri and L. L. Stillings, eds.

Balistrieri, L. S., and R. G. Blank. 2008. Dissolved and labile concentrations of Cd, Cu, Pb, and Zn in the South Fork Coeur d'Alene River, Idaho: Comparisons among chemical equilibrium models and implications for biotic ligand models. *Appl. Geochem.* 23:3355–3371.

Chanat, J. G., D. L. Moyer, J. D. Blomquist, K. E. Hyer, and M. J. Langland. 2016. Application of a weighted regression model for reporting nutrient and sediment concentrations, fluxes, and trends in concentration and flux for the Chesapeake Bay Non-tidal Water-Quality Monitoring Network, results through water year 2012. U.S. Geological Survey Scientific Investigations Report 2015–5133. http://dx.doi.org/10.3133/sir20155133.

Chess, D. 2021. Water Quality Data Summary. Presentation to the NASEM Committee. February 26, 2021.

Coats, R., J. Lewis, N. Alvarez, and P. Arneson. 2016. Temporal and Spatial Trends in Nutrient and Sediment Loading to Lake Tahoe, California-Nevada, USA. JAWRA 52(6):1347–1365.

EcoAnalysts, Inc. 2017. Couer d'Alene Lake Management Plan 2011-2015 Benthic Survey Summary Report. Prepared for Idaho Department of Environmental Quality.

Ellis, B. K., J. A. Stanford, D. Goodman, C. P. Stafford, D. L. Gustafson, D. A. Beauchamp, D. W. Chess, J. A. Craft, M. A. Deleray, and B. S. Hansen. 2011. Long-term effects of a trophic cascade in a large lake ecosystem. *Proceedings of the National Academy of Sciences* 108(3):1070–1075. https://doi.org/10.1073/pnas.1013006108.

EPA. 2020. Adaptive Management Project Management Plan for the Lower Basin of the Bunker Hill Mining and Metallurgical Complex, Shoshone County, Idaho. 20-019(E)/041320.

Gilroy, E. J., R. M. Hirsch, and T. A. Cohn. 1990. Mean-square error of regression-based constituent transport estimates. *Water Resources Research* 26(9):2069–2077.

Harrington, J. M., M. J. LaForce, W. C. Rember, S. E. Fendorf, and R. F. Rosenzweig. 1998. Phase associations and mobilization of iron and trace elements in Coeur d'Alene Lake, Idaho. *Environ. Sci. Technol.* 32(5):650–656.

Horowitz, A. J., K. A. Elrick, and R. B. Cook. 1993. Effect of mining and related activities on the sediment trace element geochemistry of Lake Coeur d'Alene, Idaho, USA. Part I: surface sediments. *Hydrol. Process.* 7:403–423.

Horowitz, A. J., K. A. Elrick, J. A. Robbins, and R. B. Cook. 1995. Effect of mining related activities on the sediment trace element geochemistry of Lake Coeur d'Alene, Idaho, USA part II: subsurface sediments. *Hydrol. Process.* 9:35–54. https://doi.org/10.1002/hyp.3360090105.

IDEQ and CDA Tribe (Idaho Department of Environmental Quality and the Coeur d'Alene Tribe). 2009. Coeur d'Alene Lake Management Plan.

IDEQ and CDA Tribe. 2020. Coeur d'Alene Lake Management Program: Total Phosphorus Nutrient Inventory, 2004–2013.

Kimball, B. A. 1997. Use of tracer injections and synoptic sampling to measure metal loading from acid mine drainage. USGS Fact Sheet 245-96. https://doi.org/10.3133/fs24596.

Kuwabara, J. S., P. F. Woods, W. M. Berelson, L. S. Balistrieri, J. L. Carter, B. R. Topping, and S. V. Fend. 2003. Importance of sediment-water interactions in Coeur d'Alene Lake, Idaho, USA: management implications. *Environmental Management* 32:348–359. https://doi.org/10.1007/s00267-003-0020-7.

Kuwabara, J. S., B. R. Topping, P. F. Woods, J. L. Carter, and S. W. Hager. 2006. Interactive effects of dissolved zinc and orthophosphate on phytoplankton from Coeur d'Alene Lake, Idaho. USGS Scientific Investigations Report 2006–5091. http://pubs.usgs.gov/sir/2006/5091.

Kuwabara, J. S., Topping, B. R., Woods, P. F., and J. L. Carter. 2007. Free zinc ion and dissolved orthophosphate effects on phytoplankton from Coeur d'Alene Lake, Idaho. *Environ. Sci. Technol.* 41(8):2811–2817. https://doi.org/10.1021/es0629231.

Lee, C. J., R. M. Hirsch, and C. G. Crawford. 2019. An evaluation of methods for computing annual water-quality loads. U.S. Geological Survey Scientific Investigations Report 2019–5084. https://doi.org/10.3133/sir20195084.

MacIntyre, S., J. P. Fram, P. J. Kushner, N. D. Bettez, W. J. O'Brien, J. E. Hobbie, and G. W. Kling. 2009. Climate-related variations in mixing dynamics in an Alaskan arctic lake. *Limnol. Oceanogr.* 54(6, part 2):2401–2417.

Marcé, R., G. George, P. Buscarinu, M. Deidda, J. Dunalska, E. de Eyto, G. Flaim, H.-P. Grossart, V. Istvanovics, M. Lenhardt, E. Moreno-Ostos, B. Obrador, I. Ostrovsky, D. C. Pierson, J. Potužák, S. Poikane, K. Rinke, S. Rodríguez-Mozaz, P. A. Staehr, K. Šumberová, G. Waajen, G. A. Weyhenmeyer, K. C. Weathers, M. Zion, B. W. Ibelings, and E. Jennings. 2016. Automatic high frequency monitoring for improved lake and reservoir management. *Environ. Sci. Technol.* 50(20):10780–10794.

Meinson, P., A. Idrizaj, P. Nõges, T. Nõges, and A. Laas. 2016. Continuous and high-frequency measurements in limnology: history, applications, and future challenges. *Environmental Reviews* 24(1). https://doi.org/10.1139/er-2015-0030.

Morra, M. J., M. M. Carter, W. C. Rember, and J. M. Kaste. 2015. Reconstructing the history of mining and remediation in the Coeur d'Alene, Idaho Mining District using lake sediments. *Chemosphere* 134:319–327. https://doi.org/10.1016/j.chemosphere.2015.04.055.

Oelsner, G. P., L. A. Sprague, J. C. Murphy, R. E. Zuellig, H. M. Johnson, K. R. Ryberg, J. A. Falcone, E. G. Stets, A. V. Vecchia, M. L. Riskin, L. A. De Cicco, T. J. Mills, and W. H. Farmer. 2017. Water-quality trends in the Nation's rivers and streams, 1972–2012: Data preparation, statistical methods, and trend results (ver. 2.0, October 2017). U.S. Geological Survey Scientific Investigations Report 2017–5006. https://doi.org/10.3133/sir20175006.

Roberts, D. C., G. C. Egan, A. L. Forrest, J. L. Largier, F. A. Bombardelli, B. E. Laval, S. G. Monismith, and S. G. Schladow. 2020. The setup and relaxation of spring upwelling in a deep, rotationally influenced lake. *Limnol. Oceanogr.* 66(4):1168–1189.

Scofield, B. D., K. Torso, S. F. Fields, and D W. Chess. 2021. Contaminant metal concentrations in three species of aquatic macrophytes from the Coeur d'Alene Lake basin, USA. *Environ. Monit. Assess.* 193:683. https://doi.org/10.1007/s10661-021-09488-y.

Smith, D. W., and R. C. Naranjo. 2021. Model Archive Summaries for Fine Sediment Particles Surrogate Regression Models, Lake Tahoe, California and Nevada: U.S. Geological Survey data release.

Spencer, C. N., B. R. McClelland, and J. A. Stanford. 1991. Shrimp stocking, salmon collapse and eagle displacement: Cascading interactions in the food web of a large aquatic ecosystem. *Bioscience* 41:14–21.

Sprague, L. A. 2001. Effects of storm-sampling frequency on estimation of water-quality loads and trends in two tributaries to Chesapeake Bay in Virginia: U.S. Geological Survey Water- Resources Investigations Report 2001–4136.

Toevs, G. R., M. J. Morra, M. L. Polizzotto, D. G. Strawn, B. C. Bostick, and S. Fendorf. 2006. Metal(loid) diagenesis in mine-impacted sediments of Lake Coeur d'Alene, Idaho. *Environ. Sci. Technol.* 40(8):2537–2543. https://doi.org/10.1021/es051781c.

USGS (U.S. Geological Survey). 2020. National Water Information System (NWIS): U.S. Geological Survey, web interface, accessed January 2022. https://waterdata.usgs.gov/nwis.

UC Davis. 2021. Tahoe: State of the Lake Report. https://tahoe.ucdavis.edu/stateofthelake.

Vander Zanden, M. J., and Y. Vadeboncoeur. 2002. Fishes as integrators of benthic and pelagic food webs in lakes. *Ecology* 83(8):2152–2161.

Ward, N. K., J. A. Brentrup, D. C. Richardson, K. C. Weathers, P. C. Hanson, R. J. Hewett, and C. C. Carey. 2022. Dynamics of the stream–lake transitional zone affect littoral lake metabolism. *Aquat. Sci.* 84:31. https://doi.org/10.1007/s00027-022-00854-7.

9

Risks of Metals Contamination in Coeur d'Alene Lake

As part of its statement of task, the committee was asked to discuss the relevance of metals release in the Coeur d'Alene (CDA) Lake to human and ecological health. This was premised on the possibility that nutrient inputs to the Lake may be increasing, which might promote growth of algae and consequently lead to low oxygen conditions in the Lake that could promote release of sediment-bound metals.

The remedy for the Bunker Hill Superfund site was based on removing sources of metals contamination and/or breaking exposure pathways linking metals to human and ecological receptors in the CDA Lake watershed. The remedy is supported by human health (Terra Graphics/URS Greiner/CH2M Hill, 2001) and ecological (CH2M Hill and URS Corp., 2001) risk assessment reports that were reviewed by a committee of the National Academies in 2005 (NRC, 2005). The U.S. Environmental Protection Agency (EPA) has not selected a remedy for CDA Lake (see Chapter 1); hence, evaluations of the Lake are not included in the published risk assessments.

This chapter specifically focuses on exposure pathways and attendant risks of metals in CDA Lake. For human health, this includes occupational and recreational exposure and exposure from fish consumption and drinking water (for those who derive drinking water from waterbodies hydrologically connected to the Lake, such as the Rathdrum Prairie Aquifer). In addition, the chapter also delves into some of the risks related to nutrient enrichment that go beyond exposure to metals, in particular the possibility of harmful algal blooms (HABs). For ecological health, the discussion is much more expansive given the Lake's complex ecology and includes risks from exposure to metals in pelagic (phytoplankton, zooplankton, fish) and benthic communities of animals and plants.

HUMAN HEALTH RISKS

As discussed in Chapter 1, the Bunker Hill Mining and Metallurgical Complex was designated as a Superfund site based on the high blood lead levels in children and contamination of the local environment by lead, arsenic, cadmium, and zinc. As part of the Superfund process, a human health risk assessment (HHRA) was conducted for the CDA basin to determine the extent of contamination, understand the potential risk to humans who come into contact with contaminated media, and provide information to support remedial activity and cleanup benchmarks (Terra Graphics/URS Greiner/CH2M Hill, 2001). The HHRA also considered the practice of subsistence lifestyles by members of the CDA Tribe. NRC (2005) provided an independent review of the progress on the remediation and the HHRA, and five-year updates of the Superfund remedy have evaluated remediation progress (the most recent being EPA Region 10, 2021).

The areas considered in the 2001 HHRA were the populated areas of the Bunker Hill Box and the upper and lower CDA basins, where nearly all historic primary mineral extraction activities and releases of primary wastes were located. A screening-level risk assessment was also completed in 1999 on 24 common-use areas along the shoreline of CDA Lake and the Spokane River (Appendix B, Terra Graphics/URS Greiner/CH2M Hill, 2001) including eight locations south of the city of Coeur d'Alene. The goal was to assess potential health risks to recreational users of beaches and picnic areas. Although the Lake's shoreline and access points were considered in the HHRA, there is no formal assessment for CDA Lake itself, even though highly contaminated secondary and tertiary particulate and dissolved contaminants have been deposited for the past 120 years in the sediments of CDA Lake.

The set of contaminants selected for the HHRA included eight metals: antimony, arsenic, cadmium, iron, lead, manganese, mercury, and zinc. The primary human health concerns are associated with arsenic, lead, and mercury. Lead is a known neurotoxin posing particular risk to sensitive subpopulations. Arsenic is a potential carcinogen in skin, bladder, kidney, lung and liver; it also poses a risk via the ingestion pathway for pre-cancer and non-cancer effects. Mercury is also considered here given its potential for human health effects and the advisory in place warning of risks from eating fish potentially contaminated with mercury. The following sections briefly summarize conclusions from the 2001 HHRA, the 2005 NRC study and the latest five-year review (EPA Region 10, 2021) as they might apply to human health exposure pathways relevant to CDA Lake.

Lead

Lead is of human health concern in the CDA basin because widely distributed lead-enriched[1] mine wastes have contaminated soils, dust, sediments, and waters that people contact, incidentally ingest, and inhale. Understanding the effects of lead on humans is essential to appreciating the risks of accidental exposures to these wastes. Because a full review of the rich literature that exists on the effects of lead toxicity in humans is beyond the scope of this report, this section briefly summarizes the state of understanding and its application to CDA Lake.

Effects of Lead in Humans

Naranjo et al. (2020) recently summarized the molecular mechanisms by which lead manifests its effects within cells:

> Multiple processes have been described in which lead has an adverse effect. . . . Lead can disturb cellular functions because it substitutes for calcium[2], and to a lesser extent, zinc, and activates processes reliant on calmodulin, a calcium-binding messenger protein. Lead also binds to the sulfhydryl group of proteins, making it particularly toxic to multiple enzymes. Lead interferes with heme production by inhibiting the enzyme delta-aminolevulinic acid dehydratase and by altering the incorporation of iron by ferrochelatase, resulting in microcytic, hypochromic anemia. The vitamin D receptor has been also described to modulate lead uptake, because it is involved in intestinal calcium absorption and calcium storage in bone. Thus, gene variants of delta-aminolevulinic acid dehydratase and the vitamin D receptor are considered susceptibility markers of lead toxicity in humans. In the liver, lead interferes with cytochrome P450 enzymes. Lead easily crosses cell membranes and exerts pro-oxidative effects within the cell with formation of reactive oxygen species, thereby activating processes of programmed cell death. Lead can also deplete intracellular glutathione, an important antioxidant.

The most notable effects of lead lie in its ability to harm the sensitive and complex processes of central nervous system development. These effects result from exposure to lead during pregnancy (on the growing fetus) and in early childhood. The degree of damage to the fetus or the growing child depends upon the length of exposure and the cumulative amount of lead in the body. Naranjo et al. (2020) note that a "cascade of neurological dysfunction" occurs with the slightest disruption of developmental processes in the central nervous systems of a growing child. At that age there are limited opportunities to repair or compensate for such changes. In particular, exposure of the developing fetus or growing child to lead can target dopamine and the hippocampus in the developing brain, both

---

[1] Pb-enriched is defined by Bookstrom et al. (2013) as concentrations in excess of 1,000 µg/g dry weight.

[2] Lead and calcium have similar physical and chemical characteristics and follow similar intracellular metabolic pathways (Potula and Kaye, 2005).

of which are involved in learning and memory. The developing child is particularly sensitive to lead exposure because the blood-brain barrier, which helps block accumulation of lead in adult brains, is immature and less effective in children. There is also an increased absorption of lead from food in the gut of children compared to adults. Normal childhood behaviors like putting dust-contaminated objects, lead-enriched dirt, or lead-contaminated toys in their mouths increases the likelihood of incidental exposure. The effects that manifest are tragic and well documented over years of study (Needleman and Bellinger, 1991; Finkelstein et al., 1998; Lidsky and Schneider, 2003; Needleman, 2004).

The effects of lead on neurological development manifest as poorer scores on behavioral and developmental tests, including IQ tests (Bellinger et al., 1987; Canfield et al., 2003; Lanphear et al., 2000; Schwartz, 1994; Wasserman et al., 1994), psychosocial morbidity (Bellinger et al., 1994; Wasserman et al., 1998) and juvenile delinquency (Needleman et al., 1996). Naranjo et al. (2020) summarized the current state of understanding. "Long-standing exposure to any amount of lead has been associated with intellectual disability in a dose-dependent manner, ranging from delay or loss of developmental milestones to reduced cognitive function and academic achievement. Other symptoms are shortened attention span, impaired executive function, delayed processing speed, and impairments in visual and verbal memory and visuospatial skills." Lead exposure, especially before age three, appears to affect learning and behavior even with only low levels of lead detected in the blood, and these effects then persist through childhood and adolescence. Effects from early exposure that appear to be manifested later in life include anxiety and depression. Aggressive, criminal, and antisocial behaviors have been described with chronic lead exposure. Renal disease, hypertension (high blood pressure), and degenerative diseases are found in adults who suffered childhood exposure to lead. Exposure to lead during development is also associated with increases in future susceptibility to nerve degeneration and the likelihood of developing Alzheimer disease later in life (Naranjo et al., 2020).

Once in the blood stream, the way lead is passed to tissues is independent of exposure route. In pregnant women, lead in the blood can cross the placenta and then the blood brain-barrier in the developing fetus. From the blood, lead can also distribute to other soft tissues and eventually to bone. It progressively accumulates in bone during exposure, where it can persist for decades (Potula and Kaye, 2005). Bone stores of lead can become dynamic and remobilize back into the blood under some conditions, especially when changes in bone turnover occur during stages of life (e.g., during pregnancy, lactation, or menopause; Potula and Kaye, 2005).

The critical concentration of lead in blood at which effects begin to manifest in children was the subject of debate when the early studies were published (Needleman, 2004) because of the difficulties with differentiating between the effects of multiple stressors. In 1991, the Centers for Disease Control and Prevention (CDC) felt the evidence was sufficient to suggest a blood lead level (BLL) of 10 µg/dL in children as the lowest level of concern. Exceeding the level of concern in a local population usually resulted in actions to reduce exposures. There is now both toxicological and epidemiological evidence that adverse health effects are associated with a BLL < 10 µg/dL (100 µg/L) (EPA Region 10, 2021). In 2012, the CDC lowered the blood lead reference value (level of concern) to < 5 µg/dL (Ruckert et al., 2021) and in 2021, the CDC further lowered it to 3.5 µg/dL (CDC, 2021). According to some literature, there is no concentration below which there are no adverse effects of lead exposure, leading some to suggest lowering the BLL of concern to about 2 µg/dL (Gilbert and Weiss, 2006). The U.S. government pledged in Healthy People 2020 to reduce mean BLLs to 1.6 µg/dL among children ages 1–5 (Digman et al., 2019). EPA has not yet changed its national lead health risk policy; thus, the remediation action objective for the CDA basin remains at a mean BLL of 10 µg/dL (EPA Region 10, 2021).

Routes of Exposure

People can be exposed to lead through ingestion, inhalation, or dermal contact. Direct lead uptake through the skin is not expected to be a major route because the most abundant forms of lead in the CDA basin are inorganic and would not efficiently penetrate the skin. Lead-enriched particles on the skin or clothing can lead to incidental ingestion, contribute to dust coatings on tools and household items, or contribute to dust that can be inhaled. In regions with a mineral extraction history, inhalation of lead-impacted ambient particulate matter in the air (e.g., blowing dust) is a significant exposure pathway. Resuspension of lead-enriched household dust may contribute to indoor exposure, with children being at higher risk compared to adults because of their higher respiratory rate, their closeness to the ground relative to adults, and their higher fractional deposition of particles in the lungs (Kastury et al., 2019). Inhalation studies with fine particulate material (≤ 0.25 μm) suggest that about one-fourth of the inhaled lead particles deposit in the lungs, and up to 95 percent of the lead on those particles is then absorbed (Kastury et al., 2019).

Compared to inhalation, incidental ingestion of soil and surface dust is a less common exposure pathway, except for children (Naranjo et al., 2020). Incidental ingestion of lead can result from hand–mouth activity typical in 1- to 4-year old children. This can be exacerbated by tracking lead-contaminated soils into the home or working in contaminated clothes and shoes, all of which can generate dust as well.

Diet and a complicated array of other factors affect intestinal lead absorption once lead is ingested. Early studies showed that, on average, about 8–18 percent of the ingested lead was absorbed into the blood and tissues of humans; 10 percent was the most frequently quoted average (NRC, 2005). But recent studies show that absorption rates may be higher (Naranjo et al., 2020). For example, data from the CDA basin show that 2.5–39 percent of the lead ingested was absorbed in the body; average absorption efficiency was 26 percent (EPA Region 10, 2021).

A major goal of the Superfund remedy was to reduce exposure of humans to lead. Box 9-1 discusses the most recent progress as related in the 2021 Five-Year Review (EPA Region 10, 2021).

Measures of Lead Exposure

Blood lead levels are the earliest reflection of exposure to lead and are the metric used to determine human exposures within the preceding 30 days. Blood lead levels have been directly related to adverse outcomes in adults and children (NRC, 2005).

Declines in average lead concentrations in domestic dust and soils in the CDA basin have been accompanied by a downward trajectory in blood lead levels of children ages 0–9. Figure 9-1 shows trends by age since the mid-1990s in OU-3 (the area of the Superfund complex outside the Box). The greatest progress in reducing blood lead levels occurred early in the remediation, but in the past ten years the decline appears to have stalled. The average blood lead levels for the 0–9 age group remain in the 2–4 μg/dL range, fluctuating around or near the latest CDC reference value (3.5 μg/dL; CDC, 2021). For OU-3, 6–22 percent of households exceeded the 5 μg/dL threshold (EPA Region 10, 2021), although the accuracy of that estimate was affected by a small sample size (Table 9-1). For comparison, the average incidence of blood lead levels in excess of 5 μg/dL among seven western states with data[3] was 2.02 ± 0.43 percent (CDC, 2021). Thus, percentage-wise, average OU-3 blood lead levels are typically two to four times higher than found on average in other western states for which data are available.

Although the blood lead levels are much improved since the 1990s, several challenges with understanding lead exposures of children in the basin remain, which could affect any future analysis specific to the Lake (as discussed in EPA Region 10, 2021).

Participation in Testing. Low participation in blood sampling was of concern in 2005 (NRC, 2005) and remains an issue (EPA Region 10, 2021). In a typical population of concern, testing 70 percent of the population at risk is ideal, although not always practical, for an accurate picture of exposure to lead. Participation rates in free blood screening events in recent years "was similar to previous years with less than a quarter of the estimated OU-3 child population participating" (EPA Region 10, 2021) and varied from 14 to 21 percent of the population in OU-3 (EPA Region 10, 2021, Table 6-13). Non participation was more concentrated in subpopulations also believed to be at higher risk, including tribal members (EPA Region 10, 2021).

Representativeness of Data. EPA Region 10 (2021) expressed concerns about representativeness of the basin blood lead surveys. Significant disparities in exposure to lead by income, race, and ethnicity have been found in numerous studies across the United States (NRC, 2005). In the CDA basin, vulnerable groups include children from low-income families, families that recreate in un-remediated areas in the basin, and children living in older housing (built before 1950; EPA Region 10, 2021). Across the nation, economically disadvantaged communities

---

[3] States with data in CDC (2021) are Colorado, Nebraska, New Mexico, Oklahoma, Oregon, Texas, and Washington.

BOX 9-1
Lead Exposure Reduction in the Coeur d'Alene Basin

The first priorities for remediation of the Bunker Hill Superfund complex were to reduce lead (Pb) concentrations in soils of homes and commercial properties within the Superfund jurisdiction. A remedial action level of ≥ 700 mg/kg in soils (lowered from the original 1,000 μg/g dw) was first established in operable unit (OU) 1 (EPA Region 10, 2021). The Institutional Controls Program set a soil disposal action level of > 350 mg/kg lead, and the clean replacement material must have < 100 mg/kg. EPA Region 10 (2021) recommended that it was "important to determine whether a community-wide soil lead level is needed for the OU-3 and if so, determine what the appropriate level is and how it would be used."

By 2002, remediation had driven geometric mean house dust lead levels below the established standard of 500 mg/kg (EPA Region 10, 2015). The 2015 Five-Year Review concluded that "community geometric mean lead concentrations in 2013 . . . range from 160 mg/kg to 288 mg/kg for vacuum samples, and from 151 mg/kg to 322 mg/kg for dust mat samples" (Table 3-1; EPA Region 10, 2015, Table 3-1). EPA Region 10 (2021) concluded that "97 percent of eligible parcels have been sampled and remediation is 93 percent complete" in the lower basin.

Although property remediation is nearly complete in the watershed and overall community geometric mean soil lead concentrations and house dust lead concentrations are lower than the action levels, some pathways of exposure still pose risks (EPA Region 10, 2021).

Remaining Unremediated Soils. Despite the impressive accomplishments of the remediation activities, "significant and pervasive" sources of unremediated metal-contaminated soils reside in the lower basin (EPA Region 10, 2021, Chapter 4). Three in every 100 lower-basin homes still have lead concentrations in their dust that exceed 1,000 mg/kg (Alta, 2020). "Approximately 6% of homes in the entire basin do not meet the current EPA risk goal due to elevated soil or dust lead concentrations, or a combination thereof" (Alta, 2020). This becomes problematic if unremediated homes are concentrated in specific ethnic or income groups. EPA Region 10 (2021) also noted that remediation in the lower basin has reduced area residents' and recreational users' contact with lead-enriched soils and sediment along creek channels, but additional hillside and floodplain areas still require access controls and/or remediation.

Recreation. One of the greatest concerns expressed by EPA Region 10 (2021) was the health risks posed from exposure to contaminated soils/sediments of people recreating in the lower basin. "There are numerous contaminated sites used for recreation throughout the Basin that span the spectrum from informally used to fairly highly developed. . . . Although recreational exposures are of relatively short duration, recent studies indicate they can affect overall Pb burdens in people. . . . the frequency of access in localized areas is greater than originally anticipated . . . pedestrian paths leading down the riverbank continue to be created and likely result in erosion and increased exposure . . . Observance of families with young children recreating in un-remediated areas dispersed across the Box and Basin has increased over the past five years." In addition "private property owners continue to establish campgrounds and recreational areas for their personal use in contaminated portions of the Coeur d'Alene River floodplain."

Recontamination. Contaminated sediments from unremediated upper basin mine sites and remobilized floodplain sediments continue to be transported downstream. Recreational sites have recontamination potential from deposition of lead-enriched particles during floods. "There are continuing issues . . . with recontamination of remediated areas after high-flow events" (EPA Region 10, 2021). The Lake is the ultimate repository for these sediments.

Long-Term Performance. EPA Region 10 (2021) noted that many remediated properties have "heavy metals-contaminated soil and water rock underneath six- or twelve-inches of clean soil, gravel and vegetation at levels classified as a hazardous substance." Sustaining the safety of remediated properties in residential and commercial areas "is dependent on several factors including private property owners maintaining soil, gravel, and vegetated barriers in compliance with cleanup criteria and Institutional Control Program requirements, and government jurisdictions maintaining public roads remediated under the Paved Roads Program and stormwater projects completed under the Remedy Protection Projects program." Developed recreational areas such as Trail of the Coeur d'Alenes also require continued maintenance.[a]

---

[a] EPA Region 10 (2021) discusses the challenges maintaining the 73-mile-long Trail of the Coeur d'Alenes, which crosses lead-enriched lands: "Education and outreach, access controls and signage, placement of temporary handwashing stations during busy summer months, and timely monitoring and maintenance of sites help to reduce exposure risks and preserve protective barriers from erosion and slope destabilization. There are continuing issues, however, with recontamination of remediated areas after high-flow events and frequent use of un-remediated areas where only signage exists. Increases in the number of families with young children recreating in un-remediated areas where signs warning of the risks are visibly posted and in numerous informal areas dispersed across the site have been observed over the past five years."

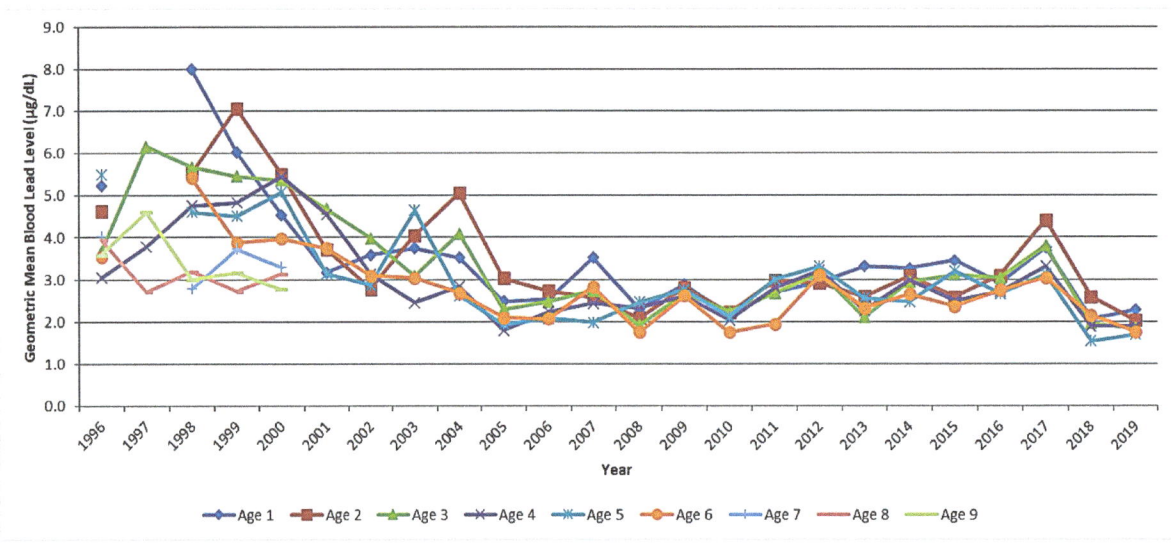

FIGURE 9-1 Geometric mean lead concentrations in blood lead levels by age in OU-3 between 1996 and 2019. SOURCE: Figure D-3 from EPA Region 10 (2021).

and Native American communities disproportionately suffer from elevated human health risks, including elevated blood lead levels (Needleman, 2004). The collapse of the mineral extraction industry by 1990 exacerbated economic problems for some people in the basin, increasing vulnerability to adverse effects from lead. NRC (2005) concluded that "American Indians who practice traditional lifestyles likely would have higher risks than other residents of the Coeur d'Alene River basin" and reviewed the importance of better understanding those risks. Concerns remain about participation being biased by a fixed-site approach to sampling and possible overrepresentation of children thought to be at lower exposure and risk (EPA Region 10, 2021).

Possible Exposure Pathways Involving CDA Lake

From what is known about lead enrichment in the Lake and the human health risks in the basin, some potential pathways of exposure relevant to the Lake itself can be identified beyond the lakeshore exposures that were the subject of the 1999 screening-level assessment. These include human exposure to contaminated lake water and lake sediments—in particular, those whose work or recreation involves incidental exposure to suspended or bed sediments in the Lake, eating fish from the Lake, and other dietary exposure pathways related to subsistence lifestyles.

TABLE 9-1 Number of Children Tested for Blood Lead Levels and Percent Exceedances of the 2012 CDC Reference Level of 5 µg/dL

Year	OU-3		Lower Basin	
	Number of children 0–9	% > 5 µg/dL	Number of children 0–9	% > 5 µg/dL
2015	94	6%	2	ND
2016	70	8%	8	38%
2017	105	22%	3	0%
2018	88	7%	5	0%
2019	84	7%	5	20%

SOURCE: EPA Region 10 (2021).

Studies of lake sediments (Horowitz et al., 1995; Harrington et al., 1998a; Toevs et al., 2006; Morra et al., 2015; EcoAnalysts, Inc., 2017) show that the highest lead concentrations are associated with finer-grained sediments in the Lake—those sediments most likely to be mobilized by disturbances like wind or recreation or to be redeposited on land during floods. As discussed in previous chapters, lead concentrations in lake sediment vary widely and range from 100 to 10,000 µg Pb/g (Bookstrom et al., 2013; Horowitz et al., 1993; EcoAnalysts, Inc., 2017). Patches of extreme lead concentrations (that exceed the action levels for soils in the upper basin Superfund site) occur near Harrison, including Carlin Bay, Rockford Bay, and other bays surrounding the confluence of the CDA River; and Cougar Bay near the head of the Spokane River (Horowitz et al., 1993; Spears et al., 2007; EcoAnalysts, Inc., 2017; Scofield et al., 2021). In contrast, annual lead sampling at the beach at Harrison showed no exceedances of EPA's lead soil target level (EPA Region 10, 2021). A systematic investigation of soils and sediments in the areas where human contact is most likely seems warranted, in order to update the 1999 screening level shoreline investigation (which showed that lead exposures from soils and sediments on the lakeshores were not universally high) and focus on where current populations are most likely to recreate.

The EPA and CDC action level for lead in drinking water is 15 µg/L (the maximum contaminant level goal is zero). CDA Lake is not a source of drinking water, and dissolved lead concentrations rarely, if ever, exceed 1 µg/L during the summer months, when incidental exposure to Lake water during recreation is most likely. Thus, incidental water ingestion during swimming in the Lake is less likely to be a significant exposure pathway than exposure to Lake sediments (EPA, 2001). Ingestion of fish fillets is also less likely to be a primary pathway of lead exposure because fish do not transport most divalent cations to their muscle tissue; rather, they trap these metals in internal organs (especially liver) and perhaps in bones. The Idaho Department of Health and Welfare (Idaho Department of Health and Welfare [IDHW], 2019) analyzed ~ 150 samples of fish fillets from a variety of edible species from the Lake and found that none of the samples exceeded 0.3 mg Pb/kg dw,[4] with all concentrations lower than consumption advisories. However, risks could be higher from consumption of whole fish from the Lake or unpeeled water potatoes from the shoreline of the Lake north of the mouth of the St. Joe River (especially to those practicing a subsistence lifestyle). No data are available on lead concentrations in whole fish, but Scofield et al. (2021) reported that the tubers of water potatoes could contain up to 30 mg Pb/kg tissue (dw).

***

Although it is possible that human health risks associated with the Lake itself are lower on average than those in the areas already covered by the HHRA to date, there is the issue of subpopulations that could be experiencing elevated exposure (as discussed by Moody and Evans, 2011). With that in mind, it seems that exposure pathways relevant to CDA Lake deserve further study, including perhaps a formal HHRA. Several suggestions from the most recent five-year review (EPA Region 10, 2021) are related to the potential lead risks inherent to CDA Lake. These include development of a shoreline-specific soil/sediment lead exposure level, updated evaluations of recreational exposures in common-use areas, better understanding of effects of flooding and flood control in lakeside recreational areas, and improved participation in blood lead testing. A thorough Lake-wide survey of lead concentrations in fine-grained fractions of shoreline soils and sediments could be a first step in determining whether and where further risk assessment would be of value, especially near areas of greatest recreational use. One study showed that "proactive mitigation approaches to cope with potential environmental degradation in lake ecosystems can have significant economic benefits to owners of lakefront properties and local communities" (Liao et al., 2016).

Arsenic

Arsenic is a naturally occurring trace element that poses a threat to human and ecosystem health, particularly when incorporated into food or water supplies. The greatest risk imposed by arsenic to human health results from contamination of drinking water. Ingestion of drinking water with hazardous levels of arsenic can lead to arsenicosis

---

[4] The most reliable metal concentrations in tissues of plants and animals are reported on a dry weight (dw) basis because water content differs among organisms and affects results reported on a wet weight (ww) basis.

and cancers of the bladder, skin, lungs and kidneys. The level of concern for arsenic in drinking water defined by both the World Health Organization and EPA is > 10 µg/L, although adverse effects have been observed at lower concentrations (Ahmad and Bhattacharya, 2019).

Within waters and sediments of CDA Lake, arsenic (As) is present primarily in one of two oxidation states: arsenate [As(V)] or arsenite [As(III)]. In solution, arsenic exists primarily as oxyanionic acids; arsenate exists as $H_2AsO_4^-$ and $HAsO_4^{2-}$, while arsenite exists as H_3AsO_3 (Goldberg and Johnston, 2001). Within soils, microbial activity has been shown to methylate arsenic, leading to chemical species such as dimethyl arsenate (Zhao et al., 2013). Another group of arsenic species, the thioarsenates, have recently been detected in flooded soils and include inorganic thioarsenates and methylated thioarsenate (Wang et al., 2020). Methylated species are usually not abundant in aqueous solutions compared to inorganic forms of arsenic (Smedley and Kinniburgh, 2002; Chen et al., 2021).

Effects of Arsenic in Humans

Numerous publications describe substantial epidemiological evidence that high concentrations of arsenic in drinking water are associated with several detrimental effects on human health, including skin lesions and cancer of the lung, bladder, kidney, and liver, among other effects (Ahmad and Bhattacharya, 2019; Fendorf et al., 2010; NRC, 1999, 2005). Upon ingestion of arsenic-bearing water or food, arsenate is taken up through phosphate transporters, while arsenite is absorbed through aquaglyceroporins (Mukhopadhyay et al., 2002). When arsenate is taken up by a cell, it is reduced to arsenite before secretion or sequestration (Mukhopadhyay et al., 2002). Arsenite, whether absorbed directly or from conversion of arsenate, may go through several methylation steps within the liver, forming monomethylarsenicals and dimethylarsenicals.

Chemical species of both arsenic oxidation states, inclusive of methylated and thiolated species, are toxic to organisms (prokaryotic and eukaryotic). Arsenate is a chemical analog to phosphate and thus may substitute for phosphate in biomolecules. In particular, it interferes with phosphate binding sites in adenosine triphosphate (ATP), resulting in the formation of adenosine diphosphate (ADP)-arsenate, which limits cellular energetics (Brazy et al., 1980; Liebl et al., 1995). Arsenite has a high affinity for thiol (-SH) groups, forming a dihydrolipoylarsenite chelate, which disrupts the structure of proteins and enzymes (Muehe and Kappler, 2014) and thus interferes with critical cellular functions that include chromosomal abnormalities, oxidative stress, altered DNA repair, altered DNA methylation, altered growth factors, cell proliferation, promotion/progression, gene amplification, and p53 gene suppression (Kitchin, 2001). Furthermore, DNA damage may be induced by methylated trivalent arsenicals that are mediated by reactive oxygen species (ROS) (Nesnow et al., 2002; Mass et al., 2001). Early onset of arsenicosis shows an erythematous flush that leads to melanosis, hyperkeratosis, and desquamation. Long-term cutaneous complications include the development of multicentric basal cell and squamous cell carcinomas (Pershagen et al., 1981), and continual exposure leads to even more serious health impacts.

Routes of Exposure

The Superfund remedy considers human exposure to arsenic in the CDA basin via a number of pathways that are the same as with lead, including ingestion of arsenic from contaminated soil and sediment in residential, commercial, and undeveloped areas; and inhalation of contaminated airborne dust generated at these locations. Ingestion of arsenic from drinking water wells is an exposure pathway less relevant to lead that has been explored in the CDA basin, as evidenced by the evaluation of private wells in OU-3 for arsenic contamination (EPA Region 10, 2021, Table 6-2). Any drinking water exposure pathway for arsenic requires that arsenic partitions from the solid phase into the aqueous phase. Processes favoring such partitioning can be grouped into four categories: (1) ion displacement, (2) desorption at pH values > 8.5, (3) reduction of arsenate to arsenite, and (4) mineral dissolution, particularly reductive dissolution of iron and manganese oxides. Processes 3 and 4 generally occur in concert (Tufano et al., 2008). Although various processes may liberate arsenic from solids, for lake systems with pH values < 8.5, a transition from aerobic to anaerobic conditions and commensurate arsenic and iron reduction appears to be a dominant means by which high concentrations of dissolved arsenic are generated (see Chapter 7 for details; Smedley and Kinniburgh, 2002; Fendorf et al., 2010).

Metrics of Arsenic Exposure

In the CDA region, measurements of arsenic exposure have not been gathered from urine or hair samples, although this is often done to characterize chronic arsenic exposure. Instead, the risks from arsenic were mainly assessed by modeling human exposures based on arsenic concentrations in environmental samples (NRC, 2005). Cancer risks and non-carcinogenic hazards were calculated on the basis of total arsenic concentrations in each area. Arsenic is naturally occurring in this region, but there is little question that soils and sediments influenced by mineral extraction activities are enriched with arsenic relative to background concentrations (Harrington et al., 1998b). It should be noted that both NRC (2005) and the most recent five-year review (EPA Region 10, 2021) emphasized that developing a human health metric for evaluation of arsenic exposures in the CDA basin was an important need.

The two main arsenic exposure pathways noted in the 2001 HHRA that present the greatest cancer risks are those associated with subsistence lifestyles among CDA Tribal members and exposure to arsenic in drinking water from private wells (EPA, 2001). For the subsistence scenarios (both traditional and modern), arsenic and iron in soil and sediment were the most notable contributors to non-cancer hazard. Non-cancer risk for non-tribal members is at an elevated level of concern solely in the Burke/Nine Mile area and only associated with well water supplying the home. The Burke/Nine Mile area had the highest neighborhood risks and hazards because of the waste pile exposures evaluated for this area. Waste piles had the highest concentrations of non-lead metals. The lower basin had the highest concentrations of arsenic and iron in soil and sediment (except for waste piles).

Possible Exposure Pathways Involving CDA Lake

Exposure pathways that involve CDA Lake directly would be if arsenic is ingested from drinking Lake waters, migration of arsenic from the Lake water into groundwater wells (see, e.g., Moore and Woessner, 2003), or migration of mobilized arsenic into major aquifers. There is some evidence that private wells in OU-3 have been evaluated for arsenic contamination (EPA Region 10, 2021, Table 6-2), but exposure pathways directly involving the Lake are not known to currently exist. However, as discussed in Chapter 1, CDA Lake is hydrologically connected to the Spokane Valley-Rathdrum Prairie Aquifer, which is a regional drinking water supply. As described in the 2011 Rathdrum Prairie Comprehensive Aquifer Management Plan (Idaho Water Resources Board[IWRB], 2011), between the outlet of CDA Lake and the confluence with the Little Spokane River, the Spokane River loses flow to the aquifer in some reaches while gaining in others. Water withdrawals from the Spokane Valley-Rathdrum Prairie Aquifer are anticipated to increase in the future as development in the region continues. As such, to the extent that in the future areas in and/or near the northern bays of CDA Lake could experience anoxic conditions and release of metals of significance to human health, such as arsenic, there exists a potential pathway for water quality in CDA Lake to impact water supplies outside the Lake area.

It is unclear what the current risks from arsenic release into bottom waters of CDA Lake are to humans. Fendorf et al. (2010) described "three environmental requirements for groundwater arsenic concentrations to increase: water saturation (which limits diffusion of atmospheric oxygen), a limited supply of sulfur, and a source of organic carbon to drive microbial dissolution of Fe oxides." Where the above conditions are met (i.e., sediments become anoxic and sulfur is consumed by FeS), there is the potential for arsenic in sediments in CDA Lake to be mobilized into porewaters, bottom waters of the Lake, or eventually groundwater (see Chapter 7). Analyzing 206 cores from CDA Lake, Harrington et al. (1998b) found that Lake sediments averaged 211 ± 11 µg As/g dw (with a range of 2–568 µg/g)—17 times higher than concentrations found in the St. Joe River sediment cores. Release of dissolved arsenic into porewaters and bottom waters from anoxic sediments has been shown to occur in CDA Lake at C6 (Figure 7-10), in the deep bend waters studied by the CDA Tribe (Chess, 2021) and in lateral lakes (Sprenke et al., 2000). Concentrations of arsenic in bottom waters at C6 can reach the human health level of concern (10 µg/L) during short periods (one or two months) of anoxia that occur annually at those locations. If anoxia were to occur in bottom waters of the deeper portions of CDA Lake (C1–C5), where arsenic concentrations in sediments are about 17 times higher than at C6 (Harrington et al., 1998b), greater release of arsenic than that seen at C6 might be expected. Sprenke et al. (2000) observed arsenic concentrations in the interstitial water of lateral lakes in the CDA basin of up to 217 mg/L, and as high as 422 mg/L in the water within the uppermost meter of lake sediment. These were sediments with 30–300 µg As/g dw, similar to sediment in CDA Lake. Thus, the large reservoir of arsenic in the sediments of CDA Lake has the potential to release high concentrations of arsenic into bottom waters and porewater if conditions change in the future. Understanding the connections between the Lake and groundwater that might provide drinking water for lakeshore populations, and beginning to monitor those sources of drinking water, could facilitate effective responses should such changes occur.

Mercury

Mercury (Hg) in aquatic systems arises from both local (e.g., mining) and global (e.g., combustion) sources, but once present in a watershed, it presents a risk to aquatic food webs and humans who consume animals from those food webs. Mercury is usually released into the environment as inorganic mercury, but the risks arise in aquatic environments because of the formation of monomethylmercury (CH_3Hg^+ or MeHg) in anoxic sediments. MeHg is primarily formed by sulfate and iron-reducing bacteria in suboxic conditions (Fitzgerald and Lamborg, 2007). Hence, MeHg production is enhanced in ecosystems with dynamic oxygen gradients, such as the water column and/or sediments of lakes, reservoirs, floodplains, and wetlands (see Figure 9-2) (Alpers et al., 2014; Beutel et al., 2020;

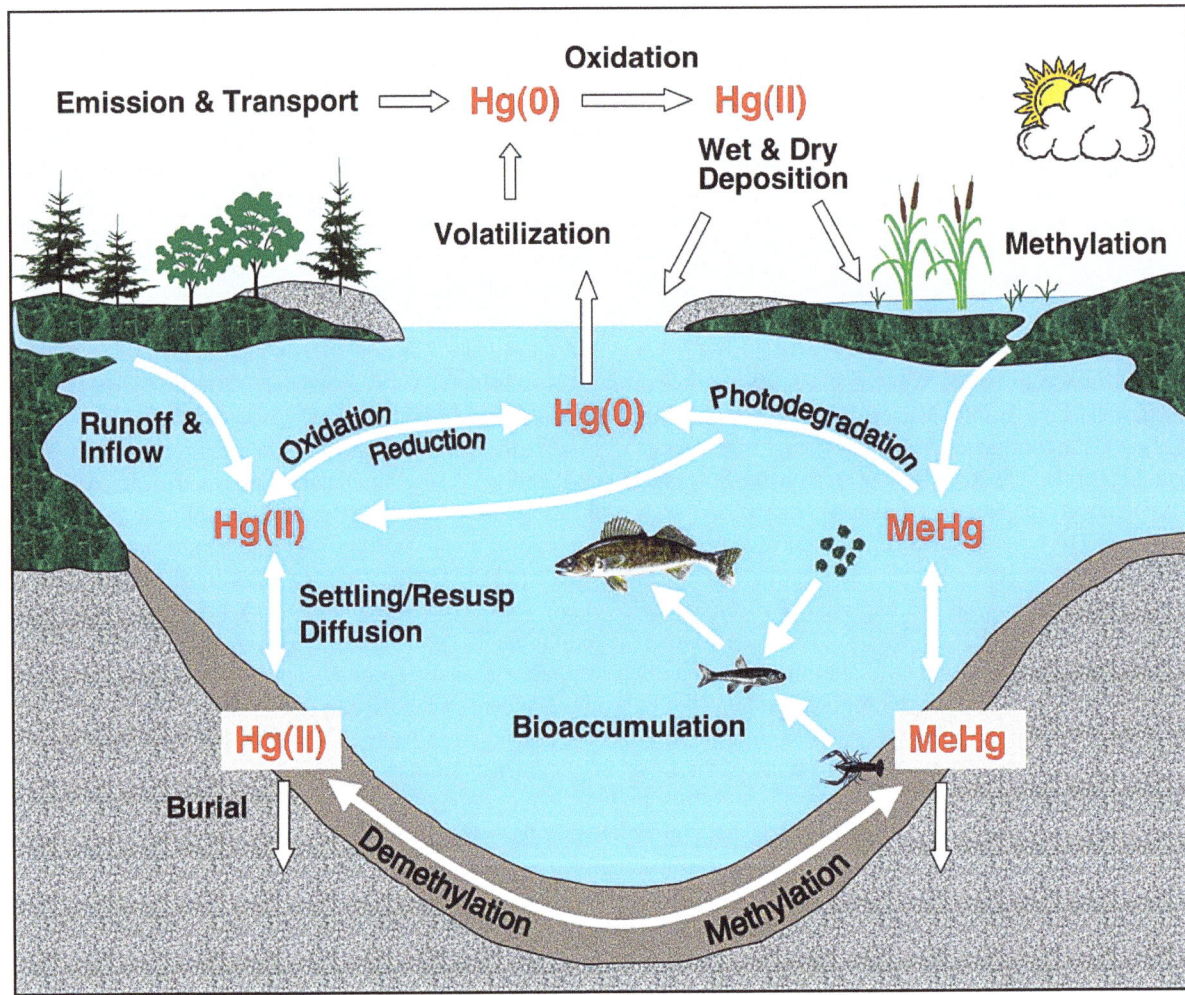

FIGURE 9-2 Mercury cycling in a lake and its watershed. Mercury emissions are transported long distances, primarily as gaseous elemental mercury [Hg(0)], oxidized in the atmosphere to reactive gaseous mercury [Hg(II)], and deposited in precipitation and by surface contact (dry deposition). Anaerobic bacteria convert a small portion of the incoming Hg(II) to methylmercury (MeHg), which is then bioconcentrated in the aquatic food chain (by a factor of $> 10^6$). Various biotic and abiotic reactions interconvert the different forms of Hg, affecting uptake, burial, and evasion back to the atmosphere. SOURCE: Engstrom (2007).

Conaway et al., 2008; Fuhrmann et al., 2021; Selin, 2009). Higher levels of mercury and longer periods of anoxia can lead to greater MeHg formation. Mercury is also highly volatile when present as elemental Hg or vaporized by combustion. Widespread mercury contamination of the atmosphere has resulted from a combination of the cumulative effect of discharges from mercury sources and local sources of combustion (e.g., coal-fired power plants).

Monomethylmercury, which accounts for < 5 percent of mercury in aqueous environments, poses the greatest threat to human and ecosystem health because it is a potent neurotoxin that biomagnifies in the food web, reaching potentially toxic concentrations in predator fish, such as bass (Benoit et al., 2003; Cossaboon et al., 2015). MeHg concentrations in high-trophic-level fish can be > 10^6 times higher than the water in which they live, and their body burden increases as they grow and age (Cossaboon et al., 2015; Lavoie et al., 2013). Thus, the primary exposure route of MeHg to humans is the consumption of fish (Sunderland, 2007; Wang, 2012), and mercury-driven fish consumption advisories exist for waterbodies throughout the 50 states (Wentz et al., 2014), including CDA Lake.

Mercury in the Coeur d'Alene Lake Region

Not surprisingly, there is mercury present in the CDA River and in CDA Lake. Mercury was used in mining (and processing) of gold and silver in the CDA watershed during the late 19th and early 20th centuries and was, undoubtedly, released into the local environment (as was typical of mining technology of the times). Bookstrom et al. (2013) listed mercury as an element that sometimes co-occurred in elevated concentrations with lead in the CDA basin. Sediment cores from the lateral lakes along the CDA River, discussed in Sprenke et al. (2000), illustrate the legacy of mercury contamination, often with the highest peaks in the subsurface (presumably deposited historically) (see Figure 9-3). Elevated total mercury concentrations in sediments, but not in food webs, were reported even earlier by Gebhardt et al. (1971): "The highest mercury concentration in water and sediment samples (among 93 locations in Idaho) were collected from the lower Coeur d'Alene River. . . . (but) fish and water samples collected above the confluence of the South Fork Coeur d'Alene River were low in mercury." The concentrations in sediments are for total mercury, which includes forms with varying levels of toxicity, not necessarily indicative of food web contamination.

In a 2009 national study conducted by EPA, 48.8 percent of lakes in the United States exceeded the EPA's recommended tissue-based mercury water quality standard of 0.3 mg/kg wet weight.[5] In all of Idaho there are 341 river miles and more than 143,000 acres of lake that are listed as being impaired by mercury. The detection of greater than 0.3 mg/kg of mercury in fish (four of 30 samples in the northern Lake, seven of 40 in the central Lake) in 2016 (IDHW, 2020) has led to CDA Lake being listed as impaired for cold water aquatic life, salmonid spawning, and primary contact recreation in 2020 (IDEQ, 2020). The highest concentrations were found in bass (0.056–0.798 mg/kg ww) and pike (0.056–0.479 mg/kg ww), both top predators. Furthermore, the CDA Lake fish consumption advisory was updated in 2020 because "[h]igh mercury levels were found in some fish species, including bass, bullhead, northern pike, panfish, and kokanee,"[6] but the data upon which such conclusions were drawn are minimal.

The relative contribution of mercury to the CDA food web coming from local contamination compared to atmospheric inputs is unknown. Bechard et al. (2009) compared mercury concentrations among six locations in Idaho in the feathers of bald eagles, another top predator prone to elevated concentrations of MeHg. Concentrations in the CDA basin were not significantly different than concentrations in the Salmon, Boise, Payette, and Snake River basins. Thus, the listing of CDA Lake for mercury contamination is not necessarily an indicator of a local source of contamination, and the degree of contamination in the food web does not indicate unusually high local concentrations of methylated mercury.

Compiled data from the U.S. Geological Survey (USGS) from the past 40 years (Figure 9-4) suggest that total mercury concentrations in the watershed may be declining (similar to the trends observed for other metals in the watershed—see Chapter 3), but the data are sparse and no statistical techniques were used to analyze trends in the data. If such declines were ever confirmed, they would coincide with decreases in atmospheric deposition in North America and Europe (Zhang et al., 2016). Locally, reduced mining activity, better source control at combustion sources, and burial of higher concentration materials in the sediment by deposition of cleaner materials could be factors contributing to declines in mercury concentration in various media.

---

[5] https://www.epa.gov/fish-tech/national-study-chemical-residues-lake-fish-tissue-results
[6] https://healthandwelfare.idaho.gov/news/fish-advisory-coeur-dalene-basin-revised-after-high-levels-mercury-found-some-species

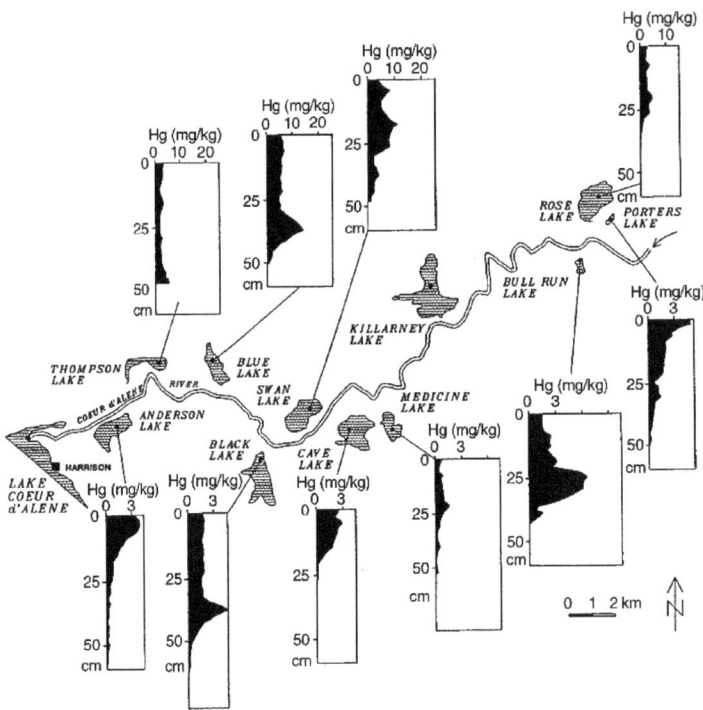

FIGURE 9-3 Total mercury concentrations in sub-bottom sediments of lateral lakes along the floodplain of the CDA River. SOURCE: Sprenke et al. (2000).

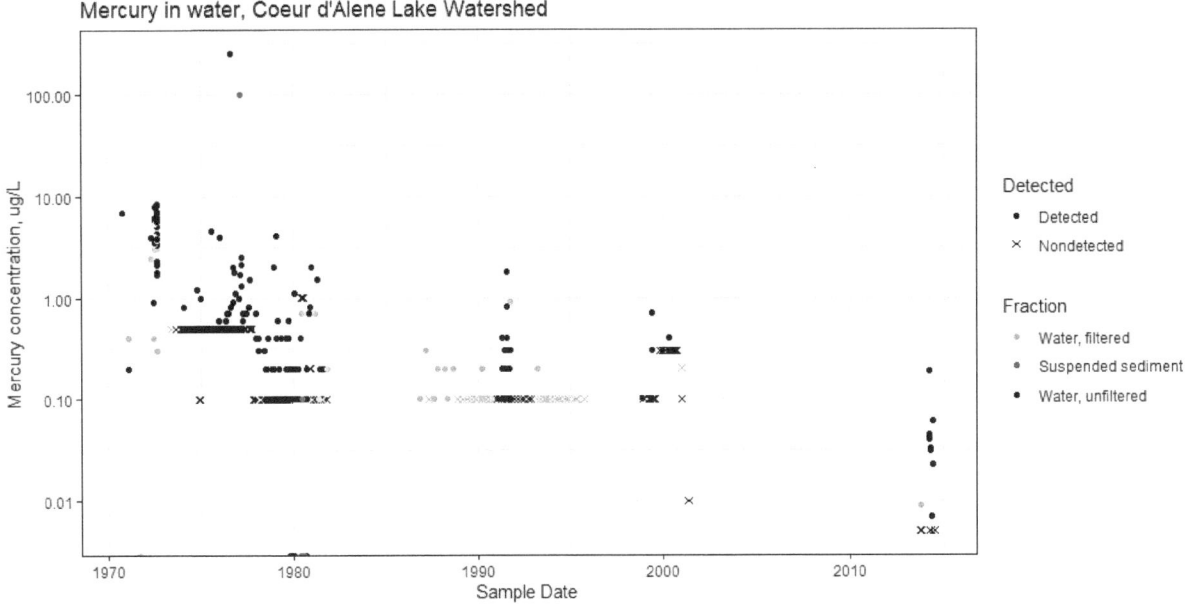

FIGURE 9-4 Trend in mercury levels in the CDA Lake watershed in water and sediment samples. Note that these results represent a search for all data in the USGS National Water Information System database with mercury-related parameter codes (any method, any media) for all subwatersheds within the CDA Lake watershed (17010301—North Fork CDA River, 17010302—South Fork CDA River, 17010303—mainstem CDA River and CDA Lake, and 17010304—St. Joe River). SOURCE: Data courtesy of the USGS.

Efforts to reduce mercury inputs to watersheds are a focus nationwide. Although more data on mercury cycling and mercury in the food web would be informative in the CDA basin, findings to date support the suggestion that the food web contamination with mercury in CDA Lake is more influenced by atmospheric sources. MeHg in some fish exceeds warning targets but is lower than some other regional exposures, which is enough to suggest that the major sources are regional (atmospheric) rather than local. However, a likely ongoing issue with regard to mercury would be increased duration or extent of anoxia in CDA Lake (or lateral lakes) leading to a greater extent of formation of MeHg. This would counteract efforts to reduce mercury inputs by having conditions that lead to greater formation of the most problematic, bioaccumulative form of mercury.

Localized Risk Associated with Algal Blooms in an Altered Chemical Environment

A concern resulting from changes in dissolved metals concentration from natural processes or remediation activities is the increase in bioproductivity of CDA Lake, leading to a eutrophic state. While nutrient and bioproductivity classification of CDA Lake is considered oligotrophic (see Chapter 5), an overabundance of bioproductivity in a localized area could negatively affect water quality and ecosystem services (Liao et al., 2016). Two potential consequences of an algae-induced eutrophic event resulting from degradation of algal detritus in CDA Lake are considered below: (1) suspended algal blooms of toxic cyanobacteria and (2) growth of filamentous algal species.

Harmful Algal Blooms

HABs involve any type of phytoplankton or periphyton blooms that cause water quality or environmental degradation (e.g., toxins, anoxia, taste and odor compounds). Toxin-producing cyanobacteria are the most common taxa that form HABs in lakes. Cyanotoxins may be toxic for aquatic and terrestrial biota and can also impair the recreational use of lakes and especially their use as drinking water sources. HABs are most common in aquatic systems with high nutrient concentrations (phosphorus and nitrogen). When cyanobacteria blooms die, they often accumulate at the surface and can appear as foam, scum, paint, or mats on a waterbody's surface. Harmful blooms are most commonly caused by cyanobacteria (of genera such as *Microcystis*, *Anabaena*, *Aphanizomenon*, *Planktothrix*, and *Cylindrospermopsis*) that produce toxins like microcystin, anatoxin-a, saxitoxin, and cylindrospermopsin. Multiple toxin exposure routes are possible, including skin contact with water containing the toxins, drinking water containing toxins, breathing in contaminated water droplets in the air, and eating fish or shellfish that contain toxins. Many of the toxins produced by HABs have no antidote, and medical intervention consists of managing symptoms. HABs present a difficult challenge to managers of aquatic ecosystems (both for source water and recreation), and their frequency is expected to expand with increases in global temperature (Paerl and Huisman, 2008, 2009; Carey et al., 2012; Paerl and Scott, 2010).

While many of the toxins that are produced by toxic cyanobacteria have been characterized, the underlying mechanisms of bloom formation (e.g., the cascade of genetic factors that trigger cyanobacteria to produce toxins) are poorly understood. In fact, while there are known species of cyanobacteria that commonly are present during a HAB event, the genes that encode toxin production may not be present or may be incomplete in different populations of the same species of cyanobacteria. Currently, the only means to determine if an algal bloom is harmful involves monitoring of toxins (commonly microcystin) in water samples or direct experimental assessments of toxicity.

Understanding the potential processes that may lead to HABs is key to predicting whether they might play a role in CDA Lake water quality in the future. One process involves production of reactive oxygen species (ROS). According to Paerl and Otten (2013), algal "blooms increase reactive oxygen species . . . proportional to the dissolved organic carbon concentration and light intensity". Reactive oxygen species (ROS) are free radicals of hydroxide ($^{\bullet}OH$) and molecular oxygen ($O_2^{\bullet-}$, superoxide) generated by natural biological and photochemical processes in high-light-intensity aquatic environments, such as fresh surface waters, that damage the cells of resident microorganisms. ROS can damage DNA, proteins, or other cellular materials or rapidly form strongly oxidizing peroxides (e.g., H_2O_2). Some cyanobacteria in fresh waters produce microcystin, which has been identified in conferring protection from ROS (Paerl and Otten, 2013). That is, microcystin binds (via thioester bonds) to key proteins to confer protection from the oxidative damage and proteolysis brought about by light (Zilliges et al., 2011), although the precise mechanism is unknown.

One might assume that the oxidative protection conferred by microcystin may allow organisms that produce this toxin to gain a competitive advantage over non-toxin-producing phytoplankton species in freshwaters. Indeed, this appears to be the case *during bloom initiation*, as these toxin-producing organisms can take advantage of highlight intensities at the top of the water column (Kardinall et al., 2007; Davis et al., 2009). However, in light-limited constructed and natural systems, there can be a gradual decrease in microcystin and microcystin producers over time during a bloom event (shown in Figure 9-5A and C) (Kardinall et al., 2007; Davis et al., 2009). That is, phytoplankton growth will eventually lead to light attenuation, favoring those species with more efficient light-harvesting capabilities, especially in wind-mixed lake systems.

In CDA Lake, high dissolved metals concentrations could both enhance ROS accumulation and confer a competitive advantage to some cyanobacterial taxa. As shown in Figure 9-5B and D, elevated concentrations of metals that occur in CDA Lake, such as cadmium and lead, can induce oxidative stress in resident organisms (Stohs and Bagchi, 1995; Nies, 1999), although the situation is more complex for zinc (Marriero et al., 2017; Xia et al., 2008). In addition, the toxins produced by some cyanobacteria can bind to metals, rendering the metals less toxic and thereby conferring another competitive advantage over non-toxin-producing, metal-sensitive algal species.

Neither dissolved organic carbon nor nutrients are especially high in the main body of CDA Lake. Thus, the likelihood of dense blooms of harmful algae under present conditions is low in the Lake overall. But occurrence of the conditions conducive to such blooms in protected localities, some littoral habitats of the Lake, or in the lateral lakes, could be enhanced by the presence of metal contamination. Monitoring such locations could provide an early warning about the beginnings of more-widespread conditions conducive to HAB formation.

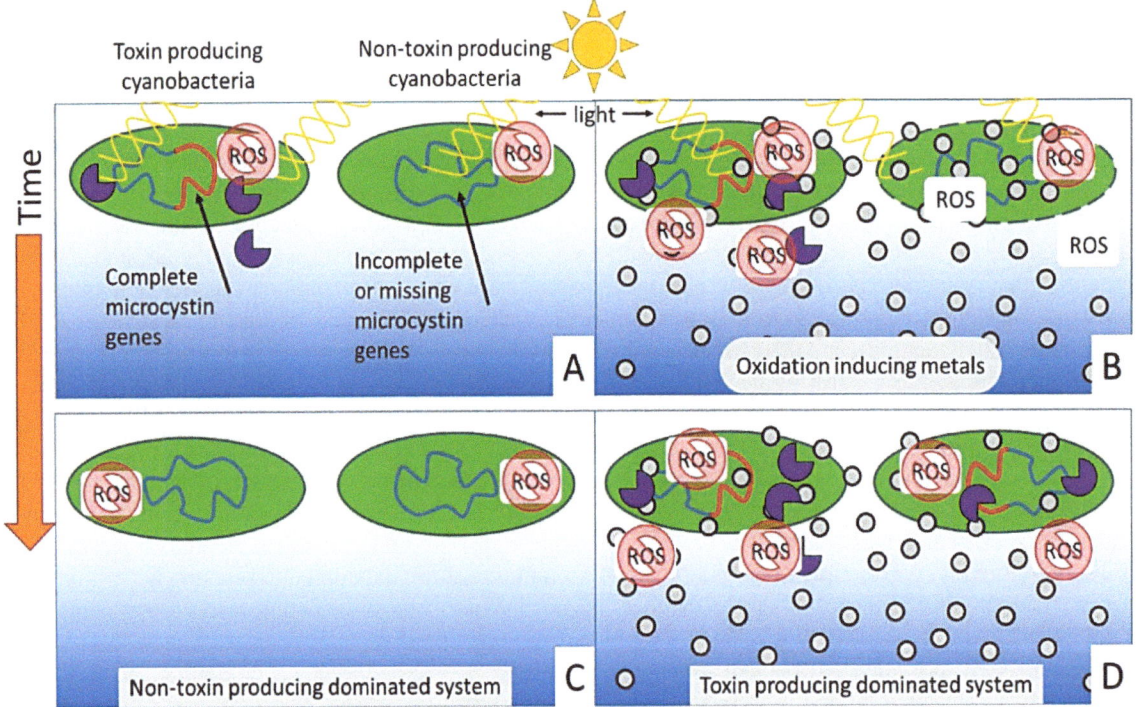

FIGURE 9-5 Succession of cyanobacteria in aquatic systems. (A) Reactive oxygen species (ROS) generated by natural biological processes or photo-reactions are neutralized (red overlay on ROS) by natural cyanobacteria defenses (glutathione, antioxidants). (B) Metal-contaminated (shown as gray circles) aquatic systems enhance ROS, which overwhelm natural cyanobacteria defenses; microcystin (purple circular sector)-generating cyanobacteria are at a competitive advantage as microcystin confers oxidative stress protection. (C) Succession and dominance of non-toxin-producing cyanobacteria in low oxidative stress systems. (D) Succession and dominance of toxin-producing over non-toxin-producing cyanobacteria in metal contaminated aquatic systems.

Filamentous Algal Blooms and Other Invasive Submerged Aquatic Vegetation

Filamentous algal blooms (FABs) and other invasive aquatic plants have been increasing within littoral, or nearshore, regions in oligotrophic lakes, like CDA Lake, despite measured low concentrations of phosphorus (Vadeboncoeur et al., 2021). FABs are typically composed of green algae (Chlorophyta) that are not usually toxic, but may provide a reservoir of carbon and nutrients to microbial communities in the sediment and increase potential for localized eutrophication. Eurasian water milfoil is a taxon of submerged aquatic vegetation with an ability to tolerate and grow in a wide range of water temperatures, depths, and turbidities—all of which contribute to its success as an invader. Liao et al. (2016) cited surveys showing that milfoil infestations prevailed in at least ten out of 28 bays of CDA Lake during the summers of 2011–2014.

Occurrence of invasive plants like milfoil and FABs in littoral areas could be an indication of a lake system trending toward a eutrophic state. A combination of factors including nutrient availability, decreased grazing by benthic communities, changes in hydrology, and warming of water temperatures due to climate change often facilitate such invasions (Vadeboncoeur et al., 2021). CDA Lake may be uniquely vulnerable to such outbreaks if the elevated concentrations of toxic metals in lake sediments are adversely affecting benthic communities that graze on early life stages of these plants.

***

In summary, the combination of ROS protection and metal sequestration could bestow a competitive advantage to toxin-producing cyanobacteria that allows them to extend successional dominance in metal-contaminated aquatic environments. However, HAB risk is likely low in the majority of CDA Lake due to a combination of factors including primarily low-nutrient, oligotrophic conditions. Occurrence of the conditions conducive to such blooms, such as an increase in temperature and/or nutrients, is more likely in the littoral habitats of the Lake or in the lateral lakes. Similarly, an event of higher probability is an increase in FAB resulting in localized degradation of water quality in nearshore Lake regions most often used by humans.

ECOLOGICAL HEALTH

Although the greatest concern for the ecological health of lakes has traditionally been eutrophication driven by excess nutrients, in CDA Lake, the ecology is threatened by profound contamination with heavy metals, including arsenic, cadmium, lead, and zinc. Much of the ecological disturbance and loss of ecosystem services in the rivers of the CDA basin has been attributed to the extreme concentrations of these metals (CH2M Hill and URS Corp., 2001). Although primary waste deposition did not occur on the lakeshore, CDA Lake is the ultimate repository for the metals mobilized from the watershed. This section explains how lake ecosystems generally respond to metal exposures, discusses benchmarks for different levels of ecological disturbance, evaluates metal stress in the CDA Lake ecosystem, and discusses the limited data available on CDA Lake ecology that reflects the risks of metal exposure.

Hierarchy of Effects of Metals in Lake Environments

Ecological responses to metal contamination are complex and difficult to separate from responses to other drivers in an ecosystem. Any adverse effects of metal(loid)s enrichment are manifested at multiple levels of biological organization (Luoma and Rainbow, 2008) as shown in the hierarchy in Figure 9-6. The first toxic effects of trace metals occur at the level of molecules within cells. With increasing availability of the toxin, effects at the molecular level become more serious and are manifested at higher levels of biological organization (McCarthy and Shugart, 1990). Thus, an early sign of metal-driven malfunction is disruption of biochemistry and the structure and function of cells.

The effects of metals on ecological receptors range from harmless to sublethal (chronic malfunctions) to lethal (toxicity). At extreme exposures, individuals can experience mortality within a few days (acute toxicity) or over

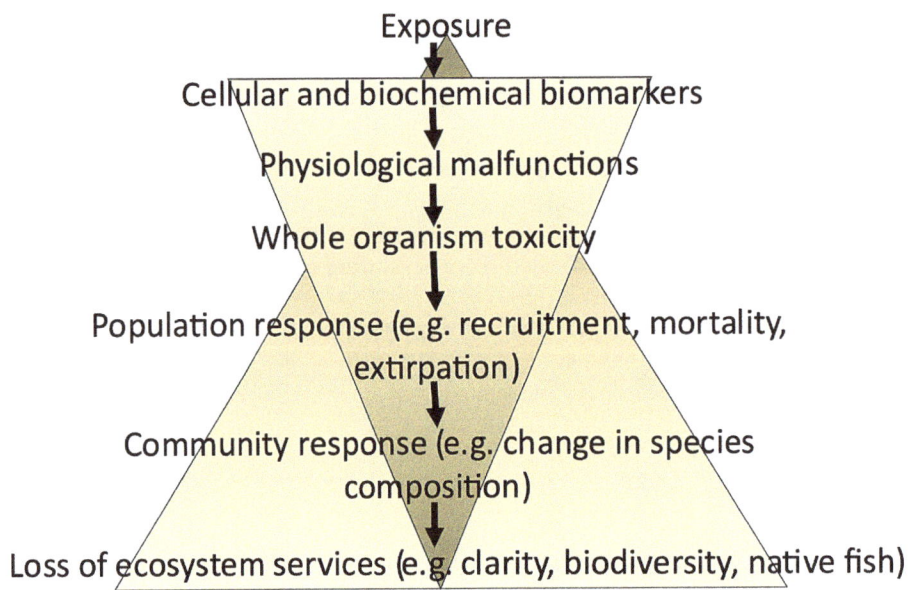

FIGURE 9-6 The hierarchy of metal effects from simpler to more complex levels of biological organization (inverse triangle) versus the degree of challenge in detecting those effects in nature (vertical triangle). SOURCE: Adapted from Luoma and Rainbow (2008).

longer periods of time (chronic toxicity). More common are adverse effects expressed as sublethal signs of malfunction, such as reduced size of individuals, poor physiological condition, changes in behavior, or changes in morphology. Metrics for determining functional disturbances include impaired reproductive success or changes in growth rates. The health of the population of a species is at risk when enough individuals are adversely affected at some life stage, such that recruitment of new individuals declines or mortality rates increase. When recruitment failures or mortality are large enough, the vulnerable population will disappear from the community (termed extirpation).

Loss of species changes the composition of the community. Starting with the most significant, metrics for community disturbance include absence of species or higher taxonomic groups most sensitive to the metals (e.g., mayflies can be the most sensitive taxa in freshwater streams and lakes—Milani et al., 2003); reduced numbers of taxa; a lower community diversity index; or exceptionally low abundance of individuals from all taxa (e.g., Clements et al., 2021). Loss of taxa from a community also can affect interactions among taxa and food webs, cascading into broader effects on the ecosystem. The ultimate effect of disturbances from contaminants is ecosystem simplification (Woodwell, 1970), degradation of ecosystem functions, and loss of ecosystem services. The greater the metal exposure, the further the cascade of responses will proceed.

In reality, ecological responses to metals exposure are not necessarily detected (or even expressed) in a linear fashion, nor do all responses necessarily lead to ecosystem simplification. Compensatory reactions can occur at every level of organization, analogous to detoxification at the molecular level. At all levels of organization, adverse effects occur when the compensatory mechanism is overwhelmed and/or when compensation imposes secondary costs.

Detecting ecological responses to metals (and distinguishing them from effects of other processes) is more difficult as one goes from simpler to more complex levels of organization (Figure 9-6). Thus, it is a challenge to detect metal-specific malfunctions with sensitivity or pinpoint cause and effect when considering population sizes, community structures, and ecological function. Lower-level responses are more readily detectable and easier to study in controlled experiments. Indeed, controlled studies to define thresholds of toxicity at different levels of organization and different life stages for different species are critical to writing the criteria embedded in governmental environmental regulations. However, the thresholds defined by single-species toxicity tests and the thresholds of metal-driven change in populations, communities, and ecosystems may not be the same (Cairns, 1986; Buchwalter et al., 2017; Clements et al., 2021).

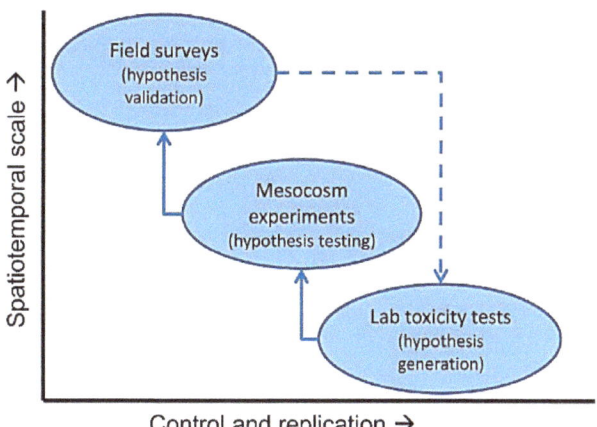

FIGURE 9-7 Linkages among laboratory toxicity testing, field experimentation (mesocosms), and field observations or surveys used in establishing a weight of evidence approach to defining ecological implications of metals. SOURCE: Buchwalter et al. (2017).

To ecologically evaluate metals in CDA Lake, a weight of evidence approach is necessary, supported by a body of work across multiple levels of organization and using different tools and approaches. Figure 9-7 shows how single species toxicity tests (laboratory bioassays) are used to generate hypotheses about the thresholds of metal toxicity, controlled experiments under field conditions (mesocosms) with multiple species can be used to test these hypotheses, and systematic field surveys or observations can be employed to validate hypotheses or generate new, testable questions. Scale and relevance to nature are lowest in toxicity tests and greatest in field studies, while control and replication are greatest in toxicity tests but lowest in field surveys.

Data on Ecological Risks from Metals in Coeur d'Alene Lake

Understanding how the CDA Lake ecosystem will change in response to changes in metal and nutrient inputs and factors like climate change requires knowing the current status of the lake at each level of biological organization, described below using the available information.

Exposure: Metal Concentrations

The first step in an assessment of ecological health and future risks in CDA Lake is determination of exposure. Zinc, cadmium, lead, and arsenic enrichment in the water column and sediments of CDA Lake is extreme by the standards of other large, oligotrophic lakes (see Chapters 1 and 6). Because water-column metal concentrations change seasonally, spatially, and year to year, mean annual metal concentrations in the water column at a few locations are not adequate indicators of ecological risk. Rather, the intensity and duration of peak metal concentrations, the period of recovery between peaks, and the timing of peaks in the life cycle of key taxa define the level of toxicity at different levels of biological organization (Groenendijk et al., 1999; CH2M Hill and URS Corp., 2001). Physiological or biochemical disturbances and even overt toxicity can result from short exposures to high concentrations of metals.

In CDA Lake, metal concentrations in the biologically active euphotic zone are highest in spring at the northern Lake monitoring locations (C1 and C4), but peak later in the year at the southern Lake location (C5). Dissolved metal concentrations during the periods of highest exposure might be the most effective metric to determine if any ecological effects might be expected, especially if such exposures last 30–60 days. Heat maps of water-column metal concentrations, such as Table 6-2, could provide a simple means of illustrating monthly changes in exposure at different locations.

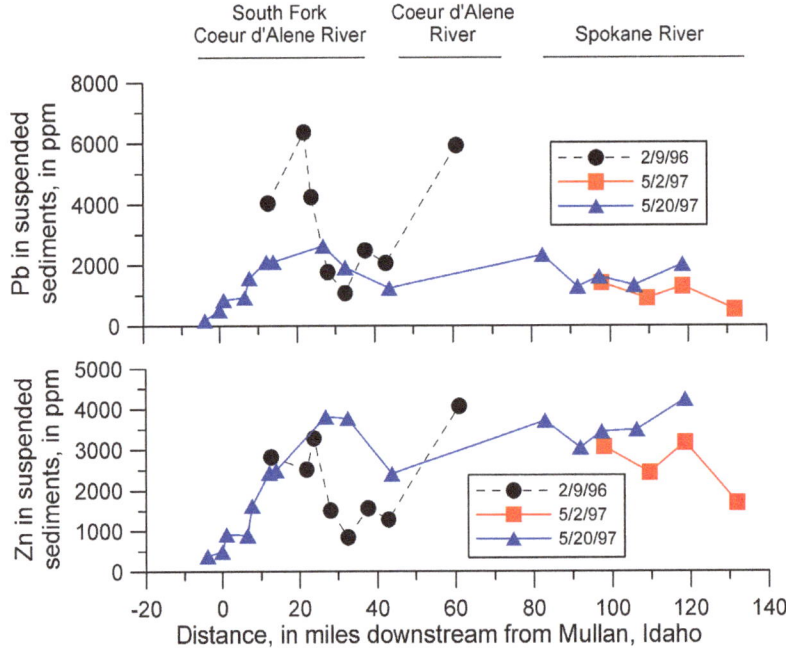

FIGURE 9-8 Lead and zinc concentrations in suspended sediment from the South Fork of the CDA River, the mainstem CDA river, and the Spokane River. SOURCE: Figure 9 in Balistrieri et al. (2002).

It is also important to consider metal concentrations other than dissolved concentrations. Suspended sediments in the Lake could be incidentally ingested by zooplankton, adsorbed onto macroflora and biofilms, or enmeshed in periphyton or microflora (e.g., aufwuchs) and thereby be a pathway for exposure for some taxa in the Lake. Balistrieri et al. (2002) found that in the watershed, metal concentrations per unit mass (the measure of relevance ecologically) of suspended material were of the same magnitude as bed sediments and soils (thousands of μg/g for lead and zinc—see Figure 9-8; tens of μg/g for cadmium and arsenic). Kuwabara et al. (2006) analyzed suspended material assumed to be primarily biogenic (defined as "phytoplankton") in June 2005 and found concentrations of zinc of about 1,300 μg/g dw at C5 and concentrations of 2,700–4,490 μg/g dw at a location near the mouth of the CDA River. Concentrations of lead at these two locations were 128–193 μg/g dw and 744–1,244 μg/g dw, respectively. Thus, the biogenic material from the surface waters of the Lake was highly enriched with zinc and lead (Kuwabara et al., 2006) but to a lesser extent than the dominantly inorganic suspended material from the watershed (Balistrieri et al., 2002). Greater understanding of seasonality, spatial distributions, and trends in metal concentrations on suspended material in the Lake is important to evaluating exposures because this material represents the base of the water-column food web.

Metal concentrations in surface sediments affect the food webs at the bottom of the Lake. Table 9-2 shows metal concentrations in surface sediments of the Lake. The spatial variability of metal concentrations in lake-bottom sediments is substantial; Horowitz et al. (1993) found lower metal concentrations at the backs of some bays and at the confluence of the St. Joe River and the highest concentrations in profundal locations. In general, however, highly enriched sediments occur Lake-wide. Metal concentrations in bottom sediments are also strongly influenced by particle size. Horowitz et al. (1993) found that the four most metal-enriched sediments were characterized by extremely fine-grained sediments (< 20 μm). More recent studies of sediment samples at five locations in the Lake showed that most of the substantial variation among locations and depths was driven by differences in particle size (EcoAnalysts, Inc., 2017), as evidenced by strong correlations of lead, cadmium, and zinc with aluminum.

TABLE 9-2 Cadmium, Lead, and Zinc Concentrations (mg/g dw) in Surface Bed Sediments from Five Different Studies at Different Times

Year Determined	Cd	Pb	Zn	Reference
1989*	56	1,800	3,500	Horowitz et al., 1993
1994*		3,820	2,995	Harrington et al., 1998a
2002**	22.9	3,780	3,250	Toevs et al., 2006
2010**	25	3,850	3,326	Morra et al., 2015
2011-2015***	3-34	231-2,520	569-2,690	EcoAnalysts, Inc., 2017
Carlin Bay****	218	10,557	14,900	EcoAnalysts, Inc., 2017

NOTES: *Median lake-wide, **one location, ***transects from 5-m water column depth to 30-m depth, respectively. **** Carlin Bay, north of the mouth of the CDA River, was separated from the rest because it represents the highest concentrations observed in the more recent studies.

Bioavailability: Waterborne Metals

Exposure is a function of the concentration of bioavailable[7] metal, not total metal. Hence, understanding both total concentrations in the environment and the bioavailability of each metal in that environment is required to evaluate ecological risks (e.g., CH2M Hill and URS Corp., 2001; Balistrieri and Blank, 2008; EcoAnalysts, Inc., 2017). The bioavailability of dissolved metals is determined by speciation (i.e., the distribution of a metal among different ligands or forms). Speciation is "a function of the water composition, presence of inorganic and organic ligands that bind with the metals, the presence of competing ions, and the properties of the metals themselves" (Smith et al., 2015). For example, lead forms strong inorganic and organic complexes, leaving little uncomplexed lead (low free ion concentrations), and is thus of low bioavailability. Dissolved zinc and cadmium form relatively weak inorganic and organic complexes (their free ion concentrations are a higher proportion of total concentrations), such that they have greater bioavailability than dissolved lead (Smith et al., 2015).

Speciation of a dissolved metal is difficult to measure directly but can be approximated by a variety of computation and analytical techniques. The most advanced method of applying speciation concepts to metal toxicity is the Biotic Ligand Model (BLM). The BLM determines uptake at a generic biological receptor (termed the biotic ligand) using equilibrium speciation/competition calculations that consider effects of interactions between the metal and major ions, inorganic ligands, and organic ligands (dissolved organic matter). It then statistically relates this uptake to outcomes of acute or chronic toxicity tests. Alternatively, diffusive gradient in thin films (DGT) is an example of a physical measurement tool designed to estimate bioavailability. All techniques available to approximate bioavailability have important limitations (e.g., Slaveykova and Wilkinson, 2005; Luoma and Rainbow, 2008) but are valuable constructs if employed within the proper context.

Dissolved metal bioavailability has not been systematically studied in CDA Lake, but studies in the watershed are informative. Balistrieri and Blank (2008) compared BLM calculations and DGT measurements to dissolved concentrations of cadmium, lead and zinc at seven mining sites in the South Fork of the CDA River and two locations with little contamination. DGT-labile lead was as low as 16 percent of dissolved lead concentrations. Up to 30–40 percent of the total lead (in two models) appeared to be adsorbed to colloidal iron oxides or strongly complexed with organic substances of low bioavailability. Similar speciation/partitioning is likely in the lake during spring runoff when lead in the water column is at its highest concentrations. For example, the colloidal and

---

[7] Bioavailability describes the fraction of total ambient metal that an organism actually takes up from both dissolved forms and forms ingested in its diet. Metals from each pathway have a separate bioavailability. Bioaccumulation is the concentration of metal that accumulates in the tissues of an organism as a result of that combined uptake. Bioaccumulation depends upon the bioavailable metal in each pathway (diet or dissolved) summed across all pathways. When metal is bioavailable to aquatic life, it has the potential to generate adverse ecological effects (toxicity). Thus, for the purposes of determining toxicity thresholds, exposure is a function of the concentration of bioavailable metal, not total metal.

nanoparticulate forms of metals common in spring runoff from the CDA River pass through 0.45-µm filters and thus are classified as dissolved metals in CDA Lake, but their bioavailability is low compared to truly dissolved metal forms. Thus, a small but variable fraction of the lead in solution in the lake is likely bioavailable. This means that benchmarks based upon total lead in water would overprotect from lead toxicity, all other things being equal.

In contrast, nearly all the zinc and cadmium in the South Fork of the CDA River were present as potentially bioavailable free ion and inorganic complexes in most situations (Balistrieri and Blank, 2008; Smith et al., 2015) with little or no colloidal metal found. There is competition for biological uptake between major ions (represented by hardness, calculated as mg/L calcium carbonate) and zinc, so high hardness will reduce bioavailability/uptake of zinc. However, waters in the northern CDA Lake have low hardness ($\leq$ 25 mg Ca/L), so such competition is minimal. Balistrieri and Blank (2008) concluded that the bioavailability of water-column cadmium and zinc was minimally restricted by organic complexation or association with colloids, a conclusion similar to that of Kuwabara et al. (2006). Given their high concentrations and high bioavailability, cadmium and zinc are likely to be the dissolved metals of greatest toxicological concern.

Bioavailability: Diet and Sediments

The proportion of particulate or foodborne metals taken up by animals after ingestion is defined by the assimilation efficiency of the metal from that food (what proportion of the total is absorbed by the animal; Wang et al., 1996). Ingestion rate, the type of food ingested, and metal concentration in the food also affect how much metal is taken up from food. Experimental studies show that metal bioavailability to copepods feeding on marine algae is largely driven by the concentration of metal within the soluble fraction of the phytoplankton cell (Reinfelder and Fisher, 1991), with the two metals most bioavailable from ingestion of algae (other than selenium) being zinc and cadmium. Less than 10 percent assimilation efficiency was seen for metals that were bound strongly to the surface of the phytoplankton (as would be expected for lead). Zinc and cadmium concentrations in the water column of CDA Lake are high enough to suggest that zooplankton are exposed to these metals via feeding upon phytoplankton. The zooplankton would then be a pathway of exposure for pelagic-feeding fish.

Birds, fish, and other predators also feed on the organisms that live in the lake sediments. This benthic food web plays a central role in supporting higher trophic-level production in most lakes (Vander Zanden and Vadeboncoeur, 2002). Thus, disturbance of the benthic food web by the extremely high concentrations of lead and zinc in the sediments of CDA Lake could have ramifications across the entire ecosystem of the Lake.

Like all deep oligotrophic lakes, sediment-feeding fauna dominate the benthos in CDA Lake, including the oligochaetes, chironomids, and sphaerid bivalves (EcoAnalysts, Inc., 2017). As shown in Figure 9-9, these organisms can differ in the strategies by which they achieve their need for both oxygen and nutrition (Luoma and Ho, 1993; Lee et al., 2000). For example, Lee et al. (2000) showed that bioavailable metal exposures of taxa feeding at the oxic surface of the sediment column were greater than the exposure of animals feeding from deeper, anoxic sediments. In CDA Lake, this could ultimately result in lower risks of toxicity to head-down feeders like tubificid oligochaetes (the most abundant taxa in the benthos of CDA Lake) compared to sphaeriid bivalves that feed at the sediment surface (and are present in some locations but not others). EcoAnalysts, Inc. (2017) identified the benthic taxa present at five different locations in CDA Lake. Evaluation of differences in functional ecology among those taxa could be a sensitive measure of how the metal contamination of the sediments is affecting the benthos (e.g., Janssen et al., 2011).

Metals in Plant and Animal Tissue

How much metal accumulates into the tissues of plants and animals is a direct determinant of metal bioavailability. Such "biomonitoring" is a well-used alternative or a complement to geochemical measures and models for estimating metal bioavailability (Bryan and Hummerstone, 1977; Phillips and Rainbow, 1994).[8] Determining the

---

[8] A biomonitor is any organism that accumulates metal(loid) in its tissue in proportion to metal concentrations and bioavailability in its environment, thereby providing a relative measure of the bioavailability of that metal from all routes of exposure (Luoma and Rainbow, 2008).

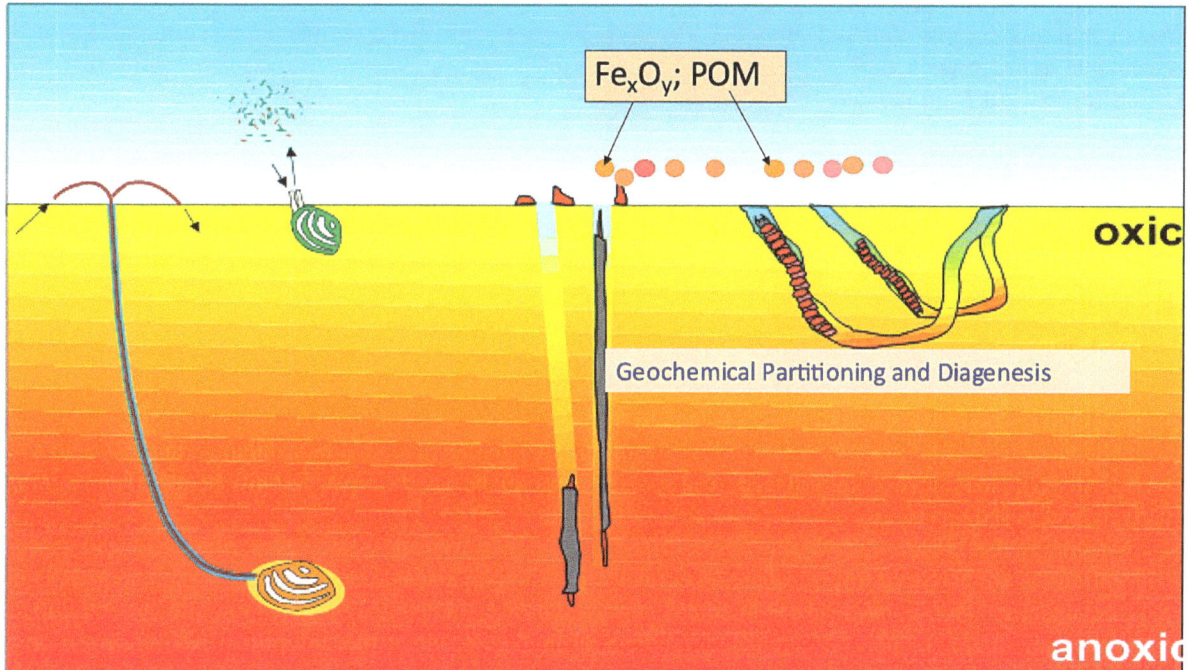

FIGURE 9-9 Different feeding strategies of benthic organisms affect their exposure to metals. Fe_xO_y are amorphous iron oxides that bind metals in CDA Lake at the oxic surface of the sediments; POM is particulate organic material that is the primary food of benthic fauna. In a typical lake, metals bound to particulate organic materials, iron oxide, and manganese oxides dominate the food of animals that feed on oxic sediments, but metal forms of lower bioavailability dominate in deeper layers (see Chapter 7; Lee et al., 2000).

amount of metal that has accumulated in taxa from nature has long been used to evaluate (1) if ecotoxicologically relevant exposures might be occurring and at what level, (2) geographic distributions of such exposures, (3) temporal variability, and more recently (4) ecosystem exposure in dose–response estimates from nature (Luoma et al., 2010; Clements et al., 2021).

There have been few attempts to determine metal bioaccumulation in the food webs of the CDA basin and CDA Lake. Scofield et al. (2021) measured arsenic, cadmium, lead, and zinc concentrations in emergent and submergent aquatic macrophytes from three lateral lakes on the CDA River, a reference location in the St. Joe River watershed, and two locations in CDA Lake (Harrison Slough and Cottonwood Bay). They found that aquatic macrophytes adsorb metals externally into their tissues during growth. The plants may be directly consumed by some species, but more importantly, they release metals back into the environment as organic detritus upon senescence and decomposition of the shed tissue (Jackson and Kalff, 1993; Weis and Weis, 2004). These metals then enter the lake food web when ingested by herbivores, omnivores, and detritivores. Scofield et al. (2021) found that arsenic and lead concentrations in all three submergent macrophytes were highest at Harrison Slough compared to the lateral lakes, reaching as high as 150 μg As/g dw and 1,250 μg Pb/g dw. Cadmium and zinc concentrations were higher in macroflora from the littoral zone of Cottonwood Bay (across the Lake from Harrison) than in the lateral lakes or Harrison Slough. Scofield et al. (2021) also determined metal concentrations in the tubers of the water potato (*Sagittaria latifolia*) from the locations above.

Water potato is a macrofloral species of cultural significance to the CDA Tribe as well as an important food item for native birds, including swans. All metal(loid) concentrations were higher in the leaves of the macrophytes than in the tubers of the water potatoes. Lead concentrations in the tubers in this study were similar to concentrations in *S. latifolia* observed in a 1999 study (up to 30 µg/g lead associated with dirt/outer skin; Audet et al., 1999); this is an indication that any decline over time in metal sediment concentrations is not yet detectable in the food web.

Birds that eat from the lake sediments (e.g., swans) or lower on the food chain have the highest dietary exposures to lead, which is not passed efficiently from prey to predators. Lead and zinc were long suspected to be a cause of widely publicized waterfowl poisonings, as indicated by analysis of soil, plant, and waterfowl specimens in the CDA region dating back to the 1950s (Chupp and Dalke, 1964). Given the high concentrations of metal(loid)s at the base of the food web, it is not surprising that elevated lead and cadmium concentrations are found in the tissues of Canada geese, American robins, and mallard (Blus et al., 1995; Henny et al., 2000; Spears et al., 2007). Ground-feeding terrestrial birds, including the American robin, song sparrow, and Swainson's thrush, are examples of riparian animals that show elevated lead concentrations in their tissues (e.g., Sample et al., 2011). Tundra swans, Canada geese, and ducks are aquatic omnivores that are exposed to lead in macrophyte tissues, detritus, and sediment incidentally ingested with their food. Beyer et al. (2000) showed that lead concentrations in the blood of ducks, swans, and goslings increased linearly with lead concentrations in food containing lead-contaminated sediments[9] from the CDA basin, illustrating that sediment lead is bioavailable via diet. Sediments were first sieved to remove artifactual lead (lead shot, sinkers etc); hence, the bioavailability resulted from mining-derived lead in sediments.

Arsenic is also bioaccumulated into at least riparian zone food webs near mine sites. Similar to lead, arsenic concentrations are higher in lower trophic level taxa. For example, arsenic concentrations in submerged plants, mosses, algae, and biofilm reach tens to hundreds of mg/kg dry weight near the Stibnite mine site in Central Idaho (Dovick et al., 2015). Tadpoles accumulated hundreds of mg/kg arsenic at the site, primarily from ingestion of algae and biofilms, while levels of arsenic in macroinvertebrates and trout were lower (maximum 30 mg/kg) than in the tadpoles. Arsenic bioaccumulation in the riparian food webs of CDA Lake have not been studied, but these could also be vulnerable to arsenic exposure.

Farag et al. (1998) measured cadmium, lead, and zinc concentrations in sediments and biofilms,[10] benthic invertebrates (primarily insect larvae), whole fish, and trout kidney at the confluence of the North and South Forks of the CDA River and in the mainstem CDA River near Harrison. They compared those concentrations to a reference location in the St. Joe River watershed. Biofilms and sediments are a food source for invertebrates and the invertebrates are a primary route of exposure for predaceous fish (Farag et al., 1998). Concentrations of metal(loid)s by all measures were patchy and site-specific, without a clear gradient between Cataldo and Harrison. Concentrations of arsenic, cadmium, and lead in invertebrates were 10–40 times higher at Harrison than in invertebrates in the St. Joe River watershed, while concentrations of zinc were 2–3 times higher; the lesser magnitude is a result of the high natural concentrations of zinc (an essential metal) in invertebrates.

Metal concentrations declined according to the ranking: sediments = biofilm > invertebrates > whole fish (Farag et al., 1998). Thus, tissue concentrations of arsenic, cadmium, zinc and lead do not increase from prey to predator (biomagnify), but all four are bioavailable to invertebrates and passed on, to a degree, to fish predators (Table 9-3). The enhanced exposure derived from the contamination in this food web is clear evidence that metal(loid)s originating from the CDA River and depositing in the Lake are not in an innocuous form unavailable to aquatic life and therefore have the potential to generate adverse ecological effects. Direct study of metal bioaccumulation in fish from the Lake itself would be an important next step in understanding the implications of this bioavailability.

---

[9] They determined that incidentally ingested sediments composed ~ 22 percent of the food eaten by swans in the field. The food in the experiments contained 22 percent sediments from the CDA basin with different lead concentrations.

[10] Biofilms consist of algae, bacteria, and associated fine detrital material attached to rocks and other substrates in water bodies. They are an important food source for an especially metal-vulnerable functional group of benthic invertebrates that scrape hard surfaces for food.

TABLE 9-3 Metal Concentrations in Two Food Sources (Sediments and Biofilm) to Three Different Taxa: Benthic Invertebrates, Perch (Whole Body) and the Kidneys of Cutthroat Trout

	Arsenic		Cadmium		Lead		Zinc	
	Ref (μg/g dw)	HSR (μg/g dw)	Ref (μg/g dw)	HSR (μg/g dw)	Ref (μg/g dw)	HSR (μg/g dw)	Ref (μg/g dw)	HSR (μg/g dw)
Sediment	5.6	171.0	0.3	25.5	57	3363	130	3895
Biofilm	9.6	149.0	2.0	21.0	38	3460	450	4543
Invertebrates	2.4	18.0	1.2	10.0	9	335	255	746
Whole Perch			0.1	1.5	<0.3	55	89	252
Kidney Cutthroat*			3.1	155.0	0.2	96	132	296

NOTE: Ref = St. Joe River basin; HSR = Harrison in the lower CDA basin. *sampled upstream of Harrison.
SOURCE: Farag et al. (1998).

Evidence from Biomarkers

Few studies have looked for biochemical signs of metal stress (biomarkers) in organisms from either the CDA watershed or CDA Lake but those that have indicate that bioaccumulated exposures are sufficient to elicit stress responses at the molecular level. Beyer et al. (2000) showed that depressed levels of δ-aminolevulinic acid dehydratase (an enzyme inhibited by lead) accompanied lead bioaccumulation in ducks fed a diet contaminated with CDA sediments. Ducks and tundra swans that bioaccumulated lead from littoral sediments in the CDA basin had blood lead levels high enough to indicate lead poisoning and also showed depressed levels of δ-aminolevulinic acid dehydratase (Blus et al., 1995, 1999; Henny et al., 2000; Spears et al., 2007). Several lines of evidence showed that the lead exposure came from the mine-contaminated soils, not lead shot (Blus et al., 1999). Osprey are also exposed to lead through their diet of fish and show depressed levels of δ-aminolevulinic acid dehydratase, although effects on reproduction rates were not detectable (Henny et al., 2000).

Biomarker studies in invertebrates and fish can yield valuable metrics in assessing both bioavailability and potential for stress in aquatic food webs. A diet of metal-contaminated invertebrates had physiological effects and led to metal accumulation in tissues of rainbow trout (Farag et al., 1994). Farag et al. (1999) fed a diet of metal-enriched invertebrates that had been collected at Pinehurst and Cataldo on the South Fork of the CDA River to cutthroat trout. They observed reduced feeding activity, "increased number of macrophage aggregates and hyperplasia of cells in the kidney, degeneration of mucosal epithelium in the pyloric caecae, and metallothionein induction" in the trout; they concluded that these effects would likely reduce growth and survival of fish in the wild. While both diet and water chemistry affected metals levels in the fish, diet was the primary route of exposure in such experiments (see, e.g., Woodward et al., 1994), and it disrupted digestive physiology, thus affecting growth and survival (e.g., Hansen et al., 2009). Even if benthic communities are not themselves disturbed, contaminated benthos can be an important pathway of potentially harmful metal exposures to their predators. No determinations of either physiological or biomarker responses in fish from different food webs (e.g., pelagic vs. benthic) in CDA Lake have been conducted. The studies with birds illustrate the potential value of adding such metrics to any comprehensive evaluation of metal effects on the ecological health of the Lake.

Whole Organism: Overt Toxicity

Wildlife provide one of the most obvious ecosystem services delivered by CDA Lake. The most overt evidence that the legacy of mineral extraction is damaging wildlife is the observations of hundreds of water bird mortalities over the years in the watershed. In the 1990s, over 600 animals (29 species of birds, including waterfowl, songbirds, and birds of prey; six species of mammals, including vole, muskrat, mink, and beaver; and amphibians and reptiles) were found sick or dead in the CDA River basin, compared to only 40 in the St. Joe River basin (Audet et al., 1999). Lead poisoning was documented in the vast majority of these animals in the Coeur d'Alene area, and over 90 percent of these poisoned animals had not ingested lead artifacts, such as shot or fishing sinkers. In the St. Joe River, lead poisoning is also observed, but it is related to ingestion of lead artifacts.

Fewer data on bird toxicity were available from CDA Lake itself until Spears et al. (2007) assessed lead concentrations in sediments and the blood of mallards and wood ducks from 22 shallow bay locations and 11 wetlands within and around the lake. Blood lead from ducks from the Lake was correlated with concentrations of lead in sediments. Blood lead concentrations exceeding thresholds for "clinical poisoning" (> 5 µg/dL) or severe clinical poisoning (> 10 µg/dL) occurred in ≥ 50 percent of the mallards from four of the eight locations sampled. Fecal samples from 19 Canada geese and three mallards indicated that these animals were "exposed to lead by ingesting contaminated lake sediment." Spears et al. (2007) calculated that adverse effects (physiological to population parameters) could occur over the range 147–944 µg/g dw lead in sediment. They also concluded that overt mortality of mallards utilizing CDA Lake could occur at lead concentrations of 1,652 mg/kg dw in sediments. These thresholds for toxicity estimated by Spears et al. (2007) are similar to those determined by Beyer et al. (2000) from experimental data with swans, where it was shown that physiological stress occurs above 530 µg Pb/g dw in sediment and that 1,800 µg Pb/g dw caused overt toxicity. Spears et al. (2007) concluded that "locations of Harrison Slough, Powderhorn Bay, Cottonwood Bay, Blackwell Island and Cougar Bay near the Spokane River outflow of CDA Lake were the areas of greatest concern for waterfowl exposure to lead-contaminated sediment." Because there is no evidence that lead concentrations in the Lake have changed detectably since the date of this study, it seems likely that similar bird exposures apply in 2022.

In summary, it appears that lead exposures of wildlife in some locations in CDA Lake can be as significant as exposures that occur in the watershed, but exposures in the Lake are patchy. In some areas, sediments are sufficiently contaminated to cause overt toxicity, and long-term adverse effects are possible. There are also areas for which contamination is insufficient to result in detectable effects. It is important to recognize that only a few locations have been studied in this large lake; no systematic picture is available. Careful management of wildlife depends upon a clearer understanding of the frequency and location of metal-driven adverse effects and studies of exposures and biomarkers in more types of birds and other wildlife. More experiments with or data on blood lead in wildlife in areas of median level exposure, or across multiple gradients of exposure, would also help clarify thresholds where sublethal effects are damaging wildlife. Clearly, remediation of the areas of extreme contamination would benefit the ecosystem services provided by wildlife for the Lake community and Idaho as a whole.

Experimental Studies of Toxicity

The toxicity of the metals of interest in CDA Lake is generally known, but toxicity testing that could be most relevant to the Lake is limited. Woods and Beckwith (1997) stated that several early direct studies of the toxicity of CDA Lake waters to fish, invertebrates and phytoplankton were cited by Savage (1986). In 1993 and 1994, two phytoplankton taxa (the diatoms *Achnanthes minutissima* and *Cyclotella stelligera*) were isolated from limnetic locations in the Lake with different zinc concentrations (Woods and Beckwith, 1997). Nominal zinc concentrations of 19.6 and 39.2 µg/L, which are typical of the Lake, both strongly inhibited growth of both species of diatoms. Later, Kuwabara et al. (2006, 2007) isolated two algal species common to the Lake, *Chlorella minutissima* and *Asterionella formosa*. The goal of the experiment was to test the hypothesis that, at high enough bioavailable concentrations, zinc suppressed cell division (growth) and inhibited utilization of the limiting nutrient phosphorus by the algae. This would allow phosphorus to accumulate intracellularly without stimulating growth. Functionally, the hypothesis meant that above some bioavailable concentration, zinc would limit the algal growth below the potential for growth defined by phosphorus. They tested only three concentrations of zinc and three concentrations of phosphorus, so the study was not designed to define a threshold for the effect. Growth in both isolates was inhibited at zinc concentrations typical of the Lake, although the chlorophyte *C. minutissima* was more affected by zinc than was the diatom isolate *A. formosa*. The adverse effect of zinc on growth rate was statistically significant for both species, but changes in cell concentration (the number of cells per volume of culturing suspension) were not statistically significant at the 95 percent confidence level for the diatom.

No experiments have been published that were designed to test potential toxicity of zinc, cadmium, lead, or arsenic to zooplankton under conditions directly relevant to the Lake. Studies determining relative sensitivities of zooplankton taxa that occur or might be expected to occur in the Lake could be of particular interest given

> **BOX 9-2**
> **Mesocosm, Multi-Species Studies of Metal Effects on a Water-Column Community**
>
> Hoang et al. (2021) present an example of how a relevant study in CDA Lake might be conducted. They focused their study on primary producers and small zooplankton because "primary producers (including phytoplankton, periphyton, and macrophytes) play central roles in aquatic ecosystem structure and function." They cited a limited multi-species testing literature that suggests phytoplankton are more sensitive to zinc than fish and invertebrates (De Schamphelaere et al., 2005; Van Sprang et al., 2009) and that "the current EPA environmental quality criterion might not be protective for primary producers and other planktonic organisms."
>
> Hoang et al. (2021) established a mesocosm containing the natural planktonic community collected with large water samples from a Florida lake, but excluded large zooplankton and small fish that feed on phytoplankton. They exposed the communities to a range of zinc concentrations (8–100 µg/L) for ~ 90 days in waters of 40–50 mg/L hardness. The phytoplankton in the control community was dominated by an equal balance of chlorophytes (green algae) and cyanobacteria (blue-green algae) over the 90 days. Their results, in brief, showed that at 8 and 20 µg/L zinc, the primary response was a decline in chlorophyte abundance and a concomitant rise in cyanobacterial populations. At 40 µg/L zinc, growth of both groups was inhibited throughout the 90 days. At 80 and 100 µg/L zinc, chlorophytes remained inhibited throughout the experiment but cyanobacterial populations began to grow after 30 days and by 90 days reached the highest concentrations in any of the treatments. Hoang et al. (2021) noted other mesocosm studies where cryptophytes rather than cyanobacteria became dominant, so the late cyanobacterial bloom in the highest treatments may not be a consistent response. They calculated a lowest detectable effect concentration to be 14 µg Zn/L for changes in group abundance and 21 µg Zn/L when chlorophyll *a* was the end point. Effects on zooplankton species richness also began in the treatment with 20 µg Zn/L.

the sensitivity of zooplankton community structure observed in other metal-enriched lakes (Yan and Strus, 1980; Marshall et al., 1983; Keller et al., 2007; Valois et al., 2010).

Extrapolations of the results of single-species toxicity tests to conclusions about the effects of metals on communities and populations have limitations. Mesocosm studies could be more relevant to CDA Lake. Such studies employ a reasonable representation of the natural community or exposure to one taxon over an entire life cycle. The most realistic studies equilibrate dissolved and dietary exposure, and capture some population or community responses (examples include Marshall et al., 1983; Cairns, 1989; Iwasaki et al., 2018; Mebane et al., 2020).

Controlled multi-species mesocosm studies with Florida flora and fauna provide one example of such an experimental approach (Hoang et al., 2021; see Box 9-2). Box 9-2 shows that complex changes occurred in phytoplankton community structure (e.g., with chlorophytes, cyanobacteria and cryptophytes) during an exposure to zinc lasting months. The most sensitive responses occurred at 8 µg Zn/L, including reduced diversity and reduced abundance of Chlorophyta (both phytoplankton and periphyton) and Chrysophyta. Cyanobacteria took over the community after 30 days of zinc exposure at extreme concentrations (80–100 µg/L). The authors concluded that the No Effects Threshold for zinc in this system was 14 µg/L (similar to the recommendation of Wong and Chau, 1990, for lakes, in general). Although results from one particular study cannot be directly extrapolated to CDA Lake, they illustrate the complexity of the phytoplankton community responses that might be expected as zinc concentrations change in the Lake. CDA Lake also seems amenable to in situ experimentation with simplified water column communities (see Marshall et al., 1983) given the north-to-south gradients in zinc exposure and the relatively high concentrations of zinc in the northern Lake.

Two mesocosm studies of zinc (or zinc and cadmium in combination) toxicity to the aquatic insect communities typical of Rocky Mountain streams are useful examples of approaches to developing concentration thresholds for water-column concentrations where rocky bottom (and perhaps related) communities began to simplify (see, e.g., Schmidt et al., 2011; Mebane et al., 2020). Such studies consistently show that mayfly populations declined then disappeared at lower-concentration metal treatments than did other taxa. Lake mayfly taxa, such as *Hexagenia* spp. (which is found in the southernmost CDA Lake locations) have been suggested (Milani et al., 2003; Reynoldson et al., 1989) as a sensitive indicator of adverse anthropogenic effects on benthic lake communities and as potential test organisms for sediment toxicity testing.

TABLE 9-4 Comparison of Sediment Quality Criteria to Measured Cadmium, Lead, and Zinc Concentrations in CDA Lake Sediments

	Arsenic	Cadmium	Lead	Zinc
	μg/g dw	μg/g dw	μg/g dw	μg/g dw
Criteria range	31–93	3.1–12.0	110–530	270–960
Median in CDA Lake		23–56	1,800–3,850	2,690–3,500
Magnitude of difference		7.4X–4.7X	7.3X–16.4X	10.0X–3.6X

NOTE: First row gives the ranges of sediment quality criteria across 15 different approaches to developing guidelines. Second row gives the range of median cadmium, lead, and zinc concentrations in CDA Lake sediments (see Table 9-2). Third row gives the magnitude by which the low and high medians estimated for CDA Lake sediments exceed the low and high criteria.

In lieu of site-specific studies, general toxicity thresholds from the literature can be used to estimate levels of concern for zinc and lead in CDA sediments. These are termed sediment quality criteria (SQG). These thresholds were mostly developed using single-species or multiple-species tests directly on sediments from different environments.

Many factors can affect the toxicity of sediments, including metal concentration, particle size, the type of organism, how that organism interacts with the sediment, and the geochemistry of the sediment (MacDonald et al., 1996). Thus, the threshold of toxicity may differ among lakes. To take this variability into account, a "weight-of evidence approach" was developed based upon a database of toxicity tests from different environments, with different levels of contamination, different geochemistry, and a variety of organisms. Studies showing adverse effects are ranked by concentration of the chemical, and then a percentile is chosen (e.g., the median concentration at which some adverse effect occurs) as the criterion (e.g., Effects Range Median, National Oceanic and Atmospheric Administration). These rankings have many limitations when used for precise predictions for any specific environment but are useful in establishing the general range within which chronic or sublethal toxicity occurs. Hubner et al. (2009) listed criteria established by 15 different approaches (first row of Table 9-4). The ranges shown in Table 9-4 include criteria used in local studies to assess the meaning of sediment metal concentrations in the CDA basin (Maret et al., 2003[11]); or in the Lake (Morra et al., 2015[12]; EcoAnalysts, Inc., 2017).

The magnitude by which the sediments of CDA Lake exceed observations from toxicity tests on sediments suggests a high likelihood of toxicity to at least some taxa and of effects on the benthic food web of the Lake. Median concentrations of cadmium, lead, and zinc in different studies exceed the median concentration expected to cause toxicity by 4- to 16-fold. In a multi-metal exposures like CDA Lake, the sum of the magnitudes by which each metal exceeds the criteria provides an estimate of the toxicity of the mixture, termed the cumulative toxic unit (assuming toxicity of different metals is additive—Maret et al., 2003). The cumulative toxic units for median cadmium, lead, and zinc in the Lake suggest that the mixture of contaminants in the sediments of the Lake exceed by 25-fold the median (not the lowest level) concentration that typically causes toxicity in a controlled test. Hubner et al. (2009) concluded that "although such criteria are not definitive measures of toxicity, they can have a high predictive ability and are a vital tool for identifying areas with potentially adverse biological effects." Clearly, CDA Lake is an environment where sediment toxicity would be expected, based upon controlled testing with sediments in general. Conducting specific studies with relevant taxa seems an important next step (EcoAnalysts, Inc., 2017).

Community Responses

Community structure, function, and productivity are driven by a wide variety of factors in lakes (see, e.g., Caires et al., 2013). Unambiguous demonstrations of community impacts for any individual stressor, including metals, are inherently challenging. Changes in food webs in the basin upstream from CDA Lake provide evidence that metals can affect food webs dramatically. The 2001 ecological risk assessment concluded: "Toxic effects of

---

[11] Maret et al. (2003) cited probably effects level (sediment criteria) of Cd 3.5, Pb 91, and Zn 315, all in μg/g dw.

[12] Morra et al. (2015) compared EPA criteria to surface sediment concentrations in a core collected from near Harrison: Cd = 2.49 vs. core 25.3; Zn = 384 vs. core 3,326; Pb 161 vs. core 3,000-4,000, all in μg/g. The cumulative toxic unit would be about 30.

contaminated sediment are believed to contribute to adverse effects on aquatic life in. . . the entire South Fork, the Coeur d'Alene River, the Spokane River, and, possibly, some parts of Coeur d'Alene Lake." Hoiland et al. (1994) reported that taxonomic richness (number of species) in the South Fork of the CDA River at Smelterville was zero (no macrofaunal life) in 1968. Life began to reappear 20 years after tailings ponds were completed in 1968. Taxa richness increased to six species in 1987 and 15 in 1991. The number of taxa in 1991 was about half the number found at reference sites on the North Fork.

Maret et al. (2003) found seven to nine EPT[13] taxa at two locations on the CDA River near Cataldo, and zero to one metal sensitive *Ephemeroptera* (mayflies) in 2000 (compared to 10 to 19 EPT taxa and three to seven mayfly species in reference areas). In 2010 the presence of zinc and cadmium was linked to lower populations of plecopterans (stoneflies) and reduced aquatic insect diversity in the CDA Lake region (Lefcort et al., 2010). Fish communities were depauperate in numerous places and some taxa typical of watersheds (sculpins) were completely missing from rivers and streams in the CDA basin (Maret and MacCoy, 2002). Thus, effects of metal contamination on the benthic and fish communities were unambiguous in the basin upstream from the Lake. The degree to which such conclusions apply to the Lake at present depends upon understanding the characteristics of the pelagic and benthic communities of the system.

Ecological Communities in Lake Water Columns

The community of organisms adapted for life in suspension in the water column of lakes, rivers, and oceans is referred to as the *plankton*, with the photosynthetic members of the plankton being the *phytoplankton*. Collectively, phytoplankton communities are formed by various species of eukaryotic photosynthetic protists ("algae") that include, at the taxonomic level of division, the Chlorophyta ("green algae," such as *Chlorella* or *Scenedesmus*), the Chryosophyta (the "golden brown algae", which include diatoms), the Cryptophyta (e.g., *Cryptomonas*, *Rhodomonas*), and the Dinophyta (e.g., *Ceratium*, *Peridinium*). Importantly, included among the phytoplankton are the photosynthetic prokaryotes, the cyanobacteria. These include very small and ubiquitous unicellular taxa (e.g., *Synechococcus*) as well as colonial taxa that can be toxic and/or fix atmospheric nitrogen (e.g., *Microcystis*, *Anabaena*).

Phytoplankton are important in forming the basis of pelagic food webs. They are consumed by filter-feeding zooplankton species that themselves form the food base of planktivorous fishes, which support piscivorous fish often prized by anglers. Phytoplankton are also key determinants of lake water quality because phytoplankton growth is often limited by nutrients such as nitrogen and phosphorus. Their growth can lead to algal blooms and decreased water quality when nutrient inputs increase, frequently from anthropogenic sources (e.g., septic leachate, fertilizer runoff). Under some conditions of increased nutrient loading, some phytoplankton species, especially cyanobacteria, can produce toxic substances harmful to fish and humans. The species composition of the phytoplankton reflects environmental conditions, with diatoms, green algae, and small unicellular cyanobacteria dominating low-nutrient, "oligotrophic" conditions and colonial cyanobacteria becoming important in eutrophic systems.

Phytoplankton Data in Coeur d'Alene Lake and Lateral Lakes

A phytoplankton community assessment of CDA Lake was performed at stations C1 (Tubbs Hill) and C4 (University Point) across several seasons from 2007 to 2017 (EcoAnalysts, Inc., 2020). The diatom *Asterionella formosa* was found to be the most abundant taxon at both stations during runoff and warm stratified periods. During the cold clear period, the potentially toxic cyanobacterium *Microcystis* was important at C1 while the diatom *Tabellaria flocculosa* and the cryptomonad *Cryptomonas* spp. were dominant at C4. The study showed that different seasons of the year harbored different phytoplankton communities. Analysis of annual data indicated that 2007–2008 was distinct from subsequent years in terms of community structure at both sites, but no consistent trends are obvious during the last ten-year interval. The relatively high abundance of the potentially toxic cyanobacterium *Microcystis* at C1 (Tubbs Hill) is of potential concern, but this is largely associated only with the cold, clear period (winter). Otherwise, the dominant taxa are diatoms, which are generally indicative of good water quality.

---

[13] EPT refers to the three orders of aquatic insects most common in cobble bottom streams and rivers (Ephemeropteral, Plecoptera, and Trichoptera). As stress increases, these three taxa groups typically decline in abundance and are replaced by other, more tolerant, groups.

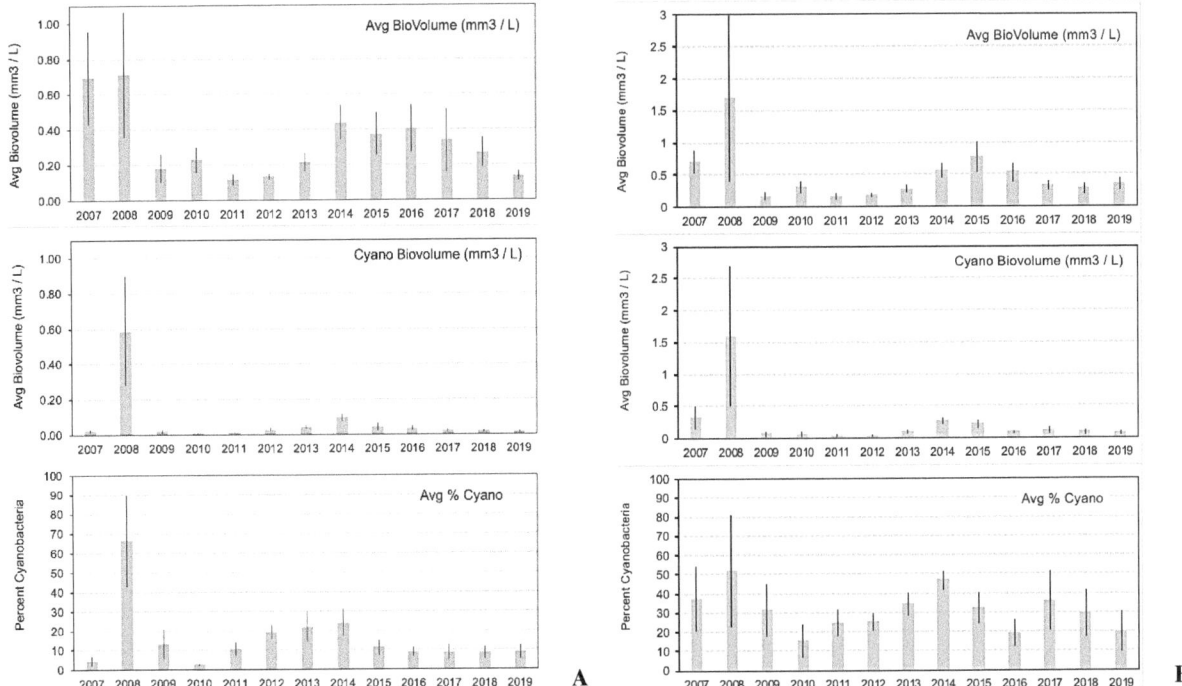

FIGURE 9-10 Dynamics of summertime (May–September) total phytoplankton biomass (top), total cyanobacteria biovolume (middle), and percentage contribution of cyanobacteria to total biovolume (bottom) for C5 (Panel A) and C6 (Panel B). NOTE: Figure 9-10B was updated after report release to reflect the full phytoplankton dataset.

Phytoplankton population and community data are also available for sites in the southern part of the Lake (C5 and C6), the lower St. Joe River, and two lateral lakes (Benewah Lake, Round Lake) for 2011–2020 (CDA Tribe and Avista Corporation, 2017, and courtesy of D. Chess, CDA Tribe). According to the report for 2011–2015 (CDA Tribe and Avista Corporation, 2017), phytoplankton cell numbers were dominated by cyanobacteria (largely small unicells). However, based on biovolume (biomass), taxonomic composition was distributed relatively evenly among major groups (cyanobacteria, diatoms, greens, and cryptophytes/chrysophytes). Site C5 had the lowest average biomass. Community differences (assessed by non-metric multidimensional scaling, a method of reducing complex multivariate data into fewer dimensions to facilitate interpretation) identifed years with higher runoff as distinct due to increases in abundances of benthic diatoms. At the genus and species level, taxa typical of eutrophic conditions (*Anabaena circinalis* and *Aphanizomenon* spp.) were abundant (based on relative biovolume) at C6 and the lateral lakes but not at C5, where diatoms (*Fragilaria crotonensis*, *Tabellaria fenestrata*), *Chlamydocapsa* spp. (green), and *Euglena* (euglenoida) were more prevalent. Ordination analysis (compression of multivariate species composition data into reduced dimensions) of phytoplankton community composition found that community composition was associated with a number of environmental variables (orthophosphorus, specific conductance, temperature, chlorophyll, nitrate, pH, fluorescence, and dissolved oxygen). Notably, little discussion of possible temporal trends within the five-year synthesis period was presented, nor were comparisons with previous years of data discussed.

Longer-term data for phytoplankton community composition were provided to the committee for 2007–2019 (courtesy of D. Chess, CDA Tribe), including summertime (May–September) total phytoplankton biovolume, cyanobacteria biovolume, and percentage of biovolume contributed by cyanobacteria.[14] As shown in Figure 9-10A, at C5 phytoplankton biovolume was high during the initial 2 years but rapidly declined after 2008, increasing to moderate levels in 2015–2016 and then declining again. Cyanobacteria biovolume was low throughout the

---

[14] Text here and throughout the chapter was edited after report release to include the full phytoplankton dataset.

observation period except in 2008 when cyanobacteria dominated the community throughout the year, contributing ~ 65 percent on average. Dominant among the cyanobacteria at C5 during 2008 was *Microcystis*, a potentially toxic taxon. Excluding 2008, cyanobacteria contributed ~ 12 percent (on average) to phytoplankton biomass across the years reported (16 percent if 2008 is included). Data for C6 indicate phytoplankton biomass levels somewhat higher than those at C5 during the same period (Figure 9-10B). Also, cyanobacteria played a more important role at C6 than at C5, contributing 31 percent to phytoplankton biomass. Important taxa at C6 included the potentially toxic forms Anabaena *circinalis*, *Microcystis* spp., *Aphanizomenon*, and *Planktothrix*. This is consistent with other metrics at C6 that indicate a more eutrophic state relative to other areas of the Lake.

Data have also been provided for epilimnetic phytoplankton for two lateral lakes (Swan Lake, Thompson Lake) in 1 year (2015) with six sampling dates between mid-July and mid-September. The data consist of biovolume concentrations (mm^3/L) obtained by microscopic identification and counting of cells combined with estimates of per-cell volumes. Identification was to the level of genus. Overall, phytoplankton biovolume was dominated by Chlorophyta (green algae such as *Euglena*, *Oocystis*, *Gloeococcus*, and *Tetraedron*; ~ 48 percent) and Cryptophyta/non-diatom Chrysophyta (*Chroomonas*, *Dinobryon*, *Komma*, and *Cryptomonas*; ~ 24 percent). Cyanobacteria (*Chroococcus*, *Synechococcus*, with some *Microcystis*) made up 13 percent of biovolume, on average, in these samples.

Overall, C6 and some of the lateral lakes are relatively productive and display some eutrophic characteristics in terms of higher cyanobacteria abundance (and nutrients and oxygen depletion), but these conditions diminish moving into the main Lake (e.g., at C5), with a transition to communities more characteristic of oligotrophic conditions in northern parts of the Lake. From 2007 to 2019 at C5, phytoplankton biovolume was highest in the early part of the observation period but has been lower in recent years. Although a single year (2008) that involved high biomass of cyanobacteria-dominated (~ 65 percent) phytoplankton was observed, C5 maintains relatively low levels of cyanobacteria overall. Based on these data, it is problematic to infer patterns in the temporal dynamics of phytoplankton communities in CDA Lake except to note that since 2007 water quality conditions in terms of phytoplankton biomass and cyanobacteria dominance seem to have improved. Nevertheless, observation of high phytoplankton biomass with cyanobacteria dominance in 2008 indicates the possibility for the Lake to support blooms of potentially harmful cyanobacteria given suitable conditions.

In their toxicity studies, Kuwabara et al. (2006) noted that the results were consistent with the observations that zinc-sensitive *C. minutissima* decreased in cell concentration away from the mouth of the CDA River, and the more tolerant *A. formosa* increased in abundance as metal concentrations increased in 2005. They concluded that "significant differences in response by the phytoplankton isolates in this study suggest that observed longitudinal shifts in phytoplankton community composition may represent a response to longitudinal gradients in solute (Zn) concentrations." They suggested implications for management of the Lake: "If dissolved Zn can be reduced in the water column from > 500 nM (i.e., current concentrations near and down stream of the Coeur d'Alene River plume) to < 3 nM (i.e., concentrations near the southern St. Joe River inlet) such that the Lake is truly phosphorus limited, management of phosphorus inputs by surrounding communities will ultimately determine the limnologic state of the lake." This implies that zinc inhibits phytoplankton growth in the Lake at present, and has become the basis for a current concern that if the presumed zinc inhibition were reduced, the Lake might be more prone to eutrophication than at present. The conclusions of Kuwabara et al. (2006, 2007) were based upon studies conducted at one point in time in a complex lake in which zinc exposures themselves are complex in both time and space. Although the experimental observations were consistent with the field collection, they did not have the benefit of the 14 years of consistent seasonal monitoring now available for the Lake (as described above) and the limnological understanding that has developed since 2005. Statistical tests of the influence of zinc on phytoplankton biomass (chlorophyll a [chl a]) performed by the committee did not support zinc inhibition (see Chapter 5). Although the committee cannot exclude the possibility of zinc inhibition of the growth of some phytoplankton taxa in the Lake, the community dynamics described above are not consistent with zinc being an important driver of phytoplankton community structure. For example, euphotic-zone zinc concentrations at C5 are below typical thresholds of toxicity in spring due to seasonal inputs from the St. Joe River, then increase dramatically to above experimentally derived thresholds in the late summer and fall. But changes in the relative abundance of *Chlorella* spp. and *A. formosa* are not coincident with those changes in zinc concentrations. This is an illustration of both the challenges in extrapolating results from limited single-species toxicity testing to the limnological processes that drive phytoplankton community dynamics in a large, complex lake and of the value of multiple lines of evidence when evaluating metal influences.

Macrobenthic Invertebrate Community Structure, Distributions, and Dynamics

Macrobenthos refers to the community of larger organisms that lives on the bottom (the *benthic habitat*) of lakes and rivers. The illuminated region of the benthic zone is referred to as the *littoral zone*, and the dark, poorly illuminated region is called the *profundal zone*. This community includes an important set of larger invertebrate animals (*macroinvertebrates*) such as the larval or adult forms of various insects (mayflies, dipterans, dragonflies, damselflies), mollusks (snails, mussels), and segmented worms (oligochaetes). Macroinvertebrates are ecologically important in the littoral zone for transferring primary production to higher trophic levels and in the profundal zone for transferring detrital materials to higher trophic levels. Macroinvertebrates can also transfer contaminants from sediments to higher trophic levels (Lavoie et al., 2013) and participate in bioturbation, which can affect the exchange of oxygen, nutrients, and metals between sediments and the water column. Macrobenthic taxa differ in their sensitivity to various environmental factors, such as oxygen or metal concentrations, and thus are often used as bioindicators of chemical conditions in rivers and lakes (Hauer and Lamberti, 2011).

The benthos of large oligotrophic and mesotrophic lakes is typically characterized by low abundances (total number of individuals) and high variability in both time and space compared to more productive lakes. Physical differences in substrate (such as grain size), instability of the substrate, differences in depth, differences in the availability of food, and other factors drive seasonal, year-to-year, and spatial variability in abundance and other community measures (White and Miller, 2008; Caires et al., 2013; Hayford et al., 2015). Organisms with broad environmental tolerances or life cycles suited to these changing conditions do best in large, deep oligotrophic or mesotrophic lakes.

In CDA Lake, highly enriched concentrations of arsenic, cadmium, lead, and zinc are an additional potentially influential variable. Sediments from Lake locations north of C5 have metal concentrations that fall within the window or exceed the window predicting toxicity according to sediment quality criteria (EcoAnalysts, Inc., 2017; Table 9-4). Such heavily contaminated conditions can select for species that avoid bioavailable metal forms, confounding broad measures of community change like abundance and even taxa richness. Thus, changes in community structure attributable to metal enrichment alone are typically difficult to distinguish from changes caused by other perturbations unless the impact on the benthos is dramatic (Nalepa and Landrum, 1988).

These challenges are evident in the studies to date of the benthos of CDA Lake. Horowitz et al. (1993, 1995) first noted that sediment cores from profundal locations in CDA Lake often included varved (layered) sediments that clearly reflect undisturbed material. Varved sediments reflect minimal sediment reworking by the benthic community, typical of a depauperate community, suggesting that was the case in at least deep lake locations in the past. The varving was sufficiently distinct that the authors used it to estimate when effects from mineral extraction activities began. Deeper layers (deposited before 1980) in the core studied by Morra et al. (2015) showed similar varving.

Benthic production supports higher trophic levels in most lakes but can often go underappreciated. In a survey across a northern temperate lake, Vander Zanden and Vadeboncoeur (2002) found that lake fish relied on the benthos for 65 percent of their food consumption and that the benthic food web played "a central role in supporting higher trophic level production and ecosystem processes in the pelagic zone." Given the anecdotal suggestion from Horowitz et al. (1993) of a depauperate benthos, it is important to have a better understanding of benthic production, benthic community structure and function, and how they are affected by the "profound enrichment" of arsenic, cadmium, lead, and zinc in the sediment of CDA Lake.

Data from four surveys (1972, 1996, 1999–2005 and 2011–2015) of macrobenthos communities in various localities within CDA Lake are available. While at coarse time resolution and involving somewhat different localities, the studies permit some insight into the status and dynamics of macrobenthos in the Lake, as described in Box 9-3.

Although the emphasis of the two most thorough benthic studies (EcoAnalysts, Inc., 2017; Kuwabara et al., 2006) differed somewhat, many results were comparable between the two. Figure 9-11 shows that the frequency distributions of density and taxa richness among samples were similar between the two studies when similar

BOX 9-3
Studies of Macrobenthos in CDA Lake Sediments

A benthic community assessment in the Lake at depths of 5–30 m for five bays (Windy, Rockford, Carlin, Beauty, and Neachen) was done during 2011–2015 (EcoAnalysts, Inc., 2017). Samples were collected from each location in different years and different months (July, September and October). No reference site from the southernmost Lake was included in the study. Abundances (density of individuals) and diversity generally decreased with depth, as is typical of most deep lakes (see, e.g., Hayford et al., 2015). Community structure was related more strongly to depth than to location (bay). Communities were generally dominated by oligochaetes and chironomids. Metal concentrations generally increased with depth, but there was no simple correlation between metal concentrations and any measure of community structure. Principal components and community similarity analyses were used to assess influences of roughly co-varying depth and metal concentration and showed a stronger relationship with depth than with metals (EcoAnalysts, Inc., 2017). The report recommended direct toxicity testing of sediments to help clarify the role of metal enrichment, as well as addition of observations from reference sites without metal contamination.

Kuwabara et al. (2003, 2006) used similar methods (ponar grab, sieve size of 500 μm) in analyses of the benthic communities in August 1999, August 2001, June 2004, and June 2005. In 1999, they noted varved cores, surface sediments indicative of iron oxide coatings, and little biological activity at a location near Harrison (Kuwabara et al., 2000). They concluded that bioturbation was unlikely to be a factor in their assessment of benthic metal flux. The more systematic benthic studies (Kuwabara et al., 2006) in 2004, 2005, and 2011 were conducted at three locations in the northern Lake (near Carlin, Rockford, and Mica Bays) and two in the southern Lake (C5 and a reference area with no metal enrichment near SJ1). Samples were collected in June in two years and in August in three years at depths of 17–38 m, except for the St. Joe River location, which was at 4-m depth. Kuwabara et al. (2003, 2006) found that "benthic macroinvertebrate densities (abundances) varied substantially both temporally and spatially," as did taxa richness (number of taxa present). The authors compared the benthic community of CDA Lake to those of ten lakes from Convict Creek basin, California, where similar study methods had been employed (Reimers et al., 1955). Macroinvertebrate densities ranged from 900 to 4,700 macroinvertebrates/m^2 in the ten oligotrophic lakes in Convict Creek basin, compared to the range of 280–5,922 at all the study sites in CDA Lake. Furthermore, all locations in CDA Lake also had substantially lower densities and lower taxa richness than the one sample from the mouth of the St. Joe River. The shallower depth and the more mesotrophic/eutrophic nature of the southern site, as well as the differences in metal contamination, could have contributed to these differences. Typically, benthic diversity is reduced by eutrophication that includes seasonal anoxia in bottom waters, as at SJ1 (Vadeboncouer et al., 2003). Thus, the low species richness at all CDA Lake sites relative to SJ1 suggests another source of stress (i.e., the high metal concentrations).

Two master's theses studied benthos in CDA Lake. Ruud (1996) surveyed macroinvertebrates in CDA Lake and compared them to uncontaminated Priest Lake. This study also considered depth variation but measured metal concentrations in the water column and in animal tissues but not in sediments. Overall, general benthic community structure was similar to that reported by EcoAnalysts, Inc. (2017) with chronomids and oligochaetes at many sites. Macroinvertebrate densities were higher than found at similar sites by Kuwabara et al. (2006). Deep-water communities did differ somewhat for the two lakes. In Priest Lake, chironomids and sphaeriinae (bivalves) were dominant at depth while in CDA Lake these taxa were rare, but nematophora ("horsehair" worms), tricladidae (e.g., planaria), and oligochaetes dominated. Animals in CDA Lake, and especially in sediments from deep water, carried overall higher metal burdens in tissues relative to those in Priest Lake.

Winner (1972) surveyed macrobenthos on three dates encompassing a period of seven months at four depths in four areas in the CDA Lake system (moving upstream to downstream): Chatcolet/Round Lakes, near Harrison, near Carlin Bay, and near the city of Couer d'Alene. The Chatcolet/Round Lakes samples in the southern Lake had lower concentrations of zinc and lead than the three locations from north of the CDA River mouth. Chironomids and oligochaetes dominated the samples. No associations were found between macroinvertebrate abundance and sediment metal concentrations. The author noted a shift from invertebrate taxa associated with eutrophic conditions in the south to those associated with more oligotrophic conditions in the north, consistent with current conditions and perceptions. Similar to other studies, unusually depauperate communities were found in some samples from the northern locations (where neither chironomids nor oligochaetes were abundant) compared to Chatcolet/Round Lakes.

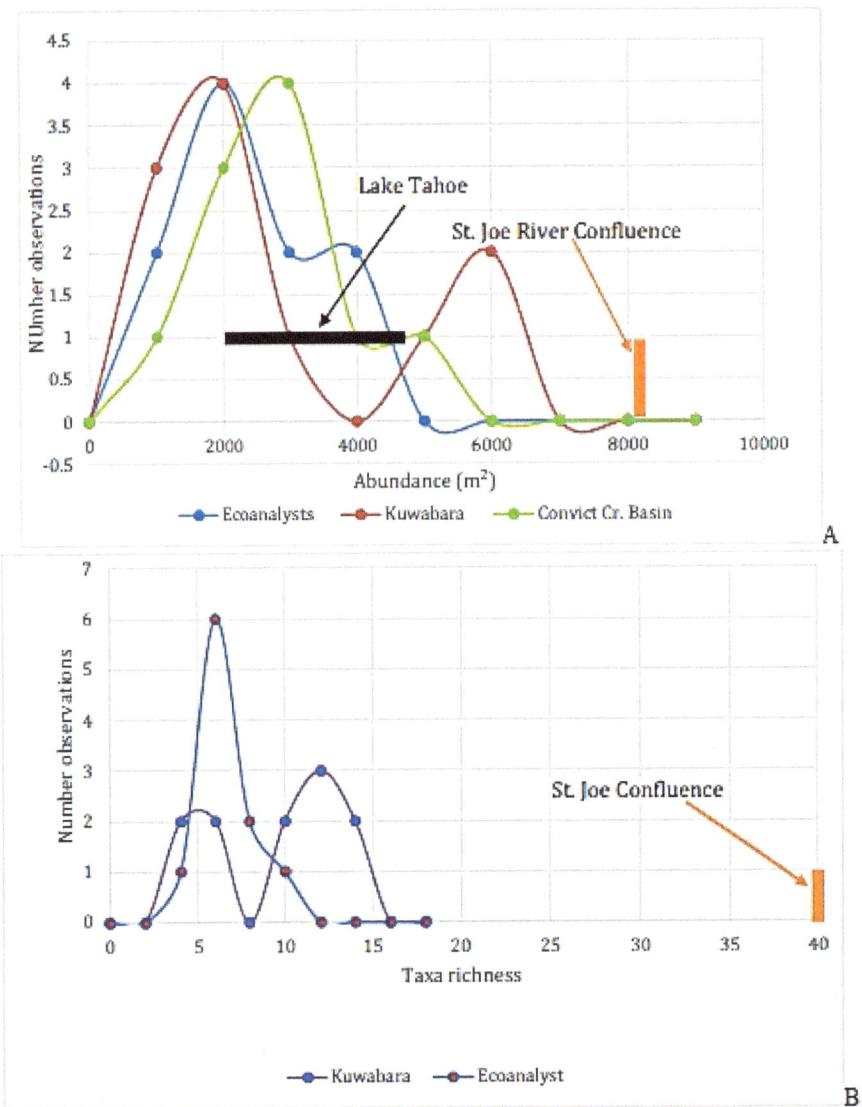

FIGURE 9-11 Frequency distribution models of (A) macroinvertebrate densities and (B) taxa richness as measured by EcoAnalysts, Inc. (2017) and Kuwabara et al. (2006). The range of mean macroinvertebrate densities observed at 0–50 m depth in Lake Tahoe in hundreds of samples in 1962–1963 and 2008–2009 (Caires et al., 2013) is also shown for perspective (black bar), as is the density and taxa richness of macroinvertebrates seen in one sample from the confluence of the St. Joe River (which was unaffected by mineral extraction activities) in CDA Lake (green bar). SOURCES: EcoAnalysts, Inc. (2017); Kuwabara et al. (2006); Caires et al. (2013).

depths were compared.[15] Figure 9-11A also compares the two CDA Lake studies to studies from oligotrophic lakes with little metal contamination—the 17 lakes in the Convict Creek basin (Reimers et al., 1955) and a study using comparable methods from Lake Tahoe (Caires et al., 2013). The main difference was the high frequency of locations with densities less than 2,000 per m^2 in CDA Lake. Caires et al. (2013) found no locations in Lake Tahoe

---

[15] Water depth influences both invertebrate density and richness in all lakes. To control for that, in Figure 9-11A only samples > 10 m depth from EcoAnalysts, Inc. (2017) were used to compare to Kuwabara et al. (2006) because the latter study included no shallow-water samples. Richness was compared between shallow-water samples in EcoAnalysts, Inc. (2017) and the location from the St. Joe River, because water depths were shallow. In the comparison with Caires et al. (2013), only samples from similar depths were compared.

with macroinvertebrate densities as low as several of the CDA Lake sites. While many factors affect invertebrate densities in oligotrophic lakes, locations as depauperate as many of those in CDA Lake are rare.

Distributions of taxa richness were reduced in the contaminated portion of CDA Lake compared to a reference location (Figure 9-11B). As noted earlier, taxa richness in benthic communities in the basin of the CDA River has recovered from zero before 1968 to about half the richness of reference sites in the most recent samplings (Hoiland et al., 1994; Maret et al., 2003). It would not be surprising if similar observations applied to the Lake.

At a qualitative level, the macrobenthos communities of CDA Lake have been consistently dominated by chironomids and oligochaetes throughout the period 1972–2015. In general, this is a community typical of large, deep oligotrophic/mesotrophic lakes. This ecological stability over the window of observation limits our ability to draw inferences about response of macrobenthos to changing metal concentrations or nutrient levels during this interval.

Locations with unusually depauperate benthic communities are characteristic of at least some deeper waters in CDA Lake. Whether the frequency of such locations is sufficient to affect benthic production overall cannot be determined from the limited data available to date. As to the role of metal enrichment, the conclusion made by EcoAnalysts, Inc. (2017) probably applies to all studies to date:

> When compared together, the decrease in community metrics and increase in metals by depth profile could not be correlated proportionately survey-wide. Thus, a combination of effects could be impacting some sites. Communities may be experiencing a metals-related effect in some deep-water locations (those from 30 m deep in Carlin Bay were markedly the least abundant and diverse in the survey); others however may be more influenced by environmental variables (such as those from 30 m deep in Neachen Bay, which had the second lowest mean abundance but also among the lowest concentrations of zinc, lead, arsenic, and cadmium).

Understanding benthic responses to either nutrients or metals could be improved by careful analysis of differences in sensitive/tolerant species and functional attributes, especially within the family Chironomidae (e.g., Caires et al., 2013) or the subclass Oligochaeta (Vivien et al., 2020). Changes in those groups can signal the causes of differences (Vivien et al., 2020) or have implications for processes such as reduced productivity (Caires et al., 2013).

Benchmarks for Different Metals

NRC (2005) suggested that an important outcome of evaluating ecological health was to establish *potential remediation goals*. In a simple world, one should be able to identify the threshold concentration beyond which each trace metal begins to elicit adverse effects in natural waters. The relationship between the concentration of trace metal in a waterbody and the influence of the resulting contaminant concentration on life in that environment depends upon detecting metal "toxicity" in nature. As is clear from the analysis above, quantifying the response of ecological health to metal enrichment, or the threshold for harmful exposures, is fraught with uncertainty and, in some cases, controversy (Luoma and Rainbow, 2008). The uncertainties stem, at least partly, from the complexity of the problem; from different perspectives and methods for determining toxicity; and the choices of data used for the benchmarks, criteria, or goals (Buchwalter et al., 2017).

Defining when remediating the legacy of mineral extraction is complete for CDA Lake may be beyond EPA's remit at present. But it is nevertheless important to define expectations of ecological health or benefits from ecosystem services that would derive from reducing metal impacts. Is remediation complete when the EPA's National Ambient Water Quality Criteria (AWQC) are met; when the Lake Management Plan (LMP) targets are met; when risks to biodiversity, native fisheries, and other ecosystem services are reduced or removed; when traditional foods are safe to eat; or when the geochemical aspects of the Lake's ecology become similar to other lakes of its size and character? Each of these goals would require a different (progressively lower) benchmark for each metal(loid). Defining transparent, explicit goals (perhaps in ecological terms) and linking those goals to a benchmark (i.e., target) is an essential step in a sustainable plan for the future of CDA Lake, even if those goals differ among interest groups. Examples of benchmarks (considering zinc only) for different expectations from lake ecology are proposed below as guides to evaluating if changes in the lake are sufficient to improve ecosystem services.

TABLE 9-5 Chronic EPA Ambient Water Quality Criteria for Dissolved Cadmium, Lead, and Zinc Corrected for Hardness Ranging from 10 to 100 mg/L $CaCO_3$

Hardness	Criteria (in µg/L) at selected water hardness (mg/L $CaCO_3$)				
	10	25	30	50	100
Cadmium	0.13	0.25	0.29	0.43	0.72
Lead	0.2	0.4	0.66	1.1	2.5
Zinc	16.7	36.2	43	65	117

SOURCES: Balistrieri et al. (2002) and EPA (2016) for cadmium.

The appropriate benchmarks to protect ecological health in a specific environment, like CDA Lake, depend upon goals designed around expectations of the ecosystem after remediation. For example, Table 1-6 shows that most large lakes in the United States have open-water dissolved zinc concentrations of < 1.0 µg/L. Van Genderen et al. (2008) found median concentrations of zinc of 2–3 µg/L in a variety of lakes from different countries in Europe. Zinc concentrations in the St. Joe River and reference areas upstream of most historic mineral extraction activities fall in a similar range, 0.2–2 µg/L (Kuwabara et al., 2007; Chapter 6). Thus, a goal of 1–3 µg /L might represent pre-mining zinc concentrations that are typical for large lakes subject to some human development on their shores. These values are close to concentrations observed in the St. Joe River near the confluence with CDA Lake. It can be assumed most such lakes provide a variety of ecosystem services and minimally disturbed ecological functions.

The simplest goal for CDA Lake is compliance with regulatory criteria. Existing criteria are designed to protect 95 percent of species and limit the most overt ecological damage. The EPA's National AWQC, from which local criteria are derived, correct the national values for local water hardness (which varies in CDA Lake from 15 to 65 mg/L $CaCO_3$; Balistrieri et al., 2002). Chronic AWQC for dissolved cadmium, lead, and zinc in the CDA basin, across a range of hardness, are presented in Table 9-5.

Most criteria, guidelines, targets, or benchmarks are defined by a single number, for simplicity, but are corrected for water quality characteristics that are themselves variable. This can result in variability in criteria among jurisdictions within the same region. For example, EPA Region 10's hardness-corrected target for zinc in the CDA basin, as described to the committee, is 58 µg Zn/L, while the LMP hardness-corrected target is 36 µg Zn/L. Presumably, the different targets reflect differences in the hardness correction between jurisdictions. DeForest and Van Genderen (2012) used a more inclusive geochemical modeling approach (the Biotic Ligand Model) to correct for local water quality conditions and calculated the appropriate range of benchmarks to be 18–27 µg Zn/L for CDA Lake. Thus, a single target concentration for cleanup is useful pragmatically, but it likely reflects jurisdictional interpretations of water quality and is not necessarily indicative of a threshold for ecological effects in a region.

In developing criteria, EPA does not allow data other than single-species toxicity testing for most metals (Buchwalter et al., 2017). Yet, compared to single-species toxicity testing, mesocosm studies can include a more realistic assemblage of species and can consider both dissolved and dietary exposures (e.g., Mebane et al., 2016), thereby providing more realistic estimates of the potential for ecological effects. In most cases, the thresholds of toxicity derived in tests that more carefully mimic nature are lower than those derived from single-species tests (Mebane et al., 2020; Clements et al., 2021)—that is, ecosystems are usually more sensitive to metals than suggested by single species toxicity testing. It has long been argued (e.g., Cairns, 1986) that combining single-species toxicity testing with mesocosm and field data might result in more realistic criteria or benchmarks than the present approach. Such criteria might better meet a goal of maximizing biodiversity and ecosystem services by minimizing simplification of the Lake community.

Hoang et al. (2021) (see Box 9-2) showed that the first effects of zinc on a mixed assemblage of phytoplankton taxa began at 8 µg Zn/L. They calculated a lowest detectable effect concentration of 14 µg Zn/L for changes in group abundance and 21 µg Zn/L when chlorophyll *a* or zooplankton toxicity was the end point in mesocosm studies of zinc toxicity. Two additional mesocosm studies of zinc (or zinc and cadmium in combination) toxicity to the aquatic insect communities typical of Rocky Mountain streams provide another example of a benchmark

TABLE 9-6 Thresholds of Water-Column Zinc Toxicity by Different Approaches

Method	Basis	Target (µg/L)	References
Single-species toxicity testing			
50 mg/L hardness	U.S. EPA target	58	EPA Region 10, 2021
25 mg/L hardness	LMP Target	36	IDEQ and CDA Tribe, 2009
Range hardness: 15–65 mg/L	Local range AWQC	21–73	Balistrieri et al., 2002
Biotic Ligand Model	Geochemistry	18–27	DeForest and Van Genderen, 2012
Mesocosm studies			
Sensitive phytoplankton	Mesocosm Expts	<14	Hoang et al., 2021
Sensitive species invertebrates	Mesocosm Expts	6–15	Mebane et al., 2020; Schmidt et al., 2011
Modern Regional Background	St. Joe River	1–3	Page 292

relevant to CDA Lake. These studies evaluated dietary and dissolved metal bioavailability in concert with evaluation of effects on the community. Both were conducted under water quality conditions typical of streams in the CDA basin. Mebane et al. (2020) observed that mayfly populations (the most sensitive taxa) declined then disappeared from the community at zinc concentrations of about 10 µg/L. Schmidt et al. (2011) showed that the bioaccumulated zinc began to result in the disappearance of sensitive mayfly species at 5.4 µg/L equilibrated in the stream, and this was associated with adverse effects in the aquatic community as a whole.

Based upon what is known to date, a variety of choices for a zinc threshold or benchmark are possible for CDA Lake, depending upon ecological and ecosystem goals (Table 9-6). Greater protection from adverse effects of zinc toxicity would result from using targets built around mesocosm studies from the literature compared to compliance with existing regulations alone.

CONCLUSIONS AND RECOMMENDATIONS

The widespread distribution of mining wastes has affected human and ecological health in the CDA region. Although metal concentrations in the Lake waters and sediments are well described and some are monitored regularly, human exposures and ecological risks associated with the Lake itself have not been the subject of comprehensive study compared to the systematic ecological and mechanistic evaluation of risks in the basin upstream of the lake. Trends in the basin indicate that human exposures to hazardous materials like lead have declined and severely damaged ecosystems in the rivers and streams have improved. But there is limited information about exposure pathways that specifically apply to the Lake.

Ultimately, a body of knowledge developed across multiple levels of biological organization will be necessary to comprehensively understand ecological implications of metal contamination in CDA Lake, as well as potential benefits from remediation into the future. Identifying present-day ecological implications of the legacy of metal contamination in CDA Lake is a first step toward addressing the future viability of ecosystem services, such as biodiversity, ecological functions, fisheries, wildlife, and support for activities ranging from recreation to a subsistence life style. The following conclusions and recommendations build from studies of the basin, considering the data that exist for the Lake and drawing analogies from other environments.

1. **Blood lead levels in children have declined in the Coeur d'Alene region but have stalled in recent years at a level about equal to the 2021 CDC level of concern (3.5 µg/dL).** Remediation of lead-contaminated lands in OU-1 and OU-2 reduced mean blood lead levels in children (aged 0–9) in the basin to below the EPA level of concern of 10 µg/dL. Yet blood lead levels within the Superfund jurisdiction are about two to four times higher than in other western states for which data are available. These estimates suffer from questions about underrepresentation in economically disadvantaged and indigenous communities, low participation in the testing programs (especially in the lower basin), and age groups studied.

2. **Assessments of human health risks specifically associated with CDA Lake would help address remaining potential pathways of lead and arsenic exposure.** Occupations that expose people to Lake water or Lake sediments, recreational exposures to water or Lake sediments, and indirect exposure to metals in the Lake via groundwater used as drinking water have not been the focus of studies to date because the Lake is outside the Superfund jurisdiction.

3. **A more comprehensive characterization of the sources of mercury in the food web of CDA Lake could benefit resource allocation decisions with regard to mercury.** Few investigations have assessed mercury biogeochemical cycling in this system or addressed the source of MeHg (local vs. regional). By evaluating mercury dynamics in CDA Lake, environmental managers will be in a better position to consider options for reducing, over time, MeHg concentrations in high-trophic-level fish.

4. **Expansion of existing monitoring to include a few sensitive nearshore environments could provide an early warning system for the onset of harmful algal blooms and expansion of nuisance-attached algae and of invasive plants.** While on a lake-wide basis, the Lake remains oligotrophic with some mesotrophy in the south, experience elsewhere suggests the first signs of changes in trophic status can occur in nearshore, local waters in the form of blooms of attached algae. Expanded lakeshore monitoring could aid in detecting those changes before they become widespread.

5. **Systematically developing a body of knowledge on how CDA Lake food webs are influenced by the legacy of mineral extraction will inform decisions about remediation and efforts to maximize ecosystem services.** High priorities for better understanding ecological processes in the Lake include (1) expanded characterization of benthic and pelagic food webs; (2) evaluation of metal exposures in key components of the food web, and (3) experiments with benthic and water-column mesocosms to identify thresholds below which the Lake ecosystem will improve. CDA Lake may be the single greatest asset in the region.

6. **Zinc concentrations in lake waters in many locations, as well as zinc and lead concentrations in many of the sediments of CDA Lake, exceed thresholds that suggest they could be toxic to some aquatic species.** However, evidence of disturbance in phytoplankton and benthic communities in CDA Lake is ambiguous. There is no evidence among existing field data of phytoplankton that supports the concept that reducing zinc concentrations in Lake waters will increase the risk of eutrophication in the Lake. Existing data from monitoring program and limited *ad hoc* studies are insufficient to clarify the contradictions in the line of evidence above, and existing data do not allow an assessment of the influence of metal contamination on the CDA Lake ecosystem.

7. **Multiple benchmarks could aid in characterizing remediation successes relevant to CDA Lake.** These would include goals for the Lake ecosystem and ecosystem services and targets for metal concentrations below which such goals could be achieved. Different goals (e.g., compliance with state law vs. avoiding effects on water column communities) will require different benchmarks. The potential benchmark for zinc in the water column that would allow the Lake to return to pre-mining reference conditions and regain lost ecological functions and ecosystem services could be as low as 2 µg/L.

REFERENCES

Ahmad, A., and P. Bhattacharya. 2019. Arsenic in Drinking Water: Is 10 µg/L a Safe Limit? *Curr. Pollution Rep.* 5:1–3. https://doi.org/10.1007/s40726-019-0102-7.

Alpers, C. N., J. A. Fleck, M. Marvin-DiPasquale, C. A. Stricker, M. Stephenson, and H. E. Taylor. 2014. Mercury cycling in agricultural and managed wetlands, Yolo Bypass, California: Spatial and seasonal variations in water quality. *Science of the Total Environment* 484:276–287. https://doi.org/10.1016/j.scitotenv.2013.10.096.

Alta (Alta Science & Engineering, Inc.). 2020. House Dust Evaluation for OU-3 of the Bunker Hill Mining and Metallurgical Complex Superfund Site. Draft Report.

Audet, D. J., L. H. Creekmore, L. Sileo, M. R. Snyder, J. C. Franson, M. R. Smith, J. K. Campbell, C. U. Meteyer, L. N. Locke, L. L. McDonald, T. L. McDonald, D. Strickland, and S. Deed. 1999. Wildlife Use and Mortality Investigation in the Coeur d'Alene Basin 1992–1997. Spokane, WA, USA: U.S. Fish and Wildlife Service.

Balistrieri, L., S. Box, A. Bookstrom, R. Hooper, and J. Mahoney. 2002. Impacts of Historical Mining in the Coeur d'Alene River Basin. Chapter 6 In: Pathways of Metal Transfer from Mineralized Sources to Bioreceptors: A Synthesis of the Mineral Resources Program's Past Environmental Studies in the Western United States and Future Research Directions. USGS Bulletin 2191. L. S. Balistrieri and L. L. Stillings, eds.

Balistrieri, L. S., and R. G. Blank. 2008. Dissolved and labile concentrations of Cd, Cu, Pb, and Zn in the South Fork Coeur d'Alene River, Idaho: Comparisons among chemical equilibrium models and implications for biotic ligand models. *Appl. Geochem.* 23:3355–3371.

Bechard, M. J., D. N. Perkins, G. S. Kaltenecker, and S. Alsup. 2009. Mercury Contamination in Idaho Bald Eagles, *Haliaeetus leucocephalus. Bull. Environ. Contam. Toxicol.* 83:698. https://doi.org/10.1007/s00128-009-9848-8.

Bellinger, D., A. Leviton, C. Waternaux, H. Needleman, and M. Rabinowitz. 1987. Longitudinal analyses of prenatal and postnatal lead exposure and early cognitive development. *N. Engl. J. Med.* 316:1037–1043. https://doi.org/10.1056/NEJM198704233161701.

Bellinger, D., A. Leviton, E. Allred, and M. Rabinowitz. 1994. Pre-and postnatal lead exposure and behavior problems in school-aged children. *Environ. Res.* 66:12–30.

Benoit, J., C. C. Gilmour, A. Heyes, R. P. Mason, and C. L. Miller. 2003. Geochemical and biological controls over methylmercury production and degradation in aquatic ecosystems. Pp. 262–297 In: Biogeochemistry of Environmentally Important Trace Elements, ACS Symposium Series No. 835. Y. Cai and O. C. Braids (eds.). Washington DC: American Chemical Society.

Beutel, M., B. Fuhrmann, G. Herbon, A. Chow, S. Brower, and J. Pasek. 2020. Cycling of methylmercury and other redox-sensitive compounds in the profundal zone of a hypereutrophic water supply reservoir. *Hydrobiologia* 847:4425–4446. https://doi.org/10.1007/s10750-020-04192-3.

Beyer, W. N., D. J. Audet, G. H. Heinz, D. J. Hoffman, and D. Day. 2000. Relation of waterfowl poisoning to sediment lead concentrations in the Coeur d'Alene River Basin. *Ecotoxicology* 9(3):207–218. https://doi.org/10.1023/A:1008998821913.

Blus, L. J., C. J. Henny, D. J. Hoffman, and R. A. Grove. 1995. Accumulation in and effects of lead and cadmium on waterfowl and passerines in northern Idaho. *Environ. Pollut.* 89(3):311–318.

Blus, L. J., C. J. Henny, D. J. Hoffman, L. Sileo, and D. J. Audet. 1999. Persistence of high lead concentrations and associated effects in tundra swans captured near a mining and smelting complex in Northern Idaho. *Ecotoxicology* 8(2):125–132. https://doi.org/10.1023/A:1008918819661.

Bookstrom, A. A., S. E. Box, R. S. Fousek, J. C. Wallis, H. Z. Kayser, and B. L. Jackson. 2013. Baseline, Historic and Background Rates of Deposition of Lead-Rich Sediments on the Floodplain of the Coeur d'Alene River, Idaho. USGS Open-File Report 2004-1211. http://pubs.usgs.gov/of/2004/1211/.

Brazy, P. C., R. S. Balaban, S. R. Gullans, L. J. Mandel, and V. W. Dennis. 1980. Inhibition of renal metabolism: relative effects of arsenate on sodium, phosphate, and glucose transport by the rabbit proximal tubule. *J. Clin. Invest.* 66:1211–1221.

Bryan, G. W., and L.G. Hummerstone. 1977. Indicators of heavy-metal contamination in the Looe estuary (Cornwall) with particular regard to silver and lead. *J. Mar. Biol. Ass. UK* 57:75–92.

Buchwalter, D. B., W. H. Clements, and S. N. Luoma. 2017. Modernizing water quality criteria in the United States: A need to expand the definition of acceptable data. *Environ. Toxicol. Chem.* 36:285–291.

Caires, A. M., S. Chandra, B. L. Hayford, and M. E. Wittman. 2013. Four decades of change: dramatic loss of zoobenthos in an oligotrophic lake exhibiting gradual eutrophication. *Freshwater Science* 32(3): 692–705.

Cairns, J. 1986. What is meant by validation of predictions based on laboratory toxicity tests? *Hydrobiologia* 137:271–278.

Cairns, J. 1989. Foreword. *Hydrobiologia* 188–189:1–5.

Canfield, R. L., D. A. Kreher, C. Cornwell, and C. R. Henderson. 2003. Low-level lead exposure, executive functioning, and learning in early childhood. *Child Neuropsychology* 9:1:35–53. https://doi.org/10.1076/chin.9.1.35.14496.

Carey, C. C., B. W. Ibelings, E. P. Hoffmann, D. P. Hamilton, and J. D. Brookes. 2012. Eco-physiological adaptations that favour freshwater cyanobacteria in a changing climate. *Water Research* 46(5):1394–1407.

CDC (Centers for Disease Control and Prevention). 2021. National Childhood Blood Lead Surveillance Data.

Chen, C., B. Yang, Y. Shen, J. Dai, Z. Tang, P. Wang, and F.-J. Zhao. 2021. Sulfate addition and rising temperature promote arsenic methylation and the formation of methylated thioarsenates in paddy soils. *Soil Biol. Biochem.* 154 (108129). https://doi.org/10.1016/j.soilbio.2021.108129.

Chess, D. 2021. Metals and Nutrients in Anoxic Hypolimnions. Presentation to the NASEM Committee. May 4, 2021.

Chupp, N. R., and P. D. Dalke. 1964. Waterfowl Mortality in the Coeur d'Alene River Valley, Idaho. *J. Wildl. Manage.* 28(4):692. https://doi.org/10.2307/3798784.

CH2M Hill and URS Corp. 2001. Final Ecological Risk Assessment. Coeur d'Alene Basin Remedial Investigation / Feasibility Study. Prepared by URS and CH2M HILL for EPA. Contract No. 86-W-98-228. Work Assignment No. 027-RI-CO-102Q. 1,820 pp. May 18, 2001.

Clements, W. H., D. B. Herbst, M. I. Hornberger, C. A. Mebane, and T. M. Short. 2021. Long-term monitoring reveals convergent patterns of recovery from mining contamination across 4 western US watersheds. *Freshwater Science* 40(2):407–426. https://doi.org/10.1086/714575.

Conaway, C. H., F. J. Black, T. M. Grieb, S. Roy, and A. R. Flegal. 2008. Mercury in the San Francisco Estuary. Pp. 29–54 *In:* Reviews of Environmental Contamination and Toxicology, volume 194. D. M. Whitacre (Ed.). Springer, New York, NY. https://doi.org/10.1007/978-0-387-74816-0_2.

Cossaboon, J. M., P. M. Ganguli, and A. R. Flegal. 2015. Mercury offloaded in Northern elephant seal hair affects coastal seawater surrounding rookery. *PNAS* 112:12058–12062. https://doi.org/10.1073/pnas.1506520112.

Coeur d'Alene Tribe and Avista Corporation. 2017. Coeur d'Alene Reservation Five-Year Synthesis 2011–2015 Monitoring Data. Spokane River Hydroelectric Project FERC Project no. 2545. February 2017.

Davis, T. W., D. L. Berry, G. L. Boyer, and C. J. Gobler. 2009. The effects of temperature and nutrients on the growth and dynamics of toxic and non-toxic strains of *Microcystis* during cyanobacteria blooms. *Harmful Algae* 8(5):715–725.

DeForest, D. K., and E. J. Van Genderen. 2012. Application of U.S. EPA guidelines in a bioavailability-based assessment of ambient water quality criteria for zinc in freshwater. *Soc. Environ. Toxicol. Chem.* 31(6):1264–1272.

De Schamphelaere, K. A. C., S. Lofts, and C. R. Janssen. 2005. Bioavailability models for predicting acute and chronic toxicity of zinc to algae, daphnids, and fish in natural surface waters. *Environmental Toxicology and Chemistry* 24:1190–1197.

Digman, T., R. D. Kaufman, L. LeStourgeon, and M. G. Brown. 2019. Control of lead sources in the United States, 1970-2017: Public health progress and current challenges to eliminating lead exposure. *J. Public Health Manag. Pract.* 25 (Suppl 1 Lead Poisoning Prevention):S13–S22. https://doi.org/10.1097/PHH.0000000000000889.

Dovick, M. A., T. R. Kulp, R. S. Arkle, and D. S. Pilliod. 2015. Bioaccumulation trends of arsenic and antimony in a freshwater ecosystem affected by mine drainage. *Environ. Chem.* 13(1):149. https://doi.org/10.1071/EN15046.

EcoAnalysts, Inc. 2017. Coeur d'Alene Lake Management Plan 2011-2015 Benthic Survey Summary Report. Prepared for Idaho Department of Environmental Quality.

EcoAnalysts, Inc. 2020. Plankton Community Assessment at Tubbs Hill and University Point Sampling Stations 2007–2017. Prepared for Idaho Department of Environmental Quality. Project Number ID-JW0028.

Engstrom, D. R. 2007. Fish respond when the mercury rises. PNAS 104:16394–16395. https://doi.org/10.1073/pnas.0708273104.

EPA Region 10. 2015. Fourth Five-Year Review Report for Bunker Hill Superfund Site, Shoshone and Kootenai Counties, Idaho.

EPA Region 10. 2021. Fifth Five-Year Review Report for the Bunker Hill Mining and Metallurgical Complex Superfund Facility (Bunker Hill Superfund Site). U.S. Environmental Protection Agency, Region 10. 130 pp.

EPA. 2001. Prepared by TerraGraphics Environmental Engineering, Inc., URS Greiner in association with CH2M HILL. Human Health Risk Assessment for the Coeur d'Alene Basin Extending from Harrison to Mullan on the Coeur d'Alene River and Tributaries: Remedial Investigation/Feasibility Study. 2,500 pp.

EPA. 2016. Aquatic Life Ambient Water Quality Criteria Cadmium–2016. EPA-820-R-16-002. Washington, DC: EPA Office of Water.

Farag, A. M., C. J. Boese, H. L. Bergman, and D. F. Woodward. 1994. Physiological changes and tissue metal accumulation in rainbow trout exposed to foodborne and waterborne metals. *Environ. Toxicol. Chem.* 13(12):2021–2029. https://doi.org/10.1002/etc.5620131215.

Farag, A. M., D. F. Woodward, J. N. Goldstein, W. Brumbaugh, and J. S. Meyer. 1998. Concentrations of metals in sediments, biofilm, benthic macroinvertebrates, and fish associated with mining waste in the Coeur d'Alene River basin, Idaho. *Archives of Environmental Contamination and Toxicology* 34:119–127.

Farag, A. M., D. F. Woodward, W. Brumbaugh, J. N. Goldstein, E. MacConnell, C. Hogstrand, and F. T. Barrows. 1999. Dietary effects of metals-contaminated invertebrates from the Coeur d'Alene River, Idaho, on cutthroat trout. *Trans. Am. Fish. Soc.* 128(4):578–592.

Fendorf, S., H. A. Michael, and A. van Geen. 2010. Spatial and temporal variations of groundwater arsenic in South and Southeast Asia. *Science* 328:1123–1127. https://doi.org/10.1126/science.1172974.

Finkelstein, Y., M. E. Markowitz, and J. R. Rosen. 1998. Low-level lead-induced neurotoxicity in children: an update on central nervous system effects. *Brain Research Reviews* 27(2):168–176.

Fitzgerald, W., and C. H. Lamborg. 2007. Geochemistry of mercury in the environment. Pp. 107–148 *In:* Environmental Geochemistry, Treatise on Geochemistry. B. S. Lollar and K. K. Turekian (Eds.). Oxford: Elsevier.

Fuhrmann, B.C., M. W. Beutel, P. A. O'Day, C. Tran, A. Funk, S. Brower, J. Pasek, and M. Seelos. 2021. Effects of mercury, organic carbon, and microbial inhibition on methylmercury cycling at the profundal sediment-water interface of a sulfate-rich hypereutrophic reservoir. *Environmental Pollution* 268:115853. https://doi.org/10.1016/j.envpol.2020.115853.

Gebhardts, S., F. Shields, and S. O'Neal. 1971. Mercury Levels in Idaho Fishes and Aquatic Environments, 1970–71. State of Idaho Department of Fish and Wildlife and Department of Health. https://collaboration.idfg.idaho.gov/FisheriesTechnicalReports/Res-Gebhards1971%20Mercury%20Levels%20in%20Idaho%20Fishes%20and%20Aquatic%20Environments.pdf.

Gilbert, S. G., and B. Weiss. 2006. A rationale for lowering the blood lead action level from 10 to 2 µg/dL. *NeuroToxicology* 27:693–701.

Groenendijk, D., B. van Opzeeland, L. M. Dioniso Pires, and J. F. Postma. 1999. Fluctuating life-history parameters indicating temporal variability in metal adaptation in riverine chironomids. *Archives of Environmental Contamination and Toxicology* 37:175–181. https://doi.org/10.1007/s002449900503.

Goldberg, S., and C. T. Johnston. 2001. Mechanisms of arsenic adsorption on amorphous oxides evaluated using macroscopic measurements, vibrational spectroscopy, and surface complexation modeling. *Journal of Colloid and Interface Science* 234(1):204–216. https://doi.org/10.1006/jcis.2000.7295.

Hansen, J. A., J. Lipton, P. G. Welsh, and B. McConnel. 2009. Reduced growth of rainbow trout (*Oncorhynchus mykiss*) fed a live invertebrate diet pre-exposed to metal-contaminated sediments. *Environmental Toxicology* 23(8):1902–1911. https://doi.org/10.1897/02-619.

Harrington, J. M., M. J. LaForce, W. C. Rember, S. E. Fendorf, and R. F. Rosenzweig. 1998a. Phase associations and mobilization of iron and trace elements in Coeur d'Alene Lake, Idaho. *Environ. Sci. Technol.* 32(5):650–656. https://doi.org/10.1021/es970492o.

Harrington, J. M., S. E. Fendorf, and R. F. Rosenzweig. 1998b. Biotic generation of arsenic(III) in metal(loid)-contaminated freshwater lake sediments. *Environ. Sci. Technol.* 32(16):2425–2430. https://doi.org/10.1021/es971129k.

Hauer, F. R., and G. A. Lamberti. 2011. Methods in Stream Ecology. Academic Press. 896 pp.

Hayford, B. L., A. M. Caires, S. Chandra, and S. F. Girdner. 2015. Patterns in benthic biodiversity link lake trophic status to structure and potential function of three large, deep lakes. *PLoS ONE* 10(1):e0117024. https://doi.org/10.1371/journal.pone.0117024.

Henny, C. J., L. J. Blus, D. J. Hoffman, L. Sileo, D. J. Audet, and M. R. Snyder. 2000. Field evaluation of lead effects on Canada geese and mallards in the Coeur d'Alene River Basin, Idaho. *Arch. Environ. Contam. Toxicol.* 39(1):97–112. https://doi.org/10.1007/s002440010085.

Hoang, T. C., J. M. Brausch, M. F. Cichra, D. J. Phlips, C. Van Genderen, and G. M. Rand. 2021. Effects of zinc in an outdoor freshwater microcosm system. *Environ. Toxicol. Chem.* 40:2051–2070. https://doi.org/10.1002/etc.5050.

Hoiland, W. K., F. W. Rabe, and R. C. Biggam. 1994. Recovery of macroinvertebrate communities from metal pollution in the south fork and mainstem of the Coeur d'Alene River, Idaho. *Water Environ. Res.* 66(1):84–88. https://doi.org/10.2175/WER.66.1.11.

Horowitz, A. J., K. A. Elrick, and R. B. Cook. 1993. Effect of mining and related activities on the sediment trace element geochemistry of Lake Coeur d'Alene, Idaho, USA. Part I: surface sediments. *Hydrol. Process.* 7:403–423. https://doi.org/10.1002/hyp.3360070406.

Horowitz, A. J., K. A. Elrick, J. A. Robbins, and R. B. Cook. 1995. Effect of mining related activities on the sediment trace element geochemistry of Lake Coeur d'Alene, Idaho. Part II: subsurface sediments. *Hydrol. Process.* 9:35–54. https://doi.org/10.1002/hyp.3360090105.

Hubner, R., K. B. Astin, and R. J. H. Herbert. 2009. Comparison of sediment quality guidelines (SQGs) for the assessment of metal contamination in marine and estuarine environments. *J. Environ. Monitoring.* 11:713–722.

IDHW (Idaho Department of Health and Welfare). 2020. Coeur d'Alene Basin Fish Tissue Analysis and Consumption Advisory. Letter Health Consultation. https://healthandwelfare.idaho.gov/sites/default/files/2020-06/CDA_Basin_Fish_Advisory_LHC.pdf.

IDEQ and Coeur d'Alene Tribe. 2009. Coeur d'Alene Lake Management Plan. State of Idaho Department of Environmental Quality.

IDEQ. 2020. Idaho's 2018/2020 Integrated Report, Final. October 2020.

Iwasaki, Y., T. S. Schmidt, and W. H. Clements. 2018. Quantifying differences in responses of aquatic insects to trace metal exposure in field studies and short-term stream mesocosm experiments. *Environ. Sci. Technol.* 52:4378–4384. https://doi.org/10.1021/acs.est.7b06628.

IWRB (Idaho Water Resources Board), Rathdrum Prairie Aquifer Comprehensive Aquifer Management Plan. https://idwr.idaho.gov/IWRB/water-planning/CAMPs/rathdrum-prairie/, 44 pp., retrieved: March 1.

Jackson, L. J., and J. Kalff. 1993. Patterns in metal content of submerged aquatic macrophytes: The role of plant growth form. *Freshwater Biology* 29(3):351–359. https://doi.org/10.1111/j.1365-2427.1993.tb00769.x

Janssen, E. M. L., J. K. Thompson, S. N. Luoma, and R. G. Luthy. 2011. PCB-induced changes of a benthic community and expected ecosystem recovery following in situ sorbent amendment. *Environ. Toxicol Chem.* 30(8):1819–1826. https://doi.org/10.1002/etc.574.

Kardinall, W. E. A., L. Tonk, I. Janse, S. Hol, P. Slot, J. Huisman, and P. M. Visser. 2007. Competition for light between toxic and nontoxic strains of the harmful cyanobacterium *Microcystis*. *Appl. Environ. Microbiol.* 73(9):2939–2946. https://doi.org/10.1128/AEM.02892-06.

Kastury, F., E. Smith, E. Lombi, M. W. Donnelley, P. W. Cmielewski, D. W. Parsons, M. Noerpel, K. G. Scheckel, A. M. Kingston, G. R. Myers, D. Paterson, M. D. de Jonge, and A. L. Juhasz. 2019. Dynamics of lead bioavailability and speciation in indoor dust and x-ray spectroscopic investigation of the link between ingestion and inhalation pathways. *Environ. Sci. Technol.* 53:11486–11495. https://doi.org/10.1021/acs.est.9b03249.

Keller, W., N. Yan, J. Gunn, and J. Heneberry. 2007. Recovery of acidified lakes: lessons from Sudbury, Ontario, Canada. *In:* Acid Rain—Deposition to Recovery. P. Brimblecombe, H. Hara, D. Houle, and M. Novak (eds.). Dordrecht: Springer. https://doi.org/10.1007/978-1-4020-5885-1_35.

Kitchin, K. T. 2001. Recent advances in arsenic carcinogenesis: Modes of action, animal model systems, and methylated arsenic metabolites. *Toxicol. Appl. Pharm.* 172:249–61. https://doi.org/10.1006/taap.2001.9157.

Kuwabara, J. S., W. M. Berelson, L. S. Balistrieri, P. F. Woods, B. R. Topping, D. J. Steding, and D. P. Krabbenhoft. 2000. Benthic flux of metals and nutrients into the water column of Lake Coeur d'Alene, Idaho: Report of an August 1999 pilot study. U.S. Geological Survey Water-Resources Investigations Report 2000-4132. https://doi.org/10.3133/wri004132.

Kuwabara, J. S., P. F. Woods, W. M. Berelson, L. S. Balistrieri, J. L. Carter, B. R. Topping, and S. V. Fend. 2003. Importance of sediment-water interactions in Coeur d'Alene Lake, Idaho, USA: management implications. *Environmental Management* 32:348–359. https://doi.org/10.1007/s00267-003-0020-7.

Kuwabara, J. S., B. R. Topping, P. F. Woods, J. L. Carter, and S. W. Hager. 2006. Interactive effects of dissolved zinc and orthophosphate on phytoplankton from Coeur d'Alene Lake, Idaho. USGS Scientific Investigations Report 2006-5091. http://pubs.usgs.gov/sir/2006/5091.

Kuwabara, J. S., Topping, B. R., Woods, P. F., and J. L. Carter. 2007. Free zinc ion and dissolved orthophosphate effects on phytoplankton from Coeur d'Alene Lake, Idaho. *Environ. Sci. Technol.* 41(8):2811–2817. https://doi.org/10.1021/es0629231.

Lanphear, B. P., K. Dietrich, P. Auinger, and C. Cox. 2000. Cognitive deficits associated with blood lead concentrations < 10 µg/dL in U.S. children and adolescents. *Public Health Reports* 115(6):521–529. doi: 10.1093/phr/115.6.521.

Lavoie, R. A., T. D. Jardine, M. M. Chumchal, K. A. Kidd, and L. M. Campbell. 2013. Biomagnification of mercury in aquatic food webs: a worldwide meta-analysis. *Environ. Sci. Technol.* 47:13385–13394. https://doi.org/10.1021/es403103t.

Lee, B.-G., S. B. Griscom, J.-S., Lee, H. J. Choi, C.-H. Koh, S. N. Luoma, and N. S. Fisher. 2000. Influences of dietary uptake and reactive sulfides on metal bioavailability from sediments. *Science* 287(5451):282–284. https://doi.org/10.1126/science.287.5451.282.

Lefcort, H., J. Vancura, and E. L. Lider. 2010. 75 years after mining ends stream insect diversity is still affected by heavy metals. *Ecotoxicology* 19(8):1416–1425. https://doi.org/10.1007/s10646-010-0526-8.

Liao, F. H., F. M. Wilhelm, and M. Solomon. 2016. The effects of ambient water quality and Eurasian watermilfoil on lakefront property values in the Coeur d'Alene area of northern Idaho, USA. *Sustainability* 8(1):44. https://doi.org/10.3390/su8010044.

Lidsky, T. I., and J. S. Schneider. 2003. Lead neurotoxicity in children: basic mechanisms and clinical correlates. *Brain* 126(1):5–19. https://doi.org/10.1093/brain/awg014.

Liebl, B., H. Mückter, P. T. Nguyen, E. Doklea, S. Islambouli, B. Fichtl, and W. Forth. 1995. Differential effects of various trivalent and pentavalent organic and inorganic arsenic species on glucose metabolism in isolated kidney cells. *Appl. Organomet. Chem.* 9:531–540. https://doi.org/10.1002/aoc.590090706.

Luoma, S. N., and P. S. Rainbow. 2008. Metal Contamination in Aquatic Environments: Science and Lateral Management. Cambridge, UK: Cambridge University Press.

Luoma, S. N., D. J. Cain, and P. S. Rainbow. 2010. Calibrating biomonitors to ecological disturbance: a new technique for explaining metal effects in natural waters. *Integr. Environ. Assess. Manag.* 6(2):199–209. doi: 10.1897/IEAM_2009-067.1.

Luoma, S. N., and K. T. Ho. 1993. Appropriate uses of marine and estuarine sediment bioassays. Pp. 193–226 *In:* Handbook of Ecotoxicology. P. Calow (ed.). Oxford: Blackwell Scientific. https://doi.org/10.1002/9781444313512.ch11.

MacDonald, D. D., S. C. Carr, S. G. Calder, E. R. Long, and C. G. Ingersoll. 1996. Development and evaluation of sediment quality guidelines for Florida coastal waters. *Ecotoxicology* 5:253–278. https://doi.org/10.1007/BF00118995.

Maret, T. R., D. J. Cain, D. E. MacCoy, and T. M. Short. 2003. Response of benthic invertebrate assemblages to metal exposure and bioaccumulation associated with hard-rock mining in northwestern streams, USA. *J. N. Am. Benthol. Soc.* 22(4):598–620. https://doi.org/10.2307/1468356.

Maret, T. R., and D. E. MacCoy. 2002. Fish assemblages and environmental variables associated with hard-rock mining in the Coeur d'Alene River Basin, Idaho. *Trans. Am. Fish. Soc.* 131(5):865–884.

Marriero, D. D. N., K. J. C. Cruz, J. B. S. Morais, J. B. Bessera, J. S. Severo, and A. R. S. De Olivera. 2017. Zinc and oxidative stress: Current mechanisms. *Antioxidants* 6(2):24. https://doi.org/10.3390/antiox6020024.

Marshall, J. S., J. I. Parker, D. L. Mellinger, and C. Lei. 1983. Bioaccumulation and effects of cadmium and zinc in a Lake Michigan plankton community. *Can. J. Fish. Aquat. Sci.* 40:1469–1479. https://doi.org/10.1139/f83-169.

Mass, M. J., A. Tennant, B. C. Roop, W. R. Cullen, M. Styblo, D. J. Thomas, and A. D. Kligerman. 2001. Methylated trivalent arsenic species are genotoxic. *Chem. Res. Toxicol.* 14(4):355–361. https://doi.org/10.1021/tx0002511.

McCarthy, J. F., and L. R. Shugart. 1990. Biomarkers of Environmental Contamination. Boca Raton: CRC Press. 79 pp.

Mebane, C. A., T. S. Schmidt, and L. S. Balistrieri. 2016. Larval aquatic insect responses to cadmium and zinc in experimental streams. *Environ Toxicol Chem* 36:749–762. https://doi.org/10.1002/etc.3599.

Mebane, C. A., T. S. Schmidt, J. L. Miller, and L. S. Balistrieri. 2020. Bioaccumulation and toxicity of cadmium, copper, nickel, and zinc and their mixtures to aquatic insect communities. *Environmental Toxicology and Chemistry* 39(4) 812–833. https://doi.org/10.1002/etc.4663.

Milani, D., T. B. Reynoldson, U. Borgmann, and J. Kolasa. 2003. The relative sensitivity of four benthic invertebrates to metals in spiked-sediment exposures and application to contaminated field sediment. *Environmental Toxicology and Chemistry* 22(4):845–854. https://doi.org/10.1002/etc.5620220424.

Moody, S. M., and E. L. Evans. 2011. Ethical issues in using children's blood lead levels as a remedial action objective. *Am. J. Public Health* 101:S156–S160. https://doi.org/10.2105/AJPH.2011.300226.

Moore, J. N., and W. W. Woessner. 2003. Arsenic Contamination in the Water Supply of Milltown, Montana. *In:* Arsenic in Ground Water. A. H. Welch and K. G. Stollenwerk (eds.). Boston: Springer. https://doi.org/10.1007/0-306-47956-7_12.

Morra, M. J., M. M. Carter, W. C. Rember, and J. M. Kaste. 2015. Reconstructing the history of mining and remediation in the Coeur d'Alene, Idaho Mining District using lake sediments. Chemosphere 134:319–327. https://doi.org/10.1016/j.chemosphere.2015.04.055.

Muehe, E. M., and A. Kappler. 2014. Arsenic mobility and toxicity in South and Southeast Asia—a review on biogeochemistry, health and socio-economic effects, remediation and risk predictions. *Environ. Chem.* 11:483–495. https://doi.org/10.1071/EN13230.

Mukhopadhyay, R., B. P. Rosen, L. T. Phung, and S. Silver. 2002. Microbial arsenic: from geocycles to genes and enzymes. *FEMS Microbiology Reviews* 26(3):311–325. https://doi.org/10.1111/j.1574-6976.2002.tb00617.x.

Nalepa, T. F., and P. F. Landrum. 1988. Benthic invertebrates and contaminant levels in the Great Lakes: Effect, fates, and role in cycling. Pp. 77–102 *In:* Toxic Contaminants and Ecosystem Health: A Great Lakes Focus. M. S. Evans (Ed.). New York: John Wiley & Sons.

Naranjo, V. I., M. Hendricks, and K. S. Jones. 2020. Lead toxicity in children: an unremitting public health problem. *Pediatric Neurology* 113:51–55. https://doi.org/10.1016/j.pediatrneurol.2020.08.005.

Needleman, H. L., and D. Bellinger. 1991. The health effects of low-level exposure to lead. *Annual Review of Public Health* 12:111–140.

Needleman, H. L., J. A. Riess, M. J. Tobin, G. E. Biesecker, and J. B. Greenhouse. 1996. Bone lead levels and delinquent behavior. *J. Am. Med. Assoc.* 275:363–69.

Needleman, H. 2004. Lead Poisoning. *Annu. Rev. Med.* 55:209–22. https://doi.org/10.1146/annurev.med.55.091902.103653.

Nesnow, S., B. C. Roop, G. Lambert, M. Kadiiska, R. Mason, W. R. Cullen, and M. J. Mass. 2002. DNA damage induced by methylated trivalent arsenicals is mediated by reactive oxygen species. *Chem. Res. Toxicol.* 15:1627–1634. https://doi.org/10.1021/tx025598y.

Nies, D. H. 1999. Microbial heavy-metal resistance. *Appl. Microbiol. Biotechnol.* 51:730–750. https://doi.org/10.1007/s002530051457.

NRC (National Research Council). 1999. Arsenic in Drinking Water. Washington, DC: National Academy Press.

NRC. 2005. Superfund and Mining Megasites: Lessons from the Coeur d'Alene River Basin. Washington, DC: National Academy Press. https://doi.org/10.17226/11359.

Paerl, H. W., and J. Huismann. 2009. Climate change: a catalyst for global expansion of harmful cyanobacterial blooms. *Environmental Microbiology Reports* 1(1):27–37. https://doi.org/10.1111/j.1758-2229.2008.00004.x.

Paerl, H. W., and J. Huisman. 2008. Blooms like it hot. *Science* 320(5872):57–58. https://doi.org/10.1126/science.1155398.

Paerl, H. W., and T. G. Otten. 2013. Blooms bite the hand that feeds them. *Science* 342(6157):433–434. Paerl, H. W., and T. G. Otten. 2013. Blooms bite the hand that feeds them. Science 342(6157):433–434. https://doi.org/10.1126/science.1245276.

Paerl, H. W., and J. T. Scott. 2010. Throwing fuel on the fire: synergistic effects of excessive nitrogen inputs and global warming on harmful algal blooms. *Environ. Sci. Technol.* 44(20):7756–7758. https://doi.org/10.1021/es102665e.

Pershagen, G., R. S. Braman, and M. Vahter. 1981. Arsenic. Pp. 76–146 *In:* Environmental Health Criteria 18. Geneva: World Health Organization.

Phillips, D. J. H., and P. S. Rainbow. 1994. Biomonitoring of Trace Aquatic Contaminants, 2nd ed. London: Chapman and Hall.

Potula, V., and W. Kaye. 2005. Report from the CDC: Is lead exposure a risk factor for bone loss? *Journal of Women's Health* 14(6):461–464. https://www.liebertpub.com/doi/pdf/10.1089/jwh.2005.14.461.

Reinfelder, J. R. and N. S. Fisher. 1991. The assimilation of elements ingested by marine copepods. *Science* 251:794–796. https://doi.org/10.1126/science.251.4995.794.

Reimers, N., J. Maciolek, and E. P. Pister. 1955. Limnological study of the lakes in Convict Creek Basin Mono County, California. *Fishery Bulletin 103 from Fishery Bulletin of the Fish and Wildlife Service* 56. 72 pp. https://spo.nmfs.noaa.gov/sites/default/files/pdf-content/fish-bull/fb56.14.pdf.

Reynoldson, T. B., D. W. Scholoesser, and B. A. Manny. 1989. Development of a benthic invertebrate objective for mesotrophic Great Lakes waters. *Journal of Great Lakes Res.* 15(4):669–686.

Ruckert, P. Z., R. L. Jones, J. G. Courtney, T. T. LeBlanc, W. Jackson, M. P. Karwowski, P.-Y. Cheng, P. Allwood, R. L. Svendsen, and P. N. Breysse. 2021. Update of the blood lead reference value — United States, 2021. *Morbidity and Mortality Weekly Report* 70(43). doi: 10.15585/mmwr.mm7043a4.

Ruud, D. F. 1996. A comparison of the macroinvertebrate communities of a trace elements enriched lake and an uncontaminated lake in North Idaho: The effects of mine waste contamination in Coeur d'Alene Lake. A Thesis Presented to Eastern Washington University.

Sample, B. E., J. A. Hansen, A. Dailey, and B. Duncan. 2011. Assessment of risks to ground-feeding songbirds from lead in the Coeur d'Alene Basin, Idaho, USA. *Integr. Environ. Assess. Manag.* 7(4):596–611. https://doi.org/10.1002/ieam.261.

Savage, N. L. 1986. A topical review of environmental studies in the Coeur d'Alene River-Lake system: Moscow, Idaho Water Resources Research Institute, 81 pp.

Schmidt, T. S., W. H. Clements, R. E. Zuellig, K. A. Mitchell, S. E. Church, R. B. Wanty, C. A. San Juan, M. Adams, and P. J. Lamothe. 2011. Critical tissue residue approach linking accumulated metals in aquatic insects to population and community-level effects. *Environ. Sci. Technol.* 45(16):7004–7010. https://doi.org/10.1021/es200215s.

Schwartz, J. 1994. Low-level lead exposure and children's IQ: a meta-analysis and search for a threshold. *Environmental Research* 65(1):42–55. https://doi.org/10.1006/enrs.1994.1020.

Scofield, B. D., K. Torso, S. F. Fields, and D W. Chess. 2021. Contaminant metal concentrations in three species of aquatic macrophytes from the Coeur d'Alene Lake basin, USA. *Environ. Monit. Assess.* 1993:683. https://doi.org/10.1007/s10661-021-09488-y.

Selin, N. E. 2009. Global Biogeochemical Cycling of Mercury: A Review. *Annual Review of Environment and Resources* 34:43–63. https://doi.org/10.1146/annurev.environ.051308.084314.

Slaveykova, V. I., and K. E. Wilkinson. 2005. Predicting the bioavailability of metals and metal complexes: critical review of the biotic ligand model. *Environmental Chemistry* 2(1):9–24. https://doi.org/10.1071/EN04076.

Smedley, P.L, and D. G. Kinniburgh. 2002. A review of the source, behaviour and distribution of arsenic in natural waters. *Appl. Geochem.* 17:517–568.

Smith, K. S., Balistrieri, L. S., and A. S. Todd. 2015. Using biotic ligand models to predict metal toxicity in mineralized systems. *Applied Geochemistry* 57: 55–72. https://doi.org/10.1016/j.apgeochem.2014.07.005.

Spears, B. L., J. A. Hansen, and D. J. Audet. 2007. Blood lead concentrations in waterfowl utilizing Lake Coeur d'Alene, Idaho. *Arch. Environ. Contam. Toxicol.* 52(1):121–128. https://doi.org/10.1007/s00244-006-0061-z.

Sprenke, K. F., W. C. Rember, M. L. Hoffmann, V. E. Chamberlain, S. F. Bender, and F. Rabbi. 2000. Toxic metal contamination in the lateral lakes of the Coeur d'Alene River valley, Idaho. *Environmental Geology* 39:575–586. https://doi.org/10.1007/s002540050469.

Stohs, S. J., and D. Bagchi. 1995. Oxidative mechanisms in the toxicity of metal ions. *Free Radical Biology and Medicine* 18(2):321–336.

Sunderland, E. M. 2007. Mercury exposure from domestic and imported estuarine and marine fish in the U.S. seafood market. *Environmental Health Perspectives* 115:235–242. https://doi.org/10.1289/ehp.9377.

Terra Graphics/URS Greiner/CH2M Hill. 2001. Final Human Health Risk Assessment for the Coeur d'Alene River Basin Extending from Harrison to Mullan on the Coeur d'Alene River and Tributaries, Remedial Investigation/Feasibility Study. Prepared for Idaho Department of Health and Welfare, Division of Health, Idaho Department of Environmental Quality, U.S. Environmental Protection Agency Region X, Seattle, WA.

Toevs, G. R., M. J. Morra, M. L. Polizzotto, D. G. Strawn, B. C. Bostick, and S. Fendorf. 2006. Metal(loid) diagenesis in mine-impacted sediments of Lake Coeur d'Alene, Idaho. *Environ. Sci. Technol.* 40(8):2537–2543. https://doi.org/10.1021/es051781c.

Tufano, K. T., C. W. Reyes, C. Saltikov, and S. Fendorf. 2008. Reductive processes controlling arsenic retention: Revealing the relative importance of iron and arsenic reduction. *Environ. Sci. Technol.* 42:8283–8289. https://doi.org/10.1021/es801059s.

Vadeboncoeur, Y., E. Jeppesen, M. J. Vander Zanden, and H.-H. Schierup. 2003. From Greenland to green lakes: Cultural eutrophication and the loss of benthic pathways in lakes. *Limnol Oceanogr.* 48(4):1408–1418. https://doi.org/10.4319/lo.2003.48.4.1408.

Vadeboncoeur, Y., M. V. Moore, S. D. Stewart, S. Chandra, K. S. Atkins, J. S. Baron, K. Bouma-Gregson, S. Brothers, S. N. Francoeur, L. Genzoli, S. N. Higgins, S. Hilt, L. R. Katona, D. Kelly, I. A. Oleksy, T. Ozersky, M. E. Power, D. Roberts, A. P. Smits, O. Timoshkin, F. Tromboni, M. J. Vander Zanden, E. A. Volkova, S. Waters, S. A. Wood, and M. Yamamuro. 2021. Blue waters, green bottoms: benthic filamentous algal blooms are an emerging threat to clear lakes worldwide. *BioScience* 71(10):1011–1027. https://doi.org/10.1093/biosci/biab049.

Valois, A., W. B. Keller, and C. Ramcharan. 2010. Abiotic and biotic processes in lakes recovering from acidification: the relative roles of metal toxicity and fish predation as barriers to zooplankton re-establishment. *Freshwater Biology* 55:2585–2597. https://doi.org/10.1111/j.1365-2427.2010.02488.x.

Van Genderen, E., W. Adams, R. Cardwell, J. Volosin, R. Santore, and P. Rodriguez. 2008. An evaluation of the bioavailability and aquatic toxicity attributed to ambient zinc concentrations in fresh surface waters from several parts of the world. *Integr. Environ. Assess. Manag.* 5:426–434. https://doi.org/10.1897/IEAM_2008-082.1.

Vander Zanden, M. J., and Y. Vadeboncoeur. 2002. Fishes as integrators of benthic and pelagic food webs in lakes. *Ecology* 83(8):2152–2161.

Van Sprang, P. A., F. A. M. ver Donk, F. van Asshe, L. Regoli, and K. A. C. De Schamphelaere. 2009. Environmental risk assessment of zinc in European freshwaters: A critical appraisal. *Science of the Total Environment* 407(20):5373–5391. https://doi.org/10.1016/j.scitotenv.2009.06.029.

Vivien, R., C. Casado-Martinez, M. Lafont, and B. J. D. Ferrari. 2020. Effect thresholds of metals in stream sediments based on in situ oligochaete communities. *Environments* 7(4):31. https://doi.org/10.3390/environments7040031.

Wang, W.-X., N. S. Fisher, and S. N. Luoma. 1996. Kinetic determinations of trace element bioaccumulation in the mussel, *Mytilus edulis*. *Marine Ecology Prog. Ser.* 140:91–113.

Wang, J., C. F. Kerl, P. Hu, M. Martin, T. Mu, L. Brüggenwirth, G. Wu, D. Said-Pullicino, M. Romani, L. Wu, and B. Planer-Friedrich. 2020. Thiolated arsenic species observed in rice paddy pore waters. *Nat. Geosci.* 13:282–287. https://doi.org/10.1038/s41561-020-0533-1.

Wang, W.-X. 2012. Biodynamic understanding of mercury accumulation in marine and freshwater fish. *Advances in Environmental Research* 1:15–35. https://doi.org/10.12989/AER.2012.1.1.015.

Wasserman, G., J. H. Graziano, P. Factor-Litvak, D. Popovac, N. Morina, A. Musabegovic, N. Vrenezi, S. Capuni-Paracka, V. Lekic, E. Preteni-Redjepi, S. Hadzialjevic, V. Slavkovich, J. Kline, P. Shrout, and Z. Stein. 1994. Consequences of lead exposure and iron supplementation on childhood development at age 4 years. *Neurotoxicol. Teratol.* 16(3):233–240.

Wasserman, G. A., B. Staghezza-Jeramillo, P. Shrout, D. Popvac, and J. Graziano. 1998. The effect of lead exposure on behavior problems in preschool children. *Am. J. Public Health* 88(3):481–486. https://doi.org/10.2105/AJPH.88.3.481.

Weis, J. S., and P. Weis. 2004. Metal uptake, transport and release by wetland plants: Implications for phytoremediation and restoration. *Environment International*, 30(5), 685–700. https://doi.org/10.1016/j.envint.2003.11.002.

Wentz, D. A., M. E. Brigham, L. C. Chasar, M. A. Lutz, and D. P. Krabbenhoft. 2014. Mercury in the Nation's streams—Levels, trends, and implications. USGS Circular 1395, p. 90. https://doi.org/10.3133/cir1395.

White, D. S., and M. F. Miller. 2008. Benthic invertebrate activity in lakes: linking present and historical bioturbation patterns. *Aquatic Biology* 2(3):269–277. https://doi.org/10.3354/ab00056.

Winner, J. E. 1972. Macrobenthic communities in the Coeur d'Alene Lake system. Masters of Science degree thesis, University of Idaho. 41 pp.

Wong, P. T. K., and Y. K. Chau. 1990. Zinc toxicity to freshwater algae. *Environmental Toxicology* 5(2):167–177.

Woods, P. F., and M. A. Beckwith. 1997. Nutrient and trace-element enrichment of Coeur d'Alene Lake, Idaho. USGS Water Supply Paper 2485. https://doi.org/10.3133/wsp2485.

Woodward, D. F., W. G. Brumbaugh, A. J. Deloney, E. E. Little, and C. E. Smith. 1994. Effects on Rainbow Trout Fry of a Metals-Contaminated Diet of Benthic Invertebrates from the Clark Fork River, Montana. *Transactions Am Fish Soc.* 123(1):51–62.

Woodwell, G. M. 1970. Effects of pollution on the structure and physiology of ecosystems. *Science* 168:429–433. https://doi.org/10.1126/science.168.3930.429.

Xia, T., M. Kovocick, M. Long, L. Madler, B. Gilbert, H. Shi, J. I. Yeh, J. L. Zink, and A. E. Nel. 2008. Comparison of the mechanism of toxicity of zinc oxide and cerium oxide nanoparticles based on dissolution and oxidative stress properties. *ACS Nano* 2(10):2121–2134. https://doi.org/10.1021/nn800511k.

Yan, N., and R. Struss. 1980. Crustacean zooplankton communities of acidic, metal-contaminated Lakes near Sudbury, Ontario. *Canadian Journal of Fisheries and Aquatic Sciences* 37(12):2282–2293. https://doi.org/10.1139/f80-275.

Zhang, Y., D. J. Jacob, H. M. Horowitz, and E. M. Sunderland. 2016. Observed decrease in atmospheric mercury explained by global decline in anthropogenic emissions. *PNAS* 113(3):526–531. https://doi.org/10.1073/pnas.1516312113.

Zhao, F.-J., E. Harris, J. Yan, J. Ma, L. Wu, W. Liu, S. P. McGrath, J. Zhou, and Y.-G. Zhu. 2013. Arsenic methylation in soils and its relationship with microbial *arsM* abundance and diversity, and as speciation in rice. *Environ. Sci. Technol.* 47(13):7147–7154. https://doi.org/10.1021/es304977m.

Zilliges, Y., J.-C. Kehr, S. Meissner, K. Ishida, S. Mikkat, M. Hagemann, A. Kaplan, T. Börner, and E. Dittmann. 2011. The cyanobacterial hepatotoxin microcystin binds to proteins and increases the fitness of *Microcystis* under oxidative stress conditions. *PLOS ONE* 6(3):e17615. https://doi.org/10.1371/journal.pone.0017615.

10

Future Water Quality Considerations

In Chapters 3 through 7, the committee evaluated past and current water quality in the Coeur d'Alene (CDA) basin and CDA Lake and its sediments, with a focus on observed trends in discharge, and nutrient and metal loading and concentrations. Some changes in climate are already apparent in the CDA region, and more are expected as the 21st century progresses, some of which may alter the trends observed by the committee. This final chapter reviews the recent climate history of the CDA region as well as climate projections that have been made through 2100. It then considers how future changes in climate, population, and land use could affect the trends noted in Chapters 3 through 7. Finally, the committee extends recent trends in metals concentration into the future to determine how much time will be needed before metals concentrations in CDA Lake fall below certain ecological thresholds.

CLIMATE CHANGE

Climate (temperature and precipitation) influences the water quality of CDA Lake through its effects on the magnitude, frequency, and timing of inflows to the Lake system. The climate interacts with the topography, geology, and land cover of the CDA watershed in complex and nonlinear ways to give rise to the hydrologic regimes that characterize the inflows to the Lake. In addition, a changing climate affects lake water temperature and consequently stratification, and it could change the frequency and intensity of forest fires in the region, with substantial impacts on lake water quality.

The first three sections below focus on (1) the magnitude of air temperature warming, (2) the amount and timing of precipitation delivered to the watershed, and (3) the extent to which climate warming will shift the delivery of precipitation from snow to rain. For these three climate factors, the historical patterns of the past several decades are examined below, followed by descriptions of projected changes in hydroclimate that will influence inflows to and conditions in the Lake into the future.

The committee originally sought climate projections for the interior Pacific Northwest from the University of Washington's Climate Impacts Group (Mauger, 2021). That information was then supplemented with more recent projections compiled by the committee using data from the University of Idaho's Northwest Knowledge Network data repository,[1] specifically for a rectangular region from −117.5°W to −114.5°W longitude and 46.5°N to 48.5°N latitude—an area that encompasses the entirety of the CDA watershed. The projections of temperature and precipitation are from the Multivariate Adaptive Constructed Analogs (MACA) V2 downscaled climate change

---

[1] https://climate.northwestknowledge.net/MACA/GDP.php

dataset (Abatzoglou and Brown, 2012). The MACA V2 dataset is an ensemble of statistically downscaled outputs from 20 global climate and Earth system models that has been bias-corrected to historical observations of climate to a spatial resolution of approximately 4 km. The downscaled climate data are available for two of the Intergovernmental Panel on Climate Change (IPCC) Representative Concentration Pathway (RCP) scenarios, RCP4.5 and RCP8.5, and both scenarios are shown below. The RCP4.5 and RCP8.5 scenarios correspond to an additional 4.5 and 8.5 W/m^2, respectively, of anthropogenic radiative forcing by the end of the 21st century (Taylor et al., 2012).

The climatology of precipitation and temperature over the region of interest was examined at two future time horizons, 2040–2060 and 2080–2099, and compared with historical precipitation and temperature patterns from the period 1985–2005. Specifically, the seasonal cycle of monthly average daytime maximum and minimum temperatures and monthly average precipitation volume, both averaged over the entire spatial region, were computed. Variability across all 20 climate model predictions was examined by computing the standard deviation of monthly average daily maximum temperature, daily average precipitation, and precipitation volume across the climate models.

Air Temperature

Unlike many other regions of the western United States, the CDA basin has experienced relatively mild warming in the recent past. As shown in Figure 1-10, air temperature in the CDA basin has been increasing over the past 40 years by about 0.4°F (0.22°C) per decade (UC Merced Climate Toolbox[2]). Zhang et al. (2018) similarly noted a slow warming trend in the region; from 1913 to 2015, the annual winter and spring (November–May) air temperature for the upper Spokane basin (of which CDA Lake is a part) averaged 3.5 ± 1.1°C and showed a marginally significant but weak (r^2 = 0.04) warming trend of 0.07 ± 0.03°C per decade.

Projections

The degree of projected warming in the CDA region varies significantly based on the associated emissions scenario examined. As shown in Figure 10-1A, during the 1985–2005 historical period, average daily maximum temperature is lowest in December, at just less than 0°C, and is greatest in July and August, at approximately 26°C. The RCP4.5 scenario suggests that during the 2040–2060 period, monthly average daily maximum temperature in December will be approximately 1.5°C, while the RCP8.5 scenario suggests that December average daily maximum temperatures in the region will be closer to 2°C. The dataset suggests that averaged over the region, summertime monthly average daily maximum temperatures in July and August will be close to 28°C in the RCP4.5 scenario and nearly 30°C in the RCP8.5 scenario.

As expected, these increases in air temperature are more profound for 2080–2099. Model projections suggest that December monthly average daily maximum temperatures would be slightly more than 2°C in the RCP4.5 scenario and slightly less than 5°C in the RCP8.5 scenario. Meanwhile, in the peak summer months of July and August, monthly average daily maximum temperatures would be approximately 30°C under the RCP4.5 scenario, and closer to 34°C under the RCP8.5 scenario.

The projections were investigated to see whether warming is uniformly distributed throughout the course of the year or whether there are temporal patterns in projected warming. For example, warming that is preferentially distributed in the winter might lead to more rapid loss of snow water storage during the winter, while preferential warming in the summer period might be important for lake thermodynamic processes and heat exchange at the Lake–atmosphere interface. During the 2040–2060 period, warming in monthly average daily maximum temperature ranges from approximately 1.5 to 2.8°C in the RCP4.5 scenario, with May having the smallest warming in monthly average daily maximum temperatures and January having the largest warming. During this same period, the RCP8.5 model projections suggests that warming in monthly average daily maximum temperature will range from 2 to 3°C (Figure 10-1B).

Shown in Figures 10-1C and D, average daily minimum temperatures are also of interest to the future of water quality in the Lake, particularly because winter and spring nighttime low temperatures influence the degree

---

[2] https://climatetoolbox.org/

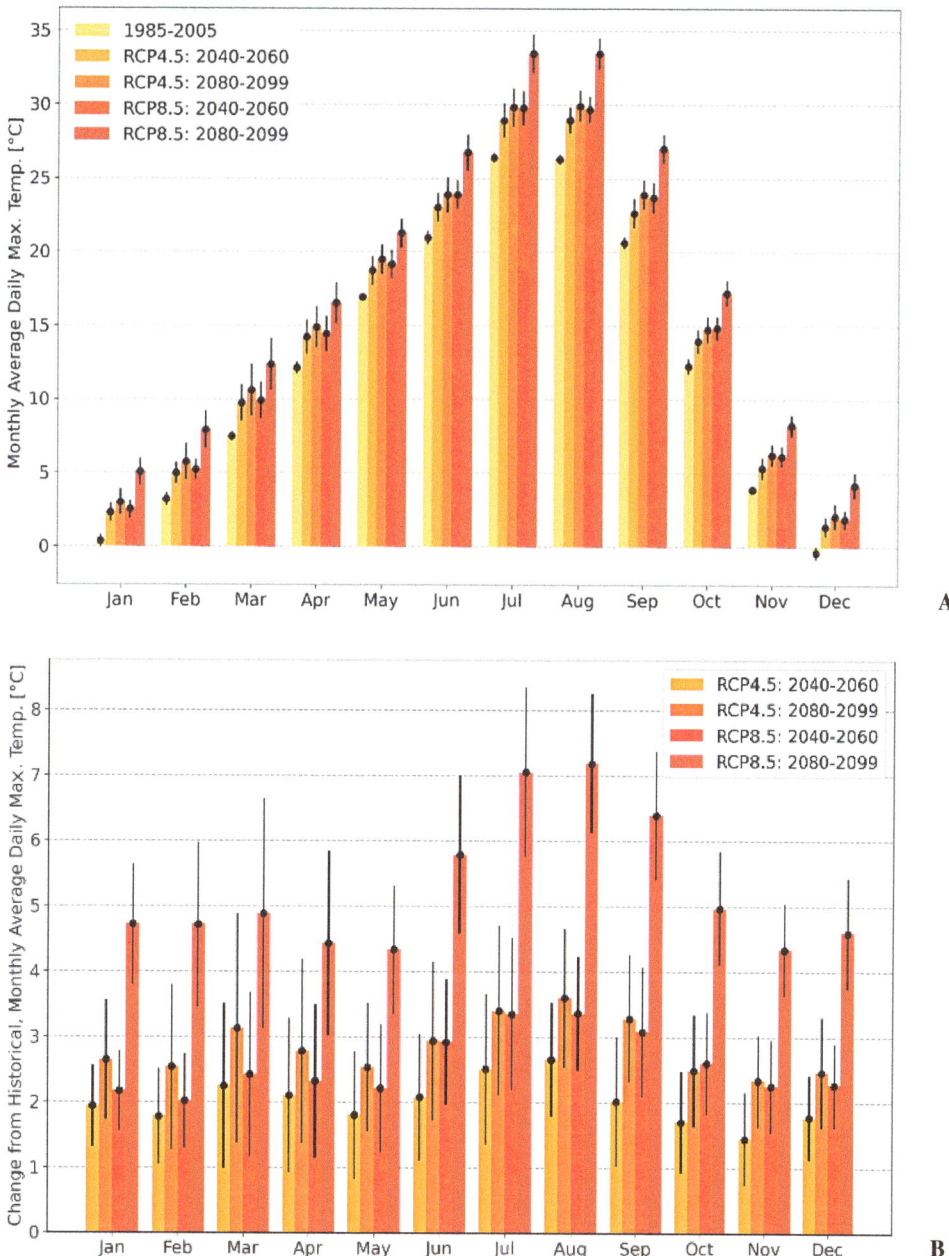

FIGURE 10-1 Projected changes in air temperature for the CDA basin. (A) Monthly average daily maximum temperature for three time periods (1985–2005, 2040–2060, and 2080–2099) and two greenhouse gas emissions scenarios (RCP 4.5 and RCP 8.5). (B) The projected change from the historical monthly average daily maximum temperature. (C) Monthly average daily minimum temperature for the same three time periods. (D) The projected change from the historical monthly average daily minimum temperature. SOURCE: Data courtesy of the MACA V2 dataset and graphed by the committee.

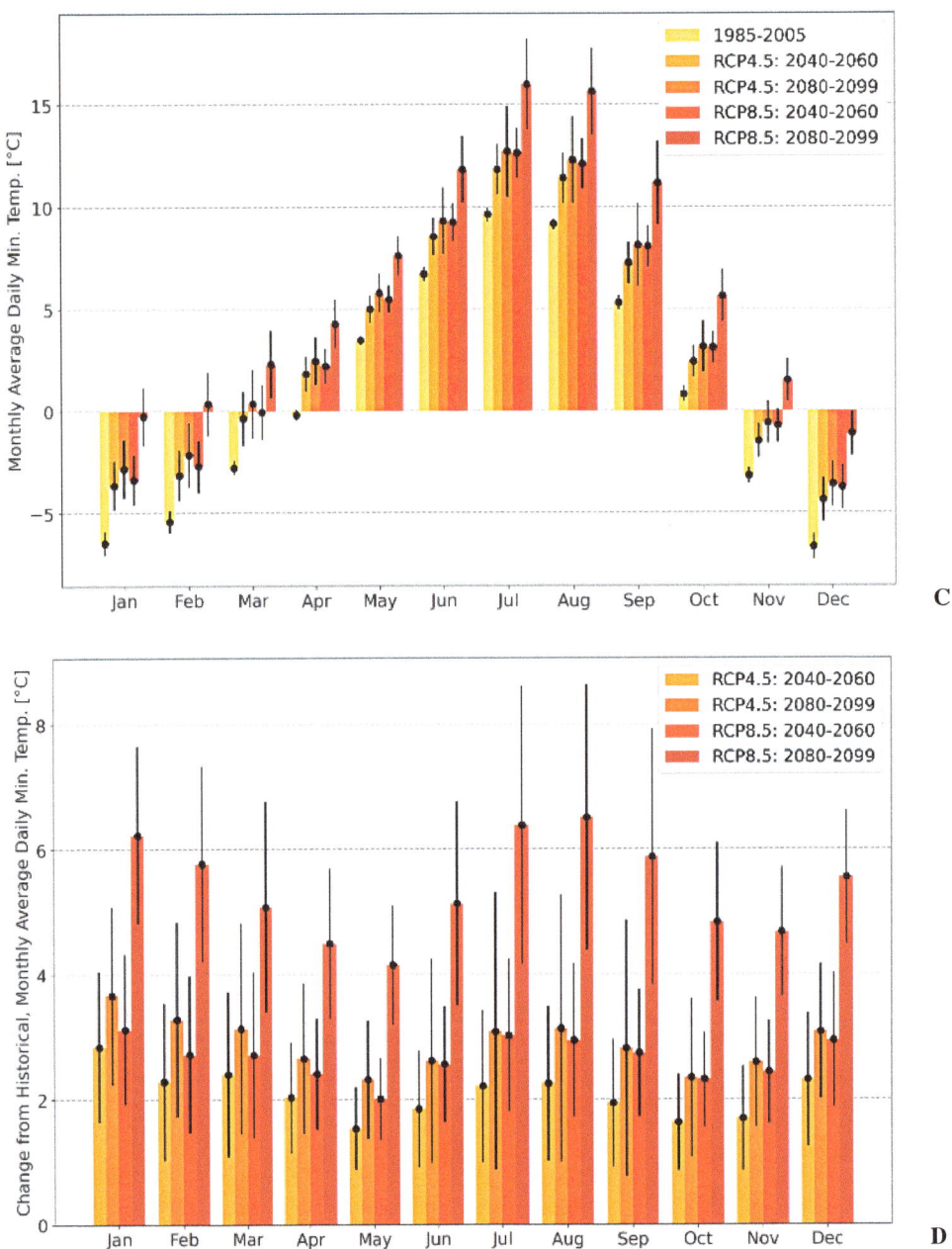

FIGURE 10-1 Continued

and extent of refreezing of snowpacks and can affect the occurrence of disturbances such as mountain pine beetle outbreaks (Preisler et al., 2012). Historical (1985–2005) monthly average daily minimum temperatures are lowest in December and January and, across the region, average approximately −7°C, while this value is highest in July, at slightly less than 10°C. For the 2040–2060 period, model projections suggest that the monthly average daily minimum temperature in December would be approximately −4°C in the RCP4.5 scenario and slightly warmer yet in the RCP8.5 scenario. July monthly average daily minimum temperatures, meanwhile, are projected to be approximately 12 and 13°C in the RCP4.5 and RCP8.5 scenarios, respectively. By the 2080–2099 horizon,

December monthly average daily minimum temperatures are projected to be approximately −3°C (RCP4.5) and −1°C (RCP8.5). The July monthly average daily minimum temperatures are projected to be approximately 12.5°C (RCP4.5) and 16°C (RCP8.5). These projected changes in average daily minimum and maximum temperatures in the region, particularly those associated with the RCP8.5 scenario, at century's end would be consistent with a "low- to no-snow future" that is of broader concern and study across the western United States (Siirila-Woodburn et al., 2021).

Precipitation

Unlike air temperature, annual precipitation within the CDA Lake watershed, although highly variable, has shown no trends in the past 40 years (see Figure 1-7).

Projections

As shown in Figure 10-2, projected changes in precipitation in the region are relatively small compared with projected changes in temperature. There is, moreover, a greater degree of variability across climate model projections. In terms of mean annual precipitation, projections suggest a small increase in total precipitation throughout the year for both the RCP4.5 and RCP8.5 scenarios and for the 2040–2060 and 2080–2099 periods. The largest change in these projections is an increase of approximately 10 percent in mean annual precipitation by the end of the 21st century. Changes in precipitation are not uniformly distributed across the year, with increases in precipitation being largest in the winter and spring months (November–April) and smallest in the summer. It should be underscored that when variability across climate models is plotted on top of the average projected change in mean monthly precipitation, almost all scenarios and time periods span zero change. This indicates disagreement in both the magnitude and direction of precipitation change among the ensemble of models considered. This analysis, however, primarily reports on projected changes in seasonal precipitation climatology.

The class of climate models underlying the MACA V2 dataset is known to exhibit large uncertainties in simulated extreme precipitation (Kharin et al., 2013), which may be of most relevance to future hydrological conditions in the CDA region because such events will drive the mobilization of contaminants stored in riverbeds, riverbanks, and floodplains. Therefore, more focused attention to which regional climate change conditions could give rise to more extreme events using, for instance, climate storyline techniques (e.g., Shepherd et al., 2018) may provide information that is more actionable for managers and planners. Work by Warner et al. (2015) suggests that those events with a daily precipitation in the top 1 percent of historical daily precipitation volumes are likely to become 5–34 percent more intense by the 2080s in the Pacific Northwest region.

The temperature and precipitation projections discussed above for the CDA region are not as dramatic as for the rest of the Pacific Northwest (Mauger, 2021). Indeed, there is not yet evidence that the megadrought currently affecting the West (Williams et al., 2022a) is affecting Idaho, and it is not reflected in the precipitation record for the CDA watershed. There is, however, considerable uncertainty about whether the megadrought will expand northward into the north central region of Idaho.

Rain versus Snow

As discussed in Chapter 1, historically the bulk of the precipitation to the CDA region has arrived as snow, which has exerted a fundamental control on the hydrologic regimes of the tributaries to CDA Lake. As might be expected given the increase in air temperature, in the past 40 years, the CDA region has experienced a trend of less and less water being stored as snow (Figure 1-12).

Projections

An important aspect of changes in temperature regimes within the watershed is the degree to which increases in temperature drive shifts in precipitation phase from snow to rain. Of particular interest in the context of climate

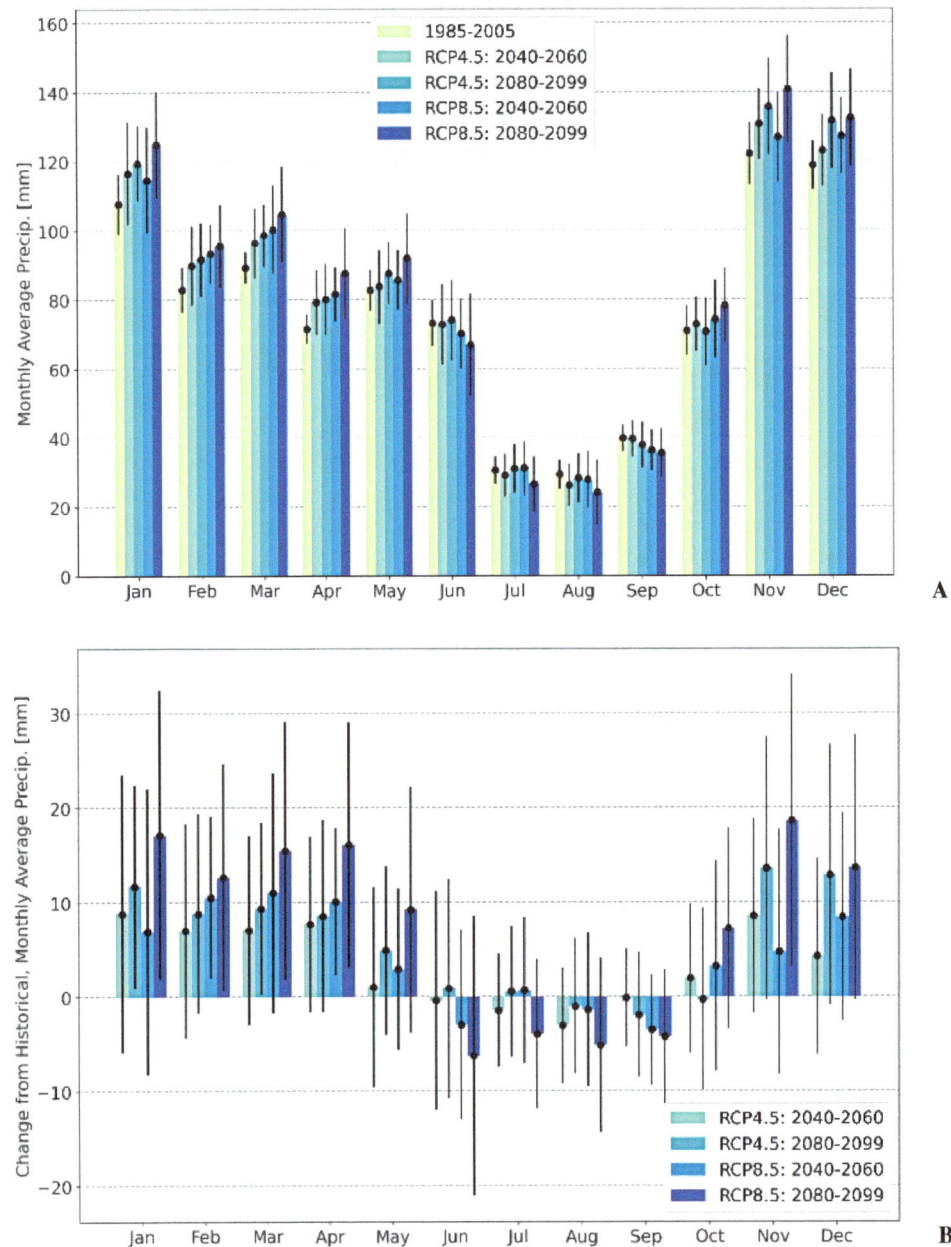

FIGURE 10-2 Projected changes in precipitation in the CDA basin. (A) Monthly average precipitation for three time periods (1985–2005, 2040–2060, and 2080–2099) and two greenhouse gas emissions scenarios (RCP 4.5 and RCP 8.5). (B) The projected change from the historical monthly average precipitation. SOURCE: Data courtesy of the MACA V2 dataset and graphed by the committee.

change are future potential shifts in the elevation of the rain–snow transition. In a set of modeling experiments, downscaled outputs from the third Coupled Model Intercomparison Project (CMIP3) were used as input to the Variable Infiltration Capacity model, version 4.0.7. Using a temperature threshold approach to partition precipitation into rain versus snow, the model simulates the response of the processes of snow accumulation, retention, and release. Examining model projections for a moderate global greenhouse gas emissions scenario, the model suggests that, depending on the climate model used as input, peak snow water equivalent in the Pacific Northwest could decrease by 27–79 percent by the 2040s, with an average of a 51 percent reduction in snow water equivalent across all models. By the 2080s, the modeling experiments suggest that snow water equivalent could decrease by 44–96 percent depending on the specific climate model inputs to the Variable Infiltration Capacity model, with an average of a 73 percent reduction in peak snow water equivalent. These potential reductions in the amount of water stored in the snowpack have potentially profound implications for runoff hydrographs, as discussed in the following section.

Streamflow

The climatic drivers of precipitation and temperature lead to the hydrologic response within the CDA Lake basin of streamflow. Historical streamflow data in the CDA region are available from many U.S. Geological Survey (USGS) stream gages, some dating back 100 years. These data were analyzed by the committee to understand both the trends in mean events and in extreme events. **Figure 3-6 in Chapter 3 showed that streamflow entering CDA Lake over the past three and a half decades has been remarkably free of trends.** This is true for average annual flow rates, for annual low flows, and for annual high flows. In addition, there does not appear to be a shift in the timing of runoff over the past 35 years. All these findings stand in sharp contrast with other parts of the western United States, which have generally seen substantial declines in streamflow and a shift of the center-of-mass of annual runoff to earlier in the year (see, e.g., Dudley et al., 2017; Huang et al., 2022).

Box 10-1 considers whether high-flow events in the CDA basin will get larger or smaller in the future. The upshot is that the future pattern of high flows from the CDA River (and hence the flux of lead and phosphorus, and to a lesser extent cadmium and zinc) entering the Lake is highly uncertain both in magnitude and direction.

Projections

The MACA V2 downscaled climate change dataset does not support predictions of hydrologic trends like streamflow. However, model-based studies performed for the larger Pacific Northwest region by the Climate Impacts Group using downscaled outputs of the CMIP3 suite of climate models have yielded streamflow hydrographs from the Variable Infiltration Capacity model. These simulations were presented to the committee in early 2021 (Mauger, 2021) and are summarized below. Box 10-2 describes how these projections could be updated to become more specific to the CDA region and to use a larger suite of hydrologic models.

The projected hydrographs (Mauger, 2021) suggested the following about the occurrence of extreme flows and the timing of flows in the Pacific Northwest. First, the model projections for low flows as measured by the minimum average annual streamflow that can be expected to occur for a seven-day period once every ten years (7Q10) are that **7Q10 would decrease by about 1 percent by the 2040s and 2 percent by the 2080s.** However, across all model simulations, future changes in 7Q10 varied from −5 percent to +2 percent by the 2040s and −6 percent to +2 percent by the 2080s, indicating uncertainty in the simulations in terms of the potential direction of change. Peak flows, as measured by the 100-year return interval event (Q100), were projected to increase by approximately 39 percent in the 2040s and 61 percent in the 2080s. But again, the modeling scenarios exhibit large ranges of variability in the change of Q100, with all of the model simulations suggesting a range of change to Q100 of −6 percent to +98 percent by the 2040s and +33 percent to 146 percent in the 2080s. While these projections are consistent with the unremarkable historical record for all CDA Lake

BOX 10-1
Will High-Flow Events in the CDA Region Get Larger or Smaller?

The committee looked more closely at the high-discharge events on the CDA River because of the very significant role that these events play in the transport of lead and phosphorus to the Lake, and to a lesser degree for zinc and cadmium. The projections in Figure 10-2 suggest an increase in precipitation for the CDA watershed, on the order of a 10 percent increase in many of the wetter months of the year. However, it is not possible at this time to translate those precipitation changes into changes in discharge and particularly changes in high flows.

There are several complicating factors. One is that, due to expected temperature increases, the size of the snowpack in the watershed can be expected to diminish in the coming years, as seen in Figure 1-12. A common mechanism for large floods in the CDA watershed is rain-on-snow events, where the total runoff from the event is a combination of a significant portion of the snowpack melting and the rainfall itself. Frozen ground also contributes to the size of large floods in the late winter and early spring, because when an intense rain falls on frozen soils, a very large fraction of that rainfall runs off rather than infiltrating into the soil. Because snowpack is expected to be smaller in the future, and the extent and duration of frozen ground are expected to be reduced, a given large rainfall event might result in a smaller amount of flood flow than would have been the case for a similar rainstorm under the climate conditions of today or of previous decades. However, if the climate change was such that a given precipitation event that would have fallen predominantly as snow would now be falling as rain, then the resulting high-flow event would be higher in a future warmer climate. The topic of changing rain-on-snow event patterns is an active area of research, and findings will be important to CDA Lake water quality (see Beniston and Stoffel, 2016; López-Moreno et al., 2021; Brandt et al., 2022).

Hence, there are two opposing drivers of change in high-flow events. The regional estimates of precipitation suggest increased total amounts of rain, and in general climate science is suggesting an increase in the intensity of precipitation across many environments. So these ideas point in the direction of larger high-flow events. But the projected declines in snow water equivalent and frozen ground are drivers that point in the direction of smaller high-flow events.

The committee looked at data from the past 30 years on high flows in the CDA River to see if there are indications of trends that are already taking place; such an empirical examination might help inform our understanding of what may be happening. Note that Chapter 3 looked at a combined record of inflow from both the CDA River and the St. Joe River. For this analysis, the committee focused on the CDA River because of the concern that high-flow events specifically on the CDA River could lead to increased transport of lead and phosphorus into the Lake.

A discharge of 200 m^3/s was used as a threshold for high flow on the CDA River. This is a discharge that has, over recent decades, been exceeded on about 10 percent of the days (i.e., on average it happens about 36 days per year). The choice of this number as a threshold is illustrated by looking at Figure 10-1-1, which shows total lead concentrations as a function of discharge for the period 2010–2020 for the CDA River near Harrison.

From the lowest discharge values up to about 150–200 m^3/s, there is virtually no relationship between concentration and discharge. Above that discharge, the data indicate the initiation of mobilization of bed sediments, and the concentrations rise sharply (and can be modeled statistically as a power function in which concentration varies in proportion to $Q^{3.3}$). The relation of lead *flux* to discharge would be in proportion to $Q^{4.3}$ power, because flux is concentration multiplied by discharge (thereby adding a value of 1 to the exponent). Hence, if discharges on all days above 200 m^3/s were to increase by 10 percent at some time in the future, fluxes on those days would increase by about 50 percent. This is a large change, and it suggests that knowing something about future trends in these higher flows is important to understanding trends in lead transport to CDA Lake. At present, climate and hydrology projections are probably not able to produce reliable predictions of change in this upper part of the discharge distribution, even though there is some confidence that overall precipitation might increase over the century by about 10 percent.

A few exploratory graphics suggest what has happened with high flows in the past few decades, as an indicator of the kind of trajectory the hydrologic system is on. The first of these is a graph of daily discharge for the CDA River near Harrison, which only shows the days with discharge greater than 200 m^3/s (Figure 10-1-2A). Another way to look at the history of high discharge for the CDA River is to look at the time series of annual maximum day discharge (Figure 10-1-2B), as well as annual maximum seven-day periods (Figure 10-1-2C). All three results indicate that recent history, which has already been subject to climate change, is moving in the direction of *lower* magnitude and frequency of high-flow events. The consequences of how the changes in high-flow magnitude and frequency due to climate will play out in this watershed could be significant for both lead and phosphorus transport to the Lake. It will be very important to continue to collect the discharge data, to analyze it regularly for trends in high flow, and to stay abreast of the climate and hydrologic change literature.

FUTURE WATER QUALITY CONSIDERATIONS

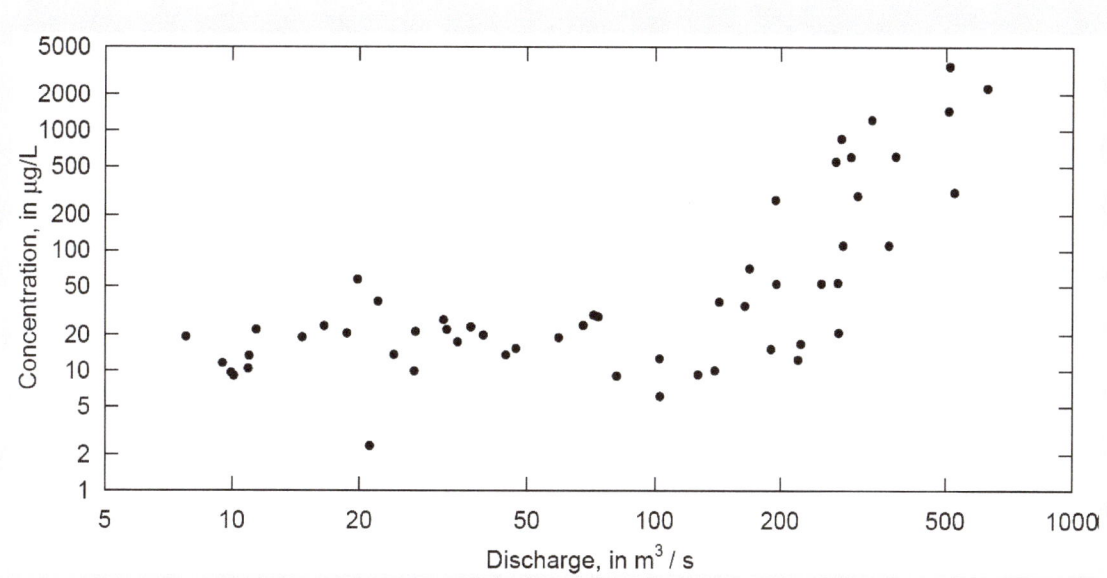

FIGURE 10-1-1 Scatterplot showing the relationship of total lead concentrations as a function of the daily mean river discharge on the sampled day. Data shown are for the CDA River near Harrison and all from water years 2010–2020. SOURCE: Data courtesy of USGS and plotted by the committee.

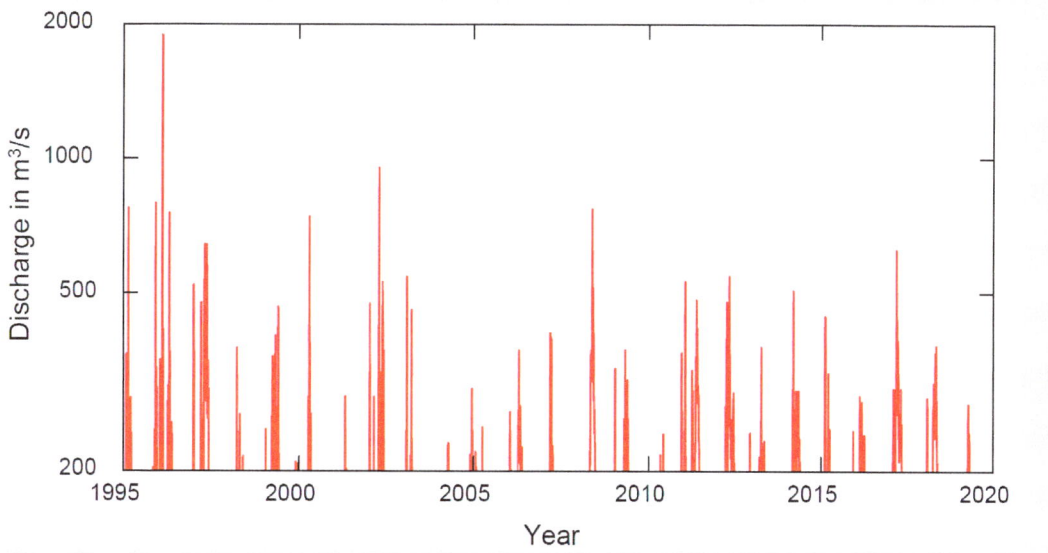

FIGURE 10-1-2 (A) Panel shows, in red, the portion of the daily hydrograph above 200 m³/s for the CDA River near Harrison, water years 1996–2020. (B) Time series of annual one-day maximum discharge for the CDA River near Harrison, by water year 1996–2020; curve is a loess smooth of the time series. (C) Time series of annual seven-day maximum discharge for the CDA River near Harrison, by water year 1996–2020; curve is a loess smooth of the time series. SOURCE: Data courtesy of USGS and plotted by the committee.

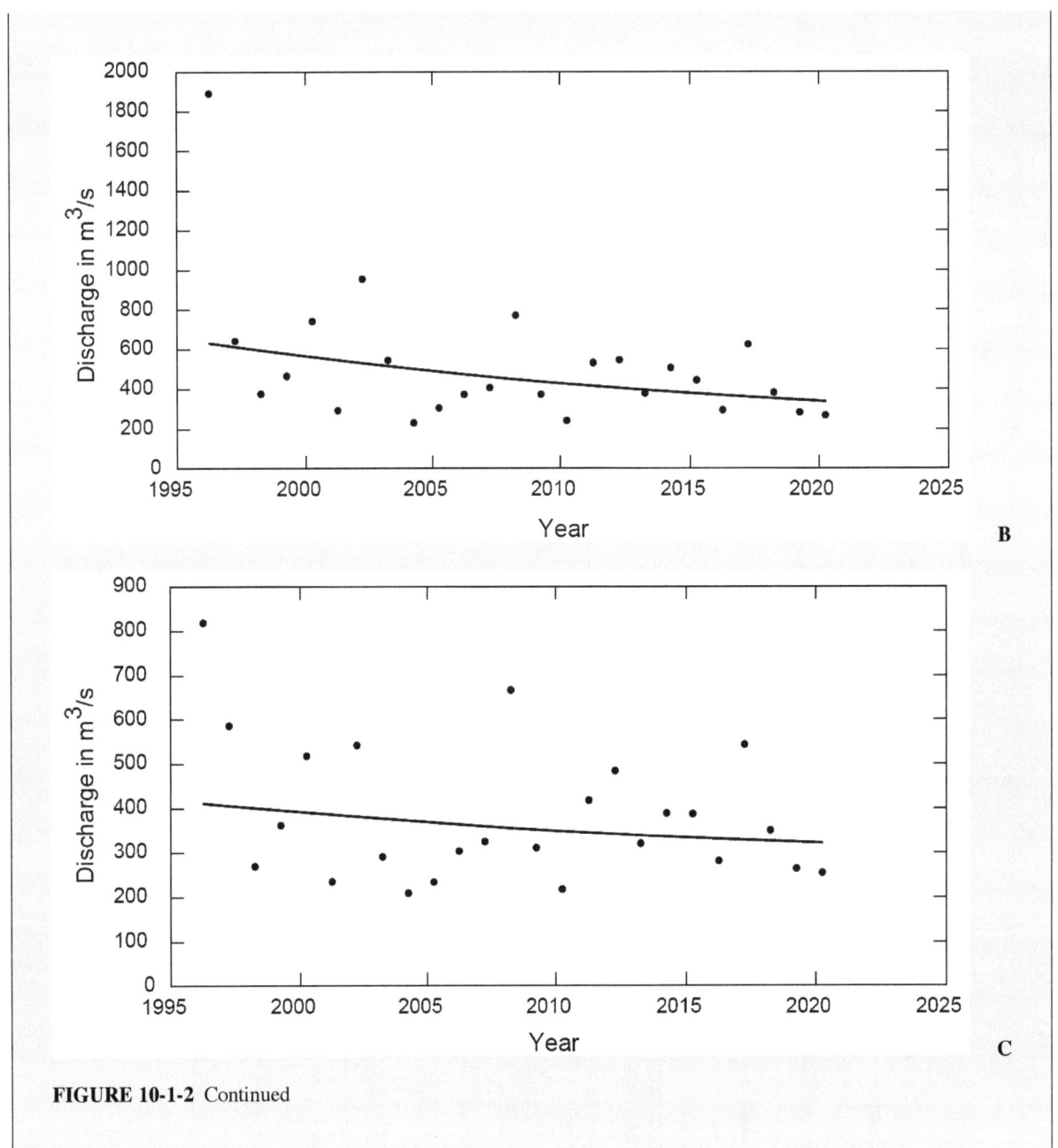

FIGURE 10-1-2 Continued

> **Box 10-2**
> **Updating the Streamflow Projections**
>
> The Computational Hydrology Research Group at the University of Washington, in collaboration with the Oregon Climate Change Research Institute at Oregon State University, developed an ensemble of streamflow projections out to 2100.[a] These simulations, which encompass the entirety of the Pacific Northwest, include three locations within the CDA watershed: the CDA River at the city of Coeur d'Alene, the St. Joe River at Calder, and the North Fork of the CDA River above Shoshone Creek near Prichard. Projected streamflow is available at these locations at a temporal resolution of one day.
>
> Furthermore, this suite of simulations encompasses improvements to the models and techniques used to derive the climate forcings to the hydrologic system, as well as to the models that simulate the hydrologic response to those forcings, relative to the previous streamflow analyses reported above for the Northwest Region. To create this dataset, climate change projections from the Fifth Coupled Model Intercomparison Project (CMIP5; Taylor et al., 2012) were used to force a suite of hydrologic models calibrated to historical data. The climate forcing data corresponds to output from ten different Global Climate Models (GCMs) within the CMIP5 dataset, along with two Representative Concentration Pathways meant to capture an intermediate degree of future warming (RCP4.5) and a more extreme scenario of future warming (RCP8.5). Data from these ten GCMs and two warming scenarios were downscaled to a 1/16° spatial resolution (approximately 6 km) using two alternative downscaling techniques—a bias-corrected, spatial-disaggregation (BCSD) approach (Wood et al., 2004) and the multivariate adaptive constructed analogs (MACA) approach (Abatzoglou and Brown, 2012). Outputs from a smaller subset of the ten GCMs were also downscaled to 18 km using a regional climate model, with further downscaling to 1/16° spatial resolution performed using the BCSD approach. These climate forcings (10 GCMs, 2 RCP scenarios, 2–3 downscaling methods) were then used as input to four alternative hydrologic models at the same 1/16° spatial resolution. Three of the four models consisted of configurations of the Variable Infiltration Capacity (VIC) model (Liang et al., 1994) that differed in the calibrated parameters. A fourth model was a calibrated version of the Precipitation-Runoff Modeling System (PRMS) model (Leavesley et al., 1983). A single hydraulic routing model (Hamman et al., 2017) was used to route runoff simulated by each hydrologic model through the stream network.
>
> To the extent that there are clear hydrologic metrics that are informative about future water quality conditions in CDA Lake, these streamflow projections are a potentially valuable dataset for future analysis because they encompass a range of alternative future climate projections and hydrologic model configurations. As such, they could be analyzed as an ensemble, allowing for quantification of the uncertainty in future streamflow metrics of interest associated with uncertainty in the climate and hydrologic model parameters.
>
> ---
> [a] https://www.hydro.washington.edu/CRCC/

inflows shown in Figure 3-6, the important message is that large increases in flood flows during this century are possible but that there is a high degree of uncertainty. If these changes do take place, they would have negative implications for water quality in the Lake (related to lead and phosphorus inputs). This heightens the importance of taking remedial measures in the lower basin that will limit the mobilization of sediments high in lead and/or phosphorus that have been deposited on the floodplain and channel of the lower CDA River.

A far more consistent result was seen for trends in the center of timing in the streamflow hydrographs, as projected by the Climate Impacts Group (Mauger, 2021). The center of timing coincides with the day of the water year associated with 50 percent of the annual volume of streamflow being exceeded. In the Pacific Northwest, the historic center of timing of the annual hydrograph occurred on approximately April 15 in the 1980s. By the 2040s, the Climate Impact Group projects that the center of timing would occur earlier, with dates ranging from March 4 to March 31, with an average date of occurrence of March 18. By the 2080s, the center of the hydrograph would occur between February 10 and March 13, with an average across all of the climate models considered of March 3. This earlier occurrence of the center of the runoff hydrograph is consistent with both a number of other modeling studies in the region and more broadly the western United States, and it is consistent with warmer winters leading to increases in the fraction of precipitation occurring as rain rather than snow. It is also consistent with the model-based suggestions of large reduction in the peak snow water equivalent in the Pacific Northwest region during the same periods of time. The historical record of the half-volume date shown in Figure 3-6D does not show any particularly large shift in hydrograph timing. This suggests that streamflow timing shifts documented in other parts of the western United States may not apply as readily to the CDA watershed.

Aquatic Thermal Regimes

One of the major impacts of climate change on CDA Lake could be increases in lake temperature that would have a variety of biological, chemical, physical, and social effects both in the Lake and in the basin. Increased lake temperatures suggest that the duration of lake stratification will lengthen, which could promote longer periods of net oxygen consumption in bottom waters affecting both oxygen levels and pH at the Lake bottom and in underlying sediments (see Chapter 5).

The committee investigated trends in lake water temperature specifically for the upper layer of the Lake (epilimnion) during the times of years (summer) when temperatures are highest and the impacts of increasing temperature would likely be most acute. Using the same dataset and methodology as for the dissolved oxygen trend analysis in Chapter 5, the committee analyzed the profile data collected at station C4 to characterize the mean monthly temperatures. The monthly average surface water (1–7 m) and bottom water (31–40 m) temperatures show that CDA Lake at this station is usually isothermal from December to March (Figure 10-3). After March, the surface temperatures rapidly warm from an average of 3.4 ± 1.1 °C to an average of 21.6 ± 1.0 °C in August, with rapid surface cooling in the fall. The bottom temperatures gradually increase from an average of 3.3 ± 0.8 °C in March to an average of 7.6 ± 0.8 °C in November, then cool again when the Lake becomes isothermal in December (Figure 10-3). The winter temperatures averaged 3.4 ± 0.8 °C throughout the water column (see Figure 10-4). During August, the epilimnion was nearly uniform in the upper 7 m, with the metalimnion extending from about 10 to 18 m depth, and then fairly uniform temperatures ranging from 8.6 to 6.3°C below 20 m depth. The complete dataset is shown in Table 10-1.

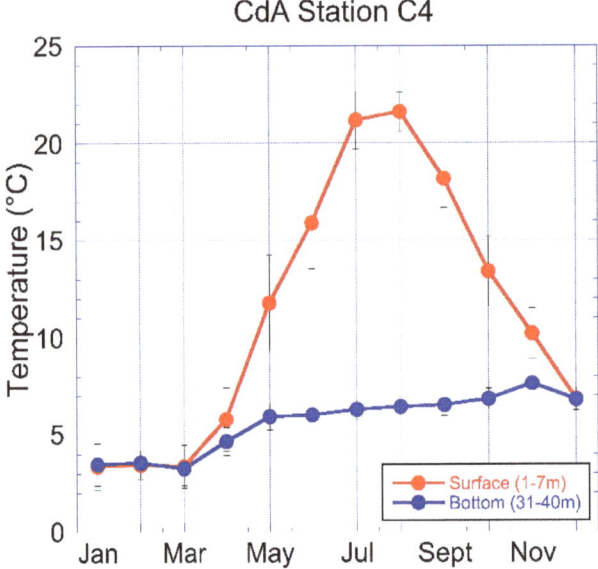

FIGURE 10-3 Monthly average ($\pm$ 1 SD) water temperatures at the C4 sampling station for the epilimnion (1–7 m) and hypolimnion (31–40 m). Data from 1991 to 2020 (see Table 10-1). SOURCE: Data courtesy of IDEQ and graphed by the committee.

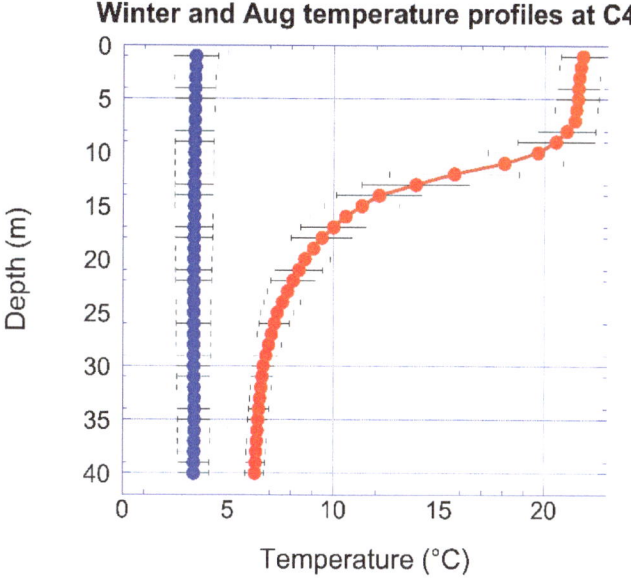

FIGURE 10-4 The average (± SD) thermal profiles at station C4 during the winter isothermal period (January to March—blue) and the peak of thermal stratification in August (red). Data from 1991 to 2020 (see Table 10-1). SOURCE: Data courtesy of IDEQ and graphed by the committee.

Trends in water temperature near the surface (1–7 m in depth) were evaluated from the set of monthly mean temperatures in Table 10-1. The Seasonal Kendall test was used to evaluate the trends in individual months and the overall trends. Because the months of January and February had so few values, the analysis was confined to March–December. Figure 10-5 shows the magnitude of the trend slope over the 30-year period 1991–2020. Note that for all ten months (except May), the estimated slopes were positive (i.e., warming). The significance of the individual monthly trends is indicated with an asterisk. The only trend that was significant was August (p = 0.003); none of the other months had p < 0.1. The overall Seasonal Kendall test for the ten months considered had an upward trend across all months of 0.035°C/yr (or an increase of 1.1 degrees over the full record). The significance level for the test was p = 0.006. Thus, it is clear that surface water temperatures have been increasing at station C4. It is reasonable to assume that this is related to the general trend toward rising air temperatures, and this can be expected to increase in the future based on the expected increase in global greenhouse gas forcing of the climate.

An identical analysis was conducted on the 1991–2020 monthly averages for temperature at 31–40 m at C4. The result was a very slight decrease in temperature over time, which was not significantly different from zero. The individual months had a mixture of downward and upward trends, but none of the individual months had trends that were significantly different from zero. From this, one can conclude that at least up to the present time, the water temperatures in the hypolimnion of C5 have not changed in any systematic pattern. These temperatures at depth are important controls on processes of benthic exchange (of metals or nutrients) between the lake bed and the water column. It may be the case that these temperatures will rise in the future, but this change has not happened yet.

There are multiple potential impacts of overall trends of lake warming. First, increased temperature leads to increased rates of biological metabolism, both primary production and respiration, making it difficult to predict impacts on overall ecosystem metabolism (e.g., heterotrophy versus autotrophy). Warmer temperatures in surface waters will likely increase the strength and duration of thermal stratification, further raising risks of hypolimnetic oxygen depletion. These changes in temperature will also alter the behavior of river plumes as they enter the Lake.

TABLE 10-1 Mean Monthly Surface Water Temperatures (1–7 m) for C4 for the Months and Years When Temperature Profiles Were Collected

C4	Jan	Feb	Mar	Apr	May	Jun	Jul	Aug	Sept	Oct	Nov	Dec
1991	2.1		2.4	4.6	9.2	14.2	19.7	21.9	18.9	16.0	9.3	6.2
1992	4.5		5.6	8.6	13.7	20.5	21.3	20.3	16.1	14.2	10.6	6.4
1993												
1994												
1995								19.0	19.2	11.1		
1996						14.1	22.2	20.9	17.2	12.0		
1997							18.9	20.9	17.9	12.6		
1998							21.8	23.7	21.0	16.4		
1999							20.2	21.0	17.7	12.4		
2000							21.7	21.2	15.1	12.2		
2001							21.0	21.4	17.9	11.4		
2002								20.3	17.3	10.4		
2003										14.1		6.7
2004		3.2		9.4	12.6	16.0	22.4	22.4		13.9		7.7
2005	3.6			4.5	13.3	14.8	21.4		19.9	15.2		5.8
2006		3.4	2.5		8.4	16.0	22.0	21.7				
2007						15.2	23.6	20.5	18.4	12.8		6.9
2008			2.7	4.7	9.9	13.1	20.1	21.5		11.8		
2009				4.8	14.0	17.2	22.1			13.4		
2010			4.5	5.3	7.6	13.1	20.4	21.8	16.5			7.1
2011		2.7		3.4	7.1	14.6	17.5	21.4		17.5	7.7	
2012			3.1	4.1	8.5	12.9	19.3	22.5	18.0	14.0	10.9	
2013				5.7	14.2	16.9	22.8	22.9		15.8		7.4
2014			2.3	6.6	13.7	16.6	22.4	21.7	17.8			7.2
2015		4.7	5.2		13.5	20.8	23.3	21.7	17.0	14.0	11.5	
2016				6.5	13.0	16.2	20.5	22.1	18.1	13.5	11.0	
2017			2.7	6.2	13.3	16.2	22.4	22.6	20.8	12.2		7.3
2018		3.5	3.3	5.6	13.7	15.0	22.1	22.8	19.0	12.6	10.4	
2019			2.9	7.4	12.2	19.6	21.2	22.3	19.7	12.4		6.5
2020			3.5	5.1	11.8	13.1	19.0	21.7				
Monthly mean	3.4	3.5	3.4	5.8	11.8	15.9	21.2	21.6	18.2	13.4	10.2	6.8
monthly SD	1.2	0.7	1.1	1.6	2.5	2.4	1.5	1.0	1.5	1.8	1.3	0.6

NOTE: Red shading denotes higher temperatures, while blue shading denotes lower temperatures.
SOURCE: Data courtesy of IDEQ.

Finally, warmer temperatures potentially favor growth of undesirable cyanobacteria, as these taxa have higher thermal optima than more desirable taxa, such as diatoms. In terms of the physical impacts of lake warming, lake surface energetics are relatively well understood, and quantitative estimates of lake warming and stratification could be made for a range of future climates. Used in conjunction with the three-dimensional model of CDA Lake (Chapter 4) one could differentiate the impacts of lake warming on the pelagic and littoral parts of the lake as well as answer more difficult questions about future ecological and biogeochemical conditions.

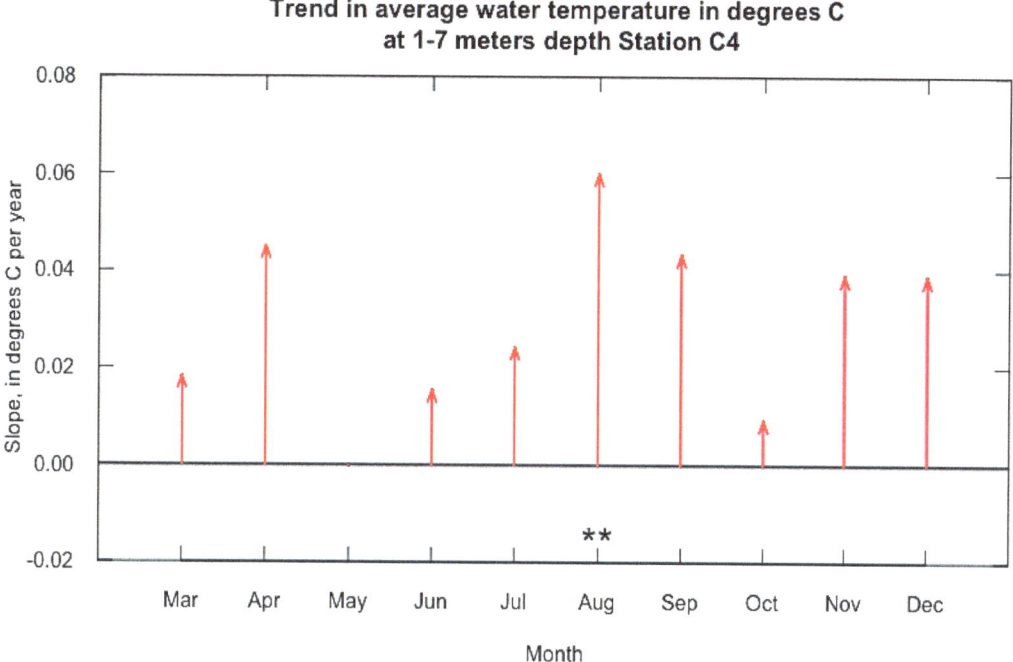

FIGURE 10-5 Trends in average surface water temperature by month for March–December, using data from 1991 to 2020. Red arrows indicate the size and direction of the trend. The asterisks indicate the statistical significance for each month (** is for $p < 0.01$; no other months had $p < 0.1$). SOURCE: Data courtesy of IDEQ and analyzed and graphed by the committee.

Potential for Increased Forest Fires and Impacts on Lake Water Quality

As with many of the effects of climate change already discussed in this chapter, there is the potential for forest fires to become more frequent and problematic in the near future, mainly due to drier conditions in the West. Abatzoglou and Williams (2016) argue that the biggest fires occur in the driest conditions, and climate change is causing drier conditions. They showed that human-caused climate change contributed to an additional 4.2 million ha of forest fire area across the western United States during 1984–2015, nearly doubling the forest fire area expected in its absence. The committee asked Erin Brooks from the University of Idaho to opine on the potential for an increase in forest fires in the CDA region that might affect water quality in CDA Lake, summarized below (Brooks, 2021).

Across the United States, successful efforts to put out fires quickly after the 1930s led to a decline in the size of fires until the early 1980s, after which the average numbers of acres burned has steadily increased (shown in Figure 10-6).

Holden et al. (2018) found that declines in summer precipitation and wetting rain days (days with precipitation ≥ 2.54 mm) are likely the primary driver of increases in wildfire area burned since 1984. Figure 10-7 shows that wetting rain days have declined, and hectares burned have increased, from 1984 to 2015, such that there is a significant correlation between these two factors over the period of investigation ($r = -0.83$). Their findings are consistent with future climate projections, which predict further decreases in summer precipitation (e.g., Figure 10-2) and longer dry periods between rain events across much of the West.

The consequences of wildfires relevant for nearby water quality are many and inherently complex, as shown schematically in Figure 10-8. First, because fires consume the canopy, duff and litter layer, and soil organic matter, the infiltration rate of burned soils decreases dramatically after a fire. This decrease in infiltration means that a greater percentage of rainfall hitting the land runs off, which increases flooding, debris flows, and sediment transport (including transport of sediment-bound lead and phosphorus). This is enhanced by the fire-induced absence

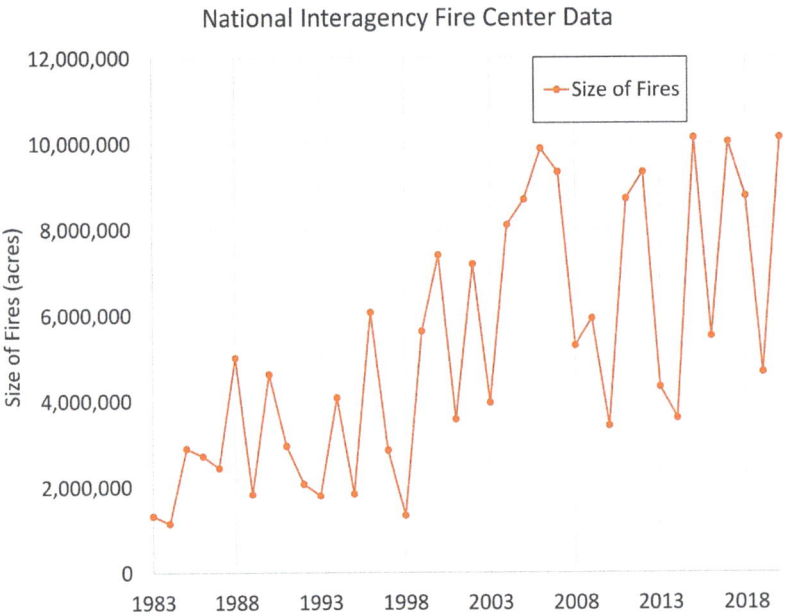

FIGURE 10-6 Acres burned in the United States from annual wildland fire statistics for federal and state agencies, 1983–2020. Note the committee does not condone graphs like this with lines connecting the dots. SOURCE: National Interagency Coordination Center. https://www.nifc.gov/fireInfo/fireInfo_stats_totalFires.html.

of surface roughness that might have served to slow flood flows. Furthermore, the absence of the canopy and litter layer means more soil is available to be eroded during precipitation events. Depending on the composition of the soil, there can be enhanced transport of contaminants to nearby waterways—a particular concern if there are high levels of phosphorus or lead adsorbed to soil particles. Given the known high metals content of soils and sediment in the CDA region, particularly in the lower basin (see Chapter 3), any increases in erodibility of the soil caused by fire could have substantial negative impacts on the loading of metals to CDA Lake.

There are other effects of fires on soil chemistry that could be a concern for water quality in nearby waterbodies. These include a potential increase in loading of dissolved phosphorus because of the burning of organic matter, increases in loading of total zinc, and increases in pH (Rust et al., 2018, 2019). According to Brooks (2021), soils in the CDA basin are andisols that present a risk of phosphorus transfer into runoff after fires. Andisols are moderately acidic soils having a high percentage of volcanic glass and allophanic properties and are characterized by high Al/Fe and high phosphorus retention. If soil pH increases following fires into a pH range of 6–7, this can lead to greater plant available phosphorus and risk of dissolved phosphorus in runoff. Brooks (2021) has found that total phosphorus in the CDA watershed is high in forested areas (although bioavailable phosphorus is higher in agricultural areas) and that there are soil phosphorus hot spots within the CDA basin (including enriched deposition environments/toe-slopes). Brooks (2021) showed results from Lake Tahoe following a forest fire, where he demonstrated a close correlation between soils with higher phosphorus and higher phosphorous export, though export was also related to precipitation, so there are other complicating factors. Finally, there is also the potential for increases in nitrogen in runoff following severe burns for many years (Rhoades et al., 2019).

There are fewer potential impacts of forest fire on drinking water in the CDA region, given that CDA River and Lake are not primary sources of drinking water for the region's population. The impacts of forest fire on aquatic habitats are unknown.

Brooks (2021) suggested that some of the water quality impacts of forest fires could be mitigated by practicing prescribed low-intensity burns to reduce fuel loads and reduce the risk of water-extractable phosphorus transport,

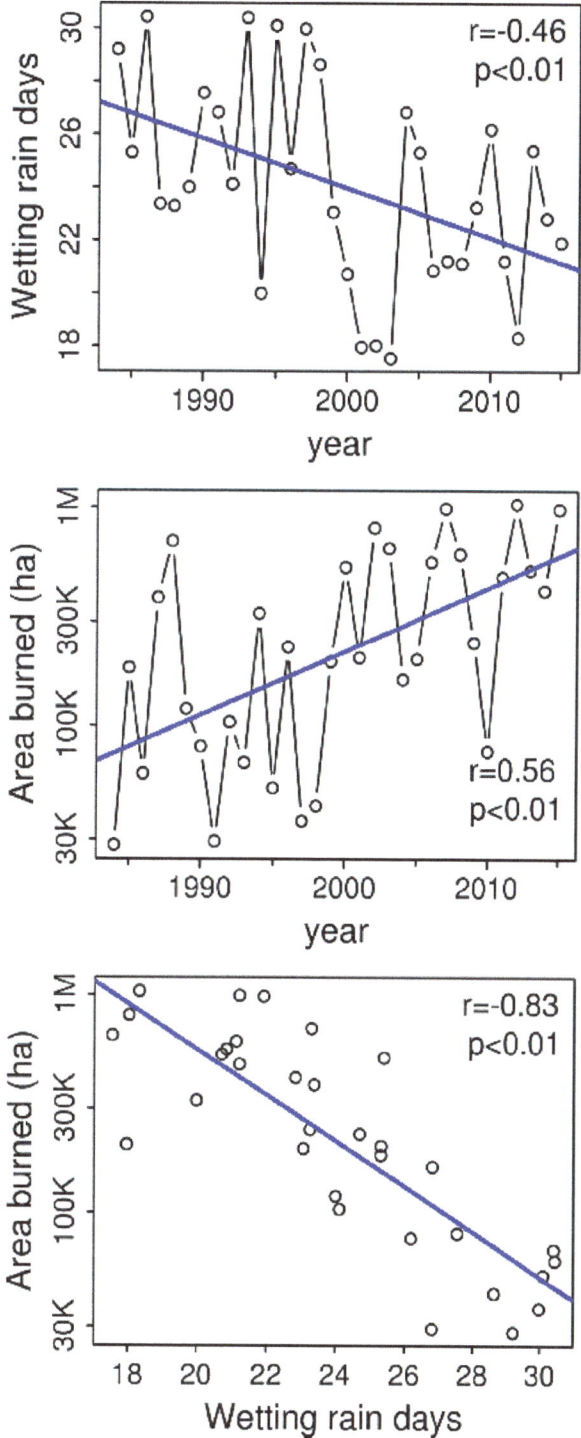

FIGURE 10-7 Linear trends in wetting rain days and area burned for forested areas in the western United States from 1984 to 2015. SOURCE: Holden et al. (2018).

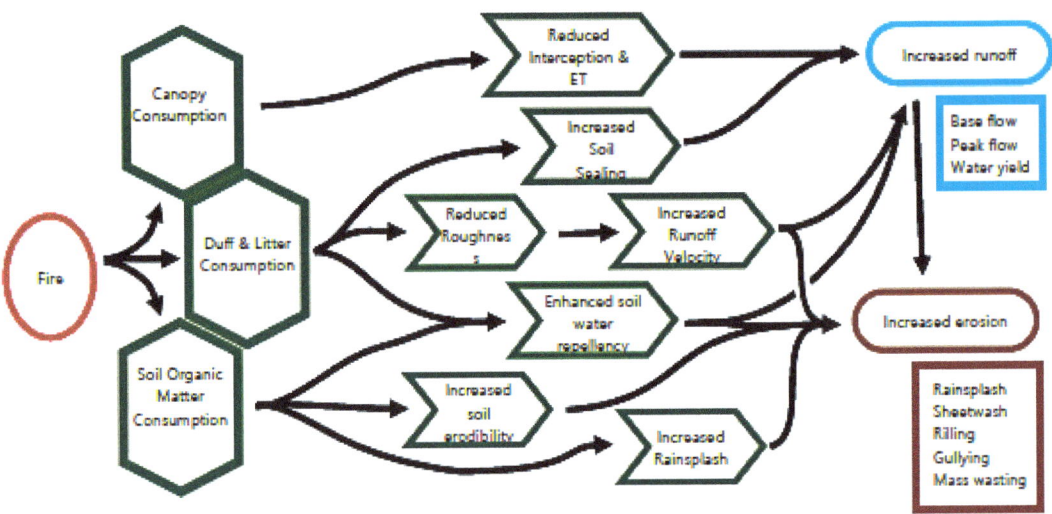

FIGURE 10-8 Hydrologic effects of forest fires subsequent to a burn. SOURCE: Brooks (2021).

if this can be done without increasing sediment transport. Because total phosphorus volatilizes at a higher temperature than carbon, such low-intensity burning should reduce organic matter while allowing phosphorus to stay in the soil. Other mitigation options in areas most likely to burn include making plans to respond to a fire quickly and protecting the ground surface with wood mulch.

The CDA region is just west of an area of Idaho and Montana that suffered more than 3 million acres burned by forest fire in 1910. Dubbed the Great 1910 Fire, the conflagration decimated the town of Wallace and other areas. Since that event, however, there have been fewer major fires in northern Idaho, especially when compared to other places in the arid parts of the Pacific Northwest. Figure 10-9 shows a map of the Pacific Northwest and fires by decade for the past 100 years, illustrating this dearth of fire activity in the CDA watershed.

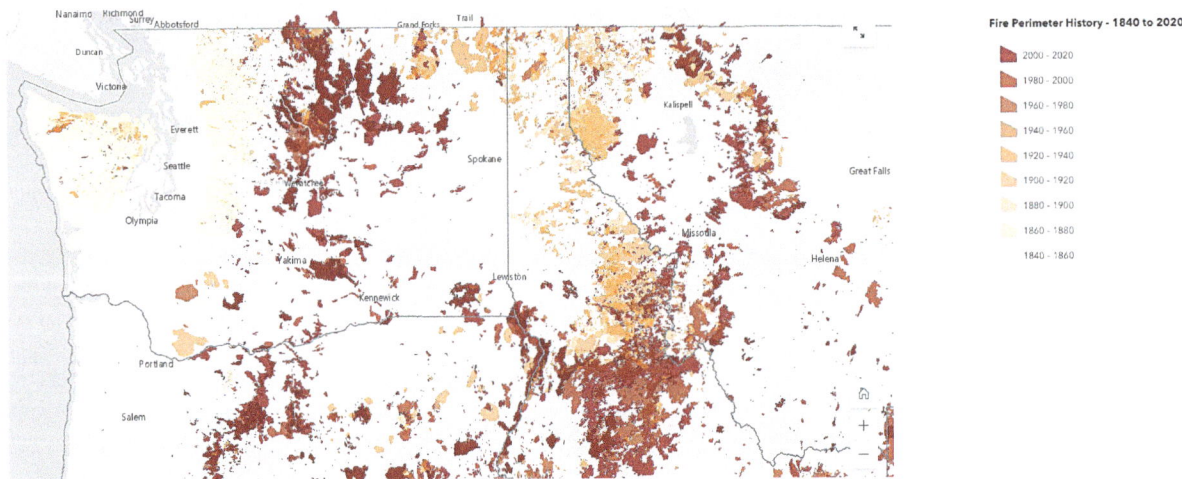

FIGURE 10-9 Fire perimeter history, by 20-year intervals in the Pacific Northwest. SOURCE: https://www.arcgis.com/apps/MapSeries/index.html?appid=6629651002db435d9df188003d790847.

Despite the lack of recent fires, Brooks (2021) argued that future wildfires should be seen as a serious risk that may increase phosphorus loading to CDA Lake, given how wildfires affect soil erosion and runoff during subsequent high-flow events. This is particularly true for fires that cover a large fraction of the lower basin's floodplain, which has unmitigated, contaminated soils. While the region has been lucky to date, climate trends for increasing air temperature in the greater Pacific Northwest region suggest that large wildfires will become more frequent.

Land Cover and Population Changes

Land cover within the CDA basin is dominated by forested land and shrubland with a small mix of agriculture, barren land, small waterbodies, wetlands, and urban development. As mentioned in Chapter 3, compared to years prior to 1992, there has been a substantial reduction in the extent of barren land in the upper basin in the past 30 years, due to remedial activities and the associated stabilization of soils. Analysis of land use trends since 2001 show relatively few changes across the CDA basin, lakeshore catchments, and St. Joe River watersheds. The percentage of forested land has decreased by less than 5 percent and the percentage of shrubland has increased by less than 2 percent (see Table 1-1). Percentage changes in developed, herbaceous, planted/cultivated, water, wetlands, and barren land cover classes have changed less than 1 percent over the 18-year time period of the committee's analysis (Table 1-1). A transition in land use to more developed land is evident within the lakeshore area, and this trend is consistent with the increased population growth in this area. For example, Kootenai County has experienced annual population growth rates between 2.4 and 2.8 percent since 2016. While the land use changes across basins are unlikely to impact nutrient and metal loads to CDA Lake, increased development within lakeshore areas could lead to local increases in solids and nutrient loads that are associated with the loss of impervious land cover and increased wastewater discharges. These potential impacts could be mitigated through low-impact development strategies and nutrient control using advanced wastewater treatment processes.

WILL CLIMATE CHANGE, POPULATION GROWTH, AND LAND USE REVERSE WATER QUALITY TRENDS IN CDA LAKE?

The committee's statement of task asked whether changes in temperature and precipitation could affect the observed trends in nutrient loading and metals concentration in CDA Lake and the rivers. Hence, the committee considered how the changes in climate discussed above, along with changes in land use and population growth, might impact the trends summarized in the concluding sections of Chapters 3 to 7. The five major effects considered relevant to the future of water quality in CDA Lake, shown schematically in Figure 10-10 as gray boxes, are (1) increased frequency and magnitude of large runoff events, (2) a forward shift in the timing of flow to the Lake, (3) warming of Lake water, (4) increased frequency and size of fires, and (5) increased lakeshore populations. (Note that these effects are directly related to, but more specific than, the precipitation and warming trends discussed at the beginning of this chapter.) Each of these five effects has the potential to reverse the observed favorable trends in metals and nutrient loading to CDA Lake (Chapter 3); timing of river flow (Chapter 4); and dissolved oxygen, nutrient, and metals concentrations in CDA Lake (Chapters 5 and 6). Thus, they may increase the risk of metals release from sediments, as described in Chapter 7.

Five Climate Changes Examined

The first effect of climate change examined is an increased frequency of large runoff events (Gray Box 1 in Figure 10-10) that would mobilize channel bed and floodplain nutrient pools, especially phosphorus, that have become increasingly stabilized as a result of watershed remediation, leading to increased nutrient loading. This could subsequently lead to increased nutrient levels in the Lake, higher phytoplankton biomass and production, and stronger hypolimnetic oxygen depletion as nutrient-stimulated production reaches bottom waters for decomposition (Arrows 1–4, 15, 16 in Figure 10-10). **The previous section on climate change highlights the uncertainty about the possibility of large runoff events increasing in magnitude and frequency in the CDA region.** This uncertainty stems partly from the concomitant likely shift from snow to rain as the predominant type of precipitation

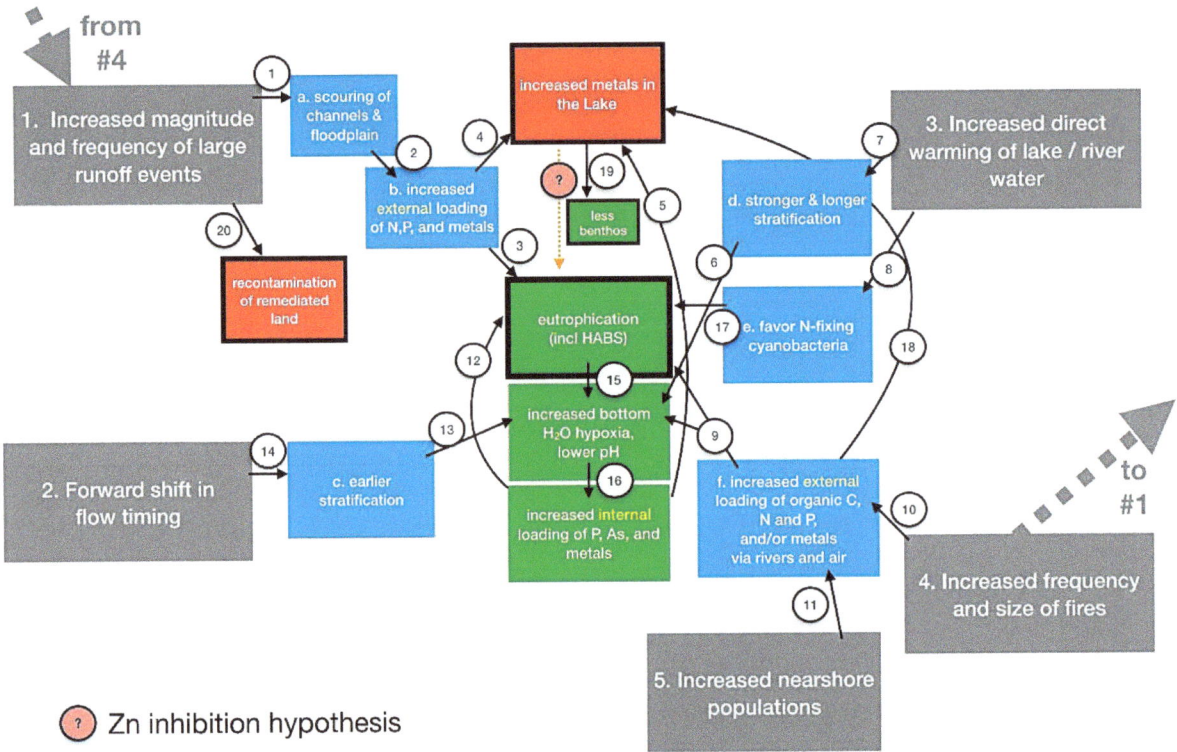

FIGURE 10-10 Potential pathways connecting five primary predicted shifts in external forcing from climate change and regional population expansion (gray boxes) that are ongoing in the CDA Lake region to outcomes of concern regarding water quality and ecological conditions (green boxes) or metal exposures and contamination (red boxes). Blue boxes indicate the mechanism of the projected impact. Note that not all connections or pathways are projected with equal degrees of confidence, as discussed in the text. The orange dashed arrow illustrates the potential connection between zinc levels in the Lake and lake eutrophication that is discussed extensively in Chapter 5.

and the recently observed decline in large events over the past 20 years (see Box 10-1 for a fuller explanation). Nonetheless, to be conservative, the committee examines the slowing or reversal of several water quality trends that could occur *if* hydrologic events become larger or more frequent in the future.

A second effect of climate change is the forward shift in flow timing (Gray Box 2), in which a greater percentage of flow from the watershed to the Lake occurs in winter and early spring than is occurring now. Although this has yet to be documented in the CDA basin (see Figure 3-6D), the projections from the Climate Impacts Group described above point to this occurring soon in the Pacific Northwest. If the center of flow timing moves earlier in the year, the St. Joe River[3] hydrograph may fall below the 1,000 cfs (28 m^3/s) discharge threshold at which the Lake stratifies to earlier in the year (Arrows 14 and 13), lengthening the period of stratification. This would then lengthen the period during which oxygen and pH could be drawn down in Lake bottom waters, with anoxia promoting mobilization of phosphorus and redox-sensitive metals in the Lake sediments and low pH promoting zinc release (Arrows 15, 16). Up to the present time, St. Joe River discharge data do not indicate that the date at which discharge falls below 28 m^3/s has shifted to earlier in the year than it was in the past.

The third climate effect considered is increased Lake temperatures (Gray Box 3). This trend, which has already been documented at C4 and is more pronounced during summer (Figure 10-5), will strengthen and lengthen

---

[3] This threshold could only be derived for the St. Joe River due to lack of continuous time series data on the CDA River (see Chapter 4). However, the threshold effect applies to both rivers, and one would expect future climate change to impact flow in both rivers in similar ways.

thermal stratification, which may further contribute to hypolimnetic oxygen depletion and likely lowering of pH (Arrows 7 and 6). There is no evidence to date of accelerating hypolimnetic oxygen depletion as a result of higher surface water temperatures, but the window of available data is narrow. Higher surface water temperatures may also favor increased microbial metabolism and the development of nitrogen-fixing cyanobacteria, contributing an additional source of nitrogen to the system (Arrows 8 and 17) (Breitbarth et al., 2006; Robarts and Zohary, 1987; Brauer et al., 2013).

The fourth climate effect that could impact CDA Lake water quality is an increased frequency and size of forest fires (Gray Box 4). Fires will mobilize sediments, metals, and nutrients, both in runoff from burned watersheds and in atmospheric transport, that could increase their loading to CDA Lake (Arrow 10; Hauer and Spencer, 1998; Spencer et al., 2003). At least over the medium term, these effects could then stimulate eutrophication and oxygen depletion (Arrow 9; McCullough et al., 2019). However, increased prevalence of smoke from regional fires may lower incident light intensity, potentially ameliorating the effects of nutrients, at least temporarily. Fire also has the potential to affect the observed trend in land use from barren to revegetated land that has accompanied the Superfund remedy. It is noted that fires can also alter hydrologic regimes and increase the risk of extreme flood events (Williams et al., 2022b) (dashed Gray Arrow from Gray Box 4 to Gray Box 1).

Finally, although not directly a result of changes in temperature and precipitation, expanding human populations and the associated increase in urban land use could slow or reverse the observed trends in water quality, particularly nutrients (Gray Box 5). This might occur because of increased inputs of nitrogen and bioavailable phosphorus from septic system leakage, localized fertilizer use, and other activities (Arrows 9 and 11).

Effects of Climate Change on Water Quality Trends

Table 10-2 summarizes how trends noted in Chapters 3 through 7 might be altered under the five climate effects discussed above. The subsequent sections elaborate on the entries in Table 10-2. Note that only those conclusions from Chapters 3 to 7 that reveal a trend (or lack thereof) are considered in Table 10-2 and the text below.

Trends in Sediment, Metals, and Phosphorus Loading to CDA Lake

The downward trends in lead inputs to CDA Lake over the past decade (Figure 3-11) are related to the readjustment of the river channel and floodplain in the lower basin. The historic mining activities and the early phases of remediation resulted in a very large influx of sediment to the upper portion of the lower CDA River, far more than what the river was able to transport downstream. However, it appears that the river channel is readjusting to this change in sediment inputs causing decreases in the transport of sediment (and attached lead) over the past decade.

Climate change over the coming decades could modify the situation based on what may happen to the magnitude and frequency of high-discharge events and based on the potential for more fires. As discussed in Box 10-1, there is a great deal of uncertainty not only in the magnitude of future changes in high flows, but even in the direction of these changes. High discharges, above about 200 m^3/s, are the primary drivers of sediment transport and hence lead transport. If the change in climate increases the frequency of these discharge conditions (currently about 36 days per year on average) and their magnitudes, they will be able to erode and transport more lead to the Lake (the transport increases with rate of discharge raised to the 4.3 power). Similarly, if winter storms that would have been snow events under past colder climate conditions become rain events in the future, there would also be an increase in the magnitude and frequency of high-discharge events. The result of both these discharge scenarios would be greater mobilization of sediment and sediment-associated contaminants like lead and phosphorus, potentially slowing or reversing the downward trends in the loading of these compounds (see rows 3.4, 3.5, and 3.7).

In contrast, if such events resulted in increased erosion and sediment transport across parts of the North and South Fork CDA watersheds not enriched in lead,[4] this could have the effect of increasing the proportion of

---

[4] The scale of contamination from the lead smelter across the landscape can be extensive (Moore and Luoma, 1990) but is unstudied in the CDA basin. Lead is vaporized and emitted as fine particulate from smelter stacks, extending the area that can be affected by fallout; but the availability for mobilization of such contamination can also decline over time as vegetation increases and matures.

TABLE 10-2 How Climate Change Will Impact Water Quality Trends Noted in Chapters 3 Through 7

	CLIMATE CHANGES AND OTHER IMPACTS OF CONCERN TO WATER QUALITY (GRAY BOXES)				
Conclusions from Chapters 3–7	1. Increased magnitude and frequency of high runoff events	2. Forward shift in timing of flow	3. Warmer lake	4. More fires	5. Increased lakeshore populations
3.1 Land cover has changed from barren to revegetated	No effect	No effect	No effect	Could burn revegetated areas	No effect
3.4 Loading of Pb has decreased	Could slow or reverse trend by mobilizing sediment and associated Pb	No effect	No effect	Could increase delivery of Pb-enriched sediments	No effect
3.5 Loading of sediment has decreased	Could slow or reverse trend by mobilizing sediment across the CDA basin	No effect	No effect	Could increase delivery of sediments	No effect
3.6 Loading of Cd has decreased	Could slow or reverse trend by mobilizing sediment and associated metals **but less effect than for Pb/P**	No effect	No effect	Could increase delivery of Cd-enriched sediments	No effect
3.6 Loading of Zn has decreased	Could slow or reverse trend by mobilizing sediment and associated metals **but less effect than for Pb/P**	No effect	No effect	Could increase delivery of Zn-enriched sediments	No effect
3.7 Loading of P has decreased	Could slow or reverse trend by mobilizing sediment and associated P	No effect	No effect	Could increase delivery of sediments and P	Could increase bioavailable P loading near the Lake
4.2 The date when St. Joe R discharge falls below 28 m³/s (1,000 cfs) comes earlier in the year	No effect	Threshold reached sooner	Enhanced	No effect	No effect
5.1 No observed trend in dissolved oxygen (DO)	No effect	May create a trend towards lower DO due to longer stratification period	May create a trend toward lower DO	May create a trend toward lower DO	May create a trend toward lower DO
5.2 [Total P] has declined over past 10 years	Trend could slow or reverse and [total P] could increase	Trend could slow or reverse and [total P] could increase	Trend could slow or reverse and [total P] could increase	Trend could slow or reverse and [total P] could increase	Trend could slow or reverse and [total P] could increase
5.3 Trends in [total N] variable, depending on lake location	[Total N] could increase	[Total N] could increase	[Total N] could increase	[Total N] could increase	[Total N] could increase

6.1 and 6.2 Downward trends in [dissolved Zn]	Trend could slow or reverse and [dissolved Zn] could increase **but less effect than for Pb**	Trend could slow or reverse and [dissolved Zn] could increase	Trend could slow or reverse and [dissolved Zn] could increase	Trend could slow or reverse and [dissolved Zn] could increase	No effect
6.3 Downward trend in [dissolved Cd] at C1 and C4	Trend could slow or reverse and [dissolved Cd] could increase **but less effect than for Pb**	Trend could slow or reverse and [dissolved Cd] could increase	Trend could slow or reverse and [dissolved Cd] could increase	Trend could slow or reverse and [dissolved Cd] could increase	No effect
6.5 [Total Pb] in the Lake has declined over the past 8–10 years	Trend could slow or reverse and [total Pb] could increase	No effect	No effect	Trend could slow or reverse and [total Pb] could increase	No effect
7.3 Risk that enhanced anoxia will release arsenic	No effect	Will increase/ enhance risk of As release	**Increase risk:** Increased DO utilization leads to increased biomass/organic carbon, which favors enhanced anoxia; lower DO leads to greater anoxia; N and P increase at sediment–water interface	Same as column 3 for a warmer lake	Same as column 3 for a warmer lake
7.4 Risk that enhanced anoxia will release P, which through feedback loop could release more arsenic	No effect	Will increase/ enhance P release	**Increase risk:** Increased DO utilization leads to increased biomass/organic carbon, which favors enhanced anoxia; lower DO leads to greater anoxia; N and P increase at sediment–water interface	Same as column 3 for a warmer lake	Same as column 3 for a warmer lake
7.5 Risk that decrease in pH will drive Zn release	No effect	Potential for increase in period over which pH can decline resulting in Zn release	**Unclear effect on risk:** N-fixing cyanobacteria could lower pH at the sediment–water interface 1. Ammonia to nitrate → decrease in pH 2. Denitrification with organic matter ($NO_3 \to N_2$) • If pH < 6.3 → pH increase • If pH > 6.3 → slight decrease in pH 3. $Fe(OH)_3$ → pH increase	Same as column 3 for a warmer lake	Same as column 3 for a warmer lake

relatively "clean" sediment compared to the lead-enriched sediment moving through the lower basin and into the Lake. Currently, inputs to the Lake from the watershed are primarily from the lower basin, much of which is high in metals and is unremediated. NRC (2005) showed that the expected dilution of pollutant loads from the North Fork did not result in decreases in lead concentrations or fluxes downstream because of recontamination in the lower basin from bed and bank sediments. Similarly, there is evidence from the Clark Fork that downstream from areas where bed contamination was removed and banks were stabilized, sediments are quickly recontaminated from unstabilized banks (Hornberger et al., 2009). How much runoff comes from uncontaminated versus unremediated areas and how much the lower basin recontaminates the CDA River need to be better understood to make predictions with greater certainty.

A third possibility, discussed in Box 10-1 but not considered in Table 10-2, is that very high discharge events may become smaller and less frequent in the future because the potential for high runoff from rain-on-snow events would diminish. This would result in a decrease in flux of sediment and lead entering the Lake (the opposite of the impacts suggested in column 1 and rows 3.4 and 3.5 in Table 10-2).

Because phosphorus delivery to the Lake is also driven by high-flow events, many of the potential outcomes for phosphorus are similar to those mentioned for lead. The exception is that if high-flow conditions become more common, the sediment delivered to the lower CDA River could become more enriched in phosphorus, even if its lead load decreased, because sources of phosphorus are ubiquitous on the landscape while the source area for lead is restricted to soils contaminated by mineral extraction activities. In the case of cadmium and zinc (column 1, the two rows labeled 3.6, in Table 10-2), the influence of increased magnitude and frequency of high-discharge events would be more limited than for lead or phosphorus. Most of the cadmium and zinc transport from the lower CDA watershed is in the dissolved form and the variations in the concentrations due to changes in discharge are smaller than for lead.

As discussed above, fire is expected to mobilize sediments and sediment-associated contaminants such as metals and nutrients, both in runoff from burned watersheds and in atmospheric transport, that could increase their loading to CDA Lake. However, the effects will differ in magnitude among regions of the watershed. For example, fire in the lower basin of the CDA River or around operable unit (OU)-1 or OU-2 might affect metals loadings, but that would not be a concern for fire in the St. Joe basin.

Trends in Dissolved Oxygen, Phosphorus, and Nitrogen Concentration within the Lake

Chapter 5 describes the recent ongoing decline in Lake total phosphorus concentrations (Table 5-6) that have accompanied decreases in phosphorus loading to CDA Lake. Likewise, dissolved oxygen conditions in the Lake are also showing signs of improvement over time (Table 5-5), while trends for total nitrogen concentration are less obvious (Table 5-7).

As depicted in Figure 10-10, a number of projected changes in climate conditions and human activity potentially put the favorable trends for dissolved oxygen and phosphorus in jeopardy. For example, increased pressures from nearshore development (Gray Box 5) threaten to increase phosphorus loading to, and concentration in, the Lake, which may drive eutrophication and hypoxia (Arrows 11 to 9 in Figure 10-10). This may then lead to possible feedback loops that could promote nutrient release from the sediments. If climate change moves the timing of peak flow to earlier in the year (Gray Box 2), the Lake may stratify earlier (Arrow 14), allowing more time for oxygen depletion to occur in the hypolimnion (Arrow 13), promoting hypoxia and subsequent nutrient release from the sediments. The same results would be likely with warmer Lake temperatures (Gray Box 3). If the frequency, magnitude, and intensity of fires increase (Gray Box 4), the same processes could be put in motion (higher nutrient loading, hypoxia and lowering of pH in Lake bottom waters, eutrophication), eventually leading to further release of nutrients from the sediments.

Trends in Metals Concentrations within the Lake

As for dissolved oxygen and nutrients, the five types of climate change considered in Figure 10-10 could also slow or reverse the favorable trends in metals concentrations observed in the Lake. First, any climate effects that increase external loads of zinc, lead, arsenic, and cadmium to the Lake (such as more frequent high-flow runoff events and fires) will slow the rate of decline in metals concentrations in the Lake. In the case of cadmium and zinc (column 1, rows 6.1, 6.2, and 6.3 of Table 10-2), the influence of these high-discharge events, if they occur, would be much more limited than for lead and phosphorus, because zinc and cadmium are more likely to travel in dissolved form. Finally, such extreme runoff events have the potential to increase the rate at which remediated lands, particularly in the floodplains of the lower basin, are recontaminated by metals, leading to greater exposure to lead (Arrow 20).

Increased frequency of fire could increase loading of lead, zinc, and arsenic from smelter-contaminated soils or once flooded soils (Arrow 10), but this effect would probably be small compared to increases in metals loading brought about by changes in hydrology subsequent to fire (Gray Arrow).

A forward shift in the timing of flow and a warmer lake, as well as population growth in lakeshore areas, can eventually lead to eutrophication and a lowering of dissolved oxygen and pH in the lake bottom waters, which could enhance internal loading of dissolved zinc and cadmium (but not lead, owing to its higher affinity for iron mineral phases and particulate organic matter). The effect could be to slow the observed decline in zinc and cadmium concentrations in the water column.

Most of the effects above will differ in magnitude by Lake location. For example, internal loading of zinc and cadmium is more important in the southern Lake than in the northern Lake, slowing the response times of the southern Lake to reduced inputs. Thus, climate changes that enhance internal loading (e.g., anoxia, reduced pH) in the southern Lake may further slow its response to other climate changes that alter external loading.

Trends in Lake Sediment Processes

Of the impacts of climate change identified in Table 10-2, increased runoff was considered too indirect to affect biogeochemical processes in the sediments because of the separation between inputs and sediment processes and because the composition of runoff water is likely to be unchanged from current conditions.

Climate changes that directly or indirectly lead to reduced dissolved oxygen concentrations in bottom waters could eventually increase porewater concentrations of arsenic and phosphorus. This includes both a forward shift in the timing of flow and warming of Lake temperatures. That is, longer periods of stratification, increased heterotrophic microbial activity, and increased nitrogen-fixing cyanobacteria within the Lake all favor increased anoxia. As organic matter, nitrogen, and phosphorus settle to the bottom of the Lake, the potential for anoxia increases at the sediment–water interface, and decreased mixing in the Lake can provide longer periods of anoxia. The presence of anoxic conditions can lead to reductive dissolution of iron hydroxides and reduction of highly sorptive As(V) to weakly sorptive As(III), both of which may lead to release of sorbed arsenic and phosphorus. Similarly, longer periods of stratification will lengthen the time that pH may be lowered, promoting release of zinc and cadmium from the sediments. As mentioned in Chapter 5, there is little evidence of anoxia, and no evidence of trends, in the main body of the Lake. However, reduced dissolved oxygen in C5 bottom waters and minor release of arsenic from C5 sediments has been observed, as well as reduced pH and concomitant elevated zinc in some bottom waters.

Increased organic matter and nutrient inputs from fires and/or increased nearshore populations will have similar effects to warming of lake water, because these changes can lead to eutrophication, anoxia, and subsequent arsenic and phosphorus release.

For zinc (and cadmium) in the sediments, the effect of the last three changes in Table 10-2 are uncertain because it is unclear if pH would go up or down during anoxia. That is, the presence of nitrogen-fixing cyanobacteria could lower pH at the sediment–water interface if ammonia is converted to nitrate or if the pH is greater than 6.3 and denitrification (nitrate to nitrogen) occurs. However, reductive dissolution of iron oxides under anoxic conditions could increase pH and offset nitrification/denitrification impacts.

FUTURE WATER QUALITY SCENARIOS

The final section of this chapter is meant to provide a sense of the time necessary to reach different stages of recovery in CDA Lake. This simplistic exercise assumes that trends in water quality observed over the past 10–30 years will continue into the future. It extrapolates how long it will take for metals concentrations to go below thresholds of ecological significance in CDA Lake.

Although mass loading is the metric necessary to identify major sources of metal, to quantify the magnitude of the contribution from each source relative to one another, and to characterize changes over time in each source as management practices are implemented (Luoma and Rainbow, 2008; see Chapter 3), it is not necessarily the best metric to consider the future of lake water quality experienced by organisms. Rather, mixing of high metal loads from the CDA River with lake water, lake sediments, and biologically generated particulate material will determine the concentration of metals in the water, suspended particulates, and sediments of CDA Lake. Exchange of metals between sediments and water are driven by concentrations, and organisms respond to metal concentrations in water and particulate material rather than loads (Luoma and Rainbow, 2008). Hence, the following section focuses on when the Lake will achieve certain concentrations of metals, if past trends are extrapolated forward.

The following analysis is constructed around a scenario that assumes that the declines in metals concentration observed in the recent past will continue and follow an exponential decay, with a rate calculated on smoothed results covering the years 2011–2020. An exponential decay provides a reasonable first approximation of what can be expected in the future and is likely to generate better estimates than a linear decay. Fitting the decline in concentrations in the Lake to an exponential decay curve assumes that the decline in the rate of input will be a consistent function of the amount of contaminated material remaining. The committee recognizes that extrapolating trends from the past decade is uncertain. However, the calculations provide a perspective on the rate of change possible in the Lake and a baseline to evaluate data against future monitoring observations.

One can express the exponential decay model as follows:

$$C_{t+k} = C_t \cdot e^{-r \cdot k}$$

where:
C_t is the concentration in year t
C_{t+k} is the concentration in year $t+k$
r is the decay constant

For example, if the concentration of a metal in some starting year (say 2020) was 10 μg/L and the decay constant was 0.05, then the exponential decay model would predict a concentration of 9.51 μg/L for the year 2021, a value of 6.06 μg/L for the year 2030, and a value of 0.82 μg/L for the year 2070. In an exponential decay model, the concentration would asymptotically approach a value of zero, and the changes over time can be expressed as some constant percentage change between years. For example, if $r = 0.05$, the percent change from any one year to the next would be a decline of 4.88 percent per year [computed as $P = 100 \cdot (e^{-r} - 1)$].

The logic of assuming an exponential decay over time, versus a linear decline, is that a linear decline would project values of concentration at some time in the future going negative, which is impossible. The exponential decay model is a common approach to describing the rates of chemical reactions (known as "first-order reactions") in which the rate of the process is proportional to the remaining amount of the chemical being evaluated. It provides a first approximation for processes that involve the mobilization of some contaminant in the environment, which, in the case of the CDA watershed and Lake system, could be either driven by chemical reactions or by transport of contaminated sediment particles.

Rates of change estimated over a decadal period were used to limit the extent of individual year-to-year changes influencing the estimate of the decay constant. Changes between any two specific years are likely to be highly influenced by the random variations driven by the specific hydrologic conditions of those specific years. Thus, the committee used linear regression of the logarithm of annual concentration values as a function of year as the means of obtaining a decay constant. Graphical evaluation of the metals data showed that they commonly conformed well to this exponential decay model over the past decade, although looking back over two to three decades does show some changes in the decay constant at earlier times in the history of the remediation process.

As present-day regional background concentrations, the committee used the concentration data for the St. Joe River at Ramsdell and the North Fork of the CDA River at Enaville. Neither of these locations had Superfund activity upstream of them and both have a small amount of historical mining and likely atmospheric inputs of metals from smelting activities, but nothing that compares to the South Fork of the CDA River.

Year When Metals Concentrations in CDA Lake Will Reach Thresholds of Ecological Significance

To determine how long will it take for values of metals concentration to fall below a specified target, assuming that current trends continue, the committee conducted four analyses for each lake site. These four analyses considered surface water data (< 20 meters); deep water data (> 20 meters); and data from two seasons: "Spring," which consists of data from March, April, and May, and "Summer," which consists of data from July, August, and September. These analyses only used data from the calendar years 2011–2020, such that everything here is based on the trend from the last ten years of the record.

The statistical method used is called "censored regression" or "survival analysis," which is appropriate when some of the data are reported as a "less than" value. This method is a more appropriate method than using values such as half the reporting limit for the censored data point. The code that was used is the cencorreg function in the R package NADA2 (see Helsel, 2011, for details). When none of the observations are censored, this method gives results that are identical to the results one would get from linear regression.

The method fits a linear regression to the logarithms of the concentration as a function of year. That fitted line is equivalent to an exponential decay over time, when expressed in terms of the actual concentration values, and it is used to make an estimate of the year in which the median concentration would fall below a specified target. In some cases, the regression line is below the target line and the result is reported as "below." If the estimated line is above the target, then the year in which the median estimate will go below the target is reported.

The analyses were run for sites C1, C4, and C5 for dissolved lead, zinc, and cadmium. At C6, it was possible to do the analysis for dissolved lead only because the zinc and cadmium data had so many non-detects that the analysis was essentially meaningless.

Table 10-3 shows the analysis for dissolved lead using the Lake Management Plan (LMP) target of 0.54 µg/L, Table 10-4 shows the analysis for dissolved zinc using the LMP target of 36 µg/L, Table 10-5 shows the analysis for dissolved zinc using the target of 15 µg/L (a potential threshold for ecological effects), Table 10-6 shows the analysis for dissolved zinc using the target of 5 µg/L (the concentration of dissolved zinc in the St. Joe River is always less than 5), and Table 10-7 shows the analysis for dissolved cadmium using the LMP target of 0.25 µg/L.

To summarize, dissolved lead and cadmium concentrations in the Lake are mostly already below the LMP targets at the Lake sites (with the exception of lead at C4, where levels should fall below the threshold very soon if trends continue). Zinc concentrations are approaching the LMP target in surface waters (see Chapter 6 for monthly details), but are above potential ecological thresholds. Zinc concentrations in bottom waters at C1 and C4 remain above LMP targets, and it will take many decades for concentrations to fall below the various thresholds. For deeper water at C5, the zinc concentrations are not following the pattern of decrease that is seen at the other sites. Trend slopes in the deeper water at C5 are not statistically significant and are generally higher than any of these thresholds. Here, the concentrations appear to be responding to internal cycling processes rather than responding to the on-going downward trends driven by the downward trends in the external loading to the Lake. The bottom water

TABLE 10-3 The Year in Which Dissolved Lead Concentrations Will Fall Below a Target of 0.54 µg/L, Assuming the Trend from the Past Ten Years Continues

Site	Surface Water		Deep Water	
	Spring	Summer	Spring	Summer
C1				
C4	2023		2025	
C5				
C6				

NOTES: A dark blue entry indicates that the median curve lies below the critical value in 2020, and the curve is declining over time. A number indicates that the median curve is declining over time, and the number represents the year in which this curve would fall below the critical value. A light blue entry indicates that the curve is, as of 2020, below the critical value and is either rising over time or falling at a rate of < 1 percent per year.

TABLE 10-4 The Year in Which Dissolved Zinc Concentrations Will Fall Below a Target of 36 µg/L, Assuming the Trend from the Past Ten Years Continues

	Surface Water		Deep Water	
Site	Spring	Summer	Spring	Summer
C1	2025		2030	2035
C4	2025		2030	2040
C5				

NOTES: A dark blue entry indicates that the median curve lies below the critical value in 2020, and the curve is declining over time. A number indicates that the median curve is declining over time, and the number represents the year in which this curve would fall below the critical value. A light blue entry indicates that the curve is, as of 2020, below the critical value and is either rising over time or falling at a rate of < 1 percent per year. A red entry indicates that the curve is, as of 2020, above the critical value and is either rising over time or falling at a rate of < 1 percent per year.

TABLE 10-5 The Year in Which Dissolved Zinc Concentrations Will Fall Below a Target of 15 µg/L, Assuming the Trend from the Past Ten Years Continues

	Surface Water		Deep Water	
Site	Spring	Summer	Spring	Summer
C1	2050	2050	2062	2069
C4	2047	2042	2053	2079
C5		2025		

NOTES: A dark blue entry indicates that the median curve lies below the critical value in 2020, and the curve is declining over time. A number indicates that the median curve is declining over time, and the number represents the year in which this curve would fall below the critical value. A light blue entry indicates that the curve is, as of 2020, below the critical value and is either rising over time or falling at a rate of < 1 percent per year. A red entry indicates that the curve is, as of 2020, above the critical value and is either rising over time or falling at a rate of < 1 percent per year.

TABLE 10-6 The Year in Which Dissolved Zinc Concentrations Will Fall Below a Target of 5 µg/L, Assuming the Trend from the Past Ten Years Continues

	Surface Water		Deep Water	
Site	Spring	Summer	Spring	Summer
C1	2080	2088	2102	2112
C4	2075	2069	2082	2129
C5		2045		

NOTES: A dark blue entry indicates that the median curve lies below the critical value in 2020, and the curve is declining over time. A number indicates that the median curve is declining over time, and the number represents the year in which this curve would fall below the critical value. A light blue entry indicates that the curve is, as of 2020, below the critical value and is either rising over time or falling at a rate of < 1 percent per year. A red entry indicates that the curve is, as of 2020, above the critical value and is either rising over time or falling at a rate of < 1 percent per year.

TABLE 10-7 The Year in Which Dissolved Cadmium Concentrations Will Fall Below a Target of 0.25 µg/L, Assuming the Trend from the Past Ten Years Continues

	Surface Water		Deep Water	
Site	Spring	Summer	Spring	Summer
C1				
C4				
C5				

Notes: A dark blue entry indicates that the median curve lies below the critical value in 2020 and the curve is declining over time. A number indicates that the median curve is declining over time and the number represents the year in which this curve would fall below the critical value. A light blue entry indicates that the curve is, as of 2020, below the critical value and is either rising over time or falling at a rate of < 1 percent per year.

observations at C5 make it clear that time to recovery in this system will be longest where internal processes are substantial. If benthic flux becomes more important in bottom waters of the northern lake as river inputs decline, one might expect the trend of decline at C1 and C4 to slow.

Year When Metals Concentrations Going into and out of CDA Lake Will Reach Thresholds of Ecological Significance

This analysis was similar to the one above, except that the committee looked at trends in flow-normalized concentrations of the metals *at two river locations:* the CDA River near Harrison and the Spokane River at the outlet of CDA Lake. The seasons are the same as defined in the exercise on the ten-year trends in lake concentrations. The fitted Weighted Regressions on Time, Discharge, and Season (WRTDS) model introduced in Chapter 3 was used to make these computations.

Dissolved Lead, Spokane River below Lake Outlet

For the spring season, the flow-normalized mean concentration in 2020 is 0.71 µg/L, and the ten-year trend is effectively zero. For the summer season, the flow-normalized mean concentration in 2020 is 0.27 µg/L, and the ten-year trend is slight upward, at +1.8 percent per year, which is not statistically significant. Critical dissolved lead concentrations in CDA Lake are 0.54 µg/L. Thus, lead releases from CDA Lake into the Spokane River are not declining, and could even be slightly increasing.

Dissolved Lead, CDA River near Harrison

For the spring season, the flow-normalized mean concentration in 2020 is 6.14 µg/L, and the ten-year trend is −4 percent per year and highly significant downward. The dissolved lead concentration would fall below the 0.54 µg/L threshold by 2082. For the summer season, the flow-normalized mean concentration in 2020 is 2.35 µg/L, and the ten-year trend is −3.7 percent per year and highly significant downward. The dissolved lead concentration would fall below the 0.54 µg/L threshold by 2061.

Dissolved Cadmium, Spokane River below Lake Outlet

For the spring season, the flow-normalized mean concentration in 2020 is 0.161 µg/L, which is already below the target of 0.25 µg/L. The ten-year trend is −2.2 percent per year and highly significant downward. For the summer season, the flow-normalized mean concentration in 2020 is 0.094 µg/L, which is already below the target of 0.25 µg/L, and the ten-year trend is −1.6 percent per year and highly significant downward.

Dissolved Cadmium, CDA River near Harrison

For the spring season, the flow-normalized mean concentration in 2020 is 0.65 µg/L, and the ten-year trend is −1.6 percent per year, which is a significant downward trend. The concentration would fall below the 0.25 µg/L LMP target in 2080. For the summer season, the flow-normalized mean concentration in 2020 is 0.69 µg/L, and the ten-year trend is −3.6 percent per year and highly significant downward. The dissolved cadmium concentration would fall below the 0.25 µg/L target in 2048. These values suggest that some trapping of cadmium in CDA Lake during the summer continues.

Dissolved Zinc, Spokane River below Lake Outlet

For the spring season, the flow-normalized mean concentration in 2020 is 39.3 µg/L, and the ten-year trend is −2.9 percent per year, a highly significant downward trend. The dissolved zinc concentration would fall below the 36 µg/L LMP target in 2024, it would cross 15 µg/L in 2054, and it would cross 5 µg/L in 2092. For the summer season, the flow-normalized mean concentration in 2020 is 20.6 µg/L and the ten-year trend is −3.1 percent per year, which is a highly significant downward trend. The concentration has already crossed below the 36 µg/L threshold, it would cross 15 µg/L in 2032, and it would cross below 5 µg/L in 2066.

Dissolved Zinc, CDA River near Harrison

For the spring season, the flow-normalized mean concentration in 2020 is 98 µg/L, and the ten-year trend is −3.2 percent per year, which is a highly significant downward trend. The concentration would cross below 36 µg/L in 2052, below 15 µg/L in 2079, and below 5 µg/L in 2113. For the summer season, the flow-normalized mean concentration in 2020 is 126 µg/L, and the ten-year trend is −3.6 percent per year, which is a highly significant downward trend. The concentration would cross below 36 µg/L in 2055, below 15 µg/L in 2080, and below 5 µg/L in 2110. Like cadmium, inputs of zinc are declining slightly faster than outputs, suggesting that some dissolved zinc continues to be retained by the Lake.

***

One can compare the results for dissolved metals in the Lake versus in the river inputs and outputs. For dissolved lead, the analysis for the Lake and for the Spokane River below the Lake indicate that dissolved lead levels have reached levels that represent a near steady-state condition and are no longer responding to the downward trend in the inputs. That steady-state condition does still show a seasonal signal, being above the 0.54 µg/L LMP target in the spring and below it in the summer.

For dissolved zinc, the analysis of the Lake results shows declines (except at C5, in the spring season at depth), and the time at which it crosses the LMP target (36 µg/L) is only a few years into the future, which applies to the Spokane River below the Lake outlet. For the inflow near Harrison, the trends are also rather steeply downward, but the levels are higher and thus the number of years before concentrations fall below the LMP target is a few decades later than the Lake or output values. Overall, the results suggest that the Lake and its outflow continue to be responding favorably to the Superfund remedy. The exception is the spring season at depth at C5, where internal loadings may play an important role rather than responses to inputs from the CDA River.

For dissolved cadmium, all of the results (Lake sites, inputs, and outputs) show an ongoing decline and levels are generally all below the 0.25 µg/L LMP target. The values seen in all of these datasets show an ongoing decline driven by the Superfund remedy and no sign of reaching some new steady-state level.

Alternative In-Lake Zinc Concentration Analysis

Another analysis was conducted by the committee, similar to the above in-lake analysis of metals concentrations that assumed the data followed an exponential decay (called the "exponential analysis"). This second analysis, called the "decadal analysis," was conducted for zinc only. The differences between monthly mean concentrations of zinc in the earliest data (2004–2008) and more recent data (2015–2020) were compared to assess the response of the Lake. Table 10-8 shows the proportional decline of zinc per decade in the photic zone and bottom waters of C4 nearest the source of zinc input, C1, and C5. For each site, the "decadal analysis" was done for means of March–April data and means of August–September data. Averaging across all locations, dissolved zinc concentrations in March–April declined 23 ± 7 percent per decade between the two time periods and concentrations in August–September declined 16 ± 4 percent per decade (−2.3 and −1.6 percent per year, respectively).

The "exponential analysis" (Tables 10-3 through 10-7) predicts a faster decline in zinc concentrations than the "decadal analysis" (Table 10-8), although both analyses suggest that zinc concentrations in the Lake are declining 1–3 percent per year in spring and summer and for surface and bottom waters. This range probably illustrates the limits of resolution of any extrapolation from the past using small datasets. Both approaches also show that at some locations and seasons LMP targets have been reached or can be reached within a decade or two, except the bottom waters at C5, which reflects the influences of hydrodynamics and internal recycling at this location.

TABLE 10-8 Proportional Decline in Monthly Geometric Mean Dissolved Zinc Concentrations in the Water Column (Photic and Bottom Waters) of C1, C4 and C5 Comparing the Periods of 2004–2008 and 2015–2020; and Years to Reach Three Benchmarks if This Rate of Decline Continued

Location	% decline/year[a]	Yrs to reach 36 µg/L	Yrs to reach 15 µg/L	Yrs to reach 5 µg/L
Photic Zone—March–April				
C4	2.4%	20	42	99
C1	1.9%	22	65	110
Bottom—March–April				
C4	3.3%	14	33	60
C1	1.9%	15	66	110
Photic Zone—August				
C4	2.1%	0	33	49
C1	1.3%	0	55	125
C5	1.7%	0	40	100
Bottom—August				
C4	1.9%	24	66	130
C1	2.1%	21	58	110
C5	1.0%	55	164	290

[a] Mean concentrations in March–April and in August–September were chosen to represent peak inputs and summer response periods, respectively. Percent decline was calculated from the ratio of mean March–April (or ratio of mean August–September) in the two time periods. Recovery time was determined by calculating a first-order exponential decline using the proportional difference as the rate constant and the concentration in 2015–2020 as time zero.

Decades to centuries will be required to reach more ambitious targets at C5. Of course, all this assumes that the present rates of decline continue into the future.

Lake Sediments as a Source of Zinc to the Lake

If remediation successfully reduces incoming metals to levels less than outgoing metals, further reduction in water column dissolved concentrations in the main body of the Lake will not only depend upon processes within the water column, but perhaps more importantly, on those that determine exchange of metals between the water column and the sediments. As discussed in Chapter 7, such exchange is largely driven by redox processes that are controlled by dissolved oxygen levels and the potential for lower pH in bottom waters that might accompany longer periods of stratification or greater productivity in surface waters. This section considers the potential for lake sediments to become a source of zinc as zinc loading to the Lake is reduced (as a result of remediation in the basin and other factors).

Kuwabara et al. (2007) estimated that dissolved zinc concentrations of about 0.2 µg/L (3 nM) would be typical of pre-mining conditions in the southern lake, which is in the range reported for other undisturbed freshwater lakes (see Table 1-6). As described in Chapter 3, the Central Treatment Plant (CTP) is removing a major, but not the only, source of zinc to the lake, and the lower basin will become a more important source of zinc as upstream remediation continues (EPA Region 10, 2021). Based on the late 2021 performance data, the CTP is currently discharging about 0.1 MT/yr (< 0.7 lb per day) of zinc, which is a small fraction of the 205 MT/yr entering the Lake (Chapter 3). Thus, further reductions at the CTP are unlikely to alter the current inputs.

The 0.2 µg/L level may be difficult to reach, given that recent average levels of dissolved zinc in the less impacted St. Joe River and North Fork of the CDA River are 1.5 and 3.5 µg/L, respectively (see Zinsser, 2020, Appendix 1). Assuming that remedial efforts in the watershed and flushing of residual zinc continue to reduce levels following current trends (3.6 percent decrease per year; see Chapter 3), it is estimated that the concentration would be 5 μg/L by 2116 and 2.4 μg/L by 2136 in the CDA River at Harrison. Using a similar analysis for the Lake, these low levels will be reached between 2045 and 2119, based on Table 10-6. Obviously, reaching the 0.2 μg/L will take even longer (beyond the year 2200 for the CDA River). This also assumes that other anthropogenic sources of zinc (e.g., tire wear, sacrificial anodes for corrosion control) do not become relevant sources that prevent background levels from being obtained once mining sources of zinc are reduced/eliminated. Even so, an evaluation of how reduced zinc inputs will affect zinc levels in the Lake is needed.

As described in previous chapters, there is a large reservoir of zinc in the sediments of CDA Lake. The Lake serves to remove zinc from the water column, and as shown in Chapter 3, fluxes leaving the Lake are lower than those entering it. If the zinc concentrations in the water column decrease, there is the potential for release of zinc from the benthic sediments to the water column. The potential for this release will depend on the concentration gradient as well as the accessibility of the zinc to the overlying water. This accessibility will be driven by resuspension (Diamond, 1995), the depth at which the zinc is present, the mineral phases with which it is associated, and any redox or pH dynamics that affect the stability of these mineral phases and the association of zinc with them. There are also likely annual/seasonal dynamics that will affect the water/sediment fluxes, with biological uptake/sorption of zinc from the upper part of the water column during periods of biological productivity and deposition to lower water column and sediment in later months (e.g., Acterberg et al., 1997). It is unclear what is the critical Lake concentration at which the sediments will become more of a source of zinc than a sink.

Reduced concentrations of zinc may also affect phytoplankton, periphyton, and zooplankton dynamics in the Lake. To prevent harmful effects on freshwater algae, a zinc concentration of 10 μg/L was recommended for the Great Lakes (Wong and Chau, 1990), and Kuwabara et al. (2007) saw minimal effects in toxicity tests with algae from CDA Lake at 1 μg Zn/L. Evidence also suggests that at levels, in the range of 1–5 μg/L, zinc can act as a nutrient and stimulate growth of algae or phytoplankton (Wong and Chau, 1990; Canli, 2005). A recent microcosm study provides insight into the potential effects of zinc over a range of concentrations (8–160 μg/L; Hoang et al., 2021; see also, Box 9-2). Concentrations as low as 8 μg/L led to reduced abundance of Chlorophyta (both phytoplankton and periphyton) and Chrysophyta and decreased the diversity. This study led to a No Observed Effects Concentration of 14 μg/L (similar to the recommendation of Wong and Chau, 1990). Concentrations of 40 μg/L and greater (similar to those present in CDA Lake) also inhibited these algae, reduced overall diversity, and led to an increase in the abundance of cyanobacteria. Continued assessment of the role of zinc in the Lake on algal abundance and diversity as nutrient levels and other factors change will be important.

As zinc levels decline, any toxic inhibitions on specific algae are likely to decrease and a more stable algal community may result. Before reaching background levels, the concentration of zinc in the range of 1–5 μg/L may serve as a nutrient to enhance algal growth and could lead to a return to greater abundance and diversity of phytoplankton, periphyton, and zooplankton. The stimulated growth and/or the more diverse community could, in turn, limit the growth of harmful cyanobacteria, which may be favored at extreme dissolved zinc concentrations (Hoang et al., 2021). Once at background zinc levels, there may be less algal growth due to zinc nutrient limitations, but other factors, particularly low phosphorus supplies, are likely to be more important. Note that based on late 2021 performance of the CTP, the plant is not a significant source of phosphorus, and almost all of the phosphorus in the CDA River is coming from nonpoint sources across the CDA watershed.

***

In summary, the calculations in this section suggest that further progress in reducing toxic metals concentrations in the euphotic zone is possible on the timescale of a few decades if the present trends can be sustained and the challenges posed by climate change and population growth can be mitigated. Better understanding of how metals are affecting the lake ecosystem is necessary to address the question of whether further reductions in zinc and cadmium concentrations in the water column of the Lake are more likely to result in ecosystem benefits than in greater risks to the ecosystem (e.g., via eutrophication). The committee suggests that estimates of the time to reach various ecological targets be regularly evaluated as conditions develop, since the drivers of those trends are likely to change (e.g., because of continued remediation, climate change, and changes in land use). Lake sediments represent an immense repository of toxic metals that will be difficult to mitigate. Better anticipating the effects of those metals would benefit from a concerted effort to fully understand the processes that lead to the releases of different metals (as well as phosphorus) from the Lake sediments and the relative rates of release at present (see Chapters 7 and 8).

CONCLUSIONS AND RECOMMENDATIONS

This chapter summarized the changes in climate likely to affect the Pacific Northwest generally and the CDA regional specifically (when local predictions were available). It then considered four changes in climate most likely to negatively impact water quality in CDA Lake, along with population growth in the nearshore watersheds, and it assessed the extent to which the trends revealed in Chapters 3 to 7 might be slowed or reversed by these changes. Because all four future climate changes, along with increased lakeshore populations, can lead to the eventual release of metals from lake sediments to the Lake (via lowering of either dissolved oxygen or pH in bottom waters), the long-term monitoring and assessment programs for both the watershed and the Lake need to be fortified and improved (see Chapter 8) to provide an early warning of deteriorating conditions. The following detailed conclusions are made about the future of water quality in CDA Lake.

1. **A major impact of climate change likely to affect water quality in the CDA region is air temperature warming as much as 2.5–3°C (4.5–5.4°F) by the year 2050, depending on the month.** Data from the CDA region over the past 30 years show warming of about 0.4°F per decade. Increases in air temperature are expected to lead to increases in Lake temperature and increased fire risk across the region.

2. **Although there are no apparent trends in precipitation in the CDA region over the past 30 years, studies in the greater Pacific Northwest suggest that precipitation events in the top 1 percent of historical daily precipitation volumes are likely to become between 5–34 percent more intense by 2080.** A shift is expected in the percentage of precipitation that falls as snow versus rain, such that by 2080 the peak snow water equivalent could decrease by an average of 73 percent. Finally, in the Pacific Northwest, the historic center of timing of the annual hydrograph is predicted to shift from April 15 in the 1980s to as early as March 4 by 2040, although there is not yet evidence of this shift over the past 30 years in the CDA region.

3. **Lake water temperatures have been increasing in surface water at station C4 over the past 30 years (although not in bottom water).** It is reasonable to assume that this warming is related to the general global trend toward rising air temperatures, and this can be expected to increase in the future based on ongoing greenhouse forcing of the climate. Responses in the lake to such warming will likely include increased rates of ecosystem metabolism and lengthening of the duration of stratification, which may extend periods of dissolved oxygen consumption in the bottom waters and Lake sediments as well as lower pH in the bottom waters.

4. **Future climate change may slow or reverse the trends in metals and phosphorus loading to CDA Lake (discussed in Chapter 3) and the trends in dissolved oxygen, phosphorus, and metals concentrations within CDA Lake (discussed in Chapters 5 and 6), and it may increase the potential for metals release from lake sediments (discussed in Chapter 7).** Those trends were based on historical data from the last 30 years and did not explicitly take future climate change into account (although it is clear that air and lake water warming have been underway during the past 30 years). The changes in climate considered by the committee were (1) increased frequency and magnitude of large runoff events, (2) a forward shift in the timing of flow to the Lake, (3) warming of lake water, and (4) increased frequency and size of fires. Of these four potential climate changes, the committee is most confident about warming of lake water, along with increases in lakeshore populations. There is not yet evidence in the CDA region of an increased frequency and magnitude of large runoff events, a forward shift in the timing of flow to the Lake, or increased frequency and size of fires, although these are predicted to occur for the Pacific Northwest.

5. **Zinc concentrations in surface waters are at or approaching the Lake Management Plan (LMP) target of 36 µg/L in some months and at some locations. If trends from the past decade continue into the future, it will take bottom waters 10 to more than 100 years to reach that target.** The slowest changes are occurring in the southern Lake, where the response to declining inputs appears to be buffered by internal inputs from bottom sediments and hydrodynamic inputs from the northern Lake. Reaching lower concentrations more reflective of background conditions, like 5 µg/L, may take as many as 500 years at C5. Dissolved lead concentrations in CDA Lake are already below the LMP target of 0.54 µg/L in the measured Lake locations, with the exception of C4 during the spring. Cadmium concentrations are also already below the LMP target of 0.25 µg/L in the three measured Lake locations (C1, C4, C5).

6. **It is unclear at what in-lake zinc concentration the Lake sediments will become more of a zinc source than a sink.** As shown in Chapter 3, CDA Lake is still a net sink for zinc. If zinc concentrations in the water column decrease owing to decreased loading from the basin, there is the potential for release of zinc from the Lake sediments to the water column, which will depend on the concentration gradient as well as the accessibility of the zinc in the sediments to the overlying water. This accessibility will be driven by resuspension, by the depth at which the zinc in present, by the mineral phases with which it is associated, and by any redox or pH dynamics that affect the stability of these mineral phases and the association of zinc with them. This conclusion could also apply to cadmium and arsenic under appropriate circumstances, but is less likely for lead.

REFERENCES

Abatzoglou, J. T., and A. P. Williams. 2016. Impact of anthropogenic climate change on wildfire across western US forests. *PNAS* 113(42):11770–11775. https://doi.org/10.1073/pnas.1607171113.

Abatzoglou, J. T., and T. J. Brown. 2012. A comparison of statistical downscaling methods suited for wildfire applications. *International Journal of Climatology* 32:772–780. https://doi.org/10.1002/joc.2312.

Acterberg, E. P., C. M. G. van den Berg, M. Boussemart, and W. Davison. 1997. Speciation and cycling of trace metals in Esthwaite Water: A productive English lake with seasonal deep-water anoxia, *Geochim. Cosmochim.* 61:5233–8253.

Beniston, M., and M. Stoffel. 2016. Rain-on-snow events, floods and climate change in the Alps: Events may increase with warming up to 4 C and decrease thereafter. *Science of the Total Environment* 571:228–236. https://doi.org/10.1016/j.scitotenv.2016.07.146.

Brandt, W. T., K. Haleakala, B. J. Hatchett, and M. Pan. 2022. A review of the hydrologic response mechanisms during mountain rain-on-snow. *Frontiers in Earth Science.* https://doi.org/10.3389/feart.2022.791760.

Brauer, V., M. Stomp, C. Rosso, S. A. M. van Beusekom, B. Emmerich, L. J. Stal, and J. Huisman. 2013. Low temperature delays timing and enhances the cost of nitrogen fixation in the unicellular cyanobacterium *Cyanothece. ISME J* 7:2105–2115. https://doi.org/10.1038/ismej.2013.103.

Breitbarth, E., A. Oschlies, and J. LaRoche. 2006. Physiological constraints on the global distribution of *Trichodesmium* – effect of temperature on diazotrophy. *Biogeosciences Discuss.* 3:779–801. https://doi.org/10.5194/bg-4-53-2007.

Brooks, E. 2021. Impacts of Wildfire on Phosphorus Transport in the CDA Basin. Presentation to the NASEM Committee. May 4, 2021.

Canli, M. 2005. The transfer of zinc in two liked trophic levels in fresh water and its effect on the reproduction of *Daphnia magna. J. Freshwater Ecol.* 20:269–276. https://doi.org/10.1080/02705060.2005.9664966.

Diamond, M. L. 1995. Application of a mass balance model to assess in-place arsenic pollution. *Environ. Sci. Technol.* 29:29–42.

Dudley, R. W., G. A. Hodgkins, M. R. McHale, M. J. Kolian, and B. Renard. 2017. Trends in snowmelt-related streamflow timing in the conterminous United States. *Journal of Hydrology* 547:208–221. https://doi.org/10.1016/j.jhydrol.2017.01.051.

EPA Region 10. 2021. Fifth five-year review for the Bunker Hill Superfund Site.

Hamman, J., B. Nijssen, A. Roberts, A. Craig, W. Maslowski, and R. Osinski. 2017. The coastal streamflow flux in the Regional Arctic System Model. *Journal of Geophysical Research: Oceans* 122:1683–1701. doi:10.1002/2016JC012323.

Hauer, F. R., and C. N. Spencer. 1998. Phosphorus and Nitrogen Dynamics in Streams Associated With Wildfire: a Study of Immediate and Long-term Effects. *International Journal of Wildland Fire* 8:183–198. https://doi.org/10.1071/WF9980183.

Helsel, D. R. 2011. Statistics for censored environmental data using Minitab and R, 2nd ed. John Wiley & Sons.

Hoang, T. C., J. M. Brausch, M. F. Cichra, D. J. Phlips, C. Van Genderen, and G. M. Rand. 2021. Effects of zinc in an outdoor freshwater microcosm system. *Environ. Toxicol. Chem.* 40:2051–2070. https://doi.org/10.1002/etc.5050.

Holden, Z. A., A. Swanson, C. H. Luce, W. M. Jolly, M. Maneta, J. W. Oyler, D. A. Warren, R. Parsons, and D. Affleck. 2018. Decreasing fire season precipitation increased recent western US forest wildfire activity. *PNAS* 115(36):E8349–E8357. https://doi.org/10.1073/pnas.1802316115.

Hornberger, M. I., S. N. Luoma, M. J. Johnson, and M. Holyoak. 2009. The influence of remediation in a mine-impacted river: do improvements upstream impact metal trends over large spatial and temporal scales? *Ecological Applications* 19(6):1522–1535. https://doi.org/10.1890/08-1529.1.

Huang, H., M. R. Fischella, Y. Liu, Z. Ban, J. V. Fayne, D. Li, K. C. Cavanaugh, and D. P. Lettenmaier. 2022. Changes in mechanisms and characteristics of western US floods over the last sixty years. *Geophysical Research Letters* 49(3):e2021GL097022. https://doi.org/10.1029/2021GL097022.

Kharin, V. V., F. W. Zwiers, X. Zhang, and M. Wehner. 2013. Changes in temperature and precipitation extremes in the CMIP5 ensemble. *Climatic Change* 119(2):345–357. https://doi.org/10.1007/s10584-013-0705-8.

Kuwabara, J. S., Topping, B. R., Woods, P. F., and J. L. Carter. 2007. Free zinc ion and dissolved orthophosphate effects on phytoplankton from Coeur d'Alene Lake, Idaho. *Environ. Sci. Technol.* 41(8):2811–2817. https://doi.org/10.1021/es0629231.

Leavesley, G. H., R. W. Lichty, B. M. Troutman, and L. G. Saindon. 1983. Precipitation-runoff modeling system-User's manual: USGS Water-Resources Investigations Report 83-4238, 207 p.

Liang, X., D. P. Lettenmaier, E. F. Wood, and S. J. Burges. 1994. A simple hydrologically based model of land-surface water and energy fluxes for general-circulation models. *Journal of Geophysical Research: Atmospheres* 99:14415–14428. doi:10.1029/94jd00483.

López-Moreno, J. I., J. W. Pomeroy, E. Morán-Tejeda, J. Revuelto, F. M. Navarro-Serrano, I. Vidaller, and E. Alonso-González. 2021. Changes in the frequency of global high mountain rain-on-snow events due to climate warming. *Environmental Research Letters* 16(9):094021. https://doi.org/10.1088/1748-9326/ac0dde.

Luoma, S. N., and P. S. Rainbow. 2008. Metal Contamination in Aquatic Environments: Science and Lateral Management. Cambridge, UK: Cambridge University Press.

Mauger, G. 2021. Climate Change Impacts in the Pacific Northwest: Implications for the Coeur d'Alene Watershed. Presentation to the NASEM Committee. May 4, 2021.

McCullough, I. M., K. Spence Cheruvelil, J.-F. Lapierre, N. R. Lottig, M. A. Moritz, J. Stachelek, and P. A. Soranno. 2019. Do lakes feel the burn? Ecological consequences of increasing exposure of lakes to fire in the continental United States. *Global Change Biology* 25(9):2841–2854. https://doi.org/10.1111/gcb.14732.

Moore, J. N., and S. N. Luoma. 1990. Hazardous wastes from large scale metal extraction: a case study. *Environ. Sci. Technol.* 24:1279–1285.

National Research Council (NRC). 2005. Superfund and Mining Megasites: Lessons from the Coeur d'Alene River Basin. Washington, DC: National Academies Press. https://doi.org/10.17226/11359.

Preisler, H. K., J. A. Hicke, A. A. Ager, and J. L. Hayes. 2012. Climate and weather influences on spatial temporal patterns of mountain pine beetle populations in Washington and Oregon. *Ecology* 93(11):2421–2434. https://doi.org/10.1890/11-1412.1.

Rhoades, C. C., A. T. Chow, T. P. Covino, T. S. Fegel, D. N. Pierson, and A. E. Rhea. 2019. The legacy of a severe wildfire on stream nitrogen and carbon in headwater catchments. *Ecosystems* 22:643–657. https://doi.org/10.1007/s10021-018-0293-6.

Robarts, R. D., and T. Zohary. 1987. Temperature effects on photosynthetic capacity, respiration, and growth rates of bloom-forming cyanobacteria. *New Zealand Journal of Marine and Freshwater Research* 21:3:391–399. https://doi.org/10.1080/00288330.1987.9516235.

Rust, A. J., S. Saxe, J. McCray, C. C. Rhoades, and T. S. Hogue. 2019. Evaluating the factors responsible for post-fire water quality response in forests of the western USA. *International Journal of Wildland Fire* 28:769–784.

Rust, A. J., T. S. Hogue, S. Saxe, and J. McCray. 2018. Post-fire water-quality response in the western United States. *International Journal of Wildland Fire* 27:203–216. https://doi.org/10.1071/WF17115.

Shepherd, T. G., E. Boyd, R. A. Calel, S. C. Chapman, S. Dessai, I. M. Dima-West, H. J. Fowler, R. James, D. Maraun, O. Martius, C. A. Senior, A. H. Sobel, D. A. Stainforth, S. F. B. Tett, K. E. Trenberth, B. J. J. M. van den Hurk, N. W. Watkins, R. L. Wilby, and D. A. Zenghelis. 2018. Storylines: an alternative approach to representing uncertainty in physical aspects of climate change. *Climatic Change* 151(3):555–571. https://doi.org/10.1007/s10584-018-2317-9.

Siirila-Woodburn, E. R., A. M. Rhoades, B. J. Hatchett, L. S. Huning, J. Szinai, C. Tague, P. S. Nico, D. R. Feldman, A. D. Jones, W. D. Collins, and L. Kaatz. 2021. A low-to-no snow future and its impacts on water resources in the western United States. *Nature Reviews Earth & Environment* 2(11):800–819. https://doi.org/10.1038/s43017-021-00219-y.

Spencer, C. N., K. Odney Gabel, and F. R. Hauer. 2003. Wildfire effects on stream food webs and nutrient dynamics in Glacier National Park, USA. *Forest Ecology and Management* 178:141–153. https://doi.org/10.1016/S0378-1127(03)00058-6.

Taylor, K. E., R. J. Stouffer, and G. A. Meehl. 2012. An Overview of CMIP5 and the experiment design. *Bull. Amer. Meteor. Soc.* 93:485–498. doi:10.1175/BAMS-D-11-00094.1.

Warner, M. D., C. F. Mass, and E. P. Salathe, Jr. 2015. Changes in winter atmospheric rivers along the North American west coast in CMIP5 climate models. *Journal of Hydrometeorology* 16(1):118–128. https://doi.org/10.1175/JHM-D-14-0080.1.

Williams, A. P., B. I. Cook, and J. E. Smerdon. 2022a. Rapid intensification of the emerging southwestern North American megadrought in 2020–2021. *Nature Climate Change* 12:232–234. https://doi.org/10.1038/s41558-022-01290-z.

Williams, A. P., B. Livneh, K. A. McKinnon, W. D. Hansen, J. S. Mankin, B. I. Cook, J. E. Smerdon, A. M. Varuolo-Clarke, N. R. Bjarke, C. S. Juang, and D. P. Lettenmaier. 2022b. Growing impact of wildfire on western US water supply. *PNAS* 119(10). https://doi.org/10.1073/pnas.2114069119.

Wong, P. T. S., and Y. K. Chau. 1990. Zinc toxicity to freshwater algae. *Tox. Assess.* 5:167-177. https://doi.org/10.1002/tox.2540050205.

Wood, A. W., L. R. Leung, V. Sridhar, and D. P. Lettenmaier. 2004. Hydrologic implications of dynamical and statistical approaches to downscaling climate model outputs. *Climatic Change* 62:189–216. doi:10.1023/B:CLIM.0000013685.99609.9e.

Zhang, H., J. E. Mu, and B. A. McCarl. 2018. Adaptation to climate change via adjustment in land leasing: Evidence from dryland wheat farms in the US Pacific Northwest. *Land Use Policy* 79:424–432. https://doi.org/10.1016/j.landusepol.2018.07.030.

Zinsser, L. M. 2020. Trends in Concentrations, Loads, and Sources of Trace Metals and Nutrients in the Spokane River Watershed, Northern Idaho, Water Years 1990-2018. USGS Scientific Investigations Report 2020-5096. https://doi.org/10.3133/sir20205096.

Acronyms

ACOE	U.S. Army Corps of Engineers
AEM3D	Aquatic Ecosystem Model
AWQC	ambient water quality criteria
BEIPC	Basin Environmental Improvement Project Commission
BLL	blood lead level
BLM	Biotic Ligand Model
BOD	biological oxygen demand
CDA	Coeur d'Alene
CDC	Centers for Disease Control and Prevention
CERCLA	Comprehensive Environmental Response, Compensation, and Liability Act
chl *a*	chlorophyll *a*
CIA	Central Impoundment Area
CMIP	Coupled Model Intercomparison Project
CTP	Central Treatment Plant
CTU	cumulative toxic unit
CWA	Clean Water Act
DGT	diffusive gradient in thin films
DO	dissolved oxygen
DOM	dissolved organic matter
EPA	U.S. Environmental Protection Agency
FAB	filamentous algal blooms
FERC	Federal Energy Regulatory Commission
HAB	harmful algal bloom
HHRA	human health risk assessment

IDEQ	Idaho Department of Environmental Quality
IDHW	Idaho Department of Health and Welfare
IDL	Idaho Department of Lands
IDWR	Idaho Department of Water Resources
LMP	Lake Management Plan
MACA	Multivariate Adaptive Constructed Analogs
NLA	National Lakes Assessment
NLCD	National Land Cover Dataset
NRCS	Natural Resources Conservation Service
NWIS	National Water Information System
OUs	operable units
PAR	photosynthetically active radiation
POM	particulate organic matter
RCP	Representative Concentration Pathway
ROD	Record of Decision
ROS	reactive oxygen species
SCM	surface complexation model
SRB	sulfate-reducing bacteria
SRP	soluble reactive phosphorus
SVRP	Spokane Valley Rathdrum Prairie
SWE	snow water equivalent
TMDL	total maximum daily load
TN	total nitrogen
TOC	total organic carbon
TP	total phosphorus
USDA	U.S. Department of Agriculture
USGS	U.S. Geological Survey
WQS	water quality standard
WRTDS	Weighted Regressions on Time, Discharge, and Season
WWTP	wastewater treatment plant

Appendix A

Coeur d'Alene Watershed Analysis Methodology for Metals and Nutrients

This appendix details the committee's methodological analysis of the history of concentration and fluxes from the Coeur d'Alene (CDA) Lake watershed described in Chapter 3. The committee's analysis uses the statistical method known as Weighted Regressions on Time, Discharge, and Season (WRTDS; Hirsch et al., 2010) to make inferences about concentration and flux, on a daily time step, based on the types of records that are typically available in the rivers of the CDA Lake watershed (typically on the order of 250 observations for each of the key contaminants at a given monitoring location over nearly three decades). WRTDS has been used extensively for many river systems in the United States, including in this watershed (see Zinsser, 2020).

The WRTDS model uses statistical smoothing in order to partition the variations in concentration into components that are related to (1) season of the year, (2) watershed hydrologic condition (characterized by the daily mean discharge on the day of sample collection), (3) long-term trend, and (4) a random component (the unexplained portion of the variation). The analysis in Chapter 3 considered six U.S. Geological Survey (USGS) monitoring locations, shown in Figure A-1 and described in Table A-1.

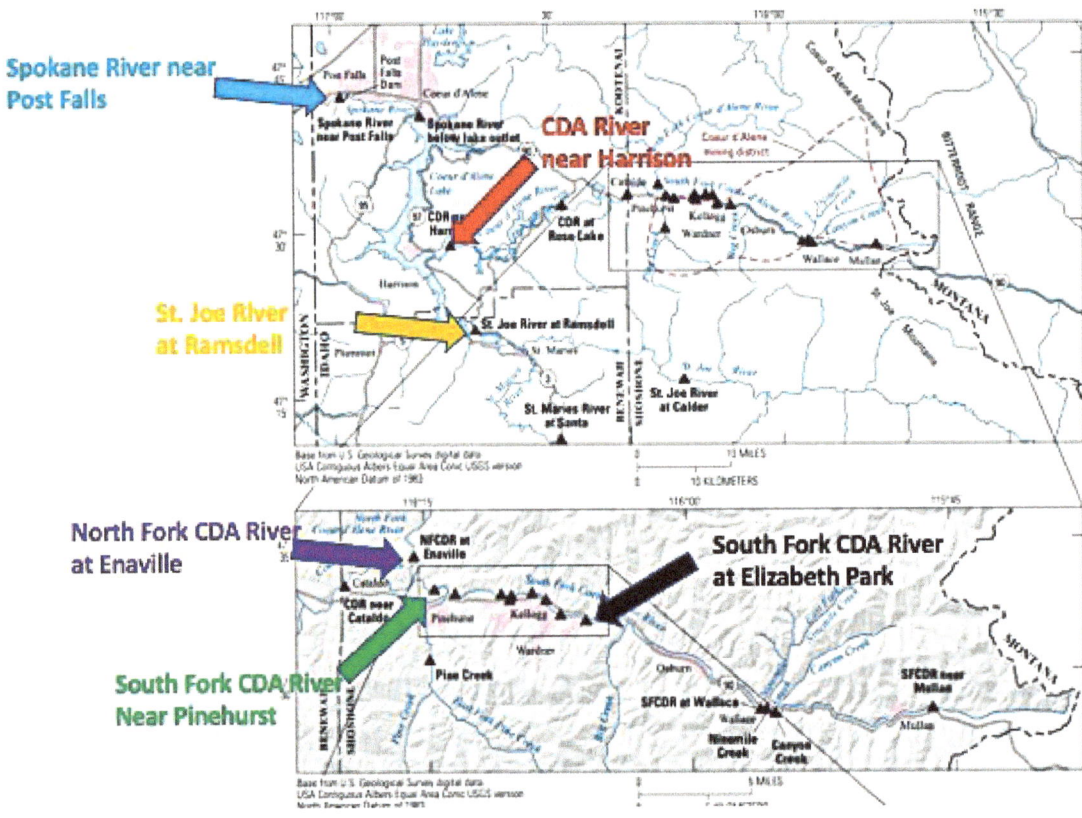

FIGURE A-1 Map of the CDA River watershed showing the locations of six river gages used in the following analysis of metals and nutrients concentration and flux trends. SOURCE: Zinsser (2020).

TABLE A-1 Station Codes for USGS Data Used in WRTDS Analysis

Gage Location	Station Code
NF CDA River at Enaville	12413000
SF CDA River at Elizabeth Park	12413210
SF CDA River near Pinehurst	12413470
CDA River near Cataldo	12413500
CDA River near Harrison	12413860[a]
St. Joe River at Ramsdell near St. Maries	12415135 (also 12415140)
Spokane River below Lake Outlet at Coeur d'Alene	12417610 (for water quality)
Spokane River near Post Falls	12419000 (for discharge)

[a]Recent record uses acoustic velocity meter discharge measurements; earlier records are based on a model that relies on stage data at Cataldo and at the Lake outlet along with channel geometry data to produce a discharge record (see discussion of this and other adjustments in Zinsser, 2020).
NOTE: NF = North Folk; SF = South Fork.

TABLE A-2 Parameter Codes for All USGS Water Quality Data Used in WRTDS Calculations

Parameter name	Parameter code
Total P	00665
Dissolved P	00671
Total N	00060
Dissolved NO_2+NO_3	00631
Suspended Sediment	80154
Total Cd	01027
Dissolved Cd	01025
Total Pb	01051
Dissolved Pb	01049
Total Zn	01092
Dissolved Zn	01090

NOTE: Cd = cadmium, N = nitrogen, NO_2 = nitrite, NO_3 = nitrate, P = phosphorus, Pb = lead, and Zn = zinc.

DESCRIPTION OF RIVER TREND CALCULATIONS

All of the river water quality trend results are developed using the USGS EGRET 3.0.7 open source software package. The R packages involved can all be downloaded for free from CRAN[1] (The Comprehensive R Archive Network). The details of the methods discussed below are described in the USGS EGRET User Guide (Hirsch and De Cicco, 2015), including the method of downloading the data from the USGS Web Service. Additionally, the Kalman filter estimates of daily fluxes use a method described by Zhang and Hirsch (2019), and further information on the Kalman filter estimates is provided at the EGRET web page for WRTDSKalman.[2] An illustrative example calculation is provided below for total phosphorus trends at the South Fork of the CDA River near Pinehurst (full list of parameter codes described in Table A-2).

The uncertainty calculations are done using the EGRETci package[3] (EGRETci 2.0.4) described by Hirsch and De Cicco (2015). Examples of these calculations are provided below. The determination of the color of the cells in the Chapter 3 summary tables (e.g., Table 3-9) is based on the likelihood results described below. The example provided below is 200 replicates, which is typical of the analyses in the report; however, for less exact results, 50 replicates could be used.

Total Phosphorus Trends Calculation Example

An illustrative example is provided here to step through calculations of total phosphorus (parameter code 00665) trends at CDA River near Pinehurst (station 12413470) from October 1988 to September 2020. R scripts, commands, and printouts are provided below, with the corresponding results shown in Figures A-2 to A-8.

---

[1] https://cran.rstudio.com/web/packages/index.html
[2] http://usgs-r.github.io/EGRET/articles/WRTDSK.html
[3] http://usgs-r.github.io/EGRETci/

Retrieve Data from USGS Web Service

```
sta <- "12413470"
param <- "00665"
startDate <- "1988-10-01"
endDate <- "2020-09-30"
Sample <- readNWISSample(sta, param, startDate, endDate)
length(Sample$Date)
Sample <- removeDuplicates(Sample)
summary(Sample)
##       Date                 ConcLow           ConcHigh            Uncen
##  Min.   :1989-07-12   Min.   :0.00400   Min.   :0.00400   Min.   :0.0000
##  1st Qu.:1998-11-01   1st Qu.:0.02100   1st Qu.:0.02000   1st Qu.:1.0000
##  Median :2001-07-30   Median :0.03300   Median :0.03150   Median :1.0000
##  Mean   :2003-03-24   Mean   :0.04276   Mean   :0.04121   Mean   :0.9526
##  3rd Qu.:2007-10-29   3rd Qu.:0.05000   3rd Qu.:0.05000   3rd Qu.:1.0000
##  Max.   :2020-09-23   Max.   :0.28000   Max.   :0.28000   Max.   :1.0000
##                                         NA's   :11
##     ConcAve            Julian           Month             Day
##  Min.   :0.00400   Min.   :50961   Min.   : 1.000   Min.   :  5.0
##  1st Qu.:0.02000   1st Qu.:54361   1st Qu.: 4.000   1st Qu.:104.8
##  Median :0.03150   Median :55362   Median : 6.000   Median :167.0
##  Mean   :0.04097   Mean   :55964   Mean   : 6.203   Mean   :173.3
##  3rd Qu.:0.05000   3rd Qu.:57644   3rd Qu.: 9.000   3rd Qu.:251.2
##  Max.   :0.28000   Max.   :62357   Max.   :12.000   Max.   :365.0
##
##     DecYear         MonthSeq        waterYear         SinDY
##  Min.   :1990    Min.   :1675    Min.   :1989    Min.   :-0.99885
##  1st Qu.:1999    1st Qu.:1787    1st Qu.:1999    1st Qu.:-0.68730
##  Median :2002    Median :1820    Median :2001    Median : 0.14154
##  Mean   :2003    Mean   :1839    Mean   :2003    Mean   : 0.05831
##  3rd Qu.:2008    3rd Qu.:1894    3rd Qu.:2008    3rd Qu.: 0.72571
##  Max.   :2021    Max.   :2049    Max.   :2020    Max.   : 1.00000
##
##      CosDY
##  Min.   :-0.9999
##  1st Qu.:-0.7925
##  Median :-0.3219
##  Mean   :-0.1872
##  3rd Qu.: 0.3593
##  Max.   : 0.9996
```

APPENDIX A

Retrieve Daily Discharge Data (code "00060")

```
Daily <- readNWISDaily(sta, "00060", startDate, endDate)
summary(Daily)
##       Date                    Q                  Julian              Month
##  Min.   :1988-10-01    Min.   :  1.642    Min.   :50677    Min.   : 1.000
##  1st Qu.:1996-09-30    1st Qu.:  3.908    1st Qu.:53599    1st Qu.: 4.000
##  Median :2004-09-30    Median :  7.702    Median :56520    Median : 7.000
##  Mean   :2004-09-30    Mean   : 15.343    Mean   :56520    Mean   : 6.523
##  3rd Qu.:2012-09-30    3rd Qu.: 19.907    3rd Qu.:59442    3rd Qu.:10.000
##  Max.   :2020-09-30    Max.   :254.852    Max.   :62364    Max.   :12.000
##
##       Day             DecYear          MonthSeq         waterYear
##  Min.   :  1.0    Min.   :1989     Min.   :1666     Min.   :1989
##  1st Qu.: 93.0    1st Qu.:1997     1st Qu.:1762     1st Qu.:1997
##  Median :184.0    Median :2005     Median :1858     Median :2004
##  Mean   :183.8    Mean   :2005     Mean   :1858     Mean   :2005
##  3rd Qu.:275.0    3rd Qu.:2013     3rd Qu.:1953     3rd Qu.:2012
##  Max.   :366.0    Max.   :2021     Max.   :2049     Max.   :2020
##
##   Qualifier                i                  LogQ                Q7
##  Length:11688       Min.   :    1      Min.   :0.4961     Min.   :  1.905
##  Class :character   1st Qu.: 2923      1st Qu.:1.3630     1st Qu.:  3.948
##  Mode  :character   Median : 5844      Median :2.0415     Median :  7.929
##                     Mean   : 5844      Mean   :2.2185     Mean   : 15.350
##                     3rd Qu.: 8766      3rd Qu.:2.9911     3rd Qu.: 20.679
##                     Max.   :11688      Max.   :5.5407     Max.   :154.974
##                                                           NA's   :6
##       Q30
##  Min.   : 2.029
##  1st Qu.: 4.227
##  Median : 9.070
##  Mean   :15.375
##  3rd Qu.:21.916
##  Max.   :88.915
##  NA's   :29
```

Retrieve the Metadata

```
INFO <- readNWISInfo(sta, param, interactive = FALSE)
INFO
##   agency_cd  site_no                             station_nm site_tp_cd
## 1      USGS 12413470 SF Coeur D Alene River nr Pinehurst, ID         ST
##     lat_va long_va dec_lat_va dec_long_va coord_meth_cd coord_acy_cd
## 1 473304.7 1161411   47.55131   -116.2363             N            1
##   coord_datum_cd dec_coord_datum_cd district_cd state_cd county_cd country
_cd
## 1          NAD83              NAD83          16       16       079
US
##             land_net_ds      map_nm map_scale_fc alt_va alt_meth_cd
## 1 SENWS32 T49N R02E B KELLOGG WEST, ID        24000   2170           M
##   alt_acy_va alt_datum_cd    huc_cd basin_cd topo_cd
## 1         10       NAVD88  17010302     <NA>       C
##                      instruments_cd construction_dt inventory_dt drain_area_v
a
## 1 YNNNYNNNNNNNNNNNNNNNNNNNNNNNNNNN              NA           NA           29
9
##   contrib_drain_area_va tz_cd local_time_fg reliability_cd gw_file_cd
## 1                   299   PST             Y           <NA>   NNNNNNNN
##   nat_aqfr_cd aqfr_cd aqfr_type_cd well_depth_va hole_depth_va depth_src_c
d
## 1        <NA>    <NA>         <NA>            NA            NA         <NA
>
##   project_no drainSqKm                             shortName staAbbrev
## 1  463015200  774.4064 SF Coeur D Alene River nr Pinehurst, ID        NA
##                                              param.nm param.
units
## 1 Phosphorus, water, unfiltered, milligrams per liter as phosphorus    mg/l
as P
##   paramShortName paramNumber constitAbbrev paStart paLong
## 1     Phosphorus       00665    Phosphorus      10     12
```

Create the eList and Estimate the WRTDS Model

```
eList <- mergeReport(INFO, Daily, Sample)
eList <- modelEstimation(eList)
##
##   estCrossVal % complete:
## 0    1    2    3    4    5    6    7    8    9    10
## 11   12   13   14   15   16   17   18   19   20
## 21   22   23   24   25   26   27   28   29   30
## 31   32   33   34   35   36   37   38   39   40
## 41   42   43   44   45   46   47   48   49   50
## 51   52   53   54   55   56   57   58   59   60
## 61   62   63   64   65   66   67   68   69   70
## 71   72   73   74   75   76   77   78   79   80
## 81   82   83   84   85   86   87   88   89   90
## 91   92   93   94   95   96   97   98   99
## Next step running  estSurfaces with survival regression:
## Survival regression (% complete):
## 0    1    2    3    4    5    6    7    8    9    10
## 11   12   13   14   15   16   17   18   19   20
## 21   22   23   24   25   26   27   28   29   30
## 31   32   33   34   35   36   37   38   39   40
## 41   42   43   44   45   46   47   48   49   50
## 51   52   53   54   55   56   57   58   59   60
## 61   62   63   64   65   66   67   68   69   70
## 71   72   73   74   75   76   77   78   79   80
## 81   82   83   84   85   86   87   88   89   90
## 91   92   93   94   95   96   97   98   99
## Survival regression: Done
multiPlotDataOverview(eList)
```

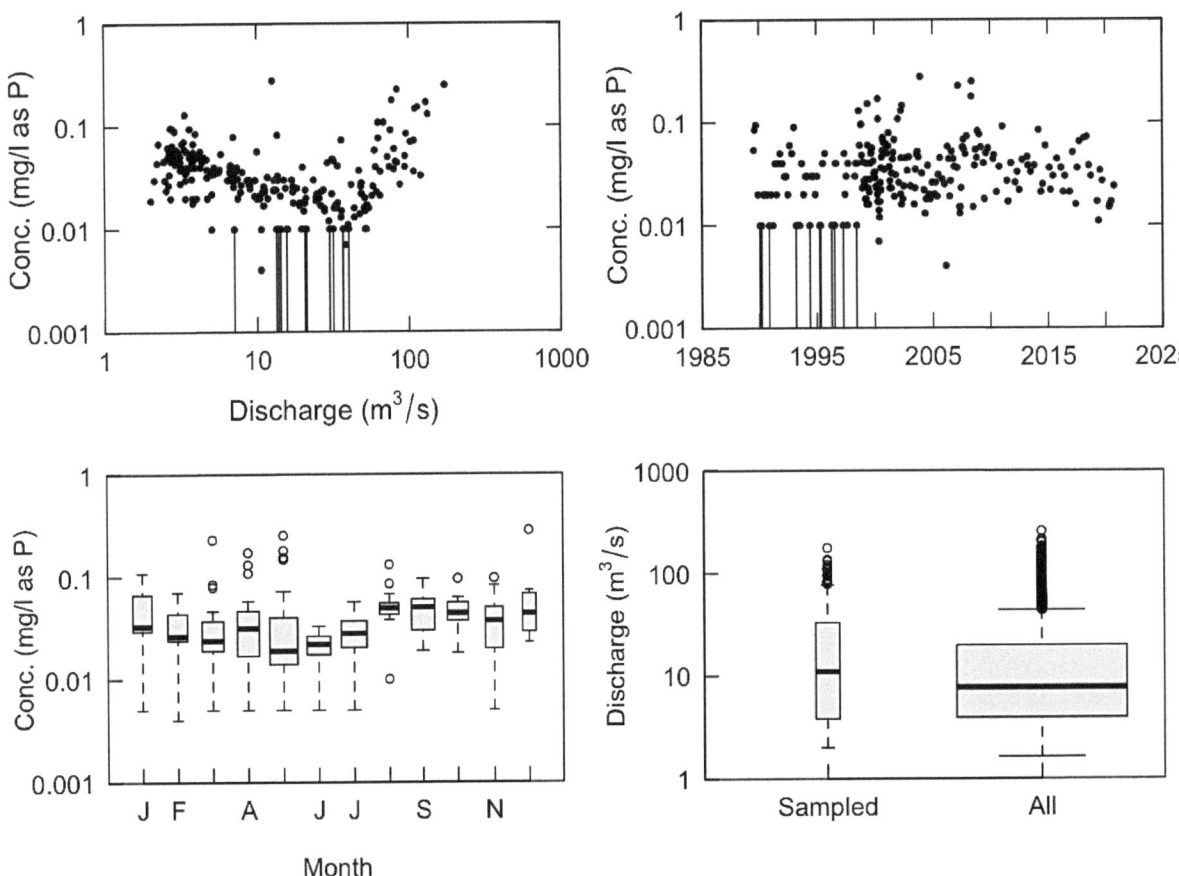

FIGURE A-2 WRTDS-generated plots of total phosphorus at the South Fork of the CDA River near Pinehurst showing discharge versus concentration (top left), concentration time series (top right), concentration binned by month of the year (bottom left), and boxplots of the discharge data (bottom right).

Calculate Error Statistics

```
errorStats(eList)
##
##   Root Mean Squared Error in natural log units =   0.557
##   Rsquared for natural log of concentration    =   0.419
##   Rsquared for natural log of flux             =   0.82
##   Standard error of estimate = 60.4 %
##   RsqLogC RsqLogF  rmse sepPercent
## 1   0.419    0.82 0.557       60.4
# this next step creates random realizations of the censored values
# they have no impact on the calculations, only impact is on graphics
eList <- makeAugmentedSample(eList)
multiPlotDataOverview(eList, randomCensored = TRUE)
```

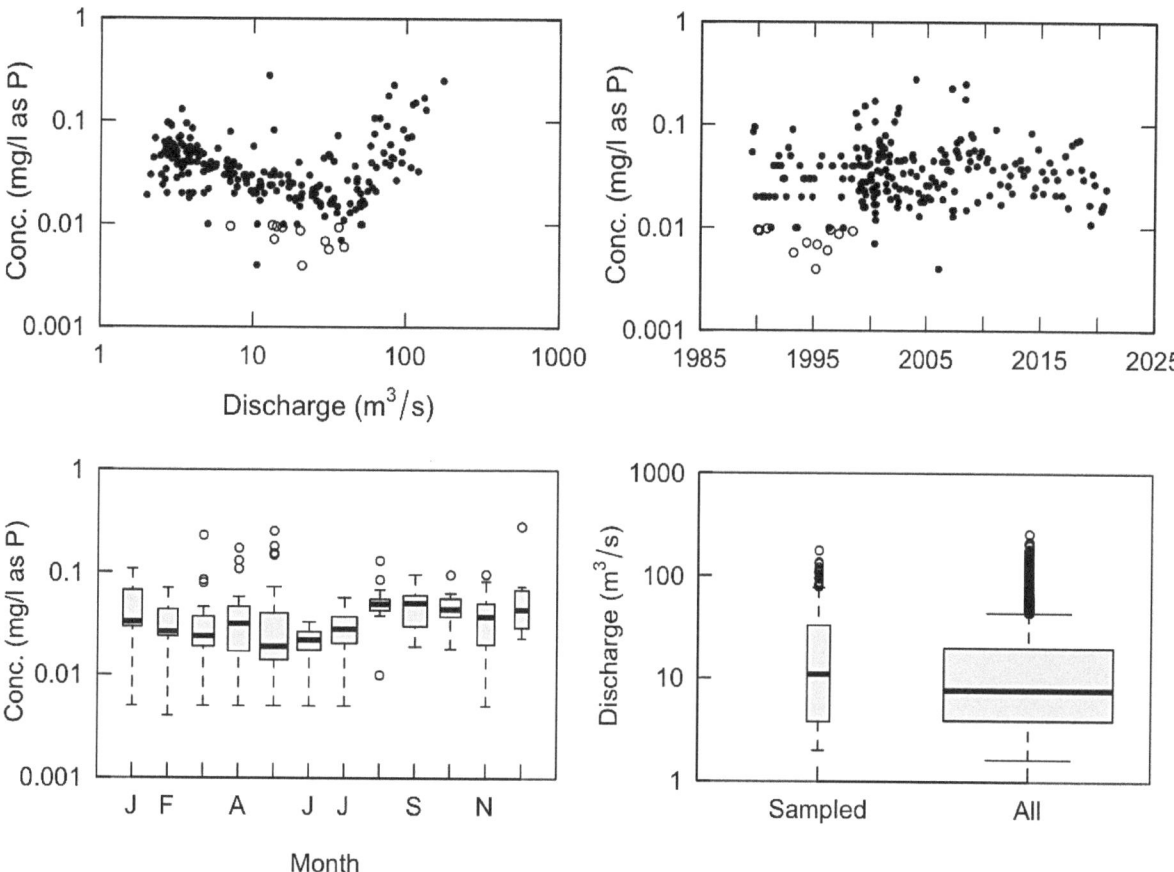

FIGURE A-3 WRTDS-generated plots of total phosphorus at the South Fork of the CDA River near Pinehurst showing discharge versus concentration (top left), concentration time series (top right), concentration binned by month of the year (bottom left), and boxplots of the discharge data (bottom right). The same plots as shown in Figure A-2, shown here with random realization of censored values.

Additional Diagnostic Figures

```
# another set of diagnostics plots
fluxBiasMulti(eList, randomCensored = TRUE)
```

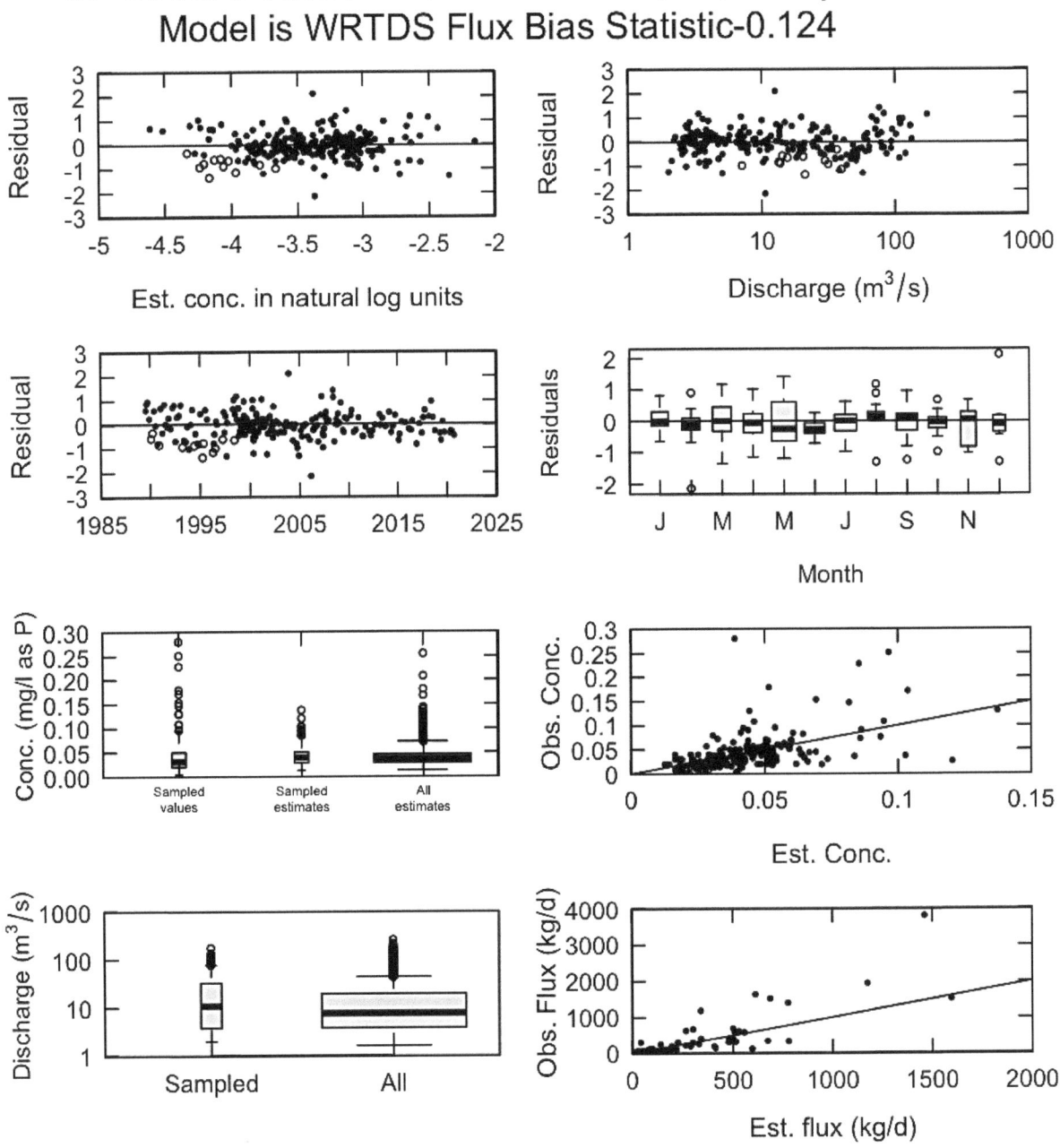

FIGURE A-4 Additional set of statistical diagnostic plots for total phosphorus trends at the South Fork (SF) of the CDA River near Pinehurst.

APPENDIX A

Kalman Filter Estimates for Concentration and Flux for Each Day

```
# now we create Kalman filter estimates for concentration and flux for each day
eList <- WRTDSKalman(eList)
## % complete:
##  0   1   2   3   4   5   6   7   8   9  10
## 11  12  13  14  15  16  17  18  19  20
## 21  22  23  24  25  26  27  28  29  30
## 31  32  33  34  35  36  37  38  39  40
## 41  42  43  44  45  46  47  48  49  50
## 51  52  53  54  55  56  57  58  59  60
## 61  62  63  64  65  66  67  68  69  70
## 71  72  73  74  75  76  77  78  79  80
## 81  82  83  84  85  86  87  88  89  90
## 91  92  93  94  95  96  97  98  99
```

Plot Trend Results and Create Summary Table of Trends (such as Table 3-9)

```
# look at trend results and annual values
# these curves make up part of figure 3-23
plotConcHist(eList, plotAnnual = FALSE, plotGenConc = TRUE)
```

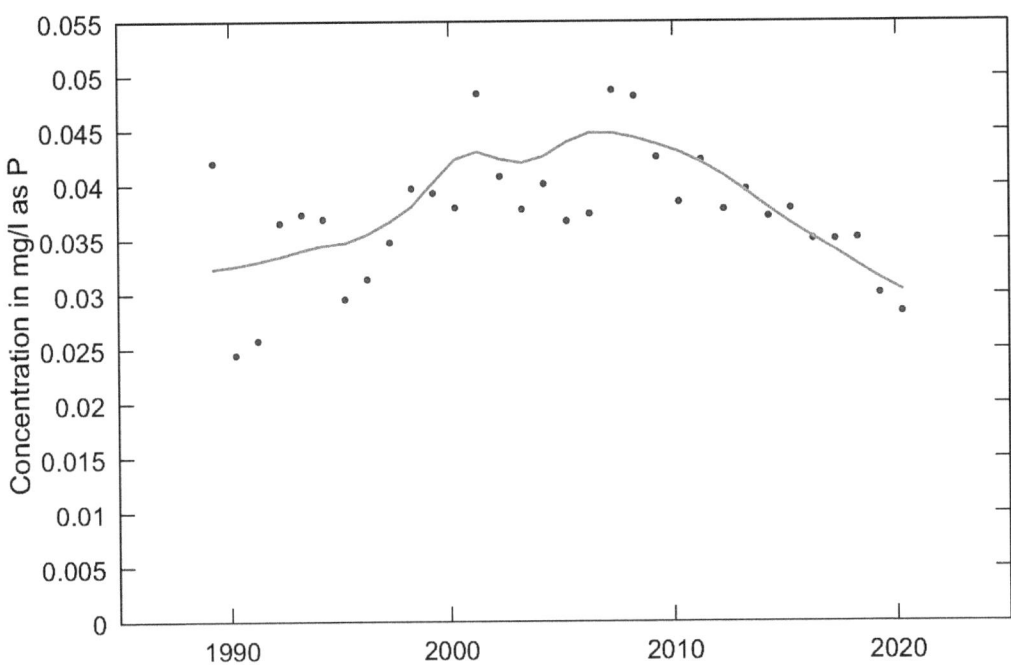

FIGURE A-5 Total phosphorus concentration as a function of water year at the South Fork of the CDA River near Pinehurst. Dots show the mean and the line shows the flow-normalized concentration.

```
plotFluxHist(eList, fluxUnit = 8, plotAnnual = FALSE, plotGenFlux = TRUE)
```

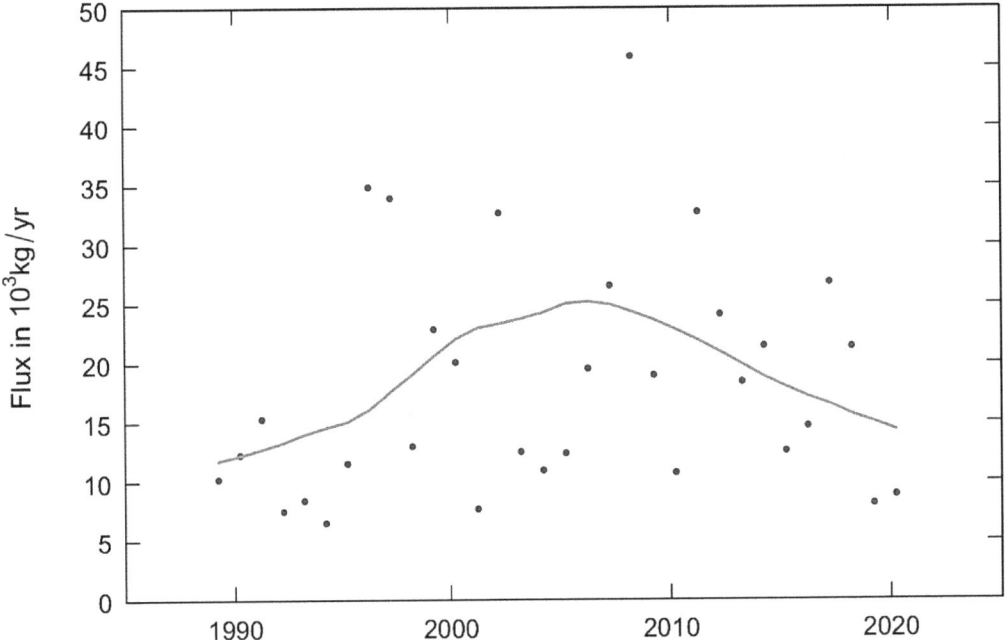

FIGURE A-6 Total phosphorus flux as a function of water year at the South Fork of the CDA River near Pinehurst. Dots show the mean and the line shows the flow-normalized flux estimate.

APPENDIX A

```
# these make up some of the values in table 3-9
tableResults(eList, fluxUnit = 8)
##
##     SF Coeur D Alene River nr Pinehurst, ID
##     Phosphorus
##     Water Year
##
##     Year    Discharge    Conc     FN_Conc      Flux     FN_Flu    GenFlux
##               cms                  mg/L                 10^3 kg/yr
##
##     1989    13.01       0.0371    0.0323       9.41     11.8      10.26
##     1990    19.67       0.0264    0.0326      12.92     12.2      12.30
##     1991    20.47       0.0265    0.0330      14.86     12.8      15.32
##     1992     8.98       0.0358    0.0334       7.15     13.3       7.54
##     1993    12.05       0.0376    0.0340       9.93     14.0       8.44
##     1994     7.84       0.0420    0.0345       7.08     14.6       6.56
##     1995    14.86       0.0318    0.0347      12.76     15.1      11.57
##     1996    23.45       0.0344    0.0355      38.24     16.1      34.91
##     1997    23.96       0.0369    0.0366      34.18     17.7      33.98
##     1998    12.39       0.0345    0.0380      11.16     19.1      13.01
##     1999    19.45       0.0390    0.0402      23.85     20.6      22.88
##     2000    16.55       0.0401    0.0424      20.42     22.1      20.11
##     2001     6.56       0.0489    0.0430       7.53     23.0       7.71
##     2002    19.36       0.0427    0.0424      32.80     23.3      32.71
##     2003    11.21       0.0407    0.0420      14.09     23.8      12.53
##     2004    11.03       0.0382    0.0426      11.15     24.3      10.96
##     2005    11.39       0.0435    0.0439      15.95     25.0      12.40
##     2006    16.66       0.0442    0.0447      24.16     25.2      19.55
##     2007    14.95       0.0470    0.0447      23.76     25.0      26.55
##     2008    18.86       0.0430    0.0443      31.73     24.3      45.92
##     2009    14.74       0.0416    0.0437      20.25     23.6      18.99
##     2010     9.53       0.0379    0.0430      10.29     22.8      10.74
##     2011    22.70       0.0435    0.0421      34.60     21.9      32.76
##     2012    20.16       0.0405    0.0409      28.01     20.9      24.10
##     2013    14.88       0.0396    0.0394      18.43     19.9      18.42
##     2014    17.14       0.0372    0.0379      20.80     18.8      21.43
##     2015    11.84       0.0394    0.0365      15.37     18.0      12.56
##     2016    14.34       0.0361    0.0352      15.80     17.2      14.65
##     2017    22.35       0.0359    0.0340      15.80     17.2      14.65
##     2018    17.67       0.0339    0.0327      28.83     16.5      26.80
##     2019    10.72       0.0317    0.0314       9.29     15.0       8.11
##     2020    12.17       0.0302    0.0303      10.07     14.3       8.84
# these also make up some of the values in table 3-9
tableChange(eList, fluxUnit = 8, yearPoints = c(1990, 2000, 2010, 2020))
##
##     SF Coeur D Alene River nr Pinehurst, ID
##     Phosphorus
##     Water Year
##
##              Concentration trends
##     time span       change      slope      change    slope
##                      mg/L       mg/L/yr      %       %/yr
##
##     1990 to 2000    0.0098     0.00098      30        3
##     1990 to 2010    0.01       0.00052      32        1.6
##     1990 to 2020   -0.0022    -7.5e-05      -6.9     -0.23
```

```
## 2000  to  2010    0.00066   6.6e-05      1.6    0.16
## 2000  to  2020    -0.012    -6e-04       -28    -1.4
## 2010  to  2020    -0.013    -0.0013      -29    -2.9
##
##
##                    Flux Trends
##    time span         change          slope         change       slope
##                    10^3 kg/yr    10^3 kg/yr /yr       %         %/yr
## 1990  to  2000       9.8            0.98            80           8
## 1990  to  2010       11             0.53            86           4.3
## 1990  to  2020       2.1            0.068           17           0.56
## 2000  to  2010       0.76           0.076           3.4          0.34
## 2000  to  2020       -7.8           -0.39           -35          -1.8
## 2010  to  2020       -8.5           -0.85           -37          -3.7
```

Total Phosphorus Trends Uncertainty Calculation Example

For the same example of total phosphorus trends at the South Fork of the CDA River near Pinehurst, calculations of uncertainty are detailed below.

Quantifying Uncertainty of the Trend Results

```
# we will set up the nBoot argument here
# to get reasonably good uncertainty estimates nBoot should be >= 200
# but to just get a very rough idea we can work with nBoot as small as 50
nBoot <- 200
# first we will look at the trends from 1990 - 2020
year1 <- 1990
year2 <- 2020
# we will not be using Generalized Flow Normalization
pairsOut <- runPairs(eList, year1, year2, windowSide = 0, oldSurface = TRUE)
##
##    SF Coeur D Alene River nr Pinehurst, ID
##    Phosphorus
##    Water Year
##
##  Change estimates  2020  minus  1990
##
##  For concentration: total change is  -0.00225 mg/L
##  expressed as Percent Change is  -6.9 %
##
##  Concentration v. Q Trend Component   -6.9 %
##        Q Trend Component               0 %
##
##
##  For flux: total change is  0.00205 million kg/year
##  expressed as Percent Change is  17 %
##
##  Concentration v. Q Trend Component   17 %
##        Q Trend Component               0 %
##
##        TotalChange    CQTC    QTC     x10     x11     x20     x22
## Conc   -0.0022     -0.0022     0    0.033   0.033   0.030   0.030
## Flux    0.0021      0.0021     0    0.012   0.012   0.014   0.014
```

```
# the lengthy output starting with "iBoot..." is just the results of each of
the bootstrap replicates
# it can be useful in those rare cases where the method runs into a numerical
issue and these replicates
pairsBootOut <- runPairsBoot(eList, pairsOut, nBoot = nBoot)
##
##   iBoot, xConc and xFlux 1 -0.0205327 0.001477626
[...]
##   Change estimates are for  2020   minus   1990
##
## Should we reject Ho that Flow Normalized Concentration Trend = 0 ? Do Not
Reject Ho
##   best estimate of change in concentration is -0.00225 mg/L
##    Lower and Upper 90% CIs -0.016482   0.006191
##   also 95% CIs -0.018901   0.008980
##   and 50% CIs -0.010393   0.000381
##   approximate two-sided p-value for Conc       0.52
##   Likelihood that Flow Normalized Concentration is trending up = 0.261   is
trending down = 0.739
##
## Should we reject Ho that Flow Normalized Flux Trend = 0 ? Do Not Reject Ho
##   best estimate of change in flux is  0.00205 10^6 kg/year
##    Lower and Upper 90% CIs -0.003983   0.007787
##   also 95% CIs -0.005889   0.008888
##   and 50% CIs -0.000616   0.003356
##   approximate two-sided p-value for Flux       0.69
##   Likelihood that Flow Normalized Flux is trending up = 0.654   is trending
down = 0.346
##
##   Upward trend in concentration is unlikely
##   Upward trend in flux is about as likely as not
##   Downward trend in concentration is likely
##   Downward trend in flux is about as likely as not
# next we will look at trends from 2000 - 2020
year1 <- 2000
pairsOut <- runPairs(eList, year1, year2, windowSide = 0, oldSurface = TRUE)
##
##      SF Coeur D Alene River nr Pinehurst, ID
##      Phosphorus
##      Water Year
##
##   Change estimates   2020   minus   2000
##
##   For concentration: total change is   -0.012 mg/L
##   expressed as Percent Change is   -28 %
##
##   Concentration v. Q Trend Component    -28 %
##        Q Trend Component                  0 %
##
##
##   For flux: total change is   -0.00776 million kg/year
##   expressed as Percent Change is   -35 %
##
##   Concentration v. Q Trend Component    -35 %
##        Q Trend Component                  0 %
##
```

```
##         TotalChange    CQTC   QTC    x10     x11    x20    x22
## Conc        -0.0120 -0.0120     0  0.042  0.042  0.030  0.030
## Flux        -0.0078 -0.0078     0  0.022  0.022  0.014  0.014
pairsBootOut <- runPairsBoot(eList, pairsOut, nBoot = nBoot)
##
##   iBoot, xConc and xFlux 1 -0.01955794 -0.009926056
[...]
##  Change estimates are for  2020  minus  2000
##
## Should we reject Ho that Flow Normalized Concentration Trend = 0 ? Reject Ho
##   best estimate of change in concentration is  -0.012 mg/L
##    Lower and Upper 90% CIs -0.02382 -0.00760
##    also 95% CIs -0.02538 -0.00606
##    and 50% CIs -0.01872 -0.01127
##   approximate two-sided p-value for Conc       0.01
## * Note p-value should be considered to be < stated value
##   Likelihood that Flow Normalized Concentration is trending up = 0.00249   is trending down = 0.998
##
## Should we reject Ho that Flow Normalized Flux Trend = 0 ? Reject Ho
##   best estimate of change in flux is -0.00776 10^6 kg/year
##    Lower and Upper 90% CIs -0.01361 -0.00398
##    also 95% CIs -0.01481 -0.00336
##    and 50% CIs -0.01006 -0.00651
##   approximate two-sided p-value for Flux       0.01
## * Note p-value should be considered to be < stated value
##   Likelihood that Flow Normalized Flux is trending up = 0.00249   is trending down = 0.998
##
##   Upward trend in concentration is highly unlikely
##   Upward trend in flux is highly unlikely
##   Downward trend in concentration is highly likely
##   Downward trend in flux is highly likely
# finally we will look at trends from 2010 - 2020
year1 <- 2010
pairsOut <- runPairs(eList, year1, year2, windowSide = 0, oldSurface = TRUE)
##
##     SF Coeur D Alene River nr Pinehurst, ID
##     Phosphorus
##     Water Year
##
##   Change estimates  2020  minus  2010
##
##   For concentration: total change is  -0.0127 mg/L
##   expressed as Percent Change is  -29 %
##
##   Concentration v. Q Trend Component    -29 %
##        Q Trend Component                  0 %
##
##
##   For flux: total change is  -0.00852 million kg/year
##   expressed as Percent Change is  -37 %
##
##   Concentration v. Q Trend Component    -37 %
##        Q Trend Component                  0 %
##
```

APPENDIX A

```
##          TotalChange   CQTC QTC    x10   x11   x20   x22
## Conc       -0.0127  -0.0127   0 0.043 0.043 0.030 0.030
## Flux       -0.0085  -0.0085   0 0.023 0.023 0.014 0.014
pairsBootOut <- runPairsBoot(eList, pairsOut, nBoot = nBoot)
##
##   iBoot, xConc and xFlux 1 -0.01958718 -0.01214326
[…]
##   Change estimates are for  2020   minus   2010
##
## Should we reject Ho that Flow Normalized Concentration Trend = 0 ? Reject Ho
##  best estimate of change in concentration is -0.0127 mg/L
##    Lower and Upper 90% CIs -0.02335 -0.00783
##   also 95% CIs -0.02633 -0.00681
##   and 50% CIs -0.01860 -0.01181
##   approximate two-sided p-value for Conc      0.01
## * Note p-value should be considered to be < stated value
##   Likelihood that Flow Normalized Concentration is trending up = 0.00249  is trending down = 0.998
##
## Should we reject Ho that Flow Normalized Flux Trend = 0 ? Reject Ho
##  best estimate of change in flux is -0.00852 10^6 kg/year
##    Lower and Upper 90% CIs -0.01443 -0.00365
##   also 95% CIs -0.01520 -0.00215
##   and 50% CIs -0.01188 -0.00728
##   approximate two-sided p-value for Flux      0.01
## * Note p-value should be considered to be < stated value
##   Likelihood that Flow Normalized Flux is trending up = 0.00249  is trending down = 0.998
##
##   Upward trend in concentration is highly unlikely
##   Upward trend in flux is highly unlikely
##   Downward trend in concentration is highly likely
##   Downward trend in flux is highly likely
```

```
# and now we will do 90% confidence intervals on the whole time series
seriesOut <- runSeries(eList, windowSide = 0, oldSurface = TRUE)
CIAnnualResults <- ciCalculations(eList, nBoot = nBoot, blockLength = 200, widthCI = 90, verbose = FALSE)
##
## Running the EGRET modelEstimation function first to have that as a baseline for the Confidence Bands
plotConcHistBoot(seriesOut, CIAnnualResults, plotAnnual = FALSE)
```

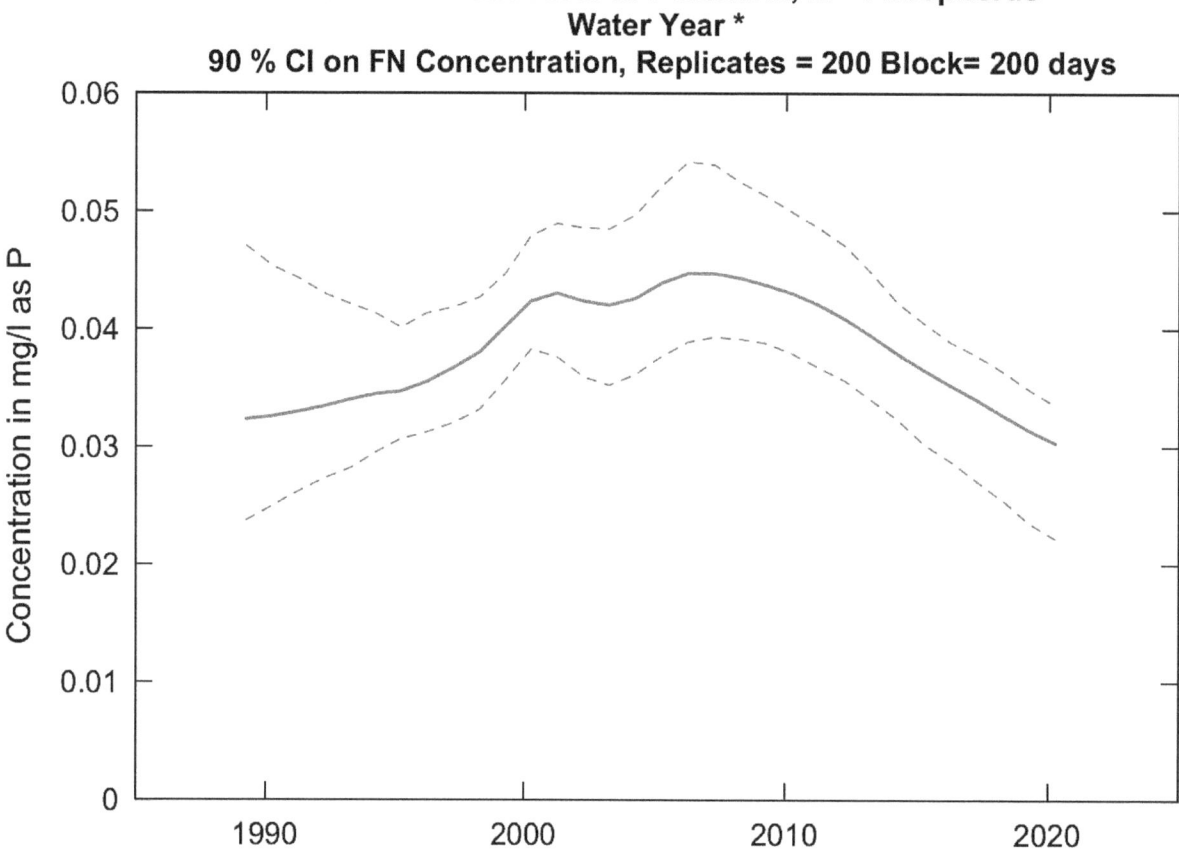

FIGURE A-7 Total phosphorus concentration at the South Fork of the CDA River near Pinehurst as a function of water year. Solid line shows the flow-normalized concentration, and the dotted lines show the 90 percent confidence interval.

```
plotFluxHistBoot(seriesOut, CIAnnualResults, plotAnnual = FALSE)
```

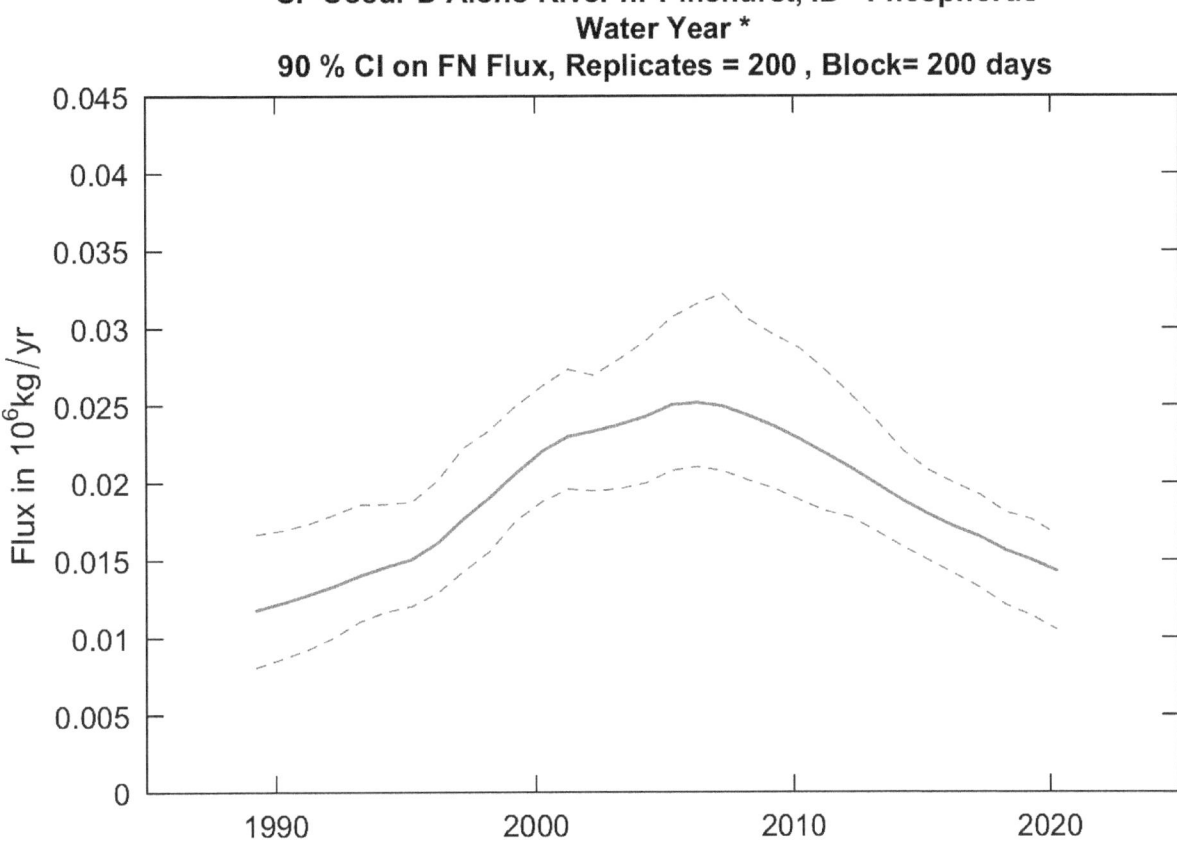

FIGURE A-8 Total phosphorus flux at the South Fork of the CDA River near Pinehurst as a function of water year. Solid line shows the flow-normalized flux, and the dotted lines show the 90 percent confidence interval.

REFERENCES

Hirsch, R. M., D. L. Moyer, and S. A. Archfield. 2010. Weighted regressions on time, discharge, and season (WRTDS), with an application to Chesapeake Bay river inputs 1. *Journal of the American Water Resources Association* 46(5):857–880.

Hirsch, R. M., and L. A. De Cicco. 2015. User guide to Exploration and Graphics for RivEr Trends (EGRET) and dataRetrieval: R packages for hydrologic data (version 2.0, February 2015): USGS Techniques and Methods Book 4, Chapter A10, 93 p. https://dx.doi.org/10.3133/tm4A10.

Zhang, Q., and R. M. Hirsch. 2019. River water-quality concentration and flux estimation can be improved by accounting for serial correlation through an autoregressive model. *Water Resources Research* 55:9705–9723. https://doi.org/10.1029/2019WR025338.

Zinsser, L. M. 2020. Trends in Concentrations, Loads, and Sources of Trace Metals and Nutrients in the Spokane River Watershed, Northern Idaho, Water Years 1990-2018. USGS Scientific Investigations Report 2020-5096.

Appendix B

In-Lake Analysis Methodology for Metals, Nutrients, and Dissolved Oxygen

This appendix details the committee's methodological trends analyses of CDA Lake data on dissolved oxygen and nutrients (Chapter 5) and metals (Chapter 6).

DISSOLVED OXYGEN TRENDS ANALYSIS METHODOLOGY

The first step in the analyses of dissolved oxygen trends in Chapter 5 was screening out problematic raw data and then standardizing the analyses within each sampling station to maximize data comparability. When analyzing the vertical profile data, the committee started by screening out any data collected on sonde upcasts (see Chapter 8). An upcast is when the sonde is retrieved from the bottom of the profile to the top while recording data. This is evident by comparing the time stamp for sonde readings to the corresponding depths. Upcast data are invalid because when the probe is being pulled upward through the water column it creates a void behind it, which gets filled with water that has been turbulently mixed and displaced upward. Conversely, downcast data are collected as the probe is being lowered through the water column; therefore, the sensors that are located on the bottom of the probe encounter laminar conditions as the sonde is lowered through the water column. Therefore, upcast data are very noisy and offset compared to downcast data. In most cases, only a small portion of the profile data was discarded. However, many profiles collected during 2016 were taken as upcasts. Upcast data should not be used, and people collecting profile data from CDA Lake in the future should be warned not to collect vertical profiles in the reverse order.

The first goal when analyzing the profile data was to characterize seasonal and depth trends in the dissolved oxygen data collected at each station. Once the times and depths with the lowest oxygen concentrations were identified, the committee used this information to design its long-term trend analyses. To look for general seasonal/depth trends in these data, the committee sorted all of the profile data collected by their Julian dates (i.e., days 1–365 within a year) and plotted all of the data collected from different years as a single annual time series. Because there was considerable year-to-year heterogeneity in sampling dates and even sampling depths, the committee binned these data into weekly and 1-meter intervals. When doing these preliminary analyses, it was evident that the script used to interpolate the data for these plots (rLakeAnalyzer[1]) created odd outcomes

---

[1] https://cran.r-project.org/web/packages/rLakeAnalyzer/rLakeAnalyzer.pdf

in regions of the overall time-depth sampling matrix that were poorly represented by real data. For example, at station C1 the majority of the vertical profiles ended at a depth of 41 m, but a small number of profiles extended down to 42–44 m (n = 6 for the entire C1 dataset). Because the deepest profile extended to 44 m, the script output estimated results for all dates down to this depth. In the vast majority of cases, the results for depths 42–44 m were based on interpolations, and some of these interpolated data were suspiciously different from the observed data at adjacent depths. Therefore, the committee changed all of the 42–44 m depths in the dataset to 41 m to eliminate strange interpolated values. Because most profiles collected at the C4 station ended at approximately 39 m, all of the observations below that depth at this station were rounded to this depth. Similarly, the deeper data collected from station C5 were rounded to 13 m, and the deeper data from station C6 were rounded to 11 m. Then the committee used the R-script (rLakeAnalyzer) to create annual time-depth dissolved oxygen isopleths for stations C1 and C4, where the most data were available, and for stations C5 and C6, where dissolved oxygen depletion was most evident. Because these plots are based on a time by depth calculated matrix of oxygen concentrations, they should not be biased by inconsistencies in the temporal and spatial frequency of sample collection.

Once the annual patterns in depth average dissolved oxygen concentrations were characterized, the committee analyzed the long-term dataset for intra-annual trends for the same sites. In this case, the data were binned by 1-meter and monthly increments. As already noted, the occasional deeper sonde readings were rounded to the most common deepest sonde reading for each sampling station. When examining the long-term vertical profile datasets, it was also apparent that ≈ 20 percent of sampling dates utilized a greatest sampling depth that was less than the previously noted most common sampling depth. For example, at the C1 station, 40 m was the most common deepest sampling depth, but in 21 percent of the profiles, the deepest depth sampled was shallower. Missing 40-m data were common before 2010 and rare afterward. Because the script used to interpolate the depth profiles is quite sensitive when the deepest depth is missing, the committee estimated the oxygen concentrations for 40 m using the next available depth, which in 75 percent of cases was either 38 or 39 m. The committee estimated the missing 40-m data by generating statistical relationships between the oxygen concentrations at lesser depths and at 40 m when data were available for both and applying this equation to estimate the 40-m values when direct observations were not available. For example, when predicting the 40-m oxygen concentrations from the 35-m data, the fit between the two datasets was strong ($r^2 = 0.96$), with predicted dissolved oxygen concentration at 40 m being 0.74 mg L^{-1} lower when the observed concentration at 35 m averaged 7.0 mg L^{-1} (the lower end of observed values) and 0.17 mg L^{-1} lower when the observed concentration at 35 m averaged 12.0 mg L^{-1} (the upper end of observed values).

These calculations generated a series of vertical dissolved oxygen profiles interpolated at 1-meter increments from the surface to the most common deepest depth sampled for each month from January 1991 to April 2021 for stations C1, C4, and C5. These profiles were then used to generate monthly matrices of surface and bottom dissolved oxygen concentrations for all of the years for which data were available. For stations C1 and C4, which were deep, the upper average was for 1–10 m and roughly represented the surface mixed layer or epilimnion when CDA Lake is stratified. The bottom layer for these stations was the deepest 10 m for which data were available. Ideally, the bottom layer average would only represent the area immediately above the sediment–water interface because the presence or absence of oxygen at this interface determines the fate of many biogeochemical reactions in the lake sediments. However, the exact depth of this interface was not known when analyzing these data, and the availability of oxygen data at the deepest depths in these profiles was not consistent. At the C5 station, which is much shallower, the surface was represented by the 1–5 m average and the bottom was represented by the 14–17 m average. The dissolved oxygen data for the C6 station were analyzed slightly differently than described above.[2] For this station, the surface averages were calculated from all data collected at depths ≤ 4 m, and the bottom averages were

---

[2] C6 data were analyzed slightly differently because this was the first station analyzed (since it had the lowest O$_2$), and because it was shallow, and using the r-script to interpolate the data would not have saved time.

calculated from all data collected at depths ≥ 8 m (i.e., 8–11 m) without any data interpolation. When two profiles were available for the same month (as was often the case for June–November since 2011), these values were averaged.

METHODS FOR IN-LAKE TREND STATISTICAL ANALYSIS

Locally Weighted Scatterplot Smoothing

One important feature of any method for evaluating long-term trends in water quality data is that the method must be able to reveal the presence of non-monotonic trends. As discussed in Chapter 3 and Appendix A, for the river data, some of the constituents of interest (specifically, lead and phosphorus) at some sites, when characterized using the Weighted Regressions on Time, Discharge, and Season method, revealed increasing trends over a decade or two, followed by decreasing trends in the most recent decade. Given that pattern in river inputs to the Lake, it is reasonable to assume that similar non-monotonic trends are possible for lake chemistry data; hence, the committee used locally weighted scatterplot smoothing.

The first step in processing the data was to plot them and determine the date at which the detection limits appear to be relatively stable, and the frequency of data collection is generally monthly or more frequent and includes both surficial water and the deeper water. All of the datasets considered here end at the end of calendar year 2020, but the starting year varies from as early as 1991 to as recent as 2004 (the record lengths are given in the tables associated with these results). The data were separated into deeper water and shallower water. For C1 and C4 (sampled by the Idaho Department of Environmental Quality), the definition of deep is > 21 meters. For C5, C6, and SJ1 (sampled by the CDA Tribe) the shallower category is denoted as euphotic zone or epilimnion (typically with sampled depths of 6 to 10 meters), and other samples are referred to as bottom water, and their sampling depths are not recorded in the database. Censored data (those observations reported as "less than" some reporting limit) were treated as values that are half the reporting limit. If the number of censored values were large, this would be an inappropriate method, but in these datasets the percentage of data points that are censored is rather low, and thus this approach makes little difference (and is certainly preferable to deleting the data or treating them as values of zero or the reporting limit). Of the 19 datasets evaluated here, 6 of them had no censored data, the median percentage of censored data was 0.7, and the maximum was 5 percent. The metals datasets from C6 and SJ1 had at least 52 percent of their data censored and most had 94–100 percent of their data censored; hence, they were eliminated from consideration for that reason.

Locally weighted scatterplot smoothing uses the loess function in R (based on Cleveland et al., 2017) to show the change in concentration as a function of time. It was applied separately to the shallow and deep datasets. The smoothing parameter (called the "span") was set to its default value of 0.75 and the fitting used a second-degree polynomial. The data were pre-processed by computing a median value for each day in the record, for both the shallow and deep water datasets, because there are often replicate samples taken on any given day. The fitting was applied to the logarithms of these daily median concentration values, and, as such, the fitted line is an estimate of the median of the data (rather than the mean). The reason the logarithms are used is to avoid oversensitivity to some of the extreme high values and to make the model residuals more nearly normal. The idea of this approach (loess on the logarithms) is to show the central tendency of the data as it changes over time.

Looking at the plotted data points, particularly from 2017 to 2020, many of the datasets suggest very steep downward trends in these final years. By design, these loess curves are designed to avoid abrupt changes in slope and tend de-emphasize what might be rapid changes in values near the start or end of the dataset. One could apply a loess smooth to the data with a substantially narrower smoothing window (lower span value). This would show the downward trends in the more recent years as being substantially steeper, but it would be intuitively difficult to argue that the many oscillations that would appear are realistic interpretations of changes in the system. This steep downward trend in the final years of some of these datasets is rather unexpected. Reevaluating the data after another one or two years have gone by will be critical to evaluating how the system is changing. The amounts of change presented in the subsequent tables are based on these loess curves and are an average of the changes in the shallow and deep waters. For example, "Change over past 10 years" is computed as follows:

$$\Delta_s = 100 \cdot \left[\frac{c_s(t_m - 10) - c_s(t_m)}{c_s(t_m - 10)}\right]$$

$$\Delta_d = 100 \cdot \left[\frac{c_d(t_m - 10) - c_d(t_m)}{c_d(t_m - 10)}\right]$$

$$\Delta = (\Delta_s + \Delta_d)/2$$

where:

Δ and Δ_d are the change in percent for the shallow and deep water, respectively;

$c_s(t_m)$ is the loess estimate of median concentration for the shallow data at time t_m, which is the maximum of the time variable in the loess model expressed in decimal years (typically about 2020.95, near the end of calendar year 2020);

$c_s(t_m - 10)$ is the estimated median at a time 10 years before t_m; and

c_d values are the loess estimates for the median concentration for the deep water data.

The loess model results are also used to provide estimates of the median at the end of 2020 for both the shallow and deep water data, in order to quantify differences among sites and between the shallow and deep water data.

Seasonal Kendall Test

The second approach uses the Seasonal Kendall test (Hirsch et al., 1982), which is widely used in environmental science applications for datasets that are expected to exhibit seasonality. The test is based on comparisons of data for each season (typically a month) from each of the years in the record and does not make comparisons across seasons. For example, it compares June values from one year to June values from another year but not June values to October values. It assumes that the probability distribution for June is different from the probability distribution for October. It is a non-parametric test, which means that it requires very minimal assumptions regarding the data. Specifically, no probability distribution is assumed for the data, no particular seasonal pattern (such as a sine wave) is assumed, and the trend is not assumed to be linear with time. However, the test is only a test for monotonic trends (meaning that a dataset with a trend that looks like a letter "U" or an inverted "U" would not be seen as having a trend). It also is not sensitive to substantial differences in the amount of trend in different seasons. For example, if six months showed upward trends and six months showed downward trends, the result might indicate that there is no trend in the record. Censored data create no particular problems for the test provided the reporting limits do not change over time (censoring is minor here, and the reporting limits do not change greatly over time). The test is performed separately on the median value for each sampled day, both for the deep water dataset and the shallow water dataset.

The implementation used in this analysis (the kendallSeasonalTrendTest in the EnvStats package in R) is documented in Millard (2013). The purpose for using it here is simply to evaluate if the observed trend evidence is stronger than what we might expect by chance alone if the true conditions were totally without trend. The results are categorized in the tables by colors, where the deep blue (for downward trends) and deep red (for upward trends) colors indicate trends that are statistically highly significant ($p < 0.01$). The lighter shades of blue and red indicate trends for which the p values lie between 0.10 and 0.01 (moderate statistical significance). Those trends that do not attain a $p < 0.05$ are shown with no shading.

Commonly, when the Seasonal Kendall test is used, the seasons identified are equivalent to the calendar months. This application departs from that in two ways. The first departure arises because sampling in the months of January and February is rather limited. Thus, data from those two months were grouped into a single two-month period while the rest of the year the periods are equivalent to the calendar months. The other departure is based on the recognition that there is typically a difference in the distributions of the data between the shallow and the deep data. This can be seen by the rather regular separation of the red and blue curves on some of the loess plots. As such, the "seasons" used here are actually defined by a particular depth group and month of the year. So, in this application, for each site and analyte pair, the number of "blocks" of data is 22. (In statistics, a "block" is a subset of the full dataset within which comparisons are made, but comparisons between blocks are not made.) The blocks are a combination of the 11 time periods and the two depth groups. The Seasonal Kendall test then performs a Mann-Kendall trend test on each of the 22 blocks of data and sums these results to form an overall test statistic, and the variance of that test statistic under the null hypothesis (the null hypothesis is that all of the 22 blocks are trend free) is the sum of the variances for each of the 22 blocks. The p-value for the test is based on the overall test statistic and its variance. It has been shown that this test statistic is well approximated by a normal distribution under the null hypothesis, regardless of the underlying distribution of the data.

The Seasonal Kendall test for each dataset is done twice: the first is on the full duration of the dataset and the second is on the last 10 years of the dataset (2011–2020). The Seasonal Kendall test does not directly inform us as to differences among the trends for the different blocks. For example, there may be trends in some months and not in others, and there could even be positive trends in some months and negative trends in others. These types of comparisons can all be made from the full set of outputs, but they were not evaluated or reported here. As part of a long-term program of trend studies to be carried out on the Lake, these kinds of differences would be a topic of interest.

In the analyses of lake dissolved oxygen (Table 5-10) and lake chlorophyll (Table 5-11), the Seasonal Kendall test was used. In the case of dissolved oxygen, the test was run on all 12 months but only for the deep water values. The slopes are based on the Theil-Sen slope estimator on the mean concentration for that month. Trend slopes are expressed as mg/L/yr, and the color codes are based on the significance of the trend for that month. These results are then summarized by the Seasonal Kendall test, and the overall slope is computed as the median of all pairwise slopes computed over all 12 months. These trend results were computed using the rkt function in the rkt R-package. The same approach was used for the shallow water chlorophyll data except for the fact that the seasons were collapsed to a total of 10. That is, because data are very scarce in the months of November through February, the months of November and December were combined into an individual season and the months of January and February were combined into an individual season, and all of the eight remaining months were defined as seasons. Here again, the rkt function was used to analyze the data.

REFERENCES

Cleveland, W. S., E. Grosse, and W. M. Shyu. 2017. Local regression models. Pp. 309–376 *In: Statistical models in S*. Routledge.

Hirsch, R. M., J. R. Slack, and R. A. Smith. 1982. Techniques of trend analysis for monthly water quality data. *Water Resour. Res.* 18(1):107–121. doi:10.1029/WR018i001p00107.

Millard, S. P. 2013. Hypothesis Tests. *In:* EnvStats. Springer, New York, NY. https://doi.org/10.1007/978-1-4614-8456-1_7.

Appendix C

Biographical Sketches of Committee Members

Samuel N. Luoma *(Chair)* is a research ecologist with the Institute of the Environment, University of California, Davis. Since 2007, he has been emeritus from the U.S. Geological Survey, he was a research associate on the faculty of The Natural History Museum, London, UK (2007–2013), and he was the first lead scientist for the CALFED Bay-Delta Program, California (2000–2003). Dr. Luoma has published more than 200 peer-reviewed papers, mostly on the fate, bioavailability, and effects of metal and metalloid contaminants in aquatic environments, along with a 2008 textbook, *Metal Contamination in Aquatic Environments: Science and Lateral Management*. His studies on the environmental implications of mining began in the late 1970s and continue today, including publications on studies in Southwest England and the Clark Fork River, Montana; consulting in South Africa and British Columbia, Canada; and review committees on Sudbury, Canada, and mountain-top mining in the southeastern United States. He has served on seven Science Advisory Board subcommittees for the U.S. Environmental Protection Agency as well as several committees of the National Academies of Sciences, Engineering, and Medicine, most recently the Committee on Progress toward Restoring the Everglades: the Fourth Biennial Review (2010–2012). He received his B.S. and M.Sc. in zoology from Montana State University and his Ph.D. in zoology from the University of Hawaii.

Robert L. Annear is a senior principal at Geosyntec Consultants and an environmental engineer with more than 20 years of experience in hydrodynamic and water quality modeling. His additional areas of professional expertise are water quality data collection and analyses, climate change impacts analyses, watershed hydrology and pollutant load modeling, sediment transport modeling, and lake limnology and ecology. While at Geosyntec he has had the opportunity to work on projects for the National Academies of Sciences, Engineering, and Medicine through the National Cooperative Highway Research Program. He has conducted numerous surface water data and model peer reviews for agencies such as the U.S. Environmental Protection Agency (EPA), Idaho Department of Environmental Quality, Oregon Department of Environmental Quality, Washington Department of Ecology, and U.S. Bureau of Reclamation. Dr. Annear has also served as a reviewer for various water resource and hydrologic journals and has participated in EPA national water quality grant review panels. Dr. Annear received his B.S. in aerospace engineering from Boston University and his M.S. and Ph.D. in civil and environmental engineering from Portland State University.

William A. Arnold is the Distinguished McKnight University Professor and the Joseph T. and Rose S. Ling Professor in the Department of Civil, Environmental, and Geo-Engineering at the University of Minnesota. His expertise is in aquatic chemistry, and his research focuses on the fate and transformation of pollutants in natural and engineered aquatic systems. Specific research areas include studying the kinetics, pathways, and mechanisms of anthropogenic chemical reactions that occur at surfaces or via photochemical processes; detecting analytes in water and sediment matrices; developing new remediation/containment techniques; assessing sources of disinfection byproducts; and using computational chemistry techniques to predict and/or explain experimental observations. Dr. Arnold has published more than 140 peer-reviewed journal articles and co-authored a *Water Chemistry* textbook. He is an associate editor for *Environmental Science and Technology Letters* and serves on the Board of Directors of the Association of Environmental Engineering and Science Professors. He served on the National Academies of Sciences, Engineering, and Medicine Committee on Future Options for Management in the Nation's Subsurface Remediation Effort. Dr. Arnold received a B.S. in chemical engineering from the Massachusetts Institute of Technology, an M.S. in chemical engineering from Yale University, and a Ph.D. in environmental engineering from the Johns Hopkins University.

Michael T. Brett is a professor in the Department of Civil and Environmental Engineering at the University of Washington. Dr. Brett's main area of scientific expertise is biological limnology, specifically food web ecology, fatty acid trophic transfer, diet estimation, lake biogeochemical modeling, model assessment, mass balance, and eutrophication control. Dr. Brett has completed several studies in the Spokane River basin, including studies on the biological availability of phosphorus in the effluents of advanced phosphorus removal wastewater treatment plants, on the modeled and observed effects of phosphorus loading on the biological limnology and oxygen inventory of Lake Spokane, and on the modeled and observed climate forcing impacts on Lake Spokane and the Spokane River. Dr. Brett currently serves on the Utah Lake Water Quality Study External Science Panel working to find a solution for chronic toxic cyanobacteria blooms in Utah Lake. He received a B.Sc. in fisheries biology at Humboldt State University, an M.Sc. in zoology at the University of Maine, and a Ph.D. in limnology from the Limnological Institute at Uppsala University, Sweden.

James J. Elser (NAS) is the Bierman Professor of Ecology at the University of Montana and since March 2016 has been director of the Flathead Lake Biological Station at Yellow Bay. Trained as a limnologist, Dr. Elser is best known for his role in developing and testing the theory of ecological stoichiometry, the study of the balance of energy and multiple chemical elements in ecological systems. Currently, Dr. Elser's research focuses most intensively on Flathead Lake as well as mountain lakes of western Montana and western China. Specific studies involve observational and experimental studies at various scales, including laboratory cultures, short-term field experiments, and sustained whole-ecosystem manipulations. He has been named a fellow of the American Association for the Advancement of Science as well as a foreign member of the Norwegian Academy of Arts and Sciences and is a recipient of the G.E. Hutchinson Award of the Association for the Sciences of Limnology and Oceanography, the world's largest scientific association dedicated to aquatic sciences. In 2019, Dr. Elser was elected to the National Academy of Sciences. He received a B.S. in biology from the University of Notre Dame, an M.S. in ecology from the University of Tennessee, and a Ph.D. in ecology from the University of California, Davis.

Scott E. Fendorf is the Huffington Family Professor of Earth Science at Stanford University, the founding chair of the Earth System Science Department, and presently the senior associate dean in the School of Earth, Energy, and Environmental Science at Stanford. He is also a senior fellow in Stanford's Woods Institute for the Environment. Dr. Fendorf is a geochemist focused on soil and water, with a specific interest in the chemical and biological processes controlling the fate of contaminants and nutrients. He has conducted fundamental experiments coupled with fieldwork across the globe to resolve the processes controlling the bioavailability and transport of metal contaminants, and his research has included metal fate and transport in the Coeur d'Alene River and Coeur d'Alene Lake. He was a member of the National Academies of Sciences, Engineering, and Medicine's Committees on Sources of Lead Contamination at or Near Superfund Sites (2016–2017) and on Bioavailability of Contaminants in Soils and Sediments (2000–2001), and he was a member of the National Academies' National Committee for Soil Science (2009–2012). He received a B.S. in soil science from California Polytechnic State University, an M.S. in soil chemistry from the University of California, Davis, and a Ph.D. in soil and environmental chemistry from the University of Delaware.

APPENDIX C

Alejandro N. Flores is an associate professor in the Department of Geosciences at Boise State University. His research focuses on understanding mountain watersheds as regional Earth systems where large-scale patterns emerge as a product of interactions between and among biophysical processes and human action. His research synthesizes numerical models of and data characterizing regional climate, ecohydrology, and human, land, and water management activities in order to assess how perturbations propagate across scales and through component systems. At Boise State, Dr. Flores is the principal investigator and director of the LEAF group, which researches the intersection of water, energy, nutrients, policy, and human activity. He is a recipient of a National Science Foundation (NSF) CAREER award and an Army Research Office Young Investigator Program award. He is a co-principal investigator on NSF's Reynolds Creek Critical Zone Observatory. He holds a B.S. and an M.S. in civil and environmental engineering from Colorado State University, and he received his Ph.D. in hydrology from the Massachusetts Institute of Technology.

Robert M. Hirsch is a research hydrologist emeritus at the U.S. Geological Survey (USGS). As a research hydrologist, the focus of his research is on the description and understanding of long-term variability and change in surface water quality and streamflow. From 1994 through May 2008, he served as the chief hydrologist of the USGS. In this capacity, Dr. Hirsch was responsible for all USGS water science programs, which encompass research and monitoring of the nation's groundwater and surface water resources, including issues of water quantity as well as quality. Dr. Hirsch has received numerous honors from the federal government and from nongovernmental organizations, including the 2006 American Water Resources Association's William C. Ackermann Medal for Excellence in Water Management and selection to be the Walter Langbein Lecturer of the American Geophysical Union in 2017, and he has twice been conferred the rank of Meritorious Senior Executive by a U.S. President. Dr. Hirsch has served on several National Academies of Sciences, Engineering, and Medicine committees, most recently the Committee to Review the New York City Watershed Protection Program. He received a B.A. in geology from Earlham College, an M.S. in geology from the University of Washington, and a Ph.D. in environmental engineering from the Johns Hopkins University.

Lynn E. Katz holds the Hussein M. Alharthy Centennial Chair in Civil Engineering at the University of Texas at Austin, serves as the director of the Center for Water and the Environment at the University of Texas at Austin, and is the associate director of the U.S. Department of Energy (DOE) Energy Frontier Research Center on Materials for Water and Energy Systems. Dr. Katz has more than 25 years of experience examining reaction phenomena at interfaces and evaluating the impact of these processes on the fate and transport of organic and inorganic contaminants in the environment. Her research has involved both fundamental and applied studies in this field and has included the development of insitu remediation and exsitu treatment processes. A major focus of research throughout her career has involved the development and application of surface complexation models for predicting metal ion fate in the environment. This research thrust was the theme of her National Science Foundation Career Award, several DOE research projects, and peer-reviewed publications. She received her B.S.E. in environmental engineering from Johns Hopkins University. She received M.S. degrees in environmental engineering and chemistry and a Ph.D. in environmental engineering, all from the University of Michigan.

James G. Moberly is an associate professor in the Chemical and Biological Engineering Department at the University of Idaho. His research focuses on discovering how microorganisms or their bioproducts can be applied to solve engineering and environmental problems, improve human health, and increase accessibility to high-quality water sources. Dr. Moberly has worked on research in the Coeur d'Alene basin during his graduate work and as a faculty member at the University of Idaho. His postdoc at Oak Ridge National Laboratory focused on determination of the biogeochemical factors and bacterial genes or pathways that produce highly toxic methylmercury. In the private sector, he performed contaminated site characterization, remediation, and treatability testing. He holds a B.S. in chemical engineering from the University of Idaho, an M.S. in chemical engineering from Washington State University, and a Ph.D. in chemical engineering from Montana State University.

S. Geoffrey Schladow currently serves as the founding director of the University of California, Davis, Tahoe Environmental Research Center, the custodian of the longest continuous multidisciplinary dataset for a large lake in the western United States, and he is also a professor of civil and environmental engineering at UC Davis. His disciplinary expertise is in environmental fluid mechanics and their interactions with water quality and ecological processes in aquatic ecosystems, using a combination of field measurements, laboratory measurements, and numerical modeling. He has studied the impacts of climate change, management interventions, and watershed activities on eutrophication, dissolved oxygen distribution, the release of nutrients and heavy metals from lake sediments, and alterations to food webs from invasive species. Dr. Schladow co-chairs the Tahoe Science Advisory Council, serves on the Technical Advisory Committee to the Blue Ribbon Committee for Clear Lake, is a member of the Salton Sea Science Committee, and is the Science Director for Chile Lagos Limpios, a nonprofit measuring and modeling the response of Patagonian lakes to climate change and land use change. He earned his Ph.D. and B.Eng. in civil engineering from the University of Western Australia and his M.Eng. in hydraulic engineering from the University of California, Berkeley.

STAFF

Laura J. Ehlers is a senior program officer for the Water Science and Technology Board of the National Academies of Sciences, Engineering, and Medicine. Since joining the National Academies in 1997, she has served as the study director for more than 27 committees, including the Committee to Review the New York City Watershed Management Strategy, the Committee on Bioavailability of Contaminants in Soils and Sediment, the Committee on Assessment of Water Resources Research, the Committee on Reducing Stormwater Discharge Contributions to Water Pollution, the Committee to Review EPA's Economic Analysis of Final Water Quality Standards for Nutrients for Lakes and Flowing Waters in Florida, and the Committee on Management of *Legionella* in Water Systems. Dr. Ehlers has periodically consulted for the EPA's Office of Research Development regarding its water quality research programs. She received her B.S. from the California Institute of Technology, majoring in biology and engineering and applied science. She earned both an M.S.E. and a Ph.D. in environmental engineering at the Johns Hopkins University.

Rachel Silvern is a program officer with the Board on Atmospheric Sciences and Climate and the Polar Research Board at the National Academies of Sciences, Engineering, and Medicine, where she has led projects on developing a framework for evaluating greenhouse gas information and machine learning and artificial intelligence to advance Earth system science. Prior to joining the National Academies, Dr. Silvern served as a California Council on Science and Technology Science Fellow, where she worked for the California State Assembly, analyzing and staffing legislation on drinking water, hazardous waste, and climate adaptation. Dr. Silvern earned a Ph.D. and an M.A. in Earth and planetary sciences from Harvard University, where she studied U.S. air quality, and a B.A. in environmental science from Barnard College.